Singularities: Formation, Structure, and Propagation

Many key phenomena in physics and engineering can be understood as singularities in the solutions to the differential equations describing them. Examples covered thoroughly in this book include the formation of drops and bubbles, the propagation of a crack, and the formation of a shock in a gas.

Aimed at a broad audience, this book provides the mathematical tools for understanding singularities and explains the many common features in their mathematical structure. Part I introduces the main concepts and techniques, using the most elementary mathematics possible so that it can be followed by readers with only a general background in differential equations. Parts II and III require more specialized methods of partial differential equations, complex analysis, and asymptotic techniques. The book may be used for advanced fluid mechanics courses and as a complement to a general course on applied partial differential equations.

J. EGGERS is Professor of Applied Mathematics at the University of Bristol. His career has been devoted to the understanding of self-similar phenomena, and he has more than 15 years of experience in teaching nonlinear and scaling phenomena to undergraduate and postgraduate students. Eggers has made fundamental contributions to our mathematical understanding of free surface flows, in particular the breakup and coalescence of drops. His work was instrumental in establishing the study of singularities as a research field in applied mathematics and in fluid mechanics. He is a member of the Academy of Arts and Sciences in Erfurt, Germany, and a Fellow of the American Physical Society and has recently been made a Euromech Fellow.

M. A. FONTELOS is a researcher in applied mathematics at the Spanish Research Council (CSIC). His scientific work has focused on partial differential equations and their applications to fluid mechanics, with special emphasis on the study of singularities and free surface flows. His main results concern the formation of singularities (or not), combining the use of rigorous mathematical results with asymptotic and numerical methods.

Cambridge Texts in Applied Mathematics

All titles listed below can be obtained from good booksellers or from Cambridge University Press. For a complete series listing, visit www.cambridge.org/mathematics.

Singularities: Formation, Structure, and Propagation

J. EGGERS
University of Bristol

M. A. FONTELOS
Consejo Superior de Investigaciones Científicas, Madrid

CAMBRIDGE
UNIVERSITY PRESS

CAMBRIDGE
UNIVERSITY PRESS

University Printing House, Cambridge CB2 8BS, United Kingdom

Cambridge University Press is part of the University of Cambridge.

It furthers the University's mission by disseminating knowledge in the pursuit of education, learning and research at the highest international levels of excellence.

www.cambridge.org
Information on this title: www.cambridge.org/9781107098411

First published 2015

Printed in the United Kingdom by Bell and Bain Ltd

A catalogue record for this publication is available from the British Library

Library of Congress Cataloguing in Publication Data
Eggers, J. (Jens G.), 1963–
Singularities : formation, structure, and propagation / J. Eggers, University of Bristol, M.A. Fontelos, Consejo Superior de Investigaciones Científicas, Madrid.
pages   cm. – (Cambridge texts in applied mathematics)
Includes bibliographical references and index.
ISBN 978-1-107-09841-1
1. Singularities (Mathematics)   I. Fontelos, M. A. (Marco Antonio)
II. Title.
QA614.58.E325   2015
514′.746–dc23
2015016866

ISBN 978-1-107-09841-1 Hardback
ISBN 978-1-107-48549-5 Paperback

To our parents:

Gisela and Hans,
Blanca and Julian.

Contents

Preface

The word "singularity" is used popularly to describe exceptional events at which something changes radically or where a new structure emerges. In the mathematical language of this book, we speak of a singularity when some quantity goes to infinity. This is usually related to the solution of a differential equation which loses smoothness in that either the unknown itself or its derivatives become unbounded at some point or region of their domain.

Very often a singularity understood in the strict mathematical sense justifies the popular use of the word, since it represents a situation or structure of special interest. For example, a singularity of the curvature lies at the center of a black hole, which is formed after the collapse of a supermassive star, and the universe itself is generally believed to have begun at a singularity. Unfortunately, the real difficulty here lies with the correct physical interpretation of the mathematical solution, which is the reason we have not been able to include examples from general relativity.

Examples of singularities discussed in this book are vortices, such as the flow around the center of a tornado, shock waves generated by the motion of a supersonic plane, caustic lines of intense brightness produced by the focusing of light, and the formation of a drop that results from the discontinuous separation of a liquid mass into two or several pieces.

Starting in the nineteenth century with the study of shock waves, singularities have been investigated on an individual basis. They have remained one of the most exciting research topics in both pure and applied mathematics. For example, two of the seven Millennium Prize problems, proposed by the Clay Foundation, were directly or indirectly related to singularities. The sixth problem was to investigate whether the Navier–Stokes equation, which describes the motion of fluids, does or does not produce any singularities. A related and hotly debated problem poses the same question for the Euler equation, which is the Navier–Stokes equation in the absence of viscosity. Both problems are still to be solved.

The third Millennium Problem, known as the Poincaré conjecture, was solved by G. Y. Perelman while studying the singularities of the partial differential equation describing Ricci flow (similar equations will be studied in Chapter 9 of this book). Perelman used his insight into the structure of these singularities, by continuing the flow across the singularity in such a way that the essential topological information was preserved.

However, few attempts have been made to present a general survey of singularities which would bring out their unifying features. In our opinion, the most important shared feature is that of self-similarity, which runs as a common thread through this book and which will be highlighted in each individual case. The significance of self-similarity and scaling was also expounded in Barenblatt's influential book [14], although the focus was not on singularities.

Self-similarity was already embodied in the similarity solution introduced in 1934 by J. Leray [138] to construct singular solutions to the Navier–Stokes equation. This posits that the solution is invariant as a function of time (or some other physical parameter), up to a change of scale. The existence of a scaling symmetry implies a dimensional reduction of the problem and reduces it to the study of the neighborhood of the singularity. This greatly simplifies the mathematical problem and makes it amenable to explicit analytical calculation. In this book we will largely ignore the important problems of existence and uniqueness but, rather, will focus on obtaining explicit solutions which can be compared to experimental data and used to explain qualitative experimental features.

Our book is intended for a broad audience of students and scientists, mainly in the areas of mathematics, physics, and engineering. It is in three parts:

• setting the scene
• formation of singularities
• persistent singularities: propagation.

The first part introduces the main concepts using elementary mathematics that can be followed by undergraduate students in their final years. The only requirement is a basic knowledge of ordinary and simple linear partial differential equations. Our aim is to introduce the fundamental ideas of blowup, self-similarity, and regularization, and to provide some essential mathematical tools such as asymptotics and matching. We introduce (or remind advanced readers of) the main concepts of continuum mechanics and develop two important tools in the study of singularities: local singular expansions and asymptotic expansions of partial differential equations. The main results and notation in vector calculus, including differential operators in curvilinear coordinate systems and an exposition of index notation, are provided in the appendices. Much of the contents of Part I will be familiar to the advanced reader but, for those who need it, it provides a preparation for most of the material to be presented in Parts II and III.

The second and third parts are more demanding and are oriented mainly toward Ph.D. students and researchers. However, advanced Masters' students and first-year graduate students in the USA will also find this material rewarding. Using explicit examples, motivated mainly by their physical interest, we explore the main themes of this book. We investigate the scaling properties of singularities as they are formed, starting from exact self-similarity and progressing to more complex forms that for example involve logarithmic corrections. We explore the structure of persistent singularities such as shocks, cusps, and vortices and finally turn to the interaction between singularities and their motion.

We wrote these two final parts of the book in the spirit of a special topics course in a postgraduate program, so that most chapters can be read independently of one another and mostly using material from the first part. We have aimed to present calculations explicitly and explain mathematical methods as we go along, but a prior knowledge of asymptotic methods such as matched asymptotic expansions and WKB methods, of basic complex variable theory, and of elements of the theory of differential equations would be helpful. Since most of the book deals with problems in continuum mechanics, some background in fluid and solid mechanics at the undergraduate level will further enhance understanding (although this is not essential since we present the main concepts and mathematical formulations in Chapter 4). We also expect the book to be useful as a "toolbox" for experienced researchers since it gathers together many ideas and techniques scattered throughout the scientific literature.

To aid self-study, we have added a number of examples to the text; these are designed to reinforce the reader's understanding of the most important concepts. The material of each chapter is supplemented by a collection of exercises of varying degree of difficulty. Some are simple extensions of material contained in the text and will be useful for self-study; others are more demanding and could be used as problems for a graduate course. Exercises which are especially demanding and explore new material have been marked with a star.

The idea of writing a unifying description of singularities was an outgrowth of the program "Singularities in mechanics: formation, propagation and microscopic description", organized with C. Josserand and L. Saint-Raymond, which took place between January and April 2008 at the Institut Henri Poincaré in

Paris. We are grateful to all participants for their input, in particular C. Bardos, M. Brenner, M. Escobedo, M. Marder, F. Merle, H. K. Moffatt, Y. Pomeau, A. Pumir, J. Rauch, S. Rica, L. Vega, T. Witten, and S. Wu. We are grateful to our editor, David Tranah of Cambridge University Press, for his encouragement to write this book and for his many suggestions along the way. We thank our colleagues and collaborators, M. Aguareles, S. Balibar, M. V. Berry, D. Bonn, M. P. Brenner, I. Cohen, S. Courrech du Pont, R. D. Deegan, L. Duchemin, T. F. Dupont, R. Evans, S. Grossmann, J. Hoppe, C. Josserand, L. P. Kadanoff, D. Leppinen, J. Li, L. Limat, J. Lister, E. Lorenceau, J. M. Martin-Garcia, G. H. McKinley, S. R. Nagel, L. M. Pismen, D. Quéré, J. H. Snoeijer, H. A. Stone, N. Suramlishvili, J. J. L. Velazquez, E. Villermaux, and C. Wagner, for their invaluable contributions toward a better understanding of singularities. The whole book was read by C. Lamstaes, who caught many errors and suggested improvements.

PART I

Setting the scene

1

What are singularities all about?

Many mathematical descriptions of natural phenomena use the language of partial differential equations (PDEs). For example, fluid flow is described by the Navier–Stokes equation for the velocity, density, and pressure inside a fluid. A typical situation is shown in Fig. 1.1: a container filled with a viscous fluid is emptying through a hole in the bottom. The flow that results deforms the interface between the fluid and the air above it, whose shape is observed by lighting the container from behind and placing a camera on the other side (the interface is axisymmetric with respect to the axis of symmetry of the hole). The light passes through the fluid but is refracted by the fluid–air interface, which appears black.

The picture on the left of Fig. 1.1 conforms with the naive expectation that the scale of deformation of the interface is set by scales imprinted on the system externally, for example the size of the sink hole in the bottom (diameter $d = 1$ mm) or the minimum distance h between the interface and the bottom. However, this expectation is contradicted by the picture on the right,

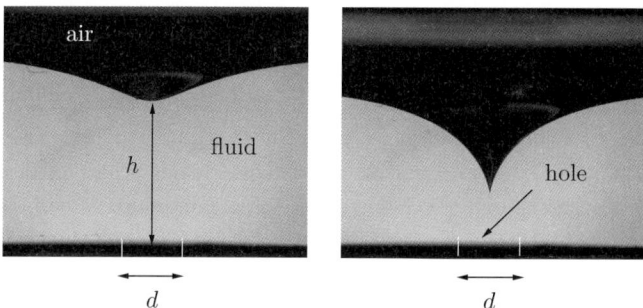

Figure 1.1 A container filled with viscous silicone oil is emptying slowly through a circular hole at the bottom [55], whose diameter $d = 1$ mm.

3

whose experimental conditions differ from those on the left only by the fact that the fluid has run out to a slightly lower mean level. Although the minimum distance h is still comparable with the hole diameter, the free surface has deformed into a very sharp tip. The size of the tip, as measured by its radius of curvature, is below the optical resolution of about 1 μm, and extrapolation of the experimental data suggests a vanishingly small value [55].

Thus the interface is no longer smooth, and the system has reached a singularity; the curvature, which involves the second derivative of the shape, diverges as the tip is approached. All examples of singularities to be discussed in this book involve quantities diverging in either space or time (so-called *blowup*) or the divergence of some derivative of the original quantities. Intuitively, this means that a local length scale of the system goes to zero. Often this is the result of nonlinearities of the problem, which couple different length scales. As in Fig. 1.1, nonlinearities serve to *focus* the flow into very small scales, although it is driven on scales which are much larger.

Near a singularity, the characteristic length scales of the solution are ultimately smaller than the microscopic length scales of the physical system it describes, such as the size of a molecule. This calls into question the assumptions made in deriving the equation (most often a PDE) from the underlying physics. As we shall see, in many cases the solution of the differential equation still presents us with a unique and physically meaningful answer to the problem we posed originally; however, a certain (in all likelihood very small) region in space or time then needs to be excluded, for which no physical prediction can be made.

In other cases the missing microscopic information needs to be supplemented by imposing additional physical conditions in order to guarantee the *uniqueness* of the solution. Finally, there are cases in which microscopic information has to be included explicitly in order to obtain a meaningful solution.

The fingerprint of a PDE

As we will see throughout this book, many nonlinear PDEs exhibit a *generic mechanism* by which singularities form. Since singularities are a local phenomenon (cf. Fig. 1.1), involving arbitrarily small length scales, their structure is usually universal, i.e. independent of the initial conditions or boundary conditions imposed over macroscopic distances. In other words, singularities are the "fingerprint" of a nonlinear PDE. They are the only thing one can say about the solution that is independent of the initial or boundary conditions and thus represents the intrinsic structure of the equation itself. Let us illustrate some key features using physical examples below.

1.1 Drop pinch-off: scaling and universality

Figure 1.2 shows a snapshot of the splash produced by a drop that has fallen into a pool of water, forming a nearly cylindrical column. The picture reveals that such a cylinder is unstable: even before the splash has had a chance to fall back into the pool the radius of the liquid column goes to zero at a point and a drop pinches off. Qualitatively, the reason is that the drop has a lower surface area than a piece of the cylinder occupying the same volume. To create an interface between fluid and air, one requires an energy equal to the surface tension γ times the surface area. Thus the system can reach a lower energy state by decaying into drops.

A point of particular interest is the moment when the volume of fluid separates into two pieces, and a drop is formed. A numerical description will run into problems at that point since such a change of topology represents a discontinuous transformation. Moreover, pinch-off events determine the structure of the resulting flow; they are the crucial moments where the solution changes qualitatively. We now focus on these singularities (to be discussed in detail in Chapter 7) which occur at the time of pinch-off, which we denote by t_0.

Scaling

The formation of a singularity is controlled by the time interval $\Delta t = t_0 - t$ to pinch-off. Note that we have defined Δt to be a positive quantity for $t < t_0$, before the singularity occurs; we will come back to this point below. The dimension of surface tension is that of an energy per unit area: $[\gamma] = \text{g/s}^2$. If the viscosity is sufficiently small to be irrelevant, the fluid motion is resisted only by inertia. Thus the other important parameter is the density ρ, representing the amount of mass being moved around, with $[\rho] = \text{g/cm}^3$. Now let us assume that the singularity is characterized by a *single* length scale, which we can take as the minimum radius $h_{\min}$. Then, dimensionally, the only possible way to represent the minimum radius is [168]:

$$h_{\min} \approx A \left(\gamma \Delta t^2 / \rho \right)^{1/3} \propto \Delta t^\alpha. \tag{1.1}$$

Thus the minimum radius behaves as a power law with scaling exponent $\alpha = 2/3$, a conclusion which has been tested experimentally with great precision; see Fig. 1.2 (right). Singularities (at which arbitrarily small length scales are produced) lack a particular length scale, so they are typically described by power laws, which are *invariant* under a change of length scale; see Section 2.4. A very simple but powerful tool to find scaling exponents is dimensional analysis, as we have just seen. A more formal exposition of this important tool is presented in Appendix C. We now illustrate it with another example.

Example 1.1 (Nuclear explosions) In 1950, G. I. Taylor [206] calculated the propagation of an intense blast wave (caused for example by a nuclear explosion) into an ambient gas. His scaling argument was that at very high intensities the only quantity measuring the strength of the explosion is the total energy E of the explosion. As far the outside temperature or pressure is concerned, they do not matter in comparison to their values inside. However, the initial density of the ambient gas atmosphere ρ_0 is relevant, since it measures the inertial resistance to the motion. Thus if we want to know the radius R of the blast at a time t, the dimensions of the relevant quantities (energy, density, length, and time) are, in terms of the units of mass, length, and time:

$$[E] = \text{g cm}^2/\text{s}^2, \quad [\rho_0] = \text{g/cm}^3, \quad [R] = \text{cm}, \quad [t] = \text{s}.$$

We can eliminate the units of mass and time between E, ρ_0, and t to find for R a unique quantity, with dimensions of length, up to a constant prefactor Γ:

$$R(t) = \Gamma \left(\frac{Et^2}{\rho_0} \right)^{1/5}. \tag{1.2}$$

This is Taylor's result for the radius of an intense blast wave; we will calculate the prefactor Γ in Exercise 11.11.

Repeating the same argument using the language of the Buckingham Π-theorem (see Appendix C), we have $k = 4$ quantities and $m = 3$ independent units. Since $k - m = 1$, we can construct a single dimensionless quantity (called a dimensionless group)

$$\Pi = \frac{R\rho_0^{1/5}}{E^{1/5}t^{2/5}}.$$

According to (C.2) we have $\phi(\Pi) = 0$ for a suitably defined function ϕ. But this is consistent only if $\phi(\Pi) \equiv \Pi - \Gamma = 0$ for some constant Γ, which once more implies (1.2). $\qquad\square$

Before we go on, let us point to the considerable subtlety that often underlies dimensional arguments. Frequently, assumptions are made on physical grounds which in general have to be confirmed by a detailed analysis. In the case of drop breakup, we have assumed that there exists a *single* length scale $h_{\min}$ characterizing the breakup. Dimensional analysis then demonstrates that the scaling exponent α is determined uniquely by the local structure of the equations alone. This scenario is known as *self-similarity of the first kind* [14].

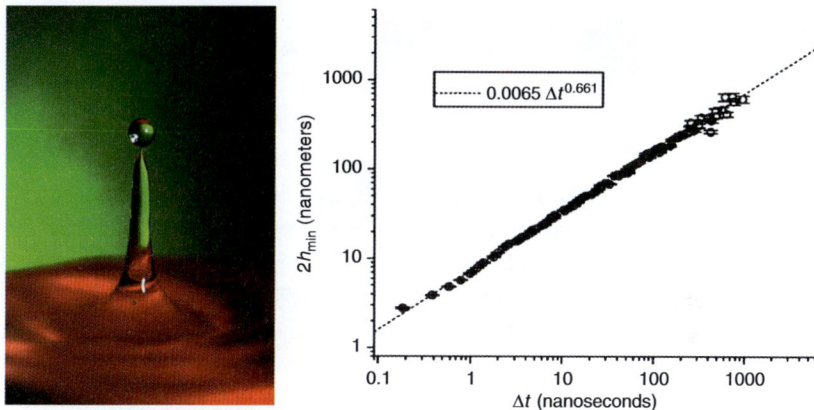

Figure 1.2 On the left, a splash of water at the moment of pinch-off. Photograph by Harold Edgerton. Copyright 2010 MIT. Courtesy of MIT Museum. On the right, the minimum diameter of a mercury drop as a function of the time interval from pinch-off. Reprinted with permission from [38]. Copyright 2004 by the American Physical Society.

However, this need not be the case, and the solution for inviscid breakup may be governed by two or more intrinsic length scales. This happens in the breakup of a two-dimensional liquid *sheet*, in which case the minimum sheet thickness and the typical width of the pinch region scale differently [39]. The ratio of these two length scales is a dimensionless number and thus cannot be fixed by dimensional analysis. As a result, the thickness and width of the sheet may shrink according to different scaling exponents, to be determined as part of the full solution to the problem. This case is known as *self-similarity of the second kind*. In general, the exponents no longer assume rational values but will become irrational numbers.

In reference to Example 1.1, the superficially very similar problem of the *focusing* of a shock wave onto a point is an example of self-similarity of the second kind [133]: the radius of the shock wave is governed by an irrational exponent, different from that in (1.2). We will return to this problem in Chapter 11 on shock waves.

Universality

Another, related, property of a singularity is the fact that its structure is insensitive to initial conditions or other aspects of the large-scale structure of the solution. This is a consequence of the fact that the singularity arises from a local balance: in drop pinch-off, the dimensionless prefactor A in (1.1) is expected to be universal since (1.1) is the solution of a nonlinear equation.

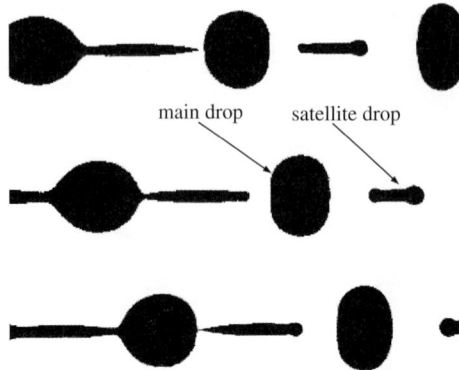

Figure 1.3 Satellite formation in a water–glycerol jet, showing a satellite drop in between two main drops. A satellite drop is the remnant of the elongated neck between two main drops [75, 146].

Any change in amplitude would disturb the balance between different terms in the equation. A detailed solution of the equation of motion shows that A has a numerical value close to 0.7; see (7.98) below.

However, the concept of universality goes much deeper and governs the entire spatial structure of a singularity. This is illustrated nicely by the breakup of a liquid jet, shown in Fig. 1.3. It is evident that there are not only main drops produced in the process, but also much smaller, so-called "satellite" drops. Such a drop is formed from an elongated neck between two adjacent main drops. The existence of the neck, in turn, is related to the fact that the profile of the jet near the point of breakup is extremely asymmetric: toward the drop it is very steep while toward the neck it is flat, forcing the neck into its slender shape.

Thus the existence of satellite drops is a consequence of the nonlinear properties of the fluid motion close to breakup, which produces the asymmetric shapes discussed above.

Drop formation of the type illustrated in Fig. 1.3 has many applications, as reviewed in for example [16]. The classical technique of ink-jet printing has been adapted to produce so-called microarrays for DNA analysis, to print integrated circuits, and to produce miniature lenses for optical applications. For all these applications it is very important to control the size of the drop accurately; thus satellite drops are detrimental to the quality since they result in at least two *different* droplet sizes. It is therefore natural to ask whether it is possible to control the excitation of the jet leading to breakup in such a way that only one type of drop is produced.

A hypothetical, more desirable, breakup configuration is illustrated schematically in Fig. 1.4, in which the profile is assumed symmetric with respect to the

Figure 1.4 A hypothetical breakup mode without satellite formation: an engineering dream that is unfulfillable.

pinch point so that breakup would occur in the middle, between two drops. In that case the neck would snap back toward each of the drops, which would receive the same amount of mass, making all drops equal. However, universality imposes that such a scenario is impossible: no matter how the jet is excited or what the initial condition may be, the pinch-off dynamics will always be similar to Fig. 1.3, *independently* of the initial conditions.

Continuation

To fully understand drop formation as shown in Fig. 1.3, one also needs to address the problem of *continuation* across the singularity, treated in Chapter 10. As a drop is formed, one proceeds from a simply connected domain (before pinch-off) to a multiply connected domain (after pinch-off), which is a discontinuous process. As a result, it is not clear whether this continuation is unique, i.e. that there is a single post-breakup solution, with which the pre-breakup solution can be continued. We will show that in the case of drop formation there is indeed only one unique continuation which does not involve discontinuities at a finite distance away from the singularity.

1.2 Stationary cusps: persistent singularities

The pinch-off singularity of a drop dominates the flow for a brief period of time, as the piece of fluid breaks into two. We now discuss another type of singularity, which is stationary and which is a generic feature for a range of parameters. As an example, consider the intensity of light generated by the reflection from a wedding ring, shown in Fig. 1.5 (left). The intensity is far from uniform but instead becomes very large along certain line-like singularities called caustics (from the Latin for "burning"). The caustic line itself has a singular tip, where it ends in a cusp. We will investigate the nature of the cusp in Chapter 14, where we show that its shape is described by the universal power law

$$y \propto x^{2/3}, \tag{1.3}$$

where x is the width of the cusp and y the distance from its apex.

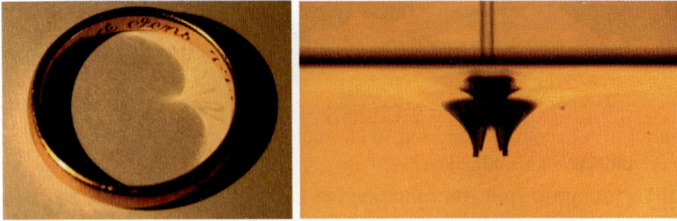

Figure 1.5 Two examples of cusp singularities, observed in very different contexts. On the left, the bright lines of a caustic produced by the focusing of wave fronts (image by Ann Eggers). On the right, a jet of viscous fluid is poured into a bath and imaged from the side, looking up toward the surface of the bath. (Reprinted with permission from [184]. Copyright 2008, AIP Publishing LLC.) The free surface ends in a cusp singularity, which lies on a circle with the jet axis at its center.

A very different problem, also treated in Chapter 14, is shown in Fig. 1.5 (right). A jet of viscous fluid falls into a bath of the same fluid, creating a strong flow. The free surface deforms into a cusp, which ends in a circular knife edge at the bottom. An analysis of this problem, which requires the solution of the viscous flow equations with a free surface, yields exactly the same cusp shape, (1.3), as for the coffee cup caustic. Thus the concept of universality may in some cases apply more broadly, connecting phenomena of very different physical origin.

One important aspect in which the two problems shown in Fig. 1.5 differ is the mechanism by which the singularity is cut off on a small scale. The motion of wave fronts, which results in the caustic, is described by a *nonlinear* equation, which *emerges* from the linear wave equation on a scale greater than the wavelength of light [21, 22]. As one approaches the caustic the light intensity, instead of diverging, dissolves into a characteristic diffraction pattern, which we will study in Section 14.5. However, the singularity at the tip of the free surface cusp is resolved by the smoothing effect of the surface tension γ. Nevertheless γ does not by itself introduce a particular length scale. The surprising result of a more detailed calculation reveals that the radius of curvature of the cusp is in fact *exponentially small* in the value of the surface tension, i.e. it is proportional to $\exp(-\gamma^{-1})$ [122]!

1.3 Shock waves: propagation

If a singularity persists for a finite period of time, the question arises how it will move in space. Among the examples considered in the third part of the book

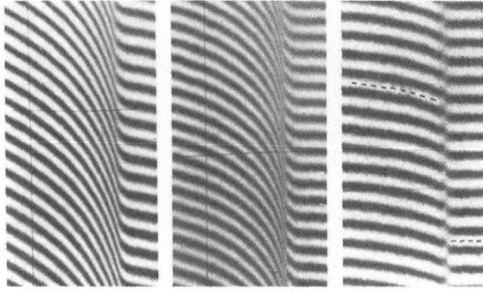

Figure 1.6 Fringe pattern showing the steepening of a wave in a gas, leading to the formation of a shock traveling from left to right. (Reproduced with permission from [98]. Copyright 1954, American Association of Physics Teachers.) The vertical position of a given fringe is proportional to the density at that point. In the rightmost picture a jump of seven fringes occurs, as indicated by the broken lines.

are shock waves (Chapter 11), vortices (Chapter 13), as well as contact lines and cracks (Chapter 15). Figure 1.6 illustrates the formation of a shock wave inside a gas-filled tube. The rapid expansion of the gas under high pressure in a section of the tube produces an abrupt change in the gas density, which propagates to the right. A fringe (bright or dark line) marks a line of constant density, so a large slope implies a rapid change in density. As the wave moves it becomes steeper and, in the rightmost picture, a discontinuity in the density is observed. Thus, between the second and the third picture, a singularity has formed at a critical time t_0. For times greater than t_0, a new persistent structure is formed (the shock), which continues to propagate to the right.

The formation of a shock is the consequence of the nonlinear properties of gas dynamics. At small amplitudes the solution is a linear wave equation, which does not develop singularities. If, however, the initial disturbance is sufficiently strong, created for example by the rapid motion of a piston or by a supersonic plane, nonlinearities lead to a progressive steepening of the wave, which ends in a singularity. The main goal is to find the properties of the shock wave (such as the magnitude of the jump) and the speed of propagation of the shock.

We will see in Chapter 11 that, to obtain a unique continuation across the singularity, additional physical conditions need to be invoked. Namely, apart from the equations of motion one has to stipulate that certain quantities are conserved across the singularity in order to find the correct laws for the motion of the shock. The motion of cracks in solids or of contact lines at the edge of a drop deposited on a solid (cf. Chapter 15) is even less universal. In these cases, specific microscopic laws have to be invoked in order to calculate the motion of the singularity, and the speed is determined from the interplay between the microscopic and macroscopic parameters.

Exercises

1.1   The motion of a viscous free surface flow is characterized by the density of the fluid, ρ, its viscosity, η, and the surface tension against the gas atmosphere, γ. The surface tension is an energy per unit area; η is determined by measuring the shear force $F = \eta U A/d$ experienced by two plates of area A which are a distance d apart; one moves at a speed U relative to the other.

   (i) Deduce the dimensions of ρ, η, and γ.

   (ii) Show that one can define length and time scales

$$\ell_\nu = \frac{\nu^2 \rho}{\gamma}, \quad t_\nu = \frac{\nu^3 \rho^2}{\gamma^2}$$

   which are intrinsic to the dynamics of free surface flow; $\nu = \eta/\rho$ is called the kinematic viscosity. The scales ℓ_ν and t_ν will play a central role in Chapter 7.

   (iii) Calculate ℓ_ν and t_ν for water and for glycerol, based on experimental values of ρ, η, and γ.

1.2   Show that the slow dripping of fluid from a circular pipette of radius r_0 is characterized by the two dimensionless numbers

$$\text{Oh} = \frac{\eta}{\sqrt{\rho r_0 \gamma}}, \quad \text{Bo} = \frac{\rho g r_0^2}{\gamma},$$

   known as the Ohnesorge and Bond numbers, respectively. Comment on their physical significance. The parameters are defined as in Exercise 1.1, and g is the acceleration of gravity.

1.3   Now assume in addition that fluid is issuing at a volume rate Q.

   (i) Calculate the mean flow rate v_0.

   (ii) Show that there is one more dimensionless parameter,

$$\text{We} = \frac{\rho r_0 v_0^2}{\gamma},$$

   known as the Weber number. What is its relation to the Reynolds number $\text{Re} = r_0 v_0/\nu$, where $\nu = \eta/\rho$?

1.4   Consider a propeller blade of a given shape characterized by its diameter d; the air in which it moves has density ρ and viscosity η.

(i) Show that the forward thrust force F on the propeller can be written as

$$F = \rho u^2 d^2 \Phi\left(\frac{Nd}{u}, \text{Re}\right),$$

where u is the forward velocity of the propeller (the velocity of the plane) and N the number of revolutions per second. The Reynolds number Re is defined as

$$\text{Re} = \frac{d^2 N}{\nu}$$

and has the form of a characteristic velocity multiplied by length and divided by ν.

(ii) How is the formula modified if the compressibility

$$\kappa = -\rho\frac{\partial p}{\partial \rho}$$

of the air is taken into account, where p is the pressure? What does this modification correspond to physically?

1.5 The Navier–Stokes equation (to be discussed in more detail in Chapter 4),

$$\rho\partial_t \mathbf{v} = -\rho(\mathbf{v} \cdot \nabla)\mathbf{v} - \nabla p + \eta\Delta\mathbf{v},$$

together with the incompressibility condition $\nabla \cdot \mathbf{v} = 0$, relates the velocity $\mathbf{v}$ and the pressure p. Here ρ is the density (which we take as being constant) and η the viscosity.

(i) Show that each term in the Navier–Stokes equation transforms in the same way under a change T of the units of time, length, and mass, as defined in Appendix C.

(ii) Show that for the solution to any flow problem the kinematic viscosity $\nu = \eta/\rho$ is the only parameter one needs to know.

1.6 A drop of dye is placed gently inside a fluid. Knowing that the diffusion constant has units $[D] = \text{m}^2/\text{s}$, show that the radius of the dyed region will grow as $t^{1/2}$ for long times. Now suppose that the fluid is stirred vigorously to produce turbulent motion. The strength of the turbulence is measured by the energy input ϵ per unit time and mass.

(i) Calculate the dimensions of ϵ.

(ii) Show that the size of the drop grows as

$$R = A\left(\epsilon t^3\right)^{1/2},$$

which is *much* faster than by diffusion; here A is a universal number [185]. This is known as Richardson's law.

1.7   In [185] Richardson also discussed the so-called Weierstrass function,

$$W(x) = \sum_{n=1}^{\infty} a^n \cos 2^n x, \tag{1.4}$$

as a model for an *individual* particle trajectory. The function W is an example of a function which is continuous but differentiable nowhere. Show that W exhibits a discrete scale invariance expressed by

$$W(2x) = a^{-1} W(x) - \cos 2x.$$

This means that, up to the large-scale modulation by $\cos(2x)$, a change of x-scale by a factor 2 can be absorbed into a change in y-scale. In the smooth case of $W(x) = x^\alpha + C$, to which exponent α would this correspond?

1.8   Consider a function $W(x)$ satisfying the following functional relation:

$$W(\gamma x) = e^{i\mu} \gamma^{2-D} W(x)$$

with $1 < D < 2$. Find all solutions of the form

$$W = x^{\alpha(\gamma)},$$

and calculate $\alpha(\gamma)$.

1.9   Consider the functional relation

$$\omega^n U(\omega) = C(n) U(\omega/k^n).$$

Show that a solution to this equation is given by

$$U(\omega) = A\omega^\nu \exp\left[-\frac{(\ln \omega)^2}{2 \ln k} \right] \exp\left(i \frac{2\pi c \ln \omega}{\ln k} \right)$$

where

$$C(n) = k^{n\nu + n^2/2} e^{2\pi i n c}$$

and

$$\nu = \Re \left\{ \frac{C'(0)}{k} \right\}, \quad c = \frac{1}{2\pi} \Im \left\{ C'(0) \right\}.$$

Discuss the behavior of the solution as $\omega \to 0$.

1.10 A cycloid is the curve traced out by a point on a circle that is rolling on a line.

(i) Show that for a circle of unit radius the cycloid through the origin is described in parametric form by

$$x = \varphi - \sin \varphi, \quad y = 1 - \cos \varphi.$$

(ii) Show that near the origin the cycloid has the shape of a cusp with scaling law (1.3).

1.11 In Chapter 5 we consider flow around a wing with profile (5.16), pictured in Fig. 5.4. Show that the rear end (trailing edge) is described by (5.17), i.e. by the same type of cusp (1.3) as before.

1.12 Imagine that two curves (such as two circles) touch at a point. Show that the shape near the point of contact is a cusp and that the width x of the cusp as a function of the distance y from the apex scales as $y \propto x^{1/2}$, which is *different* from (1.3).

2

Blowup

2.1 A scalar example

In 1960, Heinz von Foerster and his colleagues [215] argued that the simple nonlinear ordinary differential equation (ODE)

$$\dot{N} = a_0 N^2 \tag{2.1}$$

described the world population $N(t)$ as function of time. The quadratic term on the right-hand side of (2.1) implies that the *rate of growth* of the world population is ever increasing. In fact, it is seen readily that

$$N(t) = \frac{1}{a_0(t_0 - t)} \tag{2.2}$$

is a solution of (2.1) which goes to *infinity* at some finite "doomsday" time t_0 determined by the initial condition. The singular solution (2.2) is shown in Fig. 2.1, and agrees well with data up to 1960.

Example 2.1 (Chemical reaction) For another example of the same type of growth consider the following chemical reaction [192]:

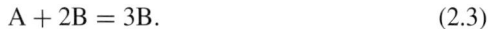

$$A + 2B = 3B. \tag{2.3}$$

It describes a reaction that requires two molecules of the chemical B to produce another molecule of the same chemical out of the chemical A. Let us assume that A is present in abundance, so that its concentration is constant, and let u be the concentration of B. Thus the rate of change of u is proportional to u^2, since two B-molecules must come together for the reaction to proceed:

$$\dot{u} = R(u) \equiv Ku^2 - ru. \tag{2.4}$$

At the same time, we allow the chemical B to decay, at a rate r, which introduces the second term on the right of (2.4). □

16

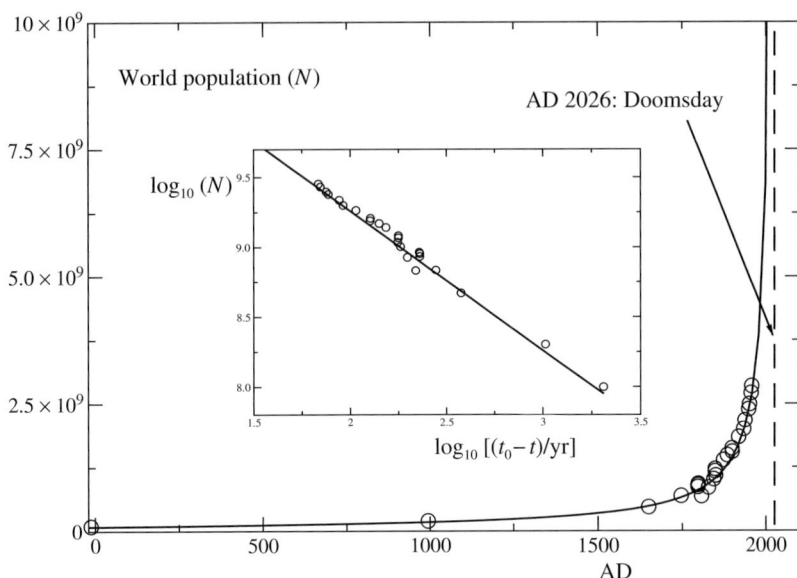

Figure 2.1 The world population N according to historical data (circles, from [215]). The solid line is a solution of (2.1) with $a_0 = 5.5 \times 10^{-12}$ years^{-1}. The population described by (2.2) is predicted to go to infinity in the year $t_0 = 2026$ AD. The inset replots the data in a doubly logarithmic form; the slope of the solid line is -1.

Of course (2.4) can be solved by separation of variables, and we will give the solution below. However, as usual, much is learned from considering special cases or rather from the limiting behavior as such a special case is approached, a method known as asymptotics. Asymptotics brings out the universal underlying structure in a potentially complicated problem. Crucially, it also works when no exact solution is available. There are two possible asymptotic behaviors of (2.4), one for small initial values of u, the other for large values of u.

Let us first consider the case that u is small, so that $u^2 \ll u$; we end up with a linear equation,

$$\dot{u} = -ru. \tag{2.5}$$

The solution to this equation is an exponential:

$$u(t) = u_0 e^{-rt}, \tag{2.6}$$

i.e. u decays to zero for long times. The assumption $u^2 \ll u$ remains consistent for all times, as u becomes ever smaller; the amplitude u_0 is determined by the initial condition.

If, however, u is sufficiently large, u^2 dominates the linear term and we return to the "doomsday equation"

$$\dot{u} = Ku^2, \tag{2.7}$$

which, on account of (2.2), is also called the hyperbolic growth equation. Let us, in the spirit of things to come, attempt to solve (2.7) by trying a power law $u = A(\Delta t)^\alpha$, where $\Delta t = t_0 - t$ and t_0 is the blowup time. Noting the negative sign in front of t, we obtain

$$\dot{u} = -A\alpha(\Delta t)^{\alpha-1}, \quad Ku^2 = KA^2(\Delta t)^{2\alpha}. \tag{2.8}$$

As the singularity is approached ($\Delta t \to 0$), the two expressions in (2.8) must balance, which means the two exponents must be the same: thus $\alpha - 1 = 2\alpha$, or $\alpha = -1$. A negative exponent implies that u indeed goes to infinity as $t = t_0$ is approached. Inserting $\alpha = -1$ into (2.8), we find $K = 1/A$ for the amplitude, and the solution becomes

$$u(t) = \frac{1}{K\Delta t}, \tag{2.9}$$

in agreement with (2.2). As $u \to \infty$, the linear term in (2.4) will become less and less important, so (2.9) is again self-consistent. The solutions (2.6) and (2.9) each correspond to a different *dominant balance* of the original equation.

Example 2.2 (General exponent) Let us consider first a generalized form of (2.7) where the nonlinearity carries an arbitrary exponent n:

$$\dot{u} = Ku^n. \tag{2.10}$$

The solution of (2.10) is

$$u = [(n-1)K\Delta t]^\alpha, \quad \alpha = -\frac{1}{n-1}, \tag{2.11}$$

which for $n > 1$ blows up in finite time.

Notice that

$$(\Delta t)^{-1/(n-1)} \ll (\Delta t)^{-1}$$

if Δt is sufficiently small, for any $n > 2$. Hence, the approach to blowup for u is faster when n is smaller. This is an apparent paradox since the reaction is stronger for large n and one would expect a faster approach to blowup in that case. However, there is no contradiction since blowup happens *earlier* for large n, provided that $u_0 = u(t = 0) > 1$: the blowup time t_0 is

$$t_0 = \frac{1}{(n-1)K}u_0^{1-n},$$

which decreases with n. In Exercise 2.9 we consider even stronger reactions leading to a logarithmic approach to the singularity. □

The singular solution (2.9) has two very important properties, which will appear again and again throughout this book: *scale invariance* and *universality*. First, scale invariance means that a change of scale in the concentration u can be absorbed into a change of time scale, keeping the same power law. For example, let us introduce a new time variable $\widetilde{\Delta t} = \Delta t/T$, so that

$$u(t) = T^{-1} \frac{1}{K \widetilde{\Delta t}}.$$

If now one introduces a new concentration variable $\tilde{u} = uT$, exactly the same law, (2.9), is recovered:

$$\tilde{u} = \frac{1}{K \widetilde{\Delta t}}. \tag{2.12}$$

Second, the singular solution (2.9) is universal, i.e. it contains no free parameters that depend on the initial condition. In particular, the constant K depends on the coefficients of (2.7) alone and is recovered in the rescaled equation (2.12). Intuitively, the diverging solution "forgets" any characteristic scale set by the initial condition, and thus picks out a universal form coming from the intrinsic balance of the equation alone. The only freedom is contained in the value of the blowup time t_0, which cannot possibly be universal, in view of time translation invariance.

The properties of scale invariance and universality should both be contrasted with the character of the solution (2.6) of the linear part of the equation. The exponential law (2.6) contains the characteristic time scale r^{-1} and is therefore not scale invariant: no change in u-scale can absorb the effect of a change in time scale. In addition, the amplitude u_0 of (2.6) is set by the initial condition and is therefore not universal.

Example 2.3 (Exponential decay) If in Example 2.2 we set $n = 1$, we obtain a linear equation. Introducing $u = A(\Delta t)^{\alpha}$, we find

$$A\alpha\,(\Delta t)^{\alpha-1} = KA\,(\Delta t)^{\alpha},$$

leading to the incompatible equation $\alpha - 1 = \alpha$ for the exponent. Hence no power law behavior can be deduced, and indeed the solution is the exponential law (2.6). This is a common fact in linear PDEs and is the reason why singularities are almost always associated with nonlinear equations. □

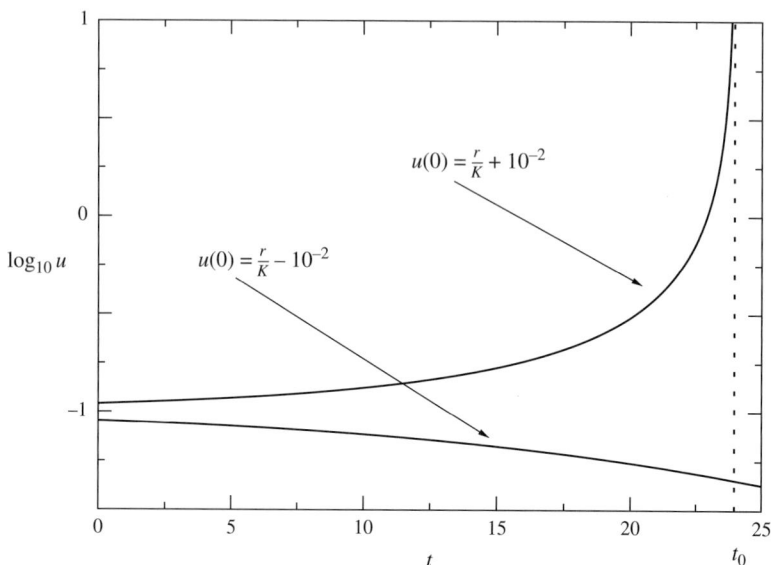

Figure 2.2 The solution of equation (2.4) as given by (2.13) with $K = 1$ and $r = 1/10$, for two different initial conditions. The slightly larger initial condition leads to blowup, the smaller to exponential decay.

One can of course recover the two possible behaviors of decay (2.6) and blowup (2.9) from the exact solution of (2.4), which is obtained by separation of variables:

$$u(t) = \frac{r}{K(1 + Ae^{rt})}, \quad \text{where} \quad A = \frac{r}{Ku(0)} - 1. \qquad (2.13)$$

This is illustrated in Fig. 2.2. If on the one hand the constant A is greater than zero, the solution decays as described by (2.6). If on the other hand $-1 < A < 0$, the denominator goes to zero eventually and the solution blows up as described by (2.9). The condition for blowup is thus that $u(0) > r/K$. With the global solution (2.9) in hand, one can determine the singularity time $t_0 = -\ln|A|/r$ in terms of the initial condition. However, the singularity time is difficult to calculate in general. In a typical PDE problem, analytical solutions are available only for the local singular behavior, and t_0 appears as a free parameter.

2.2 Crossover

Equations which contain several terms (corresponding to several competing physical effects) may exhibit solutions of very different character depending on which terms are of the same size. A balance of the left-hand side of (2.4) with the first term on the right leads to blowup, and a balance with the second term to exponential decay. The behavior near singularities is described by particular asymptotic balances, as some selected terms become large. However, the evolution resulting from one balance does not necessarily remain consistent as the singularity is approached.

To illustrate this, we consider the more general case of

$$\dot{u} = K_2 u^2 + K_3 u^3 - ru, \qquad (2.14)$$

which contains a choice of three terms to balance $\dot{u}$, corresponding to three competing physical mechanisms. For simplicity, let us assume that $r = 0$ and put

$$c = K_2 u, \quad \epsilon = \frac{K_3}{K_2}, \qquad (2.15)$$

so that (2.14) becomes

$$\dot{c} = c^2 + \epsilon c^3. \qquad (2.16)$$

We will imagine that ϵ is small, so that the second term on the right is small initially. However, as c grows the cubic term will take over eventually, corresponding to a power law with $\alpha = -1/2$, cf. (2.11). We thus have two different solutions, corresponding to the two different balances in (2.16). The balance with the cubic term leads to the solution

$$c = (2\epsilon \Delta t)^{-1/2}, \qquad (2.17)$$

describing the ultimate singular behavior. However, the balance with the quadratic term allows for another, transient, solution, i.e.

$$c = (\Delta t + \delta)^{-1}, \qquad (2.18)$$

where δ is an as yet undetermined shift in the blowup time. This shift accounts for the fact that the effective singularity time is slightly *later* than the actual singularity at $\Delta t = 0$. The crossover between (2.17) and (2.18) occurs when c^2 and ϵc^3 are of the same order, or $c = [\Delta t_c + \delta]^{-1} \approx \epsilon^{-1}$, where Δt_c is the crossover time. In addition, for one solution to match the other, (2.17) and (2.18) should be of the same order at the crossover point: $\Delta t_c + \delta \approx (2\epsilon \Delta t_c)^{1/2}$. Combining the two expressions, one finds

$$\Delta t_c \approx \epsilon/2, \quad \delta \approx \epsilon/2. \qquad (2.19)$$

This behavior is illustrated in Fig. 2.3, which shows the (easily available) exact solution of (2.14) as a solid line. The asymptotic solution (2.17) and

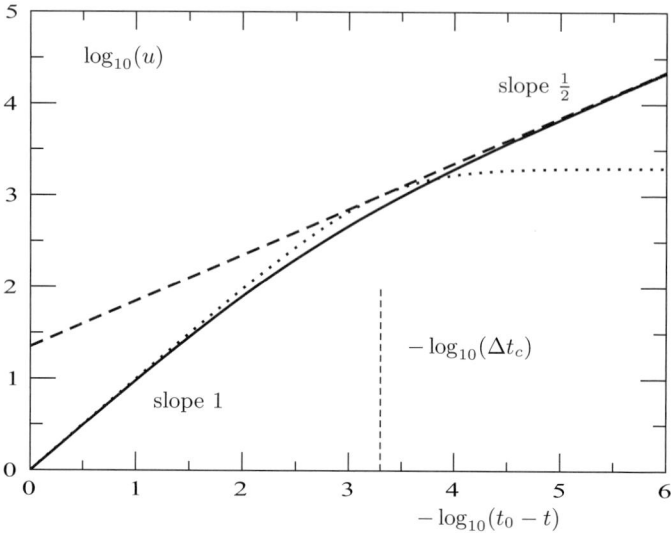

Figure 2.3 Crossover between the power laws 1 and 1/2 of the solution of (2.14) (solid line) for $\epsilon = 10^{-3}$. The limiting solutions (2.17) and (2.18) are given by the broken and dotted lines, respectively. Crossover occurs for $\Delta t_c \approx \epsilon/2$.

the transient solution (2.18) are included for comparison. We will find many more examples of crossover below, for example in describing drop breakup.

2.3 Regularization: saturation

The possibility of doomsday arriving in the year 2026 is of course an unpleasant prospect. Fortunately, there are reasons to believe that this event will not in fact occur, simply because we will run out of space or other resources necessary for procreation. One way to model this fact is to introduce a higher-order nonlinearity which has a *negative* sign into (2.1). After rescaling, and to be able to treat ϵ as a positive (small) number, we write (2.16) as

$$\dot{u} = u^2 - \epsilon u^3. \tag{2.20}$$

As can be seen in Fig. 2.4, u grows initially like a solution of (2.7) and threatens to blow up in finite time. However, as ϵu^3 comes to be of comparable magnitude, the equation converges to a stationary state defined by $\dot{u} = 0$. From the condition that the right-hand side of (2.20) vanishes, we find a saturation value of

$$u_{\text{sat}} = \frac{1}{\epsilon}. \tag{2.21}$$

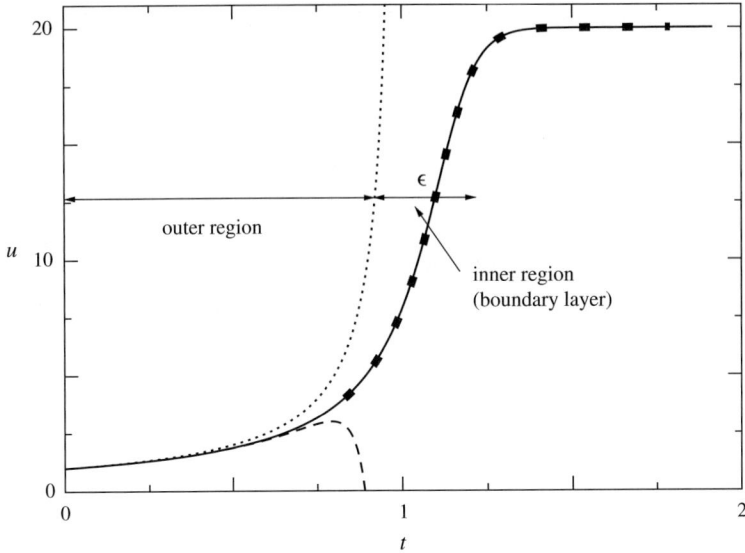

Figure 2.4 A solution of (2.20) with $\epsilon = 5 \times 10^{-2}$ and initial condition $u(0) = 1$ (solid line). The dotted line gives the outer solution, evaluated to leading order (2.22); the broken line gives the outer solution to first order, (2.30). The inner solution (2.25), with $\bar{t}_0 = \ln \epsilon$, is shown by the heavy dots.

But how does the approach to (2.21) occur? To answer this question, we introduce an important and extremely general technique known as a *matched asymptotic expansion*. It applies to any kind of *multiscale* phenomenon which contains two features taking place on two different (time) scales.

In the case of (2.20) there is the initial growth, described by

$$u_0 = \frac{1}{t_0 - t} \equiv \frac{1}{\Delta t}, \qquad (2.22)$$

which takes place over a time t_0 set by the initial condition $t_0 = 1/u_0(0)$. Of course (2.22) is the solution of (2.20) for $\epsilon = 0$, and is shown as the dotted line in Fig. 2.4. This is the large-scale or "outer" regime. Second, there is a small-scale or "inner" regime, which describes the crossover to the saturation value (2.22). The time over which crossover occurs will become shorter and shorter as ϵ decreases. To see this, note that from balancing the two terms on the right of (2.20) one finds that a typical scale of u in the transition region is $\delta u \propto 1/\epsilon$. Thus according to (2.22) the associated time scale is $\Delta t \propto 1/\delta u \propto \epsilon$. This means there is a narrow region (in time) over which the change of behavior from power law to saturation occurs. Such a region, whose thickness goes to zero with some parameter ϵ, is called a *boundary layer*. The vanishing size ϵ of the boundary layer is called the *boundary layer thickness*.

To isolate the fast crossover behavior, taking place over a short time scale ϵ, we introduce an inner variable $\bar{t} = \Delta t/\epsilon$. Since a typical u-scale is $1/\epsilon$, it is natural to introduce

$$u_{\text{in}} = \frac{1}{\epsilon} g(\bar{t}), \quad \bar{t} = \frac{\Delta t}{\epsilon}, \tag{2.23}$$

where $g(\bar{t})$ is the appropriate description for the inner region. The rescaling ensures that quantities remain of order one over the inner region. Inserting (2.23) into (2.20), we find

$$g_{\bar{t}} = -g^2 + g^3, \tag{2.24}$$

where the change in sign relative to (2.20) comes from the definition of Δt as $t_0 - t$. The solution of (2.24), by the separation of variables, is

$$\bar{t} - \bar{t}_0 = \frac{1}{g} + \ln \frac{1-g}{g} \tag{2.25}$$

and describes the shape of the profile in the inner region. To complete the inner description we have to find the unknown constant $\bar{t}_0$, which we will do by matching to the outer region.

In (2.25) the limit $\bar{t} \to -\infty$ corresponds to saturation, that is the approach of g to unity. The limit $\bar{t} \to \infty$ points toward the outer solution. The idea of the method of matched asymptotics is to make sure that the two descriptions in the outer and inner regions, respectively, are consistent. To this end, we have to formulate "matching conditions"; as a result of the matching we will be able to compute the free parameter $\bar{t}_0$ which appears in the local solution (2.25). To match, we require that the functional form of the outer solution u_{out}, to which (2.22) is a first approximation, agrees with the inner solution (2.23) in the limit $\epsilon \to 0$. This ensures that our result does not depend on the choice of the point where the two solutions are required to agree (an undesirable alternative known as patching).

On the one hand, the outer solution approaches the inner solution as $\Delta t \to 0$. On the other hand, the inner solution approaches the outer solution for $\bar{t} \to \infty$. This can be summarized as a *matching rule*:

$$\lim_{\Delta t \to 0} u_{\text{out}} \quad \text{agrees with} \quad \lim_{\bar{t} \to \infty} u_{\text{in}}. \tag{2.26}$$

To achieve this agreement formally, we expand both sides of (2.25) in power series and compare coefficients. The left-hand side is expanded for small values of Δt, the right-hand side for *large* values of $\bar{t}$. The leading approximation for the outer solution (2.22) is already of the right form: expanding (2.25) gives

$g = 1/\bar{t} + O(1/\bar{t}^2)$. Remembering that $\bar{t} = \Delta t/\epsilon$, this means that the leading-order approximations agree identically. To improve on this and to reach our goal of computing $\bar{t}_0$, we need to go to higher order in ϵ.

We set up the expansion

$$u = u_0 + \epsilon u_1 + \cdots , \tag{2.27}$$

which leads to the equation

$$\frac{du_1}{dt} = 2u_0 u_1 - u_0^3 = \frac{2u_1}{\Delta t} - \frac{1}{\Delta t^3}. \tag{2.28}$$

This linear equation is solved by standard methods, giving

$$u_1 = \frac{\ln \Delta t + C}{\Delta t^2}. \tag{2.29}$$

The constant C is determined from the initial condition. Let us choose $t_0 = u_0^{-1}(0)$ in (2.22), such that the initial condition on u is satisfied exactly: $u(0) = u_0(0)$. Then $u_1(0) = 0$ and thus $C = -\ln \Delta t$. In summary, the outer solution reads

$$u_{\text{out}} = \frac{1}{\Delta t} + \frac{\epsilon \ln[u(0)\Delta t]}{\Delta t^2}, \tag{2.30}$$

which is shown as the broken line in Fig. 2.4. Thus although the higher-order result (2.30) is a slight improvement over (2.22), it fails soon after one enters the boundary layer region. Naive perturbation theory is therefore not a useful way to describe boundary layer behavior.

In order to match to (2.30) one has to carry out the expansion of the inner solution (2.25) to the next order in $\bar{t}^{-1}$ as $\bar{t} \to \infty$. Using the leading-order expression $g = 1/\bar{t}$, the logarithm becomes

$$\ln((1-g)/g) \approx \ln((1-\bar{t}^{-1})/\bar{t}^{-1}) = \ln \bar{t} - \bar{t}^{-1} + O(\bar{t}^{-2}),$$

and thus

$$g^{-1} = \bar{t} - \bar{t}_0 - \ln \bar{t} + O\left(\bar{t}^{-1}\right).$$

Thus the expansion of the inner solution is

$$u_{\text{in}} = \frac{g}{\epsilon} \equiv \frac{1}{\epsilon}\left(\frac{1}{\bar{t}} + \frac{\bar{t}_0 + \ln \bar{t}}{\bar{t}^2}\right) = \frac{1}{\Delta t} + \epsilon \frac{\bar{t}_0 + \ln \Delta t - \ln \epsilon}{\Delta t^2}. \tag{2.31}$$

Having found expansions of the outer solution (2.30) and of the inner solution (2.31) in the same variable Δt, we can apply the matching rule (2.26). A comparison shows that the two expansions are consistent if $\ln u(0) = \bar{t}_0 - \ln \epsilon$. This means that we can determine the free coefficient $\bar{t}_0$ in the inner expansion as

$$\bar{t}_0 = \ln[\epsilon u(0)].$$

In summary, the inner solution is described by

$$\bar{t} - \ln(\epsilon u(0)) = \frac{1}{g} + \ln\frac{1-g}{g}, \qquad (2.32)$$

plotted as the heavy dots in Fig. 2.4. For $\bar{t} \to -\infty$ it approaches the asymptotic value $g = 1$, corresponding to the solution $u(t)$ for its saturation value (2.21).

To describe the approach to saturation we put $g = 1 - \delta$ and expand in δ. To leading order,

$$\frac{1}{g} \approx 1 + \delta, \qquad \frac{1-g}{g} \approx \delta,$$

and so (2.32) becomes

$$\bar{t} - \ln(\epsilon u(0)) = 1 + \ln\delta + O(\delta).$$

Thus the asymptotic value of g is approached as

$$g = 1 - \frac{1}{eu(0)\epsilon}e^{\bar{t}}, \qquad \bar{t} \to -\infty, \qquad (2.33)$$

where $e \equiv \exp(1)$. In terms of the original variable $u(t)$, the approach to saturation is described by the exponential law

$$u(t) = \frac{1}{\epsilon} - \frac{e^{\Delta t/\epsilon}}{eu(0)\epsilon^2}. \qquad (2.34)$$

Remember that saturation occurs for $\Delta t < 0$ (later than the time for which the singularity would have occurred for $\epsilon = 0$), so that u approaches $1/\epsilon$ on the time scale ϵ.

We note the slightly unusual point that the inner solution (2.32) agrees with the full solution exactly. The reason is that in deriving the inner equation (2.24) no higher-order terms in ϵ had to be neglected. There is no particular significance to this point; had we considered the more general equation

$$\dot{u} = u^2 - \epsilon u^3 - ru, \qquad (2.35)$$

this would not have been the case. The calculation is more involved but otherwise similar; see Exercise 2.4.

2.4 What is special about power laws?

We have seen that power laws

$$u = A(\Delta t)^\alpha \qquad (2.36)$$

are scale invariant. If we decide to measure time in a new unit, for example $\widetilde{\Delta t} = \Delta t/T$, the same scaling law is recovered by introducing a new dependent variable $\tilde{u} = uT^\alpha$, where

$$\tilde{u} = A(\widetilde{\Delta t})^\alpha, \tag{2.37}$$

with the same amplitude A.

However, the exponential law (2.6), which is the solution of the linear equation (2.5), is *not* scale invariant. In fact, power laws are the *only* scale-invariant functions, as we now demonstrate. To that end, consider an arbitrary law (in one dimension)

$$u(t) = f(\Delta t). \tag{2.38}$$

Scale invariance implies that a change in t-scale can be absorbed into a change in u-scale;

$$u(t)/U(T) = f(\Delta t/T), \tag{2.39}$$

for all T. The time derivative of (2.39) is

$$\dot{u}(t)/U(T) = -f'(\Delta t/T)/T. \tag{2.40}$$

Now we put $T = \Delta t$ and divide (2.40) by (2.39), which gives

$$\frac{d\ln u}{dt} = \frac{f'(1)}{f(1)}\frac{d\ln \Delta t}{dt}. \tag{2.41}$$

Integrating (2.41), we find

$$u = A(\Delta t)^\alpha, \quad \alpha = f'(1)/f(1),$$

where the amplitude A is a constant of integration. Indeed, the assumption (2.39) of scale invariance leads to a power law!

The scale invariance of the solution (2.36) is of course inherited from the scale invariance of the original differential equation, say (2.10). A more general version of (2.10) would have been

$$\dot{u} = F(u), \tag{2.42}$$

with an *arbitrary* function $F(u)$. However, near a singularity, that is for large u, one expects a single mechanism to survive as dominant. This statement corresponds to the domination of a particular power $F(u) \propto u^n$ in the limit. Our previous example for crossover (2.16) illustrates this, in that the cubic power $F(u) \propto u^3$ takes over eventually. The parameter ϵ introduces a specific scale into the problem. Only when u has become much larger than the crossover scale $u_{\mathrm{cr}} = 1/\epsilon$ is the asymptotic scale-invariant behavior observed.

Example 2.4 (Exponential law) As a counterexample, consider the equation

$$u_t = e^u \equiv F(u),\tag{2.43}$$

for which $F(u)$ is not expandable in any power law. The solution is

$$u = -\ln(t_0 - t)\tag{2.44}$$

with $t_0 = e^{-u_0}$. Thus we find an approach to blowup slower than that for any power law.   □

Exercises

2.1 To illustrate the crossover between two power laws, we considered (2.16):

$$\dot{c} = c^2 + \epsilon c^3.$$

(i) Show that an exact solution (in implicit form) is given by

$$t - \bar{t} = -\frac{1}{c} + \epsilon \ln\frac{1 + \epsilon c}{c}.$$

(ii) Calculate the singularity time t_0 for a given initial condition $c_0 = c(0)$.

2.2 Look at the crossover displayed by (2.16) using matched asymptotic expansions. The initial condition is $c(0) = c_0$. According to (2.19), the similarity form of the crossover is expected to be

$$c = \frac{1}{\epsilon}f(\xi), \quad \xi = \frac{\Delta t - \epsilon/2}{\epsilon}.$$

(i) Solve the equation for f in implicit form.
(ii) Identify the asymptotic solution for early times and the solution close to the singularity by taking appropriate limits of ξ.
(iii) Show by matching that the singularity time is

$$t_0 = c_0^{-1} + \epsilon \ln(\epsilon c_0) + O(\epsilon^2),$$

and compare it with the result of the previous exercise.

2.3 Consider a generalization of (2.16),

$$\dot{c} = c^n + \epsilon u^m,$$

where $n > 1$ and $m > n$. Show that the crossover time is

$$\Delta t_c \approx \frac{\epsilon^{(n-1)/(m-n)}}{m-1}.$$

and the time shift between the early-time and late-time asymptotics is

$$\delta \approx \frac{(m-n)\epsilon^{(n-1)/(m-n)}}{(n-1)(m-1)}.$$

2.4   In our treatment of (2.20) using matched asymptotics, the solution (2.25) for the inner dynamics is identical with the exact solution. In general this will not be the case, for example when we are considering (2.35),

$$\dot{u} = u^2 - \epsilon u^3 - ru,$$

where the linear term is included. Redo the matched asymptotic analysis for this more general model! Then compare your answer with a numerical solution of the equation.

2.5   In [72], the following differential equation is proposed for the structure function $D(r)$ which measures the kinetic energy contained in turbulent motion of size r:

$$\epsilon = \frac{3}{2} \left(v + \frac{\beta}{\epsilon} D^2(r) \right) \frac{D'(r)}{r}.$$

The boundary condition is $D(0) = 0$.
(i) Show that $D(r)$ satisfies

$$D^3(r) + \frac{3}{\beta} \epsilon v D(r) - \frac{\epsilon^2 r^2}{\beta} = 0.$$

(ii) Identify ranges of r for which D satisfies

$$D(r) = \frac{\epsilon}{3v} r^2$$

and

$$D(r) = \beta^{-1/3} (\epsilon r)^{2/3},$$

respectively.

2.6   In the case of singularities, the uniqueness of a solution is not guaranteed. As an example, the right-hand side of $\dot{x} = f(x)$ with $f(x) = \sqrt{x}$ for $x > 0$ and $f(x) = 0$ for $x < 0$ is non-differentiable at $x = 0$. Construct real solutions $x(t)$ for all $t \in \mathbb{R}$ for the initial value problem $x(0) = x_0$.
(i) Consider the cases $x_0 > 0$ and $x_0 = 0$ separately. For which case is there a unique solution?
(ii) Discuss the regularity of the solutions.

2.7   Consider the gradient dynamics $\dot{x} = -V'(x)$, with

$$V(x, a) = -x^3/3 - ax.$$

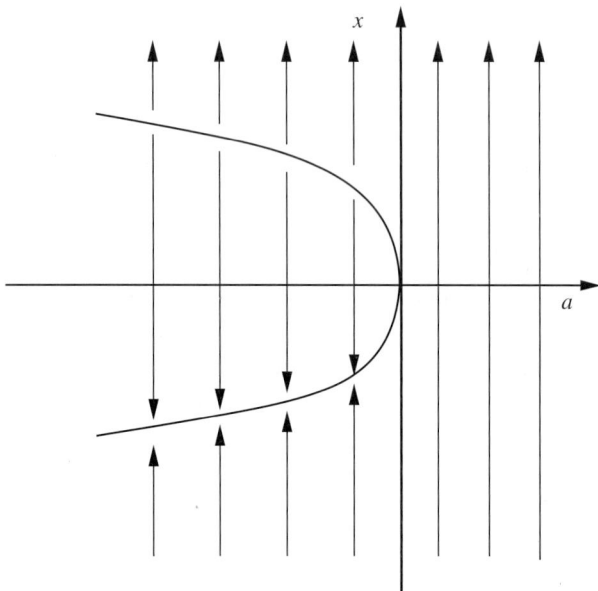

Figure 2.5 A saddle-node bifurcation.

(i) Sketch $V(x)$ for $a > 0$ and for $a < 0$. Which case contains (stable) equilibria?

(ii) Calculate the equilibrium points as function of a; see Fig. 2.5.

(iii) Show that the lower branch is linearly stable and the upper unstable; confirm the flow lines shown in the figure.

(iv) For each a, find the range of initial conditions for which a singularity occurs, for which $x \to \infty$ in finite time.

(v) What is special about the point $a = 0$?

(vi) Suppose that, for a given $a < 0$, x is initially on the lower branch. Describe what happens as a is raised slowly to positive values.

(vii) The scenario described above is known as a saddle-node bifurcation. Argue why the behavior is generic for *any* stationary solution curve that folds over to produce a vertical tangent in the ax-plane.

2.8   Consider the second-order ODE

$$\frac{d^2 y}{dx^2} + \frac{\alpha}{x}\frac{dy}{dx} + \frac{\beta}{x^2}y = 0, \quad x > 0. \tag{2.45}$$

Look for solutions of the form $y(x) = x^\gamma$ and discuss their asymptotic behavior as $x \to 0^+$. For what values of α and β are the solutions bounded? For what values α and β are the solutions highly oscillatory as $x \to 0^+$?

2.9 The following equations are such that their solutions blow up as $t \to t_0$, cf. Example 2.4. Find the asymptotic behavior of each solution close to the blowup time t_0.

(i) $u_t = \exp(\exp u)$,

(ii) $u_t = \exp(\exp(\exp \ldots (\exp x)))$, where exp occurs n times.

2.10 Consider the following ODEs:

(i) $V_\tau = -V^q$ with $q > 1$,

(ii) $V_\tau = -e^{-1/V}/V$.

Analyze the asymptotic behavior of the solutions for $\tau \to \infty$ such that $V \to 0$. This will be relevant to the analysis of Exercises 9.4 and 9.8.

2.11 According to the Malthus growth model, the rate of growth of the number of individuals $N(t)$ in a population is proportional to N, with a proportionality constant which is the difference between the birth rate a and the death rate b.

(i) Show that this simple law implies an exponential growth in $N(t)$.

(ii) A more realistic model takes into account the "fight for limited resources". This results in an extra death rate that is proportional to the numbers of encounters between individuals, which is roughly equal to the square of the number of individuals. Show that this leads to the "logistic equation"

$$\frac{dN(t)}{dt} = (a - b)N(t) - kN^2(t), \tag{2.46}$$

and to a limited maximum possible number of individuals $N(t)$.

2.12 A method for the control of a population of insects consists in releasing a certain number of infertile individuals [159]. If the number of infertile individuals n is kept constant then the following equation for the number $N(t)$ of fertile individuals can be deduced:

$$\frac{dN}{dt} = \left(\frac{aN}{N+n} - b \right) N - kN(N+n), \tag{2.47}$$

where $a, b, k > 0$ are constants.

(i) Discuss the origins of the different terms in (2.47).

(ii) Find the minimum number n of infertile individuals needed in order to eradicate the population of insects.

3

Similarity profile

3.1 The spatial structure of blowup

Most problems that interest us here are formulated in terms of PDEs. But even the hyperbolic growth equation (2.7) acquires an interesting spatial structure if the initial condition is a function of the spatial coordinate x,

$$\frac{\partial u}{\partial t} = u^2(x, t), \quad u(x, 0) = u_0(x), \tag{3.1}$$

u having been normalized to ensure that $K = 1$. Since each spatial point x evolves independently according to (3.1), it is easy to write down the solution for general initial condition $u_0(x)$ in the form (2.9):

$$u(x, t) = \frac{1}{t_s(x) - t}, \quad t_s(x) = \frac{1}{u_0(x)}. \tag{3.2}$$

Each point appears to blow up at its own local blowup time $t_s(x)$, each of which is different except in the non-generic case of (locally) *constant* initial data (in which case we return to the solution (2.9) discussed in the previous chapter). The *first* blowup t_0 occurs where u is a maximum, at $x = x_0$:

$$t_0 = \frac{1}{\max_x\{u_0\}} \equiv \frac{1}{u_0(x_0)}. \tag{3.3}$$

As seen in Fig. 3.1, the spatial structure of the solution (3.2) is nontrivial in spite of its simplicity and leads to localized blowup around the singular point (x_0, t_0).

The singular problem defined by (3.1) will turn out to share many common features with "real" PDE problems for which an exact solution is not available. We will therefore use (3.1) to investigate blowup in detail, introducing many tools used later. Of course all the information is contained already in the exact solution (3.2), a local analysis of which can be used to check our results.

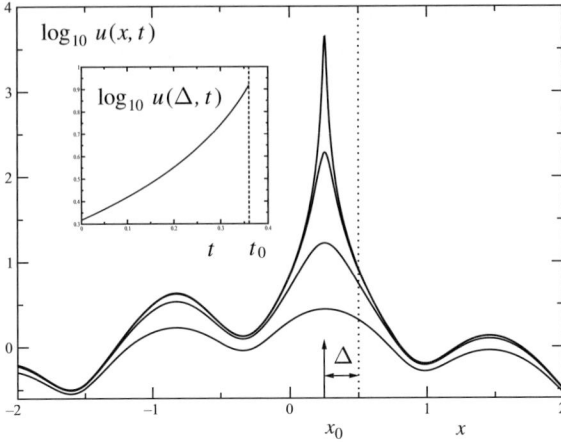

Figure 3.1 A sequence of solutions of (3.1), which are described by (3.2), for the initial condition $u_0(x) = (\sin 5x + 2)/(1 + x^2)$. The singularity occurs for $x_0 = 0.256\,414\,117\,1\dots$ and $t_0 = 0.360\,219\,531\,7$; profiles are shown for $t = 0, 0.3, 0.355$, and 0.36. For clarity, $\log_{10} u$ rather than u has been plotted. The dotted line is at $\Delta \equiv x' = 0.5 - x_0 = 0.14$. Observe that at this point the profile remains "frozen" as the singularity time is approached. It reaches a finite value in the limit $t' \to 0$; see inset.

The temporal and spatial distances from the singularity are measured by

$$t' = t_0 - t, \quad x' = x - x_0, \tag{3.4}$$

defined so as to make $t' > 0$ before the singularity; this ensures that t' can be raised to an arbitrary power. After the singularity we have $t > t_0$ and the sign of t' has to be reversed, as we will see below. As a natural extension of the previous chapter, where the growth of u was described by a power law, we try the *similarity solution*

$$u(x', t') = t'^{\alpha} f(\xi), \quad \xi = \frac{x'}{t'^{\beta}}. \tag{3.5}$$

This means that both the dependent and independent variables are rescaled by a power law as the singularity is approached. In terms of the spatial structure, (3.5) means that the profile *remains the same* as time goes by. The only thing that happens is that the coordinates are rescaled by appropriate factors.

The time derivative of the similarity solution (3.5) is calculated using the chain rule:

$$\frac{\partial u}{\partial t} = -\alpha t'^{\alpha-1} f(\xi) + t'^{\alpha} \beta \frac{x'}{t'^{\beta+1}} f'(\xi),$$

noting the extra negative sign which occurs on account of the definition of t'. Using $\xi = x'/t'^\beta$ on the right-hand side, we can eliminate x' in favor of t' and the similarity variable ξ. Thus one obtains for the left-hand side of (3.1)

$$\frac{\partial u}{\partial t} = t'^{\alpha-1}\left[-\alpha f(\xi) + \beta\xi f'(\xi)\right], \tag{3.6}$$

an expression that will be used frequently throughout this book. Two terms appear on the right-hand side of (3.6); by taking the time derivative, the power α has been reduced by one relative to (3.5). For the right-hand side of (3.1) one obtains

$$u^2 = t'^{2\alpha} f^2(\xi); \tag{3.7}$$

so, comparing powers with (3.6), it follows that $\alpha = -1$, as in the space-independent case (2.9). Combining (3.6) and (3.7) one finds the similarity equation

$$f + \beta\xi f' = f^2, \tag{3.8}$$

where β is undetermined on the basis of the local structure of the equation of motion (3.1) alone. This situation is known as *self-similarity of the second kind*. Let us follow directly with another example of self-similar behavior in which both exponents are fixed as part of the structure of the equation; this is known as *self-similarity of the first kind*.

Example 3.1 (Self-similarity of the first kind) We add a diffusion term to the right-hand side of (3.1), to obtain

$$\frac{\partial u}{\partial t} = u^2(x, t) + u_{xx}. \tag{3.9}$$

While the first term on the right describes, say, a chemical reaction which leads to localized growth, the second term tends to broaden the peak by diffusing the reactant. Accordingly, (3.9) is known as a reaction–diffusion equation.

Curiously, however, (3.9) displays a particular kind of degeneracy, which we will not study until Chapter 9 and which involves the appearance of terms logarithmic in t'. As a result, the approach to the asymptotic self-similar law proceeds on a slow logarithmic time scale. To avoid this we add another term of the same order to the right of (3.9), chosen for the sole purpose of furnishing a convenient, exactly solvable, example:

$$\frac{\partial u}{\partial t} = u^2 + u_{xx} - 2\frac{u_x^2}{u}. \tag{3.10}$$

In fact, using the transformation $u = 1/h$, (3.10) turns into the "pinch-off" problem

$$\frac{\partial h}{\partial t} = -1 + h_{xx},$$ (3.11)

which is formally linear but where $h = 0$ corresponds to the singularity.

Inserting the similarity solution (3.5) into (3.10), we find $\alpha = -1$ as before, from a balance of the time derivative with the first term on the right. The spatial derivatives are calculated using the chain rule as

$$u_x = t'^{\alpha-\beta} f'(\xi), \quad u_{xx} = t'^{\alpha-2\beta} f''(\xi),$$ (3.12)

which are formulae to be used frequently throughout this book. Hence, balancing u_{xx} with $u^2 \propto t'^{2\alpha}$ leads to $-2 = -1 - 2\beta$ or $\beta = 1/2$. The last term on the right of (3.10) is of order $t'^{\alpha-2\beta} = t'^{-2}$, which is of the same order as the other terms. Thus, for a self-similar solution to exist, the similarity function f has to satisfy

$$-f + \frac{\xi}{2} f' = f^2 + f'' - 2\frac{f'^2}{f}.$$ (3.13)

Note that the exponent $\beta = 1/2$ is fixed by a balance of terms in (3.10), corresponding to self-similarity of the first kind. □

Returning to the similarity equation (3.8) of the hyperbolic growth law, it is checked easily that a general solution is

$$f(\xi) = \frac{1}{1 + a\xi^{1/\beta}},$$ (3.14)

where a is a constant of integration. To find the parameters a and β, we must impose two conditions on (3.14). The first is a *matching condition*, which ensures that the local solution (3.14) is consistent with an outer solution which is evolving slowly on the time scale of the singularity. More formally, we demand that at any fixed distance $x' = \Delta$ from the singularity (which is at $x' = 0$) the solution must obey the matching condition

$$\lim_{t' \to 0} u(\Delta, t') = \lim_{t' \to 0} t'^{\alpha} f\left(\frac{\Delta}{t'^{\beta}}\right) = u(\Delta, 0),$$ (3.15)

where $u(\Delta, 0)$ is finite; this requirement is illustrated by the inset in Fig. 3.1. Essentially, it guarantees that the singularity only occurs at a point. The matching condition (3.15) can be converted into a boundary condition for the similarity function $f(\xi)$ as $\xi \to \pm\infty$. Namely, putting $\xi = \Delta/t'^{\beta}$, in the limit $t' \to 0$ we have

$$f(\xi) = u(\Delta, 0) t'^{-\alpha} = u(\Delta, 0) \left(\frac{\xi}{\Delta}\right)^{\alpha/\beta}.$$

The limit $t' \to 0$ corresponds to $\xi \to \pm\infty$, depending on whether Δ is positive or negative, i.e. whether (3.15) is applied to the left or the right of the singularity. In either case the singular solution must match to a static outer solution, and thus we find the boundary condition

$$f(\xi) \propto \xi^{\alpha/\beta}, \quad \xi \to \pm\infty. \tag{3.16}$$

The growth condition (3.16) is satisfied automatically by (3.14) if $a \neq 0$ but is *inconsistent* with $a = 0$.

Next we impose a *regularity condition* which demands that (3.14) be regular for all $\xi \in \mathbb{R}$. This ensures that there is no other singularity before the one we have been studying at $t = t_0$. First, $1/\beta$ must be a positive integer otherwise (3.14) would have a singularity for $\xi = 0$. Second, $1/\beta$ must be even or else (3.14) will develop a pole as the denominator goes to zero. Another consequence of this reasoning is that we must have $a > 0$ otherwise f would once more become infinite. But this means that β can only take on a discrete set of values,

$$\beta_i = \frac{1}{2(i+1)}, \quad i = 0, 1, \ldots \tag{3.17}$$

Correspondingly, there is a discretely infinite series of allowed solutions of (3.8):

$$\bar{f}_i(\xi) = \frac{1}{1 + a\xi^{2(i+1)}}, \quad i = 0, 1, 2, \ldots, \tag{3.18}$$

where we have denoted the similarity solution by an overbar. The first two similarity solutions are shown in Fig. 3.2. Another way of formulating the regularity condition (3.17) is that at the blowup point $\xi = 0$ the similarity profile $\bar{f}$ must have the regular expansion

$$\bar{f}_i(\xi) = 1 - a\xi^{2i+2} + O(\xi^{2i+4}), \quad i = 0, 1, 2, \ldots \tag{3.19}$$

The "ground state" solution $i = 0$ corresponds to a quadratic maximum; the "excited states" have maxima of quartic order or higher.

We see that the scaling exponent β is determined not by any local property of the equation of motion but by the regularity condition (3.19), which inherits its properties from the initial condition. To demonstrate this explicitly, we use the general solution (3.2) and assume that the initial condition has the regular form

$$u_0(x) = \bar{u} - Ax^{2i+2} \tag{3.20}$$

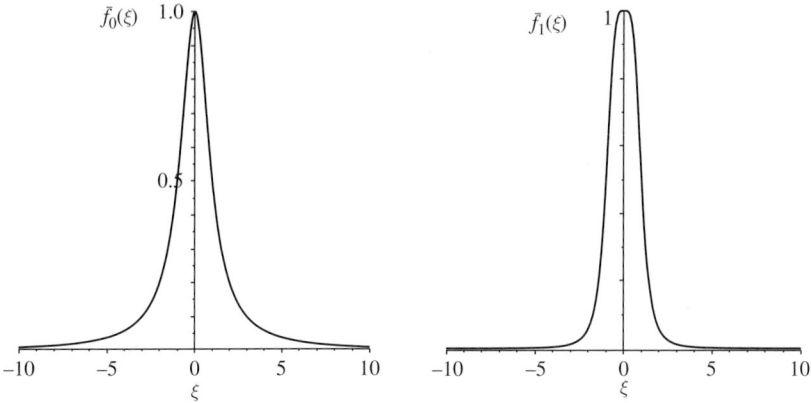

Figure 3.2 The first two similarity solutions (3.18), with the generic case $i = 0$ (left) and the first unstable solution $i = 1$ (right). The constant a has been chosen as 1. The orders of the maxima are quadratic and quartic, respectively.

near the maximum. Then blowup occurs for $t_0 = 1/\bar{u}$ at $x_0 = 0$. Expanding

$$u = \frac{1}{\left(\bar{u} - Ax^{2i+2}\right)^{-1} - t}$$

for small $x = x'$, we find

$$u = t'^{-1}\left(1 - \frac{A}{\bar{u}^2}\frac{x^{2i+2}}{t'}\right). \tag{3.21}$$

Comparison of (3.21) and (3.19) shows that the index i of the similarity solution is the corresponding value for the local maximum of the initial condition (3.20) and that

$$a = \frac{A}{\bar{u}^2}. \tag{3.22}$$

In particular, if the typical scale A of the initial condition assumes a singular limit ($A \to \infty$ or $A \to 0$) then the limit of the similarity solution (3.18) is also singular. This is a key feature of self-similarity of the second kind, as emphasized in [15, 14].

In addition, since the behavior around a generic maximum is quadratic, the generic local behavior will be given by (3.19) with $i = 0$. Thus (3.21) shows that only the ground state solution ($i = 0$) is expected to be observed. Should the local behavior (3.20) of the initial condition be non-generic ($i > 0$), then any small perturbation will generate an (albeit small) quadratic component and drive the dynamics towards the ground state solution. In Section 3.2 below we will demonstrate this more formally, showing that the ground state solution alone is stable. By contrast, the similarity equation (3.13) of Example 3.1 allows for just a single similarity solution.

Example 3.2 (Similarity solution for (3.10)) According to the matching condition (3.16), we have to find a solution to (3.13) on the real line which behaves like $f \propto \xi^{-2}$ for $\xi \to \pm\infty$. Applying the transformation $f = 1/g$, (3.13) turns into the linear equation

$$g - \frac{\xi}{2}g' = 1 - g''.$$

As can be checked by inspection, this equation has the general solution

$$g = 1 + a(\xi^2 - 2) + b\left[\sqrt{\pi}(\xi^2 - 2)\text{erfi}\left(\frac{\xi}{2}\right) - 2\xi e^{\xi^2/4}\right], \tag{3.23}$$

where

$$\text{erfi}(x) = \frac{2}{\sqrt{\pi}} \int_0^x e^{t^2}\, dt$$

is known as the imaginary error function. Clearly (and we will find this to be a typical behavior for many more examples to come) the term in square brackets on the right of (3.23) has a superexponential behavior at infinity. In fact, using formula 7.2.14 of [1], its leading asymptotic behavior is $16e^{\xi^2/4}/\xi^3$.

Thus, to find the required quadratic growth at infinity we need to have $b = 0$, and so the similarity profile is

$$\bar{f} = \frac{1}{1 + a(\xi^2 - 2)}, \tag{3.24}$$

where a is undetermined. However, for $\bar{f}$ to remain finite for all ξ we need to have $0 < a < 1/2$, otherwise $\bar{f}$ would have a pole. Let us compare this result with the sequence of similarity solutions (3.18) found for (3.1), which does not have the derivative terms on the right. There is a similar degeneracy in the amplitude a, and the first solution of the sequence is almost the same as (3.24). However, (3.24) only has a single admissible value, $\beta = 1/2$, which is determined from a balance of terms in the equation irrespective of the initial condition.

This is illustrated in Fig. 3.3, which shows the solution for an initial condition exhibiting a quartic maximum. This can be seen as the singular limit of an initial condition for which the quadratic coefficient vanishes exactly. In the case of (3.1) this would have led to the selection of a similarity solution with $\beta = 1/4$, again having a quartic maximum. In the present similarity problem

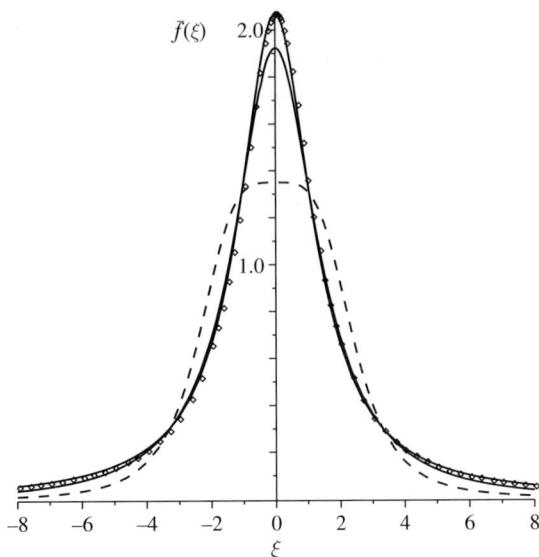

Figure 3.3 A sequence of solutions of (3.10) with initial condition $u = 1/(1 + bx^4)$ and $b = 0.016$. The solutions are rescaled to similarity variables, and the fixed point solution (3.24) is represented by dots. The broken line is the initial condition at $t' = 1.349\ldots$, and two more solutions are shown, at $t' = 0.2$ and $t' = 0.01$. For the construction of the exact solution, see Exercise 3.8.

of the first kind, however, the similarity solution (3.24) remains regular in this singular limit [14], as shown explicitly in Exercise 3.8. □

3.2 Stability

To establish which similarity solution from the set (3.18) is actually observed, one must look at the stability of each. This looks like a difficult task, since the solutions are strongly time dependent. However, there is an ingenious way around this problem [92, 93]: one introduces a new *logarithmic* time variable $\tau = -\ln t'$ in the place of t'. This means that the original problem in coordinates (x', t') is transformed to a new coordinate system (ξ, τ):

$$u(x', t') = t'^{\alpha} f(\xi, \tau). \tag{3.25}$$

Notice that $d\tau/dt' = -1/t'$, so the original equation (3.1) is transformed into

$$f_\tau = -f - \beta \xi f' + f^2. \tag{3.26}$$

The transformed equation of motion (3.26) has been written in a way that highlights the fact that for $f_\tau = 0$ (3.26) reduces to the similarity equation (3.8). Each similarity solution (3.18) is a *fixed point* of the transformed formulation (3.26), so the stability problem is now reduced to investigating the neighborhood of these fixed points.

This motivates us to call (3.26) the *dynamical system* description of the original problem. Although (3.26) is formally equivalent to (3.1), its dynamics are dominated by the fixed point. If the fixed point is stable, the solution will be attracted exponentially to it and the dynamics are determined by the fixed point alone. If there are unstable or neutral directions, it will be those which determine the dynamics. If the neutral or unstable directions are few, a problem that has infinitely many dimensions will be transformed into a manageable low-dimensional system.

To analyze the stability of the fixed point we linearize about it and look for eigenvalues. This means we need to use the ansatz

$$f(\xi, \tau) = \bar{f}(\xi) + \delta e^{\nu\tau} P(\xi), \tag{3.27}$$

where ν is the eigenvalue and δ a small parameter. If ν is negative, (3.27) tends toward the fixed point $\bar{f}$; if ν is positive, the solution is driven away from it. We insert (3.27) into (3.26), and equate terms linear in δ:

$$\nu P(\xi) = -P(\xi) - \beta\xi P'(\xi) + 2\bar{f}(\xi)P(\xi) \equiv \mathcal{L}P. \tag{3.28}$$

Using the fixed point solutions (3.18), this yields the eigenvalue equation

$$(\nu + 1)P(\xi) + \frac{\xi}{2(i + 1)}P'(\xi) = 2\frac{P(\xi)}{1 + a\xi^{2(i+1)}}. \tag{3.29}$$

The solution to this equation is found easily by separation of variables:

$$P(\xi) = \frac{\left(\xi^{2(i+1)}\right)^{1-\nu}}{\left(1 + a\xi^{2(i+1)}\right)^2}. \tag{3.30}$$

For (3.30) to be regular at the origin, we must have

$$2(i + 1)(1 - \nu) = j, \quad j = 0, 1, 2, \ldots$$

Thus, for each fixed point, labeled by i, there is an infinite discrete sequence of eigenvalues and eigenfunctions:

$$\nu_j^{(i)} = 1 - \frac{j}{2(i + 1)}, \quad P_j^{(i)}(\xi) = \frac{\xi^j}{\left(1 + a\xi^{2(i+1)}\right)^2}, \quad j = 0, 1, 2, \ldots \tag{3.31}$$

In Fig. 3.4 we show the first four eigenfunctions determining the stability of the ground state ($i = 0$) solution.

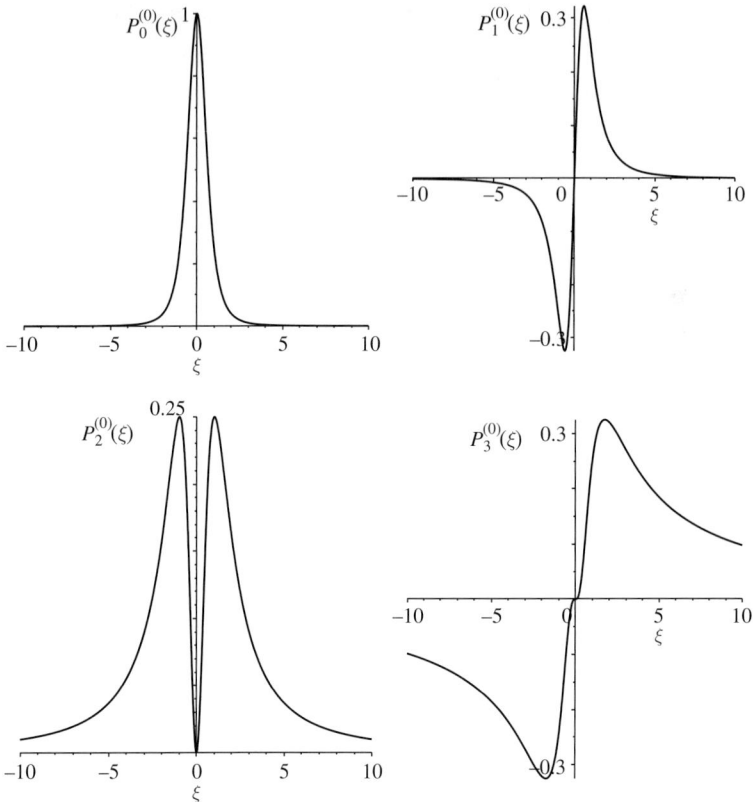

Figure 3.4 The first four eigenfunctions determining the stability of the ground
state solution ((3.18), $i = 0$); they are given by (3.31) with $j = 0, 1, 2, 3$. The
first two (at top left and right) describe the instability with respect to a shift in t_0 or
x_0, respectively. The third (at bottom left) describes an indeterminate stability with
respect to a shift in the parameter a; the fourth corresponds to a stable direction in
solution space.

Note that (3.30), (3.31) have the "ladder structure" typical of bound states,
for example in quantum mechanics: with each new value of i, two new positive eigenvalues are generated. All eigenvalues are real, and the eigenfunctions
alternate between even and odd. For the first three solutions the sequence of
eigenvalues is

$$\nu^{(0)} = 1, 1/2, 0, -1/2, \ldots \quad \text{ground state,}$$
$$\nu^{(1)} = 1, 3/4, 1/2, 1/4, 0, -1/4, \ldots \quad \text{first excited state,} \qquad (3.32)$$
$$\nu^{(2)} = 1, 5/6, 2/3, 1/2, 1/3, 1/6, 0, -1/6 \ldots \quad \text{second excited state,}$$

and so forth. Thus at first sight it looks as if *all* the similarity solutions (3.18) are unstable, since all have positive eigenvalues. We will now show that in fact there are *always* two positive eigenvalues, $\nu = 1$ and $\nu = \beta$; this is an artifact of representing the dynamics in similarity variables and does not indicate instability.

The origin of these two unstable directions in the space of solutions is easy to understand. If the initial condition is perturbed, the position of the singularity (x_0, t_0) in space and time will be perturbed as well. But this means that $x' = 0$ no longer corresponds to the singularity, and $u(x, t)$ will in fact be finite at this point as $t' \to 0$. This can only mean that the solution (which still blows up, albeit at a different point) will be driven away from the similarity solution. In the same way, an unstable direction follows from a shift in the time variable.

To calculate the unstable modes explicitly note that, owing to translational invariance,

$$u^{(\Delta)}(x', t') = t'^{\alpha} \bar{f}\left(\frac{x' + \Delta}{t'^{\beta}}\right) \equiv t'^{\alpha} f^{(\Delta)}(\xi, \tau) \tag{3.33}$$

must be a solution of (3.26) for any spatial shift Δ. Expanding in Δ we obtain

$$f^{(\Delta)}(\xi, \tau) = \bar{f}(\xi) + \Delta t'^{-\beta} \bar{f}_{\xi} + O(\Delta^2) \equiv \bar{f} + \Delta e^{\beta \tau} \bar{f}_{\xi} + O(\Delta^2). \tag{3.34}$$

Now comparing with (3.27), the term linear in Δ must be a solution of the linearized equation with $\nu = \beta$ and eigenfunction $\bar{f}_{\xi}$. Similarly, defining a time-translated solution by

$$c^{(\Delta)}(x', t') = (t' + \Delta)^{\alpha} \bar{f}\left(\frac{x'}{(t' + \Delta)^{\beta}}\right) \equiv t'^{\alpha} f^{(\Delta)}(\xi, \tau), \tag{3.35}$$

the linearized version becomes

$$f^{(\Delta)}(\xi, \tau) = \bar{f}(\xi) + \Delta e^{\tau}\left(\alpha \bar{f}_{\xi} - \beta \xi \bar{f}_{\xi}\right) + O(\Delta^2). \tag{3.36}$$

Thus now the eigenvalue is $\nu = 1$ with eigenfunction $\alpha \bar{f} - \beta \xi \bar{f}_{\xi}$. The eigenfunctions and eigenvalues stemming from translational invariance in time and space are reported in Table 3.1.

For the ground state solution $(i = 0)$ the eigenvalues predicted by this argument are 1 and 1/2, which are exactly the positive eigenvalues (3.32) of the ground state sequence. For the next solution $(i = 1)$, the eigenvalues are 1 and 1/4, which indeed appear in the sequence, but other positive eigenvalues remain. The next eigenvalue in the ground state series is zero, which comes from the fact that (3.18) is a *continuous* family of solutions. Thus changing a must point in a neutral direction, and

Table 3.1 *A summary of known eigenvalues of the stability problem for a similarity solution of the form (3.5).*

Eigenvalue	Eigenfunction	Origin
$\nu = 1$	$\alpha \bar{f} - \beta \xi \bar{f}_\xi$	time translation
$\nu = \beta$	$\bar{f}_\xi$	space translation
$\nu = 0$	$\bar{f}_a$	invariance under a-translation

$$\frac{\partial \bar{f}}{\partial a} = -\frac{\xi^{2(i+1)}}{\left(1 + a\xi^{2(i+1)}\right)^2} \equiv -P^{(i)}_{2(i+1)}(\xi), \qquad (3.37)$$

which is indeed the eigenfunction corresponding to $\nu = 0$, also listed in Table 3.1.

All the other eigenvalues, however, are *negative* for the ground state $i = 0$, and thus this solution is stable. But turning to the higher-order similarity solutions, for each step up the ladder, there are two more positive eigenvalues, which cannot be explained away. Thus all the other solutions are unstable, and a small perturbation will drive the singular solution away from the corresponding fixed point (3.18). Of course, in a problem as simple as (3.1) the origin of this structure is understood easily. We have seen that the similarity solution is selected by the behavior (3.20) of the initial condition near its maximum. If the maximum is quadratic, a perturbation will change the prefactor slightly or will result in a shift; however, the maximum will still be quadratic and the ground state similarity solution is approached. If, however, the maximum is quartic (say) then a generic perturbation will contain a quadratic component, and one will be driven away from the $i = 1$ similarity solution and cross over to the generic $i = 0$ solution.

To summarize, near a stable fixed point $\bar{f}$ the dynamics can be represented by expanding in eigenfunctions:

$$u(x', t') = t'^\alpha \left(\bar{f}(\xi) + \sum_{j=0}^{\infty} a_j t'^{-\nu_j} P_j(\xi) \right). \qquad (3.38)$$

By shifting the values of t_0, x_0, and a, the eigenfunctions corresponding to the eigenvalues 1, β, and 0 are generated. Thus by adjusting t_0, x_0, and a suitably, the prefactors a_0, a_1, and a_2 of the first three eigenfunctions can be made to vanish; see Exercise 3.9. As a result, the ground state solution can be written in the form

$$u(x', t') = t' \left(\frac{1}{1 + a\xi^2} + t'^{1/2} \sum_{j=3}^{\infty} a_j \frac{t'^{(j-3)/2} \xi^j}{(1 + a\xi^2)^2} \right), \qquad (3.39)$$

where the values of the constants a_j depend on the initial condition. In the limit $t' \to 0$, the correction terms go to zero and the fixed point is approached.

In the case of our example (3.10) of self-similarity of the first kind, the stability analysis of the similarity solution (3.24) is similar to the analysis of the ground state solution:

Example 3.3 (Stability analysis for (3.10)) For simplicity, we consider the stability of the similarity solution

$$\bar{g} = 1 + a(\xi^2 - 2) \tag{3.40}$$

for the dynamics (3.11), which is related to (3.10) by the transformation $h = 1/u$. As before, we transform to self-similar variables using a logarithmic time $\tau = -\ln t'$:

$$h = t'g(\xi, \tau), \quad \xi = x'/t'^{1/2}. \tag{3.41}$$

Then the "dynamical system" for (3.11) becomes

$$g_\tau = g - \xi g'/2 - 1 + g'', \tag{3.42}$$

whose fixed point is the similarity solution (3.40).

Inserting the ansatz

$$g(\xi, \tau) = \bar{g}(\xi) + \delta e^{\nu \tau} P(\xi) \tag{3.43}$$

into (3.42), we see that the stability is controlled by the eigenvalue equation

$$\nu P = P - \xi P'/2 + P'', \tag{3.44}$$

where ν is the eigenvalue. To find the boundary conditions to be satisfied by $P(\xi)$, we note that the time derivative

$$h_t = -\bar{g} + \xi \bar{g}_\xi/2 + \delta \left(\nu P - P + \xi P_\xi/2 \right) e^{\nu \tau}$$

must vanish for $\xi \to \pm \infty$ for the singular solution to be matchable to a static outer solution. The matching condition (3.16) corresponds to $\bar{g} \propto \xi^2$, which ensures that the first two terms, taken together, vanish. For the term proportional to δ to vanish as well, $P \propto \xi^{2-2\nu}$ for $\xi \to \pm \infty$ must be satisfied.

In Exercise 3.10 we show that solutions to (3.44) grow exponentially unless the eigenvalue satisfies $\nu_i = 1 - i/2$, with $i = 0, 1, 2, \ldots$ Solutions having these eigenvalues are $P_i(\xi) = H_i(\xi/2)$, where

$$H_n(y) = (-1)^n e^{y^2} \frac{d^n}{dy^n} e^{-y^2} \tag{3.45}$$

are the Hermite polynomials [135]. The first few eigenfunctions, obtained easily from (3.45), are listed in Table 3.2. Clearly the solutions satisfy the required

Table 3.2 *The first few eigenvalues and eigenfunctions of (3.44).*

i	v_i	$P_i(\xi)$	Parity ($\pm$)
0	1	1	$+$
1	1/2	ξ	$-$
2	0	$\xi^2 - 2$	$+$
3	$-1/2$	$\xi^3 - 6\xi$	$-$
4	-1	$\xi^4 - 12\xi^2 + 12$	$+$

boundary condition $P_i \propto \xi^i$ for $\xi \to \pm\infty$. Just as for the ground state solution of (3.8), the first three eigenvalues are $v = 1, 1/2$ and 0, corresponding to invariances under time translation, space translation, and a shift in the free parameter a. In Exercise 3.11 the reader is invited to develop an expansion analogous to (3.38) and to show that the first three terms can be eliminated using these invariances. All remaining eigenvalues are negative, which shows that the similarity solution (3.24) is stable.   □

3.3 Similarity solutions and the dynamical system

Let us review the structure that emerges from the similarity description of a singularity that occurs at a point. We consider an evolution equation of the form

$$u_t = F[u], \tag{3.46}$$

many more examples of which will appear later in the book. On the right-hand side, F is some (nonlinear) differential or integral operator that acts on u. In earlier examples we considered $F[u] = u^2$ (see (3.1)) or $F[u] = u^2 + u_{xx} - 2u_x^2/u$ (see (3.10)). Of course (3.46) could also be a system of equations (in which case u would be a vector) or the spatial variable x may have several dimensions.

The singularity occurs at a single point in space and time, x_0, t_0, and we put $t' = t_0 - t$ and $x' = x - x_0$. We will now assume that (3.46) has an exact similarity solution of the form

$$u(x,t) = t'^\alpha \bar{f}(x'/t'^\beta), \tag{3.47}$$

with appropriately chosen values of the exponents α, β. Two different cases arise here:

(a) *self-similarity of the first kind*   a similarity solution (3.47) only exists for a particular pair of values α, β;

(b) *self-similarity of the second kind*   a solution (3.47) exists for a continuous range of exponents, whose value is determined by a regularity condition.

The equation for the similarity profile $\bar{f}$ is

$$G[\bar{f}] \equiv \alpha \bar{f} - \beta \xi \bar{f}_\xi + F[\bar{f}] = 0, \qquad (3.48)$$

where $F[\bar{f}]$ is exactly the same expression as that for the original equation; the extra terms come from the time derivative (3.6).

For the similarity solution to be matchable to a time-independent outer solution, (3.6) must go to zero to leading order. Thus, for $\xi \to \pm\infty$ the similarity solution must satisfy

$$-\alpha \bar{f} + \beta \xi \bar{f}_\xi = 0, \qquad (3.49)$$

which results in the condition on $\bar{f}$ given earlier:

$$\bar{f}(\xi) \propto \xi^{\alpha/\beta}, \quad \xi \to \pm\infty. \qquad (3.50)$$

To find the similarity profile, (3.48) must be solved subject to the boundary condition (3.50).

Equation (3.47) describes only the strictly self-similar part of the evolution. Clearly, (3.47) will in general not satisfy a given initial condition, so there needs to be at least a crossover period during which the full solution approaches (3.47). As described in the previous section, this approach is studied conveniently by introducing self-similar variables $\tau = -\ln t'$ and $\xi = x'/t'^\beta$. Generalizing (3.47) to

$$u(x,t) = t'^\alpha f(\xi, \tau), \qquad (3.51)$$

(3.46) turns into the *dynamical system*

$$f_\tau = G[f] \equiv \alpha f - \beta \xi f_\xi + F[f]. \qquad (3.52)$$

Clearly, a similarity solution satisfying (3.48) is a *fixed point* of (3.47), which we denote by $\bar{f}(\xi)$. Solutions to the original PDE (3.46) for given initial data can be viewed as orbits in some infinite-dimensional phase phase. The most basic analysis of the dynamics consists in linearizing around the fixed point according to

$$f = \bar{f}(\xi) + \delta H(\xi, \tau), \qquad (3.53)$$

which results in

$$H_\tau = \mathcal{L}H, \qquad (3.54)$$

where $\mathcal{L}$ is a linear differential operator that acts on $\bar{f}$. To solve (3.54) we use the ansatz

$$H(\xi, \tau) = e^{\nu\tau} P(\xi), \tag{3.55}$$

which leads to the eigenvalue equation

$$\nu P = \mathcal{L}P. \tag{3.56}$$

To find the spectrum, (3.56) has to be supplemented with growth conditions for P at infinity. These are analogous to the condition (3.50) for the fixed point solution. The time derivative of

$$u = t'^{\alpha} \left[\bar{f}(\xi) + \delta e^{\nu\tau} P(\xi) \right] \tag{3.57}$$

reads

$$\left(-\alpha \bar{f} + \beta \xi \bar{f}_\xi \right) (t')^{\alpha-1} + \delta \left(\nu P - \alpha P + \beta \xi P_\xi \right) (t')^{\alpha-1} e^{\nu\tau},$$

which again must vanish to leading order to be matchable to a time-independent outer solution. The factor of $(t')^{(\alpha-1)}$ vanishes, according to (3.49), so the condition becomes

$$\nu P - \alpha P + \beta \xi P_\xi = 0 \tag{3.58}$$

with solution

$$P(\xi) \propto \xi^{(\alpha-\nu)/\beta}, \quad \xi \to \pm\infty. \tag{3.59}$$

Note that the eigenfunctions (3.31) behave as $\xi^{j-4(i+1)}$ at infinity, which agrees with (3.59); see Exercise 3.7.

The eigenvalue equation (3.56) with boundary condition (3.59) generally has a discrete spectrum [19]. As discussed above, there are always two eigenvalues resulting from the invariance under time and space translations, with $\nu = 1$ and $\nu = \beta$; see Table 3.1. Apart from these values, all eigenvalues must be non-positive for the fixed point to be stable. The vanishing of eigenvalues results either from a continuous symmetry or is a sign of a *slow* approach to the fixed point, which we will discuss in Chapter 9. The negative eigenvalue ν_c with the smallest absolute value determines the speed at which the self-similar solution $\bar{f}$ is approached. In critical phenomena, ν_c is known as the correction or Wegner exponent [218].

3.4 Regularization

So far we have considered only the dynamics leading up to a singularity at $t = t_0$. By adding a term which regularizes the equation, the singularity can

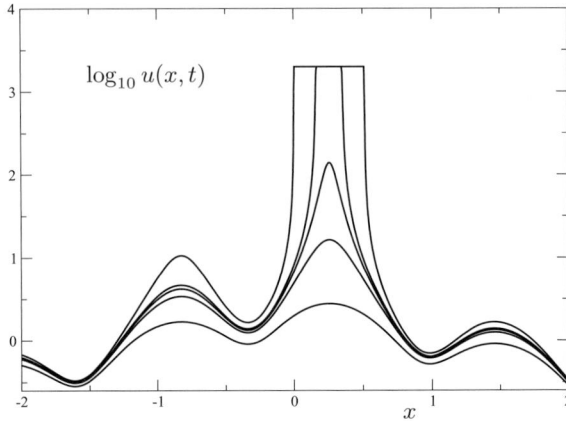

Figure 3.5 A sequence of solutions of (3.60), with initial condition $u_0(x) = (\sin 5x + 2)/(1 + x^2)$ and parameter $\epsilon = 5 \times 10^{-4}$. In the case $\epsilon = 0$, a singularity would occur for $t_0 = 0.360\,219\,531\,7$, as shown in Fig. 3.1. Profiles are shown for $t = 0, 0.3, 0.355, 0.38$, and 0.5; the first three profiles are almost identical to Fig. 3.1 but then the solution saturates around the position of the singularity $x_0 = 0.256\,414\,117\,1\ldots$

be avoided. As an example, we consider the spatially dependent version of the regularized equation (2.20):

$$\frac{\partial u}{\partial t} = u^2(x, t) - \epsilon u^3(x, t), \quad u(x, 0) = u_0(x). \tag{3.60}$$

As seen in Fig. 3.5, u first saturates in the center of the singularity. Subsequently the region where u has reached its asymptotic value spreads, as dictated by the self-similar dynamics.

The asymptotic analysis of the spatial problem is very similar to the boundary layer calculation for the temporal problem (2.20), except that we also need to estimate the size of the *spatial* region where the boundary layer resides. When $\epsilon = 0$, we return to the original equation with local similarity solution

$$u(x', t') = \frac{1}{t'} \bar{f}_0 \left(\frac{x'}{t'^{1/2}} \right) \equiv \frac{1}{t' + ax'^2}, \quad t' = t_0 - t,$$

which blows up. In addition to the previous estimates $t' = O(\epsilon)$ and $u = O(\epsilon^{-1})$, the similarity form also suggests that $x'/t'^{1/2} = O(1)$ and thus that $x' = O(\epsilon^{1/2})$. It is therefore natural to introduce

$$u(x', t') = \frac{1}{\epsilon} g \left(\eta = \frac{x'}{\epsilon^{1/2}}, \bar{t} = \frac{t'}{\epsilon} \right), \tag{3.61}$$

which generalizes (2.23). The scaling function g obeys equation (2.24), as before.

We apply the matching principle (2.26) to find a boundary condition for g. This means that

$$u(x', t') \sim \frac{1}{t' + ax'^2} \quad \text{when} \quad t' + ax'^2 \ll 1 \tag{3.62}$$

has to agree with $g(\bar{t}, \eta)$ in the limit of large arguments. This suggests that we should write the solution to (2.24) in the form

$$\bar{t} + a\eta^2 - \bar{t}_0 = \frac{1}{g} + \ln \frac{1 - g}{g}, \tag{3.63}$$

with leading-order behavior

$$g(\eta, \bar{t}) \sim \frac{1}{\bar{t} + a\eta^2} \quad \text{as} \quad \bar{t} + a\eta^2 \to \infty, \tag{3.64}$$

in agreement with (3.62). To determine $\bar{t}_0$ in (3.63) we have to expand the outer solution (3.62) to the next order in ϵ:

$$u = u_0 + \epsilon u_1 = \frac{1}{t' + ax'^2} + \epsilon u_1, \tag{3.65}$$

where u_1 satisfies

$$\frac{\partial u_1}{\partial t} = 2u_0 u_1 - u_0^3$$

with solution

$$u_1 = \frac{\ln(t' + ax'^2) + C}{(t' + ax'^2)^2}.$$

If in (3.62) the constant a and the singularity time t_0 (which appears in t') have been chosen to match the initial condition at leading order, we find that $u_1(t' = t_0) = 0$ and thus $C = -\ln(t_0 + ax'^2) \approx -\ln t_0$. Expanding g to the next order of $\bar{t} + a\eta^2$, we obtain from (3.63) as $\bar{t} + a\eta^2 \to \infty$

$$g = \frac{1}{\bar{t} + a\eta^2} + \frac{\bar{t}_0 + \ln(\bar{t} + a\eta^2)}{(\bar{t} + a\eta^2)^2}, \tag{3.66}$$

which is the analogue of (2.31). Comparing (3.66) with (3.65), we find

$$\bar{t}_0 \approx \ln \frac{\epsilon}{t_0}, \tag{3.67}$$

completing the solution at this order. To find the approach to the asymptotic value $u = 1/\epsilon$, we have to consider the limit $\bar{t} \to -\infty$ of (3.63), corresponding to $t' < 0$, while $\epsilon \to 0$. Putting $g = 1 - \delta$, this yields

$$\bar{t} + a\eta^2 - \bar{t}_0 - 1 = \ln \delta + O(\delta)$$

or

$$g = 1 - \frac{\exp(\bar{t} + a\eta^2)}{ee^{\bar{t}_0}}.$$

Thus using (3.61) and (3.67), we find

$$u(\eta, \tau) \approx \frac{1}{\epsilon} - \frac{t_0}{e\epsilon^2} \exp\left(\frac{t' + ax'^2}{\epsilon}\right), \qquad (3.68)$$

which is valid for times $t' < 0$ *after* the singularity.

The behavior of the solution is illustrated in Fig. 3.5. The initial dynamics is very similar to that of Fig. 3.1, where $\epsilon = 0$. However, for times $t > t_0$, after the blowup time of the unregularized equation, a plateau of height $u = 1/\epsilon$ forms. The position of the edge of the plateau is determined by where the correction in (3.68) is of the same order as the leading term, i.e. $e^\Delta = \epsilon$ where $\Delta \equiv (t' + ax'^2)/\epsilon$. Thus the boundary x'_b of the plateau is determined by $t' + ax'^2_b \approx \epsilon \ln \epsilon$. Since t' is negative (after the singularity), for $\epsilon \ll |t'|$ the boundary of the plateau lies at

$$x'_b \approx \pm\sqrt{\frac{-t'}{a}}, \qquad (3.69)$$

which means that the size of the plateau spreads as $|t'|^{1/2}$ in time.

The value of ϵ, however, determines how sharply the edge of the plateau is defined. The characteristic size of the crossover region δ_{cr} can be defined by the requirement that e^Δ changes by a factor of order unity over the distance δ_{cr}. Since $\Delta = \Delta_0 + 2ax'_b\delta_{cr}/\epsilon$, this implies $e^{2ax'_b\delta_{cr}/\epsilon} = O(1)$ and so

$$\delta_{cr} \sim \frac{\epsilon}{2ax'_b}. \qquad (3.70)$$

Thus the size of the crossover region goes to zero with ϵ and becomes sharper as the peak broadens.

3.5 Continuation

The regularization of a singularity, which we have just studied, is related closely to the problem of continuation; this is illustrated by the breakup of a fluid drop, as shown in Fig. 3.6. Soon after the third frame the fluid drop breaks in two, which means that a singularity of the equations of fluid motion has occurred. If we want to describe the dynamics of the two separate pieces of fluid, we must continue across the singularity. One way to achieve this is to introduce a regularization similar to (3.60), as we now illustrate. As we have

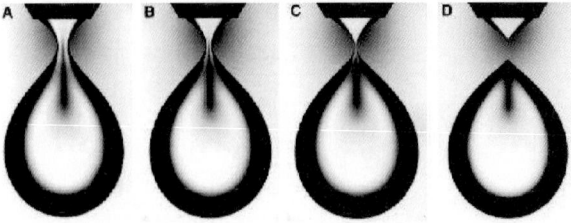

Figure 3.6 The breakup of a drop of water in an environment of much greater viscosity [63]. (Reprinted with permission from AAAS.) Between C and D the water drop breaks and separates from the nozzle.

seen before, we can transform the (unregularized) blowup problem by setting $u(x, t) = 1/h(x, t)$, so that (3.1) becomes

$$\frac{\partial h}{\partial t} = -1, \quad h(x, 0) = h_0. \qquad (3.71)$$

In Chapter 9 we will show that if $h(x, t)$ is the radial profile of a fluid neck, (3.71) does describe the breakup of one liquid inside another, in the particular limit that the drop fluid is much less viscous than its surroundings.

Equation (3.71) looks like the most harmless equation imaginable, but the interpretation of h as a radius requires that h be positive. Namely, the solution of (3.71) formally reads

$$h(x, t) = h_0(x) - t, \qquad (3.72)$$

which is illustrated by the profiles in Fig. 3.7. The time when $h(x, t)$ first goes to zero corresponds to the separation of the drop into two, and thus to a singularity t_0, as shown by the sequence on the right of Fig. 3.7.

While the part of (3.72) with $h(x, t)$ negative clearly makes no sense in terms of the original dynamics for the concentration variable u in (3.1) (and the same holds true for an interpretation as the radius of a fluid thread), (3.72) can still be interpreted as the *continuation* of (3.1) across the singularity. Namely, the solution after the singularity can be defined as the part of (3.72) for which $h(x, t)$ is positive. In that case the solution makes sense in the domain in which it is defined and represents a unique continuation from the solution before the singularity. However, after the singularity the domain is *multiply* connected, as a result of the solution having undergone a topological transition. Let us recount the two different ways to deal with the fact that the drop has split into different pieces.

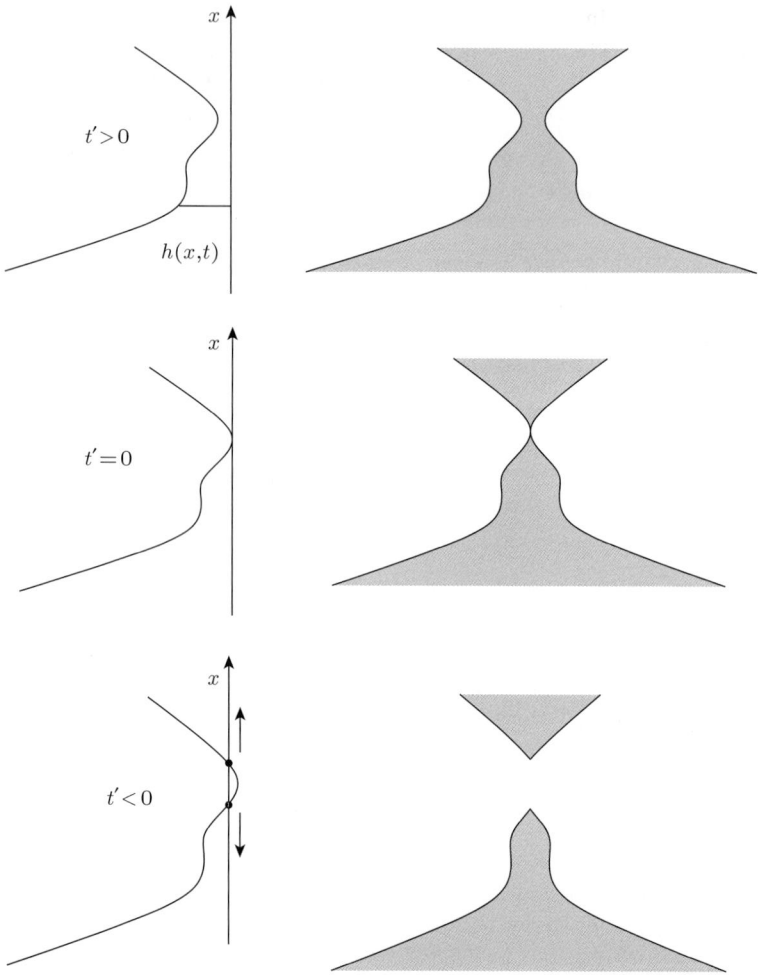

Figure 3.7 A simple model for the experimental sequence of drop pinch-off shown in Fig. 3.6. For $t' > 0$ there is a narrow neck joining the two parts of the drop (top). At $t' = 0$ the radius of the neck goes to zero (middle); for $t' < 0$ the drop consists of two pieces (bottom). Note how the dynamics is generated by a simple *shift* of the profile at a constant rate.

Sharp interface description

In the first method, the dynamics after the singularity is formulated in the physical domain alone, and the boundary between a drop and its exterior is represented by a sharp interface. In that case, apart from the dynamics in the interior of the drop itself, one needs a separate equation for the motion of the interface. In the context of numerical methods, one speaks of "front tracking" [129]. Since the condition for a boundary point x_b is $h_0(x_b) = t$, an equation of motion for the boundary point is found by differentiating with respect to time: $\dot{x}_b h'_0(x_b) = \dot{x}_b h'(x_b, t) = 1$, where a prime denotes the derivative with respect to the spatial argument. Considering for simplicity two boundary points only (cf. Fig. 3.7), we arrive at the following post-breakup dynamics:

$$\frac{\partial h}{\partial t} = -1, \quad x < x_b^{(1)} \text{ or } x > x_b^{(2)}, \tag{3.73}$$

$$\frac{\partial x_b^{(i)}}{\partial t} = \frac{1}{h'(x_b^{(i)}, t)}, \quad i = 1, 2. \tag{3.74}$$

As shown in Fig. 3.7, this is close to the physical breakup dynamics of Fig. 3.6. After breakup the equation of motion (3.73) has to be solved in two separate domains, while the boundary points are tracked using (3.74).

Diffuse interface description

The split into two or more domains brings with it the complication of having to solve several separate problems at the same time and to trace the boundaries of each domain; for example, consider the thousands of separate pieces of fluid seen in the display of a fountain. To avoid this complication and to solve the problem in a single domain, one can look at the same phenomenon using a version (3.60) of the quadratic equation regularized by a cubic term. As a result, there is no longer a singularity: the dynamics defined by (3.60) can be continued for all times and all space. Using the same transformation, $h(x, t) = 1/u(x, t)$, as before, (3.60) transforms to

$$\frac{\partial h}{\partial t} = -1 + \frac{\epsilon}{h(x, t)}, \quad h(x, 0) = h_0(x). \tag{3.75}$$

Now the "repulsive" term ϵ/h ensures that $h(x, t)$ never goes through zero but rather forms a thin "thread" of thickness ϵ.

As illustrated in Fig. 3.8, in the limit of small ϵ the regularized model (3.75) describes essentially the same problem as the continuation (3.73) of the original equation (3.71). Outside the two domains of this solution, a thin "thread" remains, whose thickness is ϵ. In fact, a 8 μm thin thread is observed to connect the two conical regions in Fig. 3.6D, owing to the presence of the water within the viscous surrounding fluid [63]. We will study a regularizing mechanism

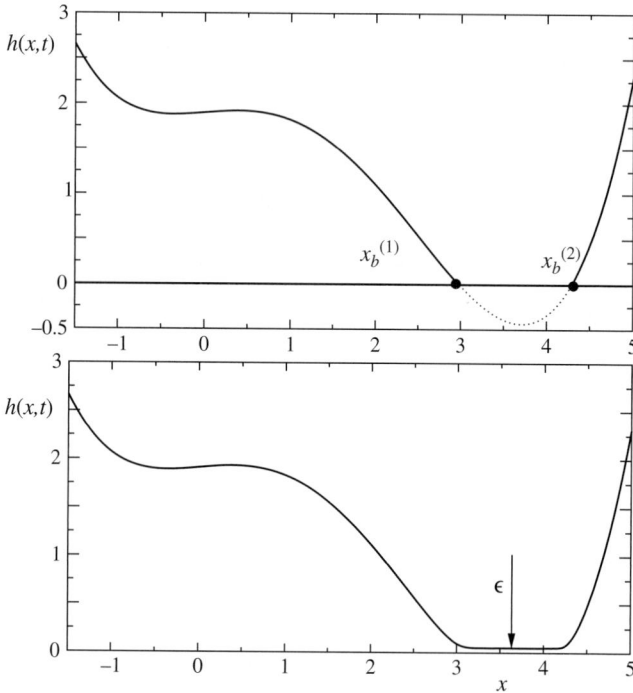

Figure 3.8 Upper panel: the profile after the singularity, as described by (3.73). The solution is defined for $x < x_b^{(1)}$ and $x > x_b^{(2)}$. Lower panel: the solution of (3.75) with the same boundary conditions and $\epsilon = 0.05$. This solution is defined for all x and tends toward the continuation in the limit $\epsilon \to 0$.

very similar to (3.75) in Chapter 10, when we consider the effect of long flexible polymers inside a fluid.

Using our earlier analysis of the asymptotic form of the solution (3.68), and $h = 1/u$, one can understand the nature of the approximation in the limit $\epsilon \to 0$. Namely, for $t' < 0$ (after the singularity), the interface thickness in the neighborhood of the front is

$$h \approx \epsilon + \frac{h_0}{e\epsilon^2} \exp\left(\frac{t' + ax'^2}{\epsilon}\right). \tag{3.76}$$

The boundaries of the thread are located at x_b', as given by (3.69); close to the singularity this amounts to the same equation as for the front motion, (3.74). The *diffuse* interface between the drop and the thread has a finite width of size (3.70), which goes to zero as ϵ goes to zero. The regularized solution approaches the continued multiply connected solution (3.73) in a pointwise manner. On the basis of the regularized solution one can "reconstruct" the fluid domains by labeling as such all regions with $h > 2\epsilon$ (say). In numerics, this procedure would be known as "front capturing" [129].

3.5.1 Similarity description

Above we addressed the problem of continuation in two different ways: first geometrically and then by regularizing the dynamics. However, a simple geometrical formulation is not always available to us, and the regularized dynamics are often complicated and not amenable to analytical treatment. It is therefore very useful to look at a third way of treating the problem, namely using a similarity description. So far we have used such a description to describe the dynamics before the singularity, but it can equally well be used to find solutions after the singularity which also carry us across the singularity. Although by virtue of the transformation $u(x, t) = 1/h(x, t)$ we could borrow our earlier results for the self-similar blowup of u, we will repeat the analysis for pedagogical reasons.

If we introduce $t' = t_0 - t$ as the time to breakup, the similarity formulation for the profile becomes

$$h = |t'| \, \phi \left(x/|t'|^{\beta} \right). \tag{3.77}$$

We have used the modulus of t' deliberately, to be able to use the expression (3.77) for $t' < 0$ as well. First we consider the time *before* breakup, for which the exponent β remains undetermined. Inserting (3.77) into (3.71), the similarity equation becomes

$$-\phi + \beta\xi\phi' = -1, \tag{3.78}$$

with solution

$$\phi = 1 + C\xi^{1/\beta}. \tag{3.79}$$

For the solution (3.79) to be regular, the exponent must be given by $1/\beta_i = 2(i + 1)$, $i = 0, 1, \ldots$, as we found before in (3.17). The interpretation is simple: $\beta = 1/2$ corresponds to a generic quadratic minimum of the initial profile, $\beta = 1/4$ corresponds to the non-generic case of a quartic minimum. As before, only the solution $i = 0$ will be stable. The constant factor C is arbitrary [63], and its value is determined by the initial conditions. In summary, before the singularity the similarity solution is

$$\phi_b = 1 + C\xi^2, \tag{3.80}$$

and the corresponding similarity profile has the form

$$h = |t'| \phi_b \left(\frac{x}{|t'|^{1/2}} \right). \tag{3.81}$$

Now we turn to the novel situation of post-breakup dynamics, for which $t > t_0$. This means that $|t'| = t - t_0$, so there is a change of sign in all the terms coming from the time derivative in the similarity equation. Thus, instead of (3.78) we have

$$\phi - \beta \xi \phi' = -1, \tag{3.82}$$

with solution

$$\phi_a = -1 + \bar{C} \xi^{1/\beta}. \tag{3.83}$$

Note that ϕ_a is no longer defined for small ξ as it would then assume negative values, which are unphysical. The exponent β and the constant $\bar{C}$ are undetermined and need to be found by *matching* to the pre-breakup solution. Namely, at a finite (but arbitrarily small) distance Δx from the pinch point, the solution must remain continuous as one passes through the singularity. In other words, the matching condition reads

$$\lim_{t' \to 0+} h_b(\Delta x, t) = \lim_{t' \to 0-} h_a(\Delta x, t). \tag{3.84}$$

Inserting the similarity solutions (3.80) and (3.82) into (3.84), we find that

$$1 + C \left(\frac{\Delta x}{t'^{1/2}} \right)^2 \sim -1 + \bar{C} \left(\frac{\Delta x}{t'^{\beta}} \right)^{1/\beta} \tag{3.85}$$

in the limit $t' \to 0$. Clearly, (3.85) can only be satisfied if $\beta = 1/2$ and $\bar{C} = C$. In other words, the matching procedure uniquely determines the post-breakup solution:

$$\phi_a = -1 + C \xi^2. \tag{3.86}$$

There are two branches of solutions to (3.86), corresponding to the two receding threads on either side of the pinch point. The tips of the threads are located at the point $h = 0$, that is at $\xi_\pm = \pm \sqrt{C}$. In terms of the physical variables, this means that the tip positions are at

$$x_{\text{tip}} = \pm \sqrt{C} |t'|^{1/2}. \tag{3.87}$$

Thus, just after the singularity the two tips recede at a relative velocity

$$\dot{x}_{\text{tip}} = \pm \frac{\sqrt{C}}{2} |t'|^{-1/2}, \tag{3.88}$$

which diverges as the singularity is approached.

Exercises

3.1 The long-time evolution of the diffusion equation

$$\frac{\partial u}{\partial t} = u_{xx} \tag{3.89}$$

is described by the similarity solution

$$u(x,t) = t^\alpha f(\xi), \tag{3.90}$$

with $\xi = x/t^\beta$. Note that we are considering the limit $t \to \infty$ rather than the usual case of a finite-time singularity $t' \to 0$.

(i) Show that $\beta = 1/2$ and derive the similarity equation

$$\alpha f - \xi f'/2 = f''. \tag{3.91}$$

(ii) Show that mass conservation

$$\int_{-\infty}^{\infty} u(x,t)dx = M \tag{3.92}$$

implies $\alpha = -1/2$.

(iii) Show that the similarity equation (3.91) can be written as

$$-(\xi f)'/2 = f''.$$

(iv) Imposing the symmetry condition $f'(0) = 0$, show that the similarity profile is given by

$$f(\xi) = \frac{M}{2\sqrt{\pi}}e^{-\xi^2/4}. \tag{3.93}$$

3.2 One can use a dynamical system to study the approach of solutions (3.89) to the asymptotic form (3.93).

(i) Introducing $\tau = \ln t$, derive the dynamical system for (3.89).

(ii) Derive the eigenvalue equation for perturbations $P(\xi)e^{\nu\tau}$ around the asymptotic solution $\bar{f}(\xi)$. Show that, with the transformation

$$v(\xi) = P(\xi)e^{-\xi^2/4},$$

the eigenvalue equation becomes

$$\nu v = v'' - \xi v/2,$$

which is identical to (3.44) up to a shift in ν. Thus show that the solutions behave asymptotically like

$$u(x,t) = t^{-1/2}\left(\bar{f}(\xi) + \sum_{j=0}^{\infty} a_j t^{-j/2} v_j(\xi)e^{-\xi^2/4}\right), \tag{3.94}$$

where

$$v(\xi) = H_i\left(\frac{\xi}{2}\right).$$

(iii) Show that the first three eigenfunctions P_0, P_1, and P_2 are generated by invariances of (3.89).

3.3 (*) Consider the similarity solution to the diffusion equation (cf. Exercise 3.1) but with *shifted* time and space coordinates [223]:

$$u^*(x,t) = \bar{t}^{-1/2}\bar{f}\left(\frac{x+x^*}{\bar{t}^{1/2}}\right),$$

where $\bar{t} = t + t^*$. We would like to determine x^* and t^* such that u^* is an optimal approximation to the solution of (3.89) with initial condition $u(x,0) = u_0(x)$.

(i) Expand $u(x,t)$ about $u^*(x,t)$:

$$u(x,t) = u^*(x,t) + \sum_{j=0}^{\infty} a_j(\bar{t})^{-(1+j)/2} v_j(\xi) e^{-\xi^2/4},$$

and show that

$$\int_{-\infty}^{\infty} [u_0(x) - u^*(x,0)]v_k(\xi)d\xi = \frac{k!\sqrt{\pi}}{(2\sqrt{t^*})^{k-1}}a_k.$$

(ii) To make the correction to $u(x,t)$ as small as possible, it is desirable that $a_0 = a_1 = a_2 = 0$. Show that this is achieved by putting

$$x^* = M_1, \quad t^* = \frac{M_2 - M_1^2}{2},$$

where

$$M_i = \int_{-\infty}^{\infty} u_0(x)x^i dx.$$

What is the physical meaning of these choices?

3.4 Verify that an exact solution to (3.89) with initial condition $u_0(x)$ can be written in the form

$$u(x,t) = \frac{1}{2\sqrt{\pi t}}\int_{-\infty}^{\infty} u_0(x')e^{-(x-x')^2/4t}dx'.$$

Thus show that

$$u(x,t) = \frac{e^{-x^2/4t}}{2\sqrt{\pi t}}\left(M_0 + \frac{xM_1}{2t} + \frac{x^2-2t}{8t^2}M_2 + \cdots\right).$$

Confirm that this agrees with (3.94), and determine the coefficients a_0, a_1, and a_2.

3.5 The *nonlinear* diffusion equation

$$\frac{\partial u}{\partial t} = \left(u^n u_x\right)_x \tag{3.95}$$

has many applications, for example as a model for the spreading of insects [159], or the spreading of a liquid in a porous medium. For $n = 3$ it describes the spreading of a viscous liquid under gravity; see Example 6.2 in Section 6.1.

(i) Show that that there exist self-similar solutions (3.90) representing the spreading of insects with total mass M. These solutions are known as Barenblatt solutions [13].

(ii) Show that the exponents are $\beta = -\alpha = 1/(n + 2)$ and derive the similarity equation for f.

(iii) Show that the self-similar profile can be written as

$$f(\xi) = \left(\frac{n}{2(n + 2)} \left(\xi_0^2 - \xi^2\right)\right)^{1/n},$$

with $f = 0$ for $\xi > \xi_0$, i.e. the solution has compact support.

(iv) Calculate ξ_0 in terms of M, using that

$$\int_0^1 \left(1 - \zeta^2\right)^a d\zeta = \frac{\sqrt{\pi}}{2} \frac{\Gamma(1 + a)}{\Gamma(3/2 + a)}.$$

3.6 If a mortality rate μ is taken into account in Exercise 3.5 then a term $-\mu u$ must be added to the right-hand side of (3.95). By changing variables into the form $\bar{u} = a(t)u$, $\bar{t} = b(t)$, with $a(t)$ and $b(t)$ chosen appropriately, show that $\bar{u}(x, \bar{t})$ also satisfies (3.95) and discuss the behavior of the solutions.

3.7 Confirm that the asymptotics of the eigenfunctions (3.31) for $\xi \to \pm\infty$ agrees with (3.59).

3.8 Consider the pinch-off problem (3.11) for two different initial conditions. According to Exercise 3.2, near pinch-off the solution is described by a similarity solution

$$h(x, t) = t'\bar{g}(\xi), \quad \xi = \frac{x}{t'^{1/2}},$$

where $\bar{g} = 1 + a(\xi^2 - 2)$.

(i) Let

$$h(x, 0) = 1 + Ax^2.$$

Show that if $0 < A < 1/2$ then h goes to zero locally and the profile corresponds to $\bar{g}$, with $a = A$.

(ii) Now let

$$h(x, 0) = 1 + Bx^4.$$

Show that h pinches off if $0 < B < 1/48$ and that the singularity time is

$$t_0 = \frac{1}{24B} - \sqrt{\frac{1}{(24B)^2} - \frac{1}{12B}}.$$

Also show that the solution again converges to $\bar{g}$, with $a = 12Bt_0$.

(iii) For both initial conditions, calculate the corrections to the similarity solution in the form (3.38).

3.9 Consider the expansion (3.38) around the ground state similarity solution

$$\bar{f}(\xi) = \frac{1}{1 + a\xi^2}.$$

Show that on replacing (x_0, t_0, a) by $(x_0 + x^*, t_0 + t^*, a + a^*)$ the first three coefficients $a_0, a_1,$ and a_2 can be made to vanish, to linear order, and compute the value of the shifts $a^*, x^*,$ and t^*.

3.10 Consider the eigenvalue equation (3.44), with boundary condition $P \propto \xi^{2-2\nu}$ for $\xi \to \pm\infty$.

(i) Show that solutions to (3.44) are either even or odd, and derive the recursion relation

$$a_{j+2} = \frac{\nu - 1 + j/2}{(j + 1)(j + 2)} a_j, \qquad (3.96)$$

where $P = \sum\limits_{j=0}^{\infty} a_j \xi^j$. Hence conclude that either $a_1 = 0$ and a_0 is nonzero (even solutions) or $a_0 = 0$ and a_1 is nonzero (odd solutions).

(ii) Show that, for large j, the coefficients behave as $a_{j+1}/a_j \approx 1/(2j)$ and thus $P \propto e^{\xi^2/4}$ *unless* $a_l = 0$ for some l. In view of the boundary condition at infinity, this means that the iteration must indeed terminate at some l.

(iii) Show that, for even solutions, the eigenvalues are $\nu = 1-l$, with $l = 0, 1, 2, \ldots$, while for odd solutions $\nu = 3/2 - l$, with $l = 1, 2, \ldots$ Use (3.96) to reproduce Table 3.2, up to an arbitrary normalization.

3.11 Consider the stability of the similarity solution (3.40) of the pinch-off problem (3.11). Show that the first three eigenvalues and eigenfunctions listed in Table 3.2 correspond to the invariances of Table 3.1. Write down an expansion of the form (3.38) for the solution $h(x', t')$, and show that the first three terms can be eliminated by an appropriate shift in t_0, x_0, and a.

3.12 Show that the porous medium equation

$$u_t = (u^{p-1} u_x)_x \tag{3.97}$$

(with $p > 1$) can be rewritten, after the transformation $u = v^{1/(p-1)}$, as

$$v_t = v_x^2 + \beta v v_{xx} \tag{3.98}$$

and determine β. Solutions to (3.97) may be compactly supported in such a way such that the boundary of the support evolves in time, i.e. the solution presents moving fronts. There is also the possibility of a waiting time phenomenon: the front remains stationary for a while before it starts to move. In order to analyze this phenomenon, consider the initial data

$$v(x, t_0) = x^\alpha.$$

(i) Find γ such that

$$v(x, t) = |t'|^\gamma f_-(\xi), \quad t < t_0$$
$$v(x, t) = |t'|^\gamma f_+(\xi), \quad t > t_0$$

are self-similar solutions to (3.98), with $t' = t_0 - t$ and $\xi = x/|t'|^{\gamma/\alpha}$.

(ii) (*) Write down the equations to be satisfied by $f_\pm$ and find a suitable transformation that makes each of them equivalent to an autonomous dynamical system of dimension 2.

(iii) (*) Perform a phase plane analysis of the dynamical system derived in (ii), and show that a unique continuation exists across the singularity [99].

3.13 (*) The Smoluchowski equation is an integro-differential equation which models the evolution of a distribution $c(x, t)$ of particles of size x that undergo processes of coagulation (i.e. particles of size y and $x - y$ merge to form a particle of size x). It describes for example the evolution of aerosols and the formation of planetesimals. This exercise shows that the concept of similarity solutions is not restricted to differential equations but can also be applied to non-local problems. The equation reads

$$c_t(x, t) = \frac{1}{2} \int_0^x K(x - y, y) c(x - y, t) c(y, t) dy$$

$$- c(x, t) \int_0^\infty K(x, y) c(y, t) dy, \tag{3.99}$$

where $K(r, s)$ is a coagulation kernel representing the probability for particles of size r and s to merge. We consider the particular case $K(r, s) = rs$.

(i) Find the exponents α and β, or a relation between them, in the similarity solution

$$c(x, t) = t'^\alpha \psi \left(\frac{x}{t'^\beta} \right)$$

of (3.99), with $t' = t_0 - t$, and derive the similarity equation to be satisfied by $\psi(\xi)$.

(ii) The similarity solution must satisfy

$$\psi(\xi) \sim \xi^{\alpha/\beta}$$

as $\xi \to 0$ (why?). Show that by choosing A, α, and β appropriately, the function

$$\psi(\xi) = \xi^{\alpha/\beta} e^{-A\xi}$$

can be made to be a self-similar solution. Is it a self-similar solution of the second kind?

4

Continuum equations

4.1 General ideas

This book deals for the most part with partial differential equations, whose basis is to view the world as a continuum. This agrees with our perception of liquids, gases, and many solids; they appear to have no characteristic structure and are perfectly isotropic: no property is linked to a particular point or direction in space. We hasten to add that the reality is often more complicated. A liquid may contain components which possess a significant microstructure, e.g. polymers. Solids are often crystalline, which means that their properties depend on the direction of observation relative to the crystal axes.

It is worth remembering the complexity of any microscopic description of a liquid, gas, or solid, even if the system is extremely small. Consider, for example, the computer simulation of a liquid jet emerging from a nozzle only 6 nm in diameter, shown in Fig. 4.1. Each molecule of the liquid is treated as a mathematical point, with a known force between any two molecules. At very small distances, the molecules repel; at larger distances, they attract. The balance between attraction and repulsion causes the molecules to condense into a liquid state. Even on a scale of a few nanometers, the jet appears essentially like a continuum.

The simulation shown in Fig. 4.1 was obtained by solving Newton's equations for each molecule, taking into account the forces exerted on it by the other molecules. This is feasible for a few hundred thousand molecules by only including interactions with molecules close to the target particle. Thus, from a microscopic point of view, we have an incredibly complicated system of non-linear ODEs. The only hope of gaining analytical insight lies in developing a continuum description. Here we will emphasize the basic physical principles on which the continuum description of fluids and solids relies, and briefly develop the equations of motion on which the rest of this book is based.

63

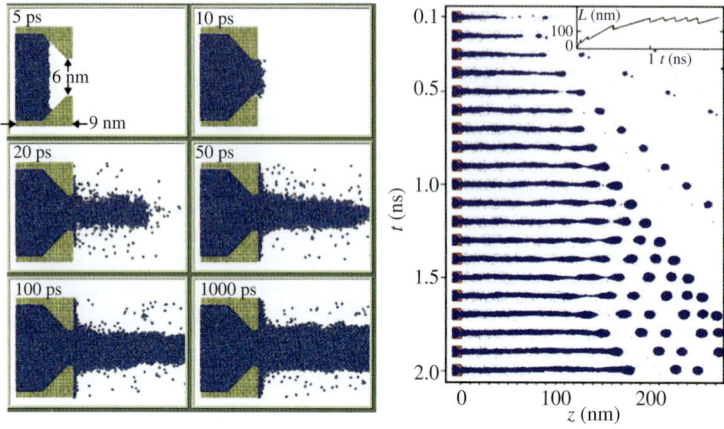

Figure 4.1 A molecular dynamics (MD) simulation of a jet of propane coming out of a gold nozzle 6 nm in diameter [158]. Reprinted with permission from AAAS.

4.2 The Navier–Stokes equation

The continuum description of liquids and gases is based on the fundamental laws of conservation of matter and of momentum. Here anything that flows is usually called a fluid and comprises both liquids and gases described by the same flow equations. The law of conservation of mass or continuity equation reads

$$\frac{\partial \rho}{\partial t} + \nabla \cdot (\rho \mathbf{v}) = 0, \tag{4.1}$$

where $\rho(\mathbf{x}, t)$ denotes the local density of the fluid at position $\mathbf{x}$ and time t and $\mathbf{v}(\mathbf{x}, t)$ is its local velocity. Integrating (4.1) over a control volume V and using Gauss' theorem, we find

$$\frac{\partial}{\partial t} \int_V \rho \, d\mathbf{x} = -\int_V \nabla \cdot (\rho \mathbf{v}) d\mathbf{x} = -\int_{\partial V} \rho \mathbf{v} \cdot d\mathbf{s}, \tag{4.2}$$

where ∂V is the surface of V. Thus the total change in mass (the left-hand side of (4.2)) equals the total flux coming in and going out through the surface of V, where $d\mathbf{s}$ is the surface element pointing in the direction of the outward normal. The *mass flux* is $\mathbf{j} = \rho \mathbf{v}$ and is thus determined by the velocity with which the particles are flowing multiplied by the mass of particles per unit volume.

A particularly simple and common case occurs when the density remains constant in space and time, as we will assume for the remainder of this chapter;

this makes the fluid incompressible. Now all spatial and temporal derivatives of ρ vanish and (4.1) becomes

$$\nabla \cdot \mathbf{v} = 0. \tag{4.3}$$

Example 4.1 (Point source) The flow around a source must be spherically symmetric, and so $\mathbf{v} = f(r)\mathbf{e}_r$, with the source at the origin. In spherical coordinates (see Appendix A for a summary of vector calculus and the notation) we have

$$\nabla \cdot \mathbf{v} = \frac{1}{r^2} \frac{\partial [r^2 f(r)]}{\partial r},$$

so (4.3) implies $f(r) = A/r^2$. The flux m from the source must equal the total flux through a sphere of radius R:

$$m = \int_{S_R} \mathbf{v} \cdot d\mathbf{s} = 4\pi R^2 f(R) = 4\pi A,$$

and so

$$\mathbf{v} = \frac{m}{4\pi r^2} \mathbf{e}_r, \tag{4.4}$$

which has a singularity at the origin. □

To formulate the conservation law for momentum, we are looking for the same structure as (4.1) except that each of the three components ρv_i of the fluid momentum must be conserved separately. Thus we are led to the equation

$$\frac{\partial \rho v_i}{\partial t} + \frac{\partial \Pi_{ij}}{\partial x_j} = 0, \tag{4.5}$$

where Π_{ij} is known as the momentum flux tensor. In (4.5) and henceforth we use the "summation convention", which stipulates that indices which appear twice are summed over (see Appendix B).

One can guess the structure of Π_{ij} by a comparison with Newtonian point mechanics, where it is known that the conservation of momentum is equivalent to Newton's law of motion $\mathbf{F} = m\mathbf{a}$, to be interpreted as the force on a small fluid "particle", and its resulting acceleration. The acceleration is given by the time derivative of the velocity of the fluid particle. Locally, the path of such a particle can be approximated as $\mathbf{x} = \mathbf{x}_0 + \mathbf{v}t$. Thus, taking the time derivative along this path, we find

$$\rho \frac{d\mathbf{v}(\mathbf{x}, t)}{dt} = \rho \left(\frac{\partial \mathbf{v}}{\partial t} + (\mathbf{v} \cdot \nabla)\mathbf{v} \right) = \mathbf{f}, \tag{4.6}$$

where $\mathbf{f}$ is the force per unit volume. The time derivative on the left, taken along a particle path, is known as the Lagrangian or material derivative. One part of $\mathbf{f}$ comes from body forces such as the force of gravity, $\rho \mathbf{g}$. However, the main problem is to find the appropriate expression for the contribution to $\mathbf{f}$ from the other particles of the fluid. Comparing (4.6) and (4.5) one finds that Π_{ij} has the structure

$$\Pi_{ij} = \rho v_i v_j - \sigma_{ij}, \tag{4.7}$$

where σ_{ij} represents the force on a fluid element.

If an infinitesimal oriented piece of surface is described by a surface element $d\mathbf{s}$, and the force on that surface is $d\mathbf{f}$, we expect the relationship between the two to be described by

$$df_i = \sigma_{ij} ds_j, \tag{4.8}$$

where σ_{ij} is called the stress tensor.

There are two types of force acting in a fluid. The first is the pressure p, which always acts in the direction of $d\mathbf{s}$. This means that the corresponding part of σ_{ij} is diagonal: $\sigma_{ij} = p\delta_{ij}$. The other part of the force is generated by viscous friction, i.e. the viscosity, of the fluid. For there to be a force, there needs to be shear, i.e. a gradient of the velocity. The general expression characterizing the amount of shearing is the symmetric part of the rate-of-deformation tensor

$$D_{ij} = \frac{\partial v_i}{\partial x_j} + \frac{\partial v_j}{\partial x_i}. \tag{4.9}$$

The antisymmetric part of D_{ij} does not contribute to the force since it corresponds to the solid body rotation of a fluid element: in that case no part of the fluid is sheared relative to another part.

Example 4.2 (Shear flow) The shear flow produced by for example two plates lying in the xz-plane and moving in the x-direction is given by

$$\mathbf{v} = (\gamma y, 0, 0), \tag{4.10}$$

where γ is the shear rate. Then the only nonzero components of D_{ij} are $D_{12} = D_{21} = \gamma$. If the lower plate is at rest and the upper is moving with speed U, we have $\gamma = U/d$, where d is the distance between the plates. $\square$

Example 4.3 (Solid body rotation) The velocity at a point $\mathbf{x}$ on a rotating solid body is given by

$$\mathbf{v} = \boldsymbol{\Omega} \times \mathbf{x}, \tag{4.11}$$

where $\boldsymbol{\Omega}$ is the angular velocity vector. Writing $v_i = \epsilon_{ijk}\Omega_j r_k$ (see Appendix B) and using the antisymmetry of ϵ_{ijk} it follows that $D_{ij} = 0$.   □

For an isotropic fluid, and in the case where only terms linear in D_{ij} are taken into account, the viscous contribution to σ_{ij} is proportional to D_{ij}, the constant of proportionality being the viscosity η. The justification for this linear approximation is that D_{ij} (which has units of inverse time) is much smaller than the inverse of a microscopic time, which is of order 10^{12} s^{-1}. This statement may no longer be true if the fluid contains for example large macromolecules, which may possess much longer relaxation times.

In summary, combining the effects of viscosity and pressure we have

$$\sigma_{ij} = -p\delta_{ij} + \eta D_{ij}. \tag{4.12}$$

Now using (4.5), together with (4.7) and (4.12), one finds the famous *Navier–Stokes equation*,

$$\partial_t \mathbf{v} = -(\mathbf{v} \cdot \nabla)\mathbf{v} - \frac{1}{\rho}\nabla p + \nu \triangle \mathbf{v} + \mathbf{g}, \tag{4.13}$$

where we have introduced the *dynamic viscosity* $\nu = \eta/\rho$. In the case of an incompressible fluid, ν is the only parameter that enters the description of fluid motion; see Exercise 1.5. The constant $\mathbf{g}$ represents external forcing by a gravity field. We reiterate that (4.13) is an expression of the conservation of the momentum $\rho\mathbf{v}$ of the fluid.

Multiplying (4.13) by $\rho\mathbf{v}$, the left-hand side is transformed into the time derivative of the kinetic energy density of the fluid, $\rho v^2/2$:

$$\rho\mathbf{v}\cdot\frac{\partial\mathbf{v}}{\partial t} = \frac{\rho}{2}\frac{\partial(\mathbf{v}\cdot\mathbf{v})}{\partial t} \equiv \frac{\rho}{2}\frac{\partial v^2}{\partial t}.$$

Using (4.13) one obtains

$$\frac{\partial}{\partial t}\frac{\rho v^2}{2} = \rho v_i \frac{\partial v_i}{\partial t} = -\rho v_i \partial_j v_j v_i - v_i \partial_i p + \eta v_i \partial_j D_{ij}$$

$$= -\rho\left(v_i \partial_i\right)\left(\frac{v^2}{2} + \frac{p}{\rho}\right) + \eta\partial_i\left(v_j D_{ij}\right) - \eta D_{ij}\partial_j v_i.$$

Since $\nabla \cdot \mathbf{v} = \partial_i v_i = 0$, all terms on the right but the last can be written as a divergence:

$$\frac{\partial}{\partial t}\frac{\rho v^2}{2} + \partial_i\left[\rho v_i \left(\frac{v^2}{2} + \frac{p}{\rho}\right) - \eta v_j D_{ij}\right] = -\eta D_{ij}\partial_j v_i. \tag{4.14}$$

This once more has the structure of a conservation law, the term in the square brackets being the *energy flux* $\mathbf{j}^{(e)}$, whose components are

$$j_i^{(e)} = v_i \left(\frac{\rho v^2}{2} + p \right) - \eta v_k \left(\frac{\partial v_i}{\partial x_k} + \frac{\partial v_k}{\partial x_i} \right). \tag{4.15}$$

The important difference from (4.1) is that another term, $-\epsilon$, appears on the right-hand side of (4.14), which cannot be written as a divergence and where we define

$$\epsilon \equiv \eta D_{ij} \partial_j v_i = \eta \left(\partial_j v_i + \partial_i v_j \right) \partial_j v_i.$$

It is easy to verify that the *rate of energy dissipation per unit volume* ϵ can be written in the form

$$\epsilon = \frac{\eta}{2} \left(\frac{\partial v_i}{\partial x_j} + \frac{\partial v_j}{\partial x_i} \right)^2, \tag{4.16}$$

which is manifestly positive. In summary, the energy balance reads

$$\frac{\partial}{\partial t} \frac{\rho v^2}{2} + \nabla \cdot \mathbf{j}^{(e)} = -\epsilon. \tag{4.17}$$

The right-hand side of (4.17) is always negative, so kinetic energy is always lost. In a closed system, such that there is no influx of energy, the total kinetic energy can only decrease. This is an important statement about the stability of the system: locally, the fluid will always return to an equilibrium state of rest. Only external driving will keep the system in motion.

Example 4.4 (Energy dissipation in shear flow) The local rate of energy dissipation in the shear flow (4.10) is $\epsilon = \eta \gamma^2$, where we have used $v_i = \gamma y \delta_{i1}$ and thus $\partial_j v_i = \gamma \delta_{i1} \delta_{j2}$. If the flow is between two plates of area A, a distance d apart, the total rate of energy dissipation becomes $E_d = A d \eta \gamma^2$.

If $\mathbf{e}_y$ is the normal to the upper plate, the shear force on the upper plate is $F_x = A \sigma_{12} = A \eta \gamma$. The plate is moving with velocity $\mathbf{v} = U \mathbf{e}_x$, so the energy input is $U F_x = U A \eta \gamma = A d \eta \gamma^2$, which equals the rate of dissipation. $\qquad \square$

4.3 Boundary conditions

To solve (4.13) we need boundary conditions. At a solid wall the fluid sticks; intuitively, the reason is that the outer solid layer can be treated as equivalent to a layer of fluid molecules held at rest in the frame of reference of the solid. Thus the fluid velocity approaches zero at the wall:

$$\mathbf{v}|_{\text{solid wall}} = 0. \tag{4.18}$$

The *force* acting on the solid can be calculated from the stress tensor, using (4.8). In the case of free surfaces, which will concern us mostly here, the situation is reversed: as the free surface moves, the velocity is unknown. Instead, the stresses are prescribed.

Namely, the condition to be met at a free surface (such as defines a fluid drop surrounded by a gas) is that there is a balance of forces as one traverses the interface. Since a surface element points in the direction of the normal **n**, this means that

$$n_i \sigma_{ij}^{\text{in}} = n_i \sigma_{ij}^{\text{out}}. \tag{4.19}$$

If there is only gas in the exterior, one can often neglect the force exerted by it, and the right-hand side of (4.19) is zero. If, however, there is a significant outer phase, the velocity also needs to be continuous across the interface:

$$\mathbf{v}^{\text{in}} = \mathbf{v}^{\text{out}}. \tag{4.20}$$

For small systems, surface forces are important relative to volume forces (such as gravity); hence the *surface tension* plays an important and even predominant role driving the flow. This extra force comes about since there is a surface (free) energy $E_s = \gamma A$ associated with the surface area A between two phases (it may be the free surface between a liquid and a gas). As the surface area increases, work has to be done against the surface tension forces. Consider the following experiment: a small spherical drop of area $A = 4\pi R^2$ is expanding as fluid is injected into it through a syringe needle. If this process is performed slowly, the flow can be neglected and the surface energy created equals the work performed by the pressure difference Δp between the interior and exterior of the drop.

Thus, if δR is a virtual displacement of the drop radius, we have

$$\gamma \delta A = 8\pi \gamma R \delta R = \Delta p \delta V = 4\pi \Delta p R^2 \delta R,$$

since $V = 4\pi R^3/3$. In other words, there is a pressure jump across the interface such that

$$\Delta p = \frac{2\gamma}{R} \tag{4.21}$$

is the pressure excess in the interior of the drop over atmospheric pressure. The coefficient $1/R + 1/R = 2/R$ is twice the mean curvature of a sphere. In the case of a general surface, $2/R$ is replaced by twice the mean curvature:

$$p^{\text{in}} - p^{\text{out}} = \gamma \kappa \equiv \gamma \left(\frac{1}{R_1} + \frac{1}{R_2} \right), \tag{4.22}$$

where R_1 and R_2 are the two principal radii of curvature [62], to be reckoned positive if the corresponding inscribed circle lies in the interior of the surface.

Example 4.5 (Mean curvature of an axisymmetric body) When $h(z)$ is the radius of an axisymmetric body with symmetry axis z, then twice the mean curvature is given by [62]

$$\kappa \equiv \frac{1}{R_1} + \frac{1}{R_2} = \frac{1}{h(1+h'^2)^{1/2}} - \frac{h''}{(1+h'^2)^{3/2}}. \qquad (4.23)$$

In Exercise 4.11 we will show how this expression is obtained from a variational problem. For $h = R = $ const, one obtains the mean curvature $\kappa = 1/R$ of a cylinder; for $h = \sqrt{R^2 - z^2}$ one obtains $\kappa = 1/R + 1/R = 2/R$, the result for a sphere. □

In terms of the free surface condition (4.19), this means that there is an additional normal force as one crosses the interface. Indeed, it is intuitive to write the free surface balance as two separate conditions for the normal and the tangential forces, obtained by multiplying (4.19) by the outward normal $\mathbf{n}$ or by one of the two linearly independent tangent vectors $\mathbf{t}$. Then, if the mean curvature is counted positive for a convex surface, we have

$$n_i \sigma_{ij}^{\text{out}} n_j - n_i \sigma_{ij}^{\text{in}} n_j = \gamma \left(\frac{1}{R_1} + \frac{1}{R_2} \right), \qquad (4.24)$$

$$n_i \sigma_{ij}^{\text{out}} t_j - n_i \sigma_{ij}^{\text{in}} t_j = 0. \qquad (4.25)$$

4.4 Free surface motion

As we have mentioned, the free surface moves with the fluid. Imagine for example a point on the surface marked with a spot of dye. Then the velocity of the dye will be precisely the fluid velocity at the location of the dye, and the dye will always remain on the surface. However, the velocity of each fluid particle on the surface contains more information than is needed to prescribe free surface motion, since any motion in a direction *tangential* to the free surface can be disregarded. This observation can be used to formulate the equation of motion for the free surface, since one can always parameterize a surface using two parameters $\boldsymbol{\xi} = (\xi_1, \xi_2)$, so that $\mathbf{x}(\boldsymbol{\xi}, t)$ describes the free surface at time t. Then, if $\mathbf{v}$ is the velocity field at $\mathbf{x}$, the equation of motion for the surface can be written as

$$\left(\frac{\partial \mathbf{x}(\boldsymbol{\xi}, t)}{\partial t} - \mathbf{v}(\mathbf{x}, t) \right) \cdot \mathbf{n} = 0, \qquad (4.26)$$

where by multiplying with the normal vector we have made sure that any motion tangential to the surface does not enter.

In many cases the formulation (4.26) is awkward since we have to introduce an extra set of coordinates for the free surface. One can get around this problem by writing the surface as an isoline of a suitably defined function of the spatial coordinates and time, $C(\mathbf{x}, t)$. In terms of $(\boldsymbol{\xi}, t)$ this means that $C(\mathbf{x}(\boldsymbol{\xi}, t), t) = C_0$. Taking the time derivative one finds that

$$\frac{\partial C}{\partial t} + \mathbf{x}_t \cdot \nabla C = 0$$

on the free surface. As ∇C points in the direction of $\mathbf{n}$, according to (4.26) one can replace $\mathbf{x}_t$ by the velocity $\mathbf{v}$, and we obtain the convection equation for the free surface,

$$\frac{\partial C}{\partial t} + \mathbf{v} \cdot \nabla C = 0. \tag{4.27}$$

Example 4.6 (Interface between two phases) Consider the boundary between an inner and an outer fluid, described by $C(\mathbf{x}, t) = C_0$. Then (4.27) can be applied in either phase, and $\partial_t C$ must be continuous across the boundary. In addition the normal vector to the interface is $\mathbf{n} = \nabla C / |\nabla C|$, so subtracting the two equations yields

$$\mathbf{n} \cdot \mathbf{v}^{\text{in}} = \mathbf{n} \cdot \mathbf{v}^{\text{out}}. \tag{4.28}$$

Thus, for purely kinematic reasons, the normal components of the velocity must be continuous across the interface. Note that, for a viscous fluid, both the normal and tangential components are continuous, according to (4.20). □

Given an initial interface, one can always find a function C which is constant along the interface. When C is evolved using (4.27), the interface can then be reconstructed from it at any given time. Note, however, that the initial definition of C is by no means unique. One particular physical interpretation of C is that of a density or concentration variable which takes different values (say C_1 and C_2) in two different phases separated by a sharp interface. The interface can then be defined as the isoline where C assumes the average between C_1 and C_2.

Example 4.7 (Graph representation) Consider the special case where a free surface has a graphical representation $h(x, y, t)$ as a function of the (Cartesian) coordinates x, y, where h represents the height above the plane $z = 0$. For C, we can now choose the particular representation

$$C(x, y, z, t) = h(x, y, t) - z,$$

so that the level set $C = 0$ defines the free surface. Applying (4.27), one finds directly that

$$\frac{\partial h}{\partial t} + \mathbf{v}_\parallel \cdot \nabla_\parallel h = v_z, \tag{4.29}$$

where $\mathbf{v} = (\mathbf{v}_\parallel, v_z)$. In Exercise 4.10 we demonstrate an analogous expression for an axisymmetric surface.   □

4.5 Special limits

Two particular limits of the Navier–Stokes equation are of interest. These correspond to the viscosity ν being either very large or very small. Since the numerical value of ν depends on the choice of unit system, the above statements need to be made in terms of a dimensionless variable, known as the Reynolds number,

$$\text{Re} = \frac{LU}{\nu}. \tag{4.30}$$

Here L is a characteristic length scale of the problem and U is a velocity scale. In a situation where the viscosity can be neglected (meaning that the Reynolds number is infinite), (4.13) simplifies to

$$\partial_t \mathbf{v} = -(\mathbf{v} \cdot \nabla)\mathbf{v} - \frac{1}{\rho}\nabla p + \mathbf{g}, \tag{4.31}$$

known as the *Euler equation*. A particular class of solutions of (4.31) can be constructed by assuming that the flow is irrotational.

4.5.1 Potential flow

A potential flow is one described by a velocity potential

$$\mathbf{v} = \nabla\phi, \tag{4.32}$$

where ϕ is a scalar. Taking the curl of (4.32), it also follows that the vorticity

$$\boldsymbol{\omega} = \nabla \times \mathbf{v} \tag{4.33}$$

is zero everywhere: the flow is irrotational. A way to measure the amount of rotation in a flow is to integrate the velocity along a closed curve C, the result of which is called the *circulation* Γ:

$$\Gamma = \oint_C \mathbf{v} \cdot d\mathbf{l}. \tag{4.34}$$

According to Stokes' theorem, the circulation can also be calculated by integrating $\boldsymbol{\omega}$ over an area A enclosed by C:

$$\int_A \boldsymbol{\omega} ds = \oint_C \mathbf{v} d\mathbf{l} = \Gamma. \tag{4.35}$$

In particular, a potential flow has no circulation. However, this applies only if the curve C can shrink to zero within the flow domain, i.e. if it does not contain any "holes"; if, however, the domain is not simply connected then there can be a circulation around the holes of the domain.

Now let us consider a closed loop $C(t)$ such that each point of the loop is convected by the flow and which is therefore time dependent in general. Kelvin's theorem (which we do not prove [133]) states that under Euler dynamics (4.31), the circulation around such a convected loop is constant:

$$\frac{d\Gamma}{dt} = 0. \tag{4.36}$$

In particular, if the flow is potential initially (and so $\Gamma = 0$ for all possible loops), it will remain potential. This statement must be qualified for solid boundaries, at which vorticity can be generated even in the limit of large Re. However, in the case of free surfaces the vorticity generated at the boundary is small, so often the potential flow assumption (4.32) is accurate.

Taking the divergence of (4.32) and using (4.3), one finds that the potential obeys *Laplace's equation* in the interior of the flow:

$$\triangle \phi = 0. \tag{4.37}$$

Example 4.8 (Potential of point source) The potential of the point source of Example 4.1 is

$$\phi = \frac{m}{4\pi r}. \tag{4.38}$$

Now

$$\triangle \frac{1}{r} = -\partial_i \frac{x_i}{r^3} = -\frac{\delta_{ii}}{r^3} + 3\frac{x_i x_i}{r^5} = -\frac{3}{r^3} + 3\frac{r^2}{r^5} = 0,$$

so (4.38) indeed satisfies (4.37). □

When supplemented with suitable boundary conditions, (4.37) can be used to calculate the velocity field. The Euler equation (4.31) is needed only to compute the pressure. To this end note that in the case of potential flow the identity (B.7) becomes $(\mathbf{v} \cdot \nabla)\mathbf{v} = -\nabla v^2 / 2$. Inserting this, as well as the representation (4.32), into (4.31) we find

$$\nabla \left(\frac{\partial \phi}{\partial t} + \frac{1}{2} |\nabla \phi|^2 + \frac{p}{\rho} + gz \right) = 0.$$

Integrating once, we deduce the *Bernoulli equation*

$$\frac{\partial \phi}{\partial t} + \frac{1}{2} |\nabla \phi|^2 + \frac{p}{\rho} + gz = C(t), \qquad (4.39)$$

where $C(t)$ depends on time only. Here we have assumed that gravity points in the negative z-direction: $\mathbf{g} = -g\mathbf{e}_z$.

With a known ϕ the pressure can now be computed from (4.39). At a free surface, the viscous boundary conditions (4.24) and (4.25) cannot both be satisfied with a velocity field that is potential. However, in the limit of vanishing viscosity the tangential force balance (4.25) is satisfied exactly, and the normal-force balance determines the pressure jump across the interface. From (4.39), we then obtain the boundary condition

$$\frac{\partial \phi^{in}}{\partial t} - \frac{\partial \phi^{out}}{\partial t} + \frac{1}{2} \left| \nabla \phi^{in} \right|^2 - \frac{1}{2} \left| \nabla \phi^{out} \right|^2 + \frac{\gamma}{\rho} \left(\frac{1}{R_1} + \frac{1}{R_2} \right) = 0. \quad (4.40)$$

Another case is that of solid boundaries, for which the viscous condition (4.18) would overdetermine the equation. If we multiply (4.18) by the normal $\mathbf{n}$ to the wall we find that

$$\mathbf{n} \cdot \mathbf{v} = (\mathbf{n} \cdot \nabla) \phi = 0. \qquad (4.41)$$

The physical interpretation of this condition is that the fluid cannot penetrate the wall, and this should hold independently of the fluid's properties. Together with Laplace's equation, the condition (4.41) on the normal derivative of the potential is enough to find a unique solution (the Neumann problem). Likewise, at the boundary between two fluids the normal component of the velocity must be continuous, as implied by (4.28).

4.5.2 Two-dimensional flow

We have seen that (4.37) embodies the condition of incompressibility. In two dimensions, there is another way of writing an incompressible velocity field: by introducing the so-called *stream function* $\psi(x, y)$. If we denote the velocity field in components by $\mathbf{v} = (u, v)$, and ψ is such that

$$u = \frac{\partial \psi}{\partial y}, \qquad v = -\frac{\partial \psi}{\partial x}, \qquad (4.42)$$

then incompressibility is verified automatically:

$$\nabla \cdot \mathbf{v} = \frac{\partial u}{\partial x} + \frac{\partial v}{\partial y} = \frac{\partial^2 \psi}{\partial y \partial x} - \frac{\partial^2 \psi}{\partial x \partial y} = 0.$$

The stream function derives its name from the fact that it is constant along lines of flow; see Exercise 4.3. Combining (4.32) and (4.42), one notes that ϕ and ψ are connected by the equations

$$u = \frac{\partial \phi}{\partial x} = \frac{\partial \psi}{\partial y}, \quad v = \frac{\partial \phi}{\partial y} = -\frac{\partial \psi}{\partial x}, \tag{4.43}$$

which are the Cauchy–Riemann equations of complex function theory, if ϕ and ψ are the real and complex parts of some complex potential

$$w = \phi + i\psi. \tag{4.44}$$

Equations (4.43) are equivalent to w being complex differentiable, and the existence of such an analytic function guarantees the existence of a corresponding flow!

Since the normal velocity vanishes at a solid wall, the flow must always be parallel to the surface: in other words, the surface is a streamline along which $\psi = \text{const}$. Thus the boundary condition to be satisfied by w is that its imaginary part is constant along solid surfaces. Once one has found a function meeting this condition which is analytic in the interior of the flow, one has automatically found a solution to the corresponding flow problem. The *complex velocity* is found by complex differentiation of the potential:

$$\frac{dw}{dz} = \frac{\partial \phi}{\partial x} + i\frac{\partial \psi}{\partial x} = u - iv; \tag{4.45}$$

it is the complex conjugate of the velocity $u + iv$.

Example 4.9 (Complex logarithm) There are two simple and important examples of complex potentials: if $w = A \ln z$ then

$$\frac{dw}{dz} = \frac{\Gamma}{z} = A\frac{x - iy}{x^2 + y^2},$$

corresponding to a velocity field

$$\mathbf{v} = \frac{m}{2\pi r}\mathbf{e}_r, \tag{4.46}$$

with $m = 2\pi A$. In analogy to Example 4.1, this represents a source or a sink of strength m, depending on whether m is positive or negative.

If the same potential is multiplied by the complex unit, we obtain $w = iA \ln z$. Then

$$\frac{dw}{dz} = i\frac{A}{z} = A\frac{ix + y}{x^2 + y^2},$$

which leads to a velocity

$$\mathbf{v} = \frac{\Gamma}{2\pi r}\mathbf{e}_\theta, \qquad (4.47)$$

with $\Gamma = -2\pi A$. Calculating (4.34) for a circle around the origin, one finds a circulation Γ, representing a vortex rotating in the clockwise ($\Gamma < 0$) or anticlockwise ($\Gamma > 0$) direction. $\qquad\square$

4.5.3 Stokes flow

The other important limit of the Navier–Stokes equation (4.13) is reached for large viscosity or small Reynolds number. This implies that the viscous term $\nu\triangle\mathbf{u}$ is much larger than the inertial or transport terms $\partial_t\mathbf{v} + (\mathbf{v}\cdot\nabla)\mathbf{v}$. The pressure term must of course be kept, as pressure gradients are driving the flow. Thus one finds the equation

$$\nabla p = \eta\triangle\mathbf{v}, \qquad (4.48)$$

with $\eta = \rho\nu$ the shear viscosity. We have omitted gravity, but a term $\rho\mathbf{g}$ can of course be added on the right. The boundary conditions are the same as for the Navier–Stokes equation, namely (4.18) for a solid wall and (4.24), (4.25) for a free surface. Equation (4.48) is known as *Stokes' equation*. It is a linear equation and will lead to smooth velocity fields when solved in a fixed smooth geometry. However, in the presence of a free surface the dynamics become nonlinear and may lead to singularities quite easily. We recall the example shown on the right of Fig. 1.5, to which we will return below.

In two dimensions, (4.48) may be given a particularly simple form using the stream function; see (4.42). First we eliminate the pressure by taking the curl of (4.48):

$$0 = \triangle\nabla\times\mathbf{v} = \triangle\left(\frac{\partial v}{\partial x} - \frac{\partial u}{\partial y}\right)\mathbf{e}_z.$$

Now inserting (4.42) for u and v, we find that

$$\triangle^2\psi = 0. \qquad (4.49)$$

Thus, in two dimensions, solving Stokes' equation is equivalent to solving the biharmonic equation.

Example 4.10 (Plane Poiseuille flow) Consider the two-dimensional flow, between two parallel plates at $y = \pm a$, driven by a constant pressure gradient $-K$ in the x-direction, parallel to the walls. By symmetry the solution is independent of x, and thus (4.49) reduces to

$$\frac{\partial^4 \psi}{\partial y^4} = 0,$$

with boundary conditions $\psi = \pm\psi_0$ and $\psi_y = 0$ at $y = \pm a$. The former conditions ensure that $v = 0$ and the latter $u = 0$, at both boundaries. Integrating, the result is

$$\psi = \frac{\psi_0}{2a^3}\left(3a^2 y - y^3\right), \tag{4.50}$$

and from (4.42) the velocity field is

$$\mathbf{v} = \frac{3\psi_0}{2a^3}\left(a^2 - y^2\right)\mathbf{e}_x.$$

Evaluating the left-hand side of (4.48), the negative pressure gradient is $-p_x = K = 3\psi_0/a^3$. $\qquad\square$

4.6 Elasticity

In elasticity, one describes the deformation of a solid body in response to forces applied to it. If we denote the initial position of any material point as $\mathbf{r}$ and its position after deformation as $\mathbf{s}(\mathbf{r}) = \mathbf{r}+\mathbf{u}(\mathbf{r})$ then the squared distance between two neighboring points $\mathbf{r}$ and $\mathbf{r} + d\mathbf{r}$ will be, after the deformation,

$$|\mathbf{s}(\mathbf{r} + d\mathbf{r}) - \mathbf{s}(\mathbf{r})|^2 \simeq \left|\sum_i \frac{\partial \mathbf{s}}{\partial x_i}dx_i\right|^2 = g_{ij}dx_i dx_j,$$

where

$$g_{ij} = \frac{\partial \mathbf{s}}{\partial x_i} \cdot \frac{\partial \mathbf{s}}{\partial x_j}. \tag{4.51}$$

The distance between the points will then be changed by an amount

$$|\mathbf{s}(\mathbf{r} + d\mathbf{r}) - \mathbf{s}(\mathbf{r})|^2 - |d\mathbf{r}|^2 = (g_{ij} - \delta_{ij})dx_i dx_j = 2\epsilon_{ij}dx_i dx_j,$$

where the matrix ϵ_{ij} is called the *strain tensor*.

Under the assumption of small deformations (linear elasticity), one finds

$$\epsilon_{ij} = \frac{1}{2}\left(\frac{\partial u_j}{\partial x_i} + \frac{\partial u_i}{\partial x_j}\right) \tag{4.52}$$

for the strain tensor. It is the analogue of the rate-of-deformation tensor (4.9) in hydrodynamics. While in elasticity forces are generated in response to deformation (and vice versa), forces in hydrodynamics come from the velocity gradients of the flow field. By requiring the material to be isotropic, one arrives at a constitutive relation very similar to (4.12):

$$\sigma_{ij} = 2\mu\epsilon_{ij} + \lambda\epsilon_{kk}\delta_{ij}. \qquad (4.53)$$

The parameters μ and λ are called the Lamé constants. Sometimes a different set of constants, namely Young's modulus E and Poisson's ratio ν, is used:

$$\mu = \frac{E}{2(1+\nu)}, \quad \lambda = \frac{E\nu}{(1+\nu)(1-2\nu)}. \qquad (4.54)$$

The free energy of the elastic body is

$$\mathcal{F} = \int_V F d^3 x, \qquad (4.55)$$

where the free energy density [132] is determined by the product of forces and deformations:

$$F = \frac{1}{2}\sigma_{ij}\epsilon_{ij} = \frac{\lambda}{2}(\epsilon_{ii})^2 + \mu\epsilon_{lk}^2 \equiv \frac{\lambda}{2}(\text{tr }\boldsymbol{\epsilon})^2 + \mu\,\text{tr}(\boldsymbol{\epsilon}^2). \qquad (4.56)$$

The stresses are recovered if one differentiates the free energy with respect to the deformations:

$$\sigma_{ik} = \frac{\partial F}{\partial \epsilon_{ik}}. \qquad (4.57)$$

The analogue of Newton's equation $\mathbf{F} = m\mathbf{a}$ now becomes

$$\rho\ddot{\mathbf{u}} = (\nabla \cdot \boldsymbol{\sigma}),$$

and inserting the stress tensor (4.53), we obtain

$$\rho\ddot{\mathbf{u}} = (\lambda + \mu)\nabla(\nabla \cdot \mathbf{u}) + \mu\Delta\mathbf{u}, \qquad (4.58)$$

an equation for the deformation field $\mathbf{u}$.

Equation (4.58) is a wave equation for a vector field $\mathbf{u}$ which admits solutions in the form of traveling waves known as elastic waves. We will be concerned exclusively with slow, quasistatic, motion of the solid in response to external forcing. In that case, and in the absence of body forces, the condition for the static equilibrium of an elastic body becomes

$$\nabla \cdot \boldsymbol{\sigma} = 0. \qquad (4.59)$$

Using (4.53), and substituting (4.54) for the Lamé constants, (4.59) can be written as

$$\nabla(\nabla \cdot \mathbf{u}) + (1 - 2\nu)\Delta\mathbf{u} = 0. \qquad (4.60)$$

Taking the divergence of (4.60) and using $\Delta = \nabla \cdot \nabla$, one finds that

$$\Delta(\nabla \cdot \mathbf{u}) = 0, \tag{4.61}$$

i.e. the divergence of the deformation vector is harmonic.

Depending on the physical situation, two different types of conditions are applied to the boundary $\partial\Omega$ of the body. If the body is forced (for example by pushing a stamp onto its surface), one is prescribing the stress or traction T_i, and the boundary condition is

$$\sigma_{ij}n_j = T_i. \tag{4.62}$$

If the body is clamped along a part of its boundary (so that there are no deformations), we must impose

$$u_i = 0. \tag{4.63}$$

Example 4.11 (Sheared plate) Consider an elastic plate of width a, which is clamped at its lower surface ($y = 0$), while a constant shear stress is applied at the upper surface in the x-direction. This means the boundary conditions at $y = 0$ are $u_x = u_y = 0$, while at $y = a$ we have $\sigma_{xy} = \tau = $ const and $\sigma_{yy} = 0$ (no normal stress). From (4.59), and since σ is independent of x, we have

$$\frac{\partial\sigma_{xy}}{\partial y} = 0, \quad \frac{\partial\sigma_{yy}}{\partial y} = 0.$$

Thus both $\sigma_{xy} = \tau$ and $\sigma_{yy} = B$ are constants and, by the boundary condition, $B = 0$. From (4.53),

$$0 = \sigma_{yy} = 2\mu\frac{\partial u_y}{\partial y} + \lambda\left(\frac{\partial u_x}{\partial x} + \frac{\partial u_y}{\partial y}\right),$$

and since there is no dependence on x and $u_y = 0$ at $y = 0$ we have $u_y = 0$ everywhere. Next,

$$\tau = \sigma_{xy} = \mu\left(\frac{\partial u_x}{\partial y} + \frac{\partial u_y}{\partial x}\right),$$

from which we obtain

$$u_x = \frac{\tau}{\mu}y,$$

which satisfies $u_x = 0$ at $y = 0$. As expected, the material is deformed in a linear way. $\qquad\square$

Note that the Stokes equation (4.48) can also be written in the form (4.59), but with the viscous stress tensor (4.12). Namely,

$$0 = \partial_i\sigma_{ij} = -\partial_j p + \eta\partial_i\left(\partial_j v_i + \partial_i v_j\right) = -\partial_j p + \eta\Delta v_j,$$

since the fluid is assumed incompressible ($\partial_i v_i = 0$). Elastic deformations, however, generally do not satisfy an incompressibility constraint, but if $\nu \to 1/2$, it follows from (4.54) that $\lambda \to \infty$; then, according to (4.53), for the stress to remain finite we must have $\epsilon_{kk} = 0$ or $\nabla \cdot \mathbf{u} = 0$. Thus if we set $p = \lambda\epsilon_{kk}$ in the limit $\nu \to 1/2$, the elastic stress tensor becomes identical to its viscous counterpart, with the deformation field $\mathbf{u}$ playing the role of the velocity field. In other words, if Poisson's ratio is $\nu = 1/2$, elastic deformations are governed by the same equation as that for the motion of a viscous fluid [30]. We will exploit this analogy later on when we compare cusp solutions on the surface of a liquid to crack tips in a solid.

In two dimensions the description of an elastic solid simplifies in various ways. The equilibrium condition (4.59), written in components, becomes

$$\frac{\partial \sigma_{xx}}{\partial x} + \frac{\partial \sigma_{xy}}{\partial y} = 0, \quad \frac{\partial \sigma_{yx}}{\partial x} + \frac{\partial \sigma_{yy}}{\partial y} = 0.$$

These equations are satisfied if the stress components are given by a single function ϕ such that

$$\sigma_{xx} = \frac{\partial^2 \phi}{\partial y^2}, \quad \sigma_{yy} = \frac{\partial^2 \phi}{\partial x^2}, \quad \sigma_{xy} = \sigma_{yx} = -\frac{\partial^2 \phi}{\partial x \partial y}; \quad (4.64)$$

ϕ is called the *Airy stress function*. One can show that this is also the most general description. Using this representation and (4.53), it follows that

$$\triangle \phi = \sigma_{xx} + \sigma_{yy} = 2(\mu + \lambda)(\epsilon_{xx} + \epsilon_{yy}).$$

However, we have

$$\epsilon_{xx} + \epsilon_{yy} = \frac{\partial u_x}{\partial x} + \frac{\partial u_y}{\partial y} = \nabla \cdot \mathbf{u},$$

which, according to (4.61), is an harmonic function. It follows that ϕ satisfies the biharmonic equation:

$$\triangle^2 \phi = 0. \quad (4.65)$$

This is the same equation as (4.49) although the meaning of the stress function ϕ is different from that of the stream function ψ. In polar coordinates the relations (4.64) between the stress tensor and the stress function become

$$\sigma_{rr} = \frac{1}{r^2}\frac{\partial^2 \phi}{\partial \theta^2} + \frac{1}{r}\frac{\partial \phi}{\partial r}, \quad \sigma_{\theta\theta} = \frac{\partial^2 \phi}{\partial r^2}, \quad \sigma_{r\theta} = -\frac{\partial}{\partial r}\left(\frac{1}{r}\frac{\partial \phi}{\partial \theta}\right). \quad (4.66)$$

Example 4.12 (Circular hole) Consider a circular hole of radius R in an infinite plate subjected to a uniform radial stress $\sigma_{rr} = \tau$ far from the hole. As a result, $\phi = \phi(r)$ depends on r only and $\triangle^2 \phi = 0$ yields

$$\Delta\phi = \frac{\partial\phi'}{\partial r} + \frac{\phi'}{r} = A_1 \ln r + A_2.$$

Integrating, the general solution for ϕ is

$$\phi = Ar^2 + Br^2 \ln r + C + D \ln r, \qquad (4.67)$$

and the corresponding radial stresses are

$$\sigma_{rr} = 2A + B + 2B \ln r + \frac{D}{r^2}.$$

From the condition at infinity, $B = 0$ and $A = \tau/2$; clearly, the additive constant C does not play a role. The walls of the hole must be stress free, and thus

$$0 = \sigma_{rr}(R) = \tau + \frac{D}{R^2},$$

so that $D = -\tau R^2$. In conclusion the stress distribution is given by

$$\sigma_{rr} = \tau - \frac{\tau R^2}{r^2}, \quad \sigma_{\theta\theta} = \tau + \frac{\tau R^2}{r^2}, \quad \sigma_{r\theta} = 0. \qquad (4.68)$$

$\square$

An even simpler description can be achieved if the deformations of a plate are out-of-plane only, i.e. $u_1 = u_2 = 0$ and u_3 depends on x and y alone. In that case the only non-vanishing components of ϵ_{ij} are ϵ_{13} and ϵ_{23}, and thus $\sigma_{ij} = 2\mu\epsilon_{ij}$. Then the equilibrium condition (4.59) reads

$$0 = \frac{\partial\sigma_{13}}{\partial x} + \frac{\partial\sigma_{23}}{\partial y} = \mu\left(\frac{\partial^2 u_3}{\partial x^2} + \frac{\partial^2 u_3}{\partial y^2}\right)$$

or

$$\Delta u_3 = 0. \qquad (4.69)$$

Thus, for a two-dimensional out-of-plane stress, u_3 plays the same role as the potential ϕ for potential flow and we have

$$\sigma_{13} = \mu\frac{\partial u_3}{\partial x}, \quad \sigma_{23} = \mu\frac{\partial u_3}{\partial y}, \qquad (4.70)$$

in analogy with (4.43). In particular, the complex methods of Section 4.5.2 can be employed to solve plane elastic problems with out-of-plane deformations.

Exercises

4.1 Calculate the two-dimensional potential flow around a circular cylinder of radius R, placed in a uniform stream $\mathbf{v} = U\mathbf{e}_x$.

(i) Show that in polar coordinates the potential $\phi = Ur\cos\theta$ corresponds to a uniform stream.

(ii) Show that

$$\phi = U\left(r + \frac{A}{r}\right)\cos\theta$$

solves (4.37); determine A such that the normal flow condition (4.41) is satisfied on the surface of the cylinder.

(iii) Compute the tangential flow v_θ on the surface of the cylinder. Thus use (4.39), without gravity, to show that the pressure distribution around the cylinder is

$$p = \frac{\rho U^2}{2}\left(1 - 4\sin^2\theta\right).$$

Show that the total force on the cylinder vanishes.

4.2 In order for the cylinder in Exercise 4.1 to experience a lift force, there must be a circulation Γ along a closed loop around it.

(i) Show that the uniform flow around a cylinder is described by the complex potential

$$w(z) = U\left(z + \frac{R^2}{z}\right),$$

either by using the result of the previous problem or by verifying the boundary conditions directly.

(ii) Demonstrate that, in complex notation, the circulation (4.34) can be calculated as

$$\Gamma = \Im\left\{\oint_C \frac{dw}{dz}dz\right\}. \tag{4.71}$$

Thus show that, by adding

$$-\frac{i\Gamma}{2\pi}\ln z$$

to the complex potential, the boundary conditions are still verified but that the circulation around the cylinder is now Γ.

(iii) Recalculate the pressure distribution on the surface of the cylinder, and show that there is a total lift force

$$\mathbf{F} = \rho U\Gamma\mathbf{e}_y$$

on the cylinder.

4.3 For a stream function in a two-dimensional flow with $u = \dfrac{\partial \psi}{\partial y}$ and $v = -\dfrac{\partial \psi}{\partial x}$, show that:

   (i) the streamlines are given by $\psi = $ const;

   (ii) $|\mathbf{v}| = |\nabla \psi|$, which implies that the flow is faster where the streamlines are closer;

   (iii) the volume flux crossing any curve from $\mathbf{x}_0$ to $\mathbf{x}_1$ is given by $\psi(\mathbf{x}_1) - \psi(\mathbf{x}_0)$.

4.4 (*) Calculate the Stokes flow around a sphere of radius R placed in a uniform stream $\mathbf{U}$; the fluid viscosity is η. To subtract the mean flow, write the velocity field as

$$\mathbf{v} = \nabla \times \mathbf{A} + \mathbf{U}.$$

Why is it possible to write the velocity perturbation as the curl of a vector ?

   (i) Assume that the vector potential $\mathbf{A}$ can be written in terms of a single scalar function $f(\mathbf{r})$:

$$\mathbf{A} = f'(\mathbf{r})\mathbf{n} \times \mathbf{U},$$

where $\mathbf{n} = \mathbf{r}/r$ is the outward normal. Noting that (4.48) can be written in the form $\Delta \nabla \times \mathbf{v} = 0$, show that f satisfies $\Delta^2 f = 0$.

   (ii) Show that f must have the form

$$f = ar + \frac{b}{r},$$

and deduce the velocity field

$$\mathbf{v} = \mathbf{U} - a\frac{\mathbf{n}(\mathbf{U} \cdot \mathbf{n}) + \mathbf{U}}{r} + b\frac{3\mathbf{n}(\mathbf{U} \cdot \mathbf{n}) - \mathbf{U}}{r^3}.$$

   (iii) Use the boundary conditions to show that $a = 3R/4$ and $b = R^3/4$.

   (iv) Calculate the stress tensor in spherical coordinates and show that the total force on the sphere is

$$\mathbf{F} = 6\pi R\eta\mathbf{U}.$$

4.5 The Navier–Stokes equation (4.13) inherits its *Galilean invariance* from the laws of mechanics. Consider fluid motion as viewed by an observer moving at a constant velocity $\mathbf{u}$. Then the spatial coordinate $\mathbf{x}$ will turn into $\tilde{\mathbf{x}} = \mathbf{x} - \mathbf{u}t$. What does this correspond to physically? What are the velocity and pressure fields $\tilde{\mathbf{v}}(\tilde{\mathbf{x}}, t)$ and $\tilde{p}(\tilde{\mathbf{x}}, t)$ in the new coordinate system? Transform the Navier–Stokes equation to the new variables $\tilde{\mathbf{v}}$ and $\tilde{p}$, and show that the equation remains invariant under the transformation.

4.6 Consider the problem of Example 4.10 but using the Navier–Stokes equation (4.13).

(i) Taking the flow field as $\mathbf{v} = u(y)\mathbf{e}_x$, show that the x-component of (4.13) reduces to

$$\eta \frac{d^2 u}{dy^2} = -K.$$

(ii) Using the boundary conditions on the plate, show that the flow is parabolic,

$$u(y) = \frac{K}{2\eta} y(h - y);$$

compare this with Example 4.10.

(iii) Compute the shear stress on the walls.

(iv) Compute the total flux (per unit length in the y-direction) through the channel.

4.7 The axisymmetric version of Example 4.10, namely the flow through a circular pipe of radius R driven by a pressure gradient $-K$ in the z-direction, is known as Poiseuille flow.

(i) Assuming that the flow field can be written in the form $\mathbf{v} = u(r)\mathbf{e}_z$, derive an equation for $u(r)$ (see Appendix A for the axisymmetric version of (4.13)).

(ii) Using the boundary condition on the surface of the pipe, and regularity for $r = 0$, find $u(r)$ in terms of K.

(iii) Show that the total flux through the pipe is

$$Q = \frac{\pi K}{8\nu} R^4,$$

which is known as the Hagen–Poiseuille law.

4.8 A flat plate in a quiescent viscous fluid starts at $t = 0$ to move with speed U. You can assume that by symmetry $\mathbf{v} = u(y)\mathbf{e}_x$.

(i) Show that the Navier–Stokes equation reduces to

$$\frac{\partial u}{\partial t} = \nu \frac{\partial^2 u}{\partial y^2}. \tag{4.72}$$

(ii) Show that (4.72) has a *similarity solution*

$$u(y, t) = f(y/\sqrt{\nu t}).$$

(iii) Using the boundary conditions $u(x, 0) = 0$ and $u(0, t) = U$ for $t > 0$ show that

$$u(y, t) = \frac{U}{\sqrt{\pi}} \int_{y/\sqrt{\nu t}}^{\infty} e^{-\xi^2/4} d\xi \equiv U \operatorname{erf}\left(\frac{y}{2\sqrt{\nu t}}\right),$$

where

$$\text{erf} = \frac{2}{\sqrt{\pi}} \int_0^x e^{-t^2} dt \tag{4.73}$$

is the error function.

4.9 A solid plate inclined at an angle α with respect to the horizontal is covered by an incompressible fluid layer of uniform thickness h. The pressure above the fluid is atmospheric. The viscosity of the fluid is η, ρ is its density, and g is the acceleration of gravity. Consider steady flow down the incline.

  (i) Taking $\mathbf{e}_x$ in the downstream direction and $\mathbf{e}_y$ perpendicular to it, assume that $\mathbf{v} = u(y)\mathbf{e}_x$. Confirm that the flow is incompressible and show that the Navier–Stokes equation is expressed by

$$0 = -\frac{\partial p}{\partial x} + \eta \frac{\partial^2 u}{\partial y^2} + \rho g \sin\alpha, \qquad 0 = -\frac{\partial p}{\partial y} - \rho g \cos\alpha.$$

  (ii) What are the boundary conditions on the plate and at the free surface?

  (iii) Show that $\partial p / \partial x = 0$ and find $u(y)$.

  (iv) Find the volume flux down the plane.

  (v) Compute the rate of energy dissipation in the fluid layer. What is the physical interpretation of the result?

4.10 Following Exercise 4.7, show that the equation of motion for an axisymmetric free surface $h(z, t)$ is (6.44).

4.11 The three-dimensional mean curvature of an axisymmetric body is given by (4.23). Show that the same result is obtained by considering the variation of the surface area

$$A = \pi \int \mathcal{A} dz \equiv 2\pi \int h \sqrt{1 + h_z^2}\, dz$$

with respect to the cross-sectional area $\pi a = \pi h^2$. In other words, show that

$$\kappa = \frac{\delta \mathcal{A}}{\delta a} - \left(\frac{\delta \mathcal{A}}{\delta a'}\right)', \tag{4.74}$$

where the primes refer to differentiation by a space variable, and thus

$$\kappa = \frac{1}{\sqrt{a + a'^2/4}} - \left(\frac{a'}{2\sqrt{a + a'^2/4}}\right)'. \tag{4.75}$$

4.12 Show that in a frame of reference that is rotating with angular velocity ω, the Navier–Stokes equation (4.13) has the form

$$\rho \left(\mathbf{v}_t + \mathbf{v} \cdot \nabla \mathbf{v} \right) = -\nabla p + \eta \Delta \mathbf{v} - 2\rho \left(\omega \times \mathbf{v} \right) \tag{4.76}$$
$$-\rho \left[\omega \times (\omega \times \mathbf{r}) \right],$$

where $2\rho \left(\omega \times \mathbf{v} \right)$ is the so-called Coriolis force and $\rho \left[\omega \times (\omega \times \mathbf{r}) \right]$ is the centrifugal acceleration.

(i) Neglecting Coriolis forces, show that the centrifugal term can be eliminated from (4.76) and thus from (4.13), by replacing p by $\Pi = p - \rho(\omega \times \mathbf{r})^2/2$.

(ii) Now consider an axisymmetric drop undergoing rigid body rotation around the z-axis. Using (i), find the ordinary differential equation to be solved by the shape $r = h(z)$ of the drop.

4.13 (*) Consider a finger of heavy fluid with density ρ penetrating into a light fluid of negligible density [123]. Close to the tip, the finger will have a shape given by

$$y = h(x, t) = y_0(t) + \kappa(t)x^2, \tag{4.77}$$

and the velocity potential in the upper fluid will be

$$\phi(x, y, t) = \phi_0(t) + U(t)y + A(t)(x^2 - y^2). \tag{4.78}$$

(i) Show that ϕ is harmonic.

(ii) Use Bernoulli's equation (4.39) to show that

$$\phi_{0,t} + \frac{p}{\rho} + \frac{1}{2}U^2 + g y_0 = 0, \tag{4.79}$$

$$A_t + 2A^2 + U_t \kappa - 2AU\kappa + g\kappa = 0, \tag{4.80}$$

$$-A_t + 2A^2 = 0. \tag{4.81}$$

(iii) Use the kinematic boundary condition to show that

$$\kappa_t = -6A\kappa$$

and thus

$$A(t) = -\frac{1}{2t}, \quad U(t) = -\frac{1}{2}gt + \frac{\alpha}{t^4}, \quad \kappa(t) = \frac{1}{3\alpha}t^3,$$

with $\alpha > 0$ a free parameter. Does the curvature at the tip of the parabola blow up in finite time?

4.14 Calculate the stress distribution around a circular hole of radius R in a plate which is stressed uniaxially in the y-direction, i.e. $\sigma_{yy} = \tau$ far from the hole.

(i) Show that the far-field stress distribution in polar coordinates reads

$$\sigma_{rr} = \frac{\tau}{2}\left(1 - \cos 2\theta\right), \quad \sigma_{\theta\theta} = \frac{\tau}{2}\left(1 + \cos 2\theta\right), \quad \sigma_{r\theta} = \frac{\tau}{2}\sin 2\theta.$$
$$(4.82)$$

Thus the problem may be viewed as the superposition of two stress distributions: a purely radial one, described by the solution (4.68), and one having an angular dependence of $\cos 2\theta$.

(ii) Show that the stress function

$$\phi = f(r)\cos 2\theta$$

describes the angular dependence of (4.82). Thus derive an equation for $f(r)$, and show that the general solution is of the form

$$f(r) = Ar^2 + Br^4 + \frac{C}{r^2} + D.$$

(iii) Find A, B, C, D using (4.82), as well as the condition that the boundary of the hole be stress free. Thus, superimposing (4.82) and the radially symmetric solution (4.68), conclude that

$$\sigma_{rr} = \frac{\tau}{2}\left(1 - \frac{R^2}{r^2}\right) + \frac{\tau}{2}\left(1 + \frac{3R^4}{r^4} - \frac{4R^2}{r^2}\right)\cos 2\theta,$$

$$\sigma_{\theta\theta} = \frac{\tau}{2}\left(1 + \frac{R^2}{r^2}\right) - \frac{\tau}{2}\left(1 + \frac{3R^4}{r^4}\right)\cos 2\theta,$$

$$\sigma_{r\theta} = -\frac{\tau}{2}\left(1 - \frac{3R^4}{r^4} + \frac{2R^2}{r^2}\right)\sin 2\theta.$$

4.15 Calculate the stress distribution in a plate around a circular hole of radius R for an out-of-plane stress. The plate is subject to a constant shear stress $\sigma_{23} = \tau$ far from the hole. Use the analogy with potential flow to show that the complex potential w with $u_3 = \Re\{w\}$ is given by

$$w = -i\tau\left(z - \frac{R^2}{z}\right),$$

and that

$$\sigma_{13} = -\tau \sin 2\theta, \quad \sigma_{23} = \tau\left(1 + \cos 2\theta\right).$$

4.16 (*) Show that solutions of the biharmonic equation (4.49) in two dimensions can be written in the form

$$\psi(x, y) = \Re\{\varphi(z) + \bar{z}\chi(z)\},$$

where $\varphi(z)$ and $\chi(z)$ are complex analytic functions (with argument $z = x + iy$, $\bar{z} = x - iy$ being the complex conjugate of z). This is known as the Goursat representation.

Hint: in terms of $z = x + iy$ one can write

$$\Delta\psi = 2\frac{\partial^2}{\partial z \partial \bar{z}}\psi$$

and hence

$$\Delta^2\psi = 4\frac{\partial^4\psi}{\partial z^2 \partial \bar{z}^2} = 0,$$

so that $\partial^2\psi/\partial z^2$ is linear in $\bar{z}$ with coefficients that depend on z.

4.17 (*) Equation (4.58) is a wave equation for a vector field $\mathbf{u}$. Hence, in absence of external forces $\mathbf{f}$, it admits solutions in the form of traveling waves.

(i) Show that there exist transverse and longitudinal waves that move at velocities

$$c_l = \sqrt{\frac{E(1-v)}{\rho(1+v)(1-2v)}}, \qquad c_t = \sqrt{\frac{E}{2\rho(1+v)}},$$

respectively (cf. [132]).

(ii) If the elastic body has a free surface, show that there exist surface waves, called Rayleigh waves, that propagate with a velocity c_R which is determined from the root of

$$\left(2 - \frac{c_R^2}{c_t^2}\right)^2 - 4\left(1 - \frac{c_R^2}{c_t^2}\right)^{1/2}\left(1 - \frac{c_R^2}{c_l^2}\right)^{1/2} = 0.$$

5
Local singular expansions

In this chapter we describe the local behavior produced by singularities of boundaries. The equations themselves are linear equations of a very benign sort (the Laplace equation (4.37) or the double Laplacian (4.49)), which in the case of smooth boundaries would lead to solutions which are themselves smooth. We discuss two different situations which arise generically for regions bounded by surfaces. In the first case the boundary has a corner which traces out a line in three-dimensional space. Since the corner is associated with a vanishing length scale the curvature of this line is irrelevant to the local behavior near the corner, and the problem is effectively two dimensional. The other generic case is that of a conical tip.

In each case the problem lacks a characteristic length scale, so we expect the solutions to be scale invariant as a function of the distance r from the singularity. The power of the method lies in the fact that, to leading order, the solution can be determined without reference to a solution of the global problem: the local behavior is universal.

5.1 Potential flow in a corner

Let us analyze the inviscid potential flow around a sharp corner. As argued above, the geometry can be described locally as a two-dimensional corner of opening angle α; see Fig. 5.1. We have to solve Laplace's equation (4.37) in the interior of the corner and impose boundary conditions (4.41) on the solid surface of the wedge. In practice the corner will be rounded, on some length scale ϵ. However, in the limit $\epsilon \to 0$ we expect the solution to be self-similar, which means we are looking for solutions of the form

$$\phi = r^\lambda f(\theta), \tag{5.1}$$

89

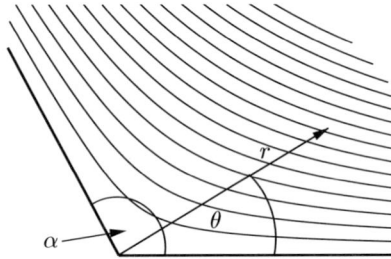

Figure 5.1 The flow inside a corner of opening angle α. Streamlines defined by $\psi(r, \theta) = \psi_0 = $ const are shown.

where r and θ are polar coordinates, see Fig. 5.1. The Laplacian becomes

$$\Delta\phi = \frac{1}{r}\frac{\partial}{\partial r}\left(r\frac{\partial \phi}{\partial r}\right) + \frac{1}{r^2}\frac{\partial^2 \phi}{\partial \theta^2} = r^{\lambda-2}\left[\lambda^2 f(\theta) + f''(\theta)\right]$$

so that (4.37) is satisfied if

$$\lambda^2 f(\theta) + f''(\theta) = 0, \tag{5.2}$$

with general solution

$$f(\theta) = A \sin \lambda\theta + B \cos \lambda\theta. \tag{5.3}$$

The azimuthal component of the velocity is

$$v_\theta = \frac{1}{r}\frac{\partial \phi}{\partial \theta} = r^{\lambda-1} f'(\theta);$$

hence the normal velocity at the walls vanishes (see (4.41)) if the homogeneous boundary conditions

$$f'(0) = 0, \quad f'(\alpha) = 0 \tag{5.4}$$

are satisfied. Since

$$f'(\theta) = A\lambda \cos \lambda\theta - B\lambda \sin \lambda\theta,$$

we find from the first boundary condition in (5.4) that $A = 0$ and, from the second boundary condition, it follows that

$$\sin \lambda\alpha = 0. \tag{5.5}$$

Thus the scaling exponent λ assumes the discrete set of values $\lambda_n = n\pi/\alpha$, with $n = \pm 1, \pm 2, \ldots$, and the potential is given by

$$\phi = r^{\lambda_n} \cos \lambda_n\theta. \tag{5.6}$$

We have disregarded the case $n = 0$, which leads to a constant potential. The exponents in (5.6) are not determined by dimensional analysis but follow from the solution of an eigenvalue problem; this is another example of self-similarity of the second kind.

If, however, a characteristic angular velocity is imposed, the problem turns into an example of self-similarity of the first kind.

Example 5.1 (Hinge flow) Consider potential flow in a corner, as shown in Fig. 5.1. Imagine that the corner is hinged and that the plate at $\theta = \alpha$ is closing at an angular velocity $\dot{\alpha} = -\Omega$. Thus the boundary condition at the upper plate is now that the azimuthal velocity is

$$v_\theta = r^{\lambda-1} f'(\theta) \equiv -r\Omega;$$

the solution (5.3) is otherwise the same. It follows that the exponent is $\lambda = 2$ and the second boundary condition in (5.4) is replaced by $f'(\alpha) = -\Omega$. Thus hinge flow is a solution to an inhomogeneous problem, whereas before we were seeking a homogeneous "eigensolution". From the first boundary condition in (5.4) we once more have $A = 0$, but now $B = \Omega/(2\sin 2\alpha)$ so that the velocity potential in the wedge is

$$\phi = \frac{r^2\Omega}{2\sin 2\alpha} \cos 2\theta, \tag{5.7}$$

which solves the problem. Note that the exponent $\lambda = 2$ is fixed by the dimensions $[\phi] = \mathrm{cm}^2/\mathrm{s}$ and $[\Omega] = 1/\mathrm{s}$, as is customary for self-similarity of the first kind. In Exercise 5.1 we will consider the singularity that occurs in (5.7) for $\alpha \to \pi/2$. $\qquad\square$

Coming back to the solution (5.6) of the *homogeneous* problem, below we will analyze which exponent is generic by considering a dominant balance near the tip. Since the problem is linear, the real local behavior will be a superposition of solutions of the form (5.6). However, if the index n becomes negative, the solution will be more and more singular. A minimum requirement to be fulfilled by any solution is that the total kinetic energy be finite. Otherwise, since energy is a conserved quantity, there is no hope of this solution ever being realized. Let S_δ be the spherical shell between two spheres of radius δ and $\delta/2$, both centered on the tip of the wedge. Then the energy in the shell is

$$E_{\mathrm{kin}} = \frac{\rho}{2} \int_{S_\delta} |\mathbf{v}|^2 \, d\mathbf{x} = \frac{\rho}{2} \int_{\delta/2}^{\delta} \int_0^{\alpha} |\nabla\phi|^2 \, r \, dr \, d\theta.$$

Using (5.6),

$$
|\nabla\phi|^2 = \left(\frac{\partial\phi}{\partial r}\right)^2 + \frac{1}{r^2}\left(\frac{\partial\phi}{\partial\theta}\right)^2 = \lambda_n^2 r^{2\lambda_n-2},
$$

and so the kinetic energy in the shell becomes

$$
E_{\text{kin}} = \frac{\rho}{2}\alpha\lambda_n^2 \int_{\delta/2}^{\delta} r^{2\lambda_n-1}\,dr = \frac{\rho\alpha\lambda_n}{4}\left(1 - 2^{-2\lambda_n}\right)\delta^{2\lambda_n}. \tag{5.8}
$$

For the kinetic energy in the shell not to make a diverging contribution as $\delta \to 0$, the scaling exponent of δ must clearly be non-negative; this means that n must be positive.

In summary, the local description of the solution is

$$
\phi = \sum_{n=1}^{\infty} a_n r^{n\pi/\alpha} \cos\frac{n\pi\theta}{\alpha}, \tag{5.9}
$$

which is a linear superposition of all solutions with finite energy. In fact (5.9) is a standard Fourier cosine series: any solution can be written in the form (5.9) with unique values of the coefficients a_n. Namely, suppose that we know the potential ϕ at a certain distance r_0 from the corner: $\phi_0(\theta) = \phi(r_0, \theta)$. Then we can compute the coefficients a_n using the formula

$$
a_n = \frac{2}{\alpha} r_0^{-n\pi/\alpha} \int_0^{\alpha} \phi_0(\theta) \cos\frac{n\pi\theta}{\alpha}\,d\theta.
$$

Similar results hold true whenever the angular problem (which is (5.2) in the case of Laplace's equation) can be written as a Sturm–Liouville problem [139]. Most linear equations of mathematical physics belong in this category.

So far, (5.9) is just a convenient way of writing the local solution and it seems one has not gained much. However, near the corner the infinite sum (5.9) is dominated by the contribution with the smallest power. In other words one can predict, without having solved the (in general very difficult) global problem, that to leading order the *local* solution must have the *universal* form

$$
\phi = a_1 r^{\pi/\alpha} \cos\frac{\pi\theta}{\alpha}. \tag{5.10}
$$

The corresponding behavior of the velocity field $\mathbf{v} = \nabla\phi$ is given by

$$
v_r = \frac{\partial\phi}{\partial r} = ar^{(\pi-\alpha)/\alpha} \cos\frac{\pi\theta}{\alpha}, \quad v_\theta = \frac{1}{r}\frac{\partial\phi}{\partial\theta} = -ar^{(\pi-\alpha)/\alpha} \sin\frac{\pi\theta}{\alpha}, \tag{5.11}
$$

where $a = \pi a_1/\alpha$. In polar coordinates the velocity components and the stream function ψ (see (4.42)) are related by

$$v_r = \frac{1}{r}\frac{\partial \psi}{\partial \theta}, \quad v_\theta = -\frac{\partial \psi}{\partial r} \tag{5.12}$$

and, integrating (5.12), we find

$$\psi = a_1 r^{\pi/\alpha} \sin\frac{\pi\theta}{\alpha}.$$

The isolines of ψ are shown in Fig. 5.1; they are the flow lines as the fluid turns the corner.

The flow (5.11) describes the generic behavior near the singularity, unless for some particular reason the amplitude a vanishes (see below). If $\alpha > \pi$, the velocity becomes infinite near the corner. This might be interpreted as implying that the corners at the exterior of buildings are very windy places! To determine the coefficient a_1 (as well as any of the higher orders in the expansion (5.9)) one has to solve the full problem, which is possible only in some very special cases.

5.2 Potential flow around a two-dimensional airfoil

An example in which the singular flow around a sharp corner plays a crucial role is the physics of flight. In Fig. 5.2 we show the potential flow around an airplane wing, which has a cusp at the trailing edge, where $\alpha = 2\pi$. As predicted by the analysis of the previous section, the flow near the corner is singular as the flow lines turn the corner. This is an undesirable situation for flight, where one would like the flow to leave the edge of the wing smoothly. This can be achieved by adding a circulation to the flow, chosen to make the coefficient a_1 in (5.10) vanish. The flow with circulation then produces an upward (lift) force on the wing.

To be able to perform calculations analytically and to find an exact solution for the flow around the wing, we use the Joukowsky wing shown on the right of Fig. 5.3. According to Section 4.5.2, we need to find an analytic function $w(z)$ in the exterior of the wing with $\Im\{w(z)\} = \text{const}$ on the surface of the wing. The wing is constructed cleverly to be the image of the circle parameterized by φ (see Fig. 5.3),

$$\zeta = -\mu + (1+\mu)e^{i\varphi}, \tag{5.13}$$

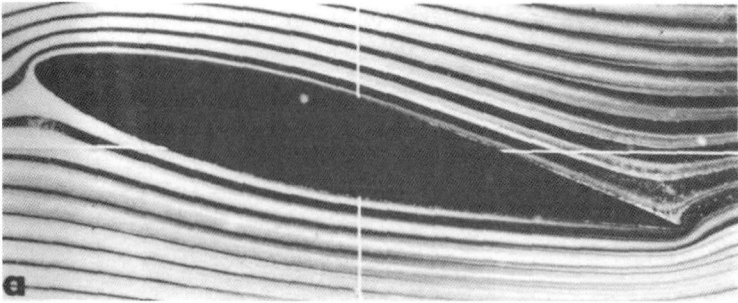

Figure 5.2 Potential flow around a wing at an angle of attack of 13° [219, 212]; note the singularity at the trailing edge. The flow was produced in the creeping-flow limit between two narrowly spaced glass plates (a so-called Hele-Shaw cell). In the next chapter we show that such a flow is potential; here we have the special case where the total circulation around the wing vanishes.

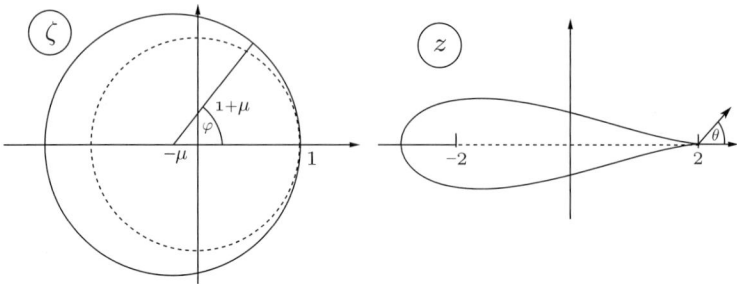

Figure 5.3 A simple mathematical model for a wing (right) is found by mapping to a circle, using the Joukowsky transformation (5.14). The center of the circle is at $\zeta = -\mu$, and its radius is $1 + \mu$.

with radius $R = 1 + \mu$ and centered at $\zeta = -\mu$, under the conformal mapping

$$z = f(\zeta) = \zeta + \frac{1}{\zeta}. \qquad (5.14)$$

This so-called *Joukowsky transformation* maps the unit circle (centered in the origin) in the ζ-plane onto a line in the z-plane between -2 and 2 (broken lines) and larger circles onto ellipses. It is thus fairly intuitive that an *off-center* circle which just touches the unit circle at $\zeta = 1$ (see Fig. 5.3 (left)) is mapped onto wing that is a rounded but with a cuspoidal corner at $z = 2$.

Since (5.14) is conformal, $w(f(\zeta))$ (which we denote as $w(\zeta)$ for brevity) must be an analytic function in the exterior of the circle and $\Im\{w(\zeta)\}$ must be constant on the surface of the circle. We have thus reduced our problem to the much simpler problem of finding the solution for the flow around a circle; see Exercise 4.1 [2]. In order to achieve lift, as mentioned above it is necessary to include a circulation Γ around the wing (see Exercise 4.2); so, in the ζ-plane, the potential reads

$$w(\zeta) = U\left(\zeta + \frac{R^2}{\zeta}\right) - \frac{i\Gamma}{2\pi}\ln\zeta.$$

To address the problem of flight it is of interest to consider the general case of an arbitrary angle of attack β between the incoming flow (with speed U) and the wing; this can be achieved by a complex rotation, replacing ζ by $\zeta e^{-i\beta}$. Taking into account that the circle is shifted by a distance μ to the left, and the radius is now $R = 1 + \mu$ (see Fig. 5.3 (left)), the solution in the ζ-plane is [2]

$$w(\zeta) = U\left((\zeta + \mu)e^{-i\beta} + \frac{(1+\mu)^2}{\zeta + \mu}e^{i\beta}\right) - \frac{i\Gamma}{2\pi}\ln(\zeta + \mu). \qquad (5.15)$$

Mapping the boundary of the circle (5.13) using (5.14), the contour of the wing is given by

$$z = -\mu + (1+\mu)e^{i\varphi} + \frac{1}{(1+\mu)e^{i\varphi} - \mu}. \qquad (5.16)$$

As can be seen in Fig. 5.3, the trailing edge of the wing is at $z = 2$. Expanding (5.16) in a power series around $\varphi = 0$ (cf. Exercise 1.11), one finds the shape near the trailing edge in the form

$$x = 2 - (1+\mu)^2\varphi^2, \quad y = (1+\mu)^2\mu\varphi^3. \qquad (5.17)$$

Thus the edge of the wing has the form of a cusp, whose width y scales as $x^{3/2}$.

In terms of the analysis in Section 5.1, a cusp corresponds to an opening angle $\alpha = 2\pi$. The generic behavior of the velocity near the trailing edge is thus given by (5.11) with $\alpha = 2\pi$. Defining θ as in Fig. 5.3 (right), we also need to replace θ in (5.11) by $\theta - \pi$, giving

$$v_r \approx ar^{-1/2}\sin\frac{\theta}{2}, \quad v_\theta \approx ar^{-1/2}\cos\frac{\theta}{2}, \qquad (5.18)$$

so the velocity generically becomes infinite. However, rather than "turning the corner" at very high speed, it is found experimentally that the flow prefers to separate smoothly from the sharp trailing edge, producing a singularity of lower order. Is it possible to make the coefficient a vanish? To answer this question one has to solve the full problem, as determined by the conformal mapping (5.14).

The complex velocity field around the wing is determined from

$$u - iv = \frac{dw}{dz} = \frac{dw}{d\zeta}\left(\frac{dz}{d\zeta}\right)^{-1}$$
$$= \left[U\left(e^{-i\beta} - \frac{(1+\mu)^2}{(\zeta+\mu)^2}e^{i\beta}\right) - \frac{i\Gamma}{2\pi(\zeta+\mu)}\right]\left(1 - \frac{1}{\zeta^2}\right)^{-1}. \quad (5.19)$$

To analyze the flow field near the trailing edge, we put $z = 2 + re^{i\theta}$; see Fig. 5.3 (right). The transformation (5.14) can be put in the more convenient form

$$\frac{z-2}{z+2} = \left(\frac{\zeta-1}{\zeta+1}\right)^2,$$

from which it follows that points near the trailing edge are parameterized by $\zeta = 1 + r^{1/2}e^{i\theta/2}$. If one inserts this expression into (5.19) and expands to leading order in r, one finds

$$u - iv = -\frac{ir^{-1/2}}{4\pi(1+\mu)}\left[\Gamma + 4\pi U(1+\mu)\sin\beta\right]e^{-i\theta/2}. \quad (5.20)$$

Note that this is precisely of the anticipated form (5.18), with

$$a = \frac{\Gamma + 4\pi U(1+\mu)\sin\beta}{4\pi(1+\mu)}.$$

The requirement that this coefficient be zero is the famous Kutta condition,

$$\Gamma = -4\pi U(1+\mu)\sin\beta. \quad (5.21)$$

As seen in Fig. 5.4, this corresponds to a flow leaving the sharp trailing edge of the wing smoothly. While the circulation in the original potential flow formulation is arbitrary, the condition that the velocity field should remain finite selects a particular value of Γ. This observation is very important for the theory of flight, although we are not going to expand much on this here. Namely, one can show that the total lift force on *any* two-dimensional body is given by

$$F_{\text{lift}} = -\rho U\Gamma,$$

a result known as Blasius' theorem. Using the value of the circulation (5.21) determined from the Kutta condition, we arrive at

$$F_{\text{lift}} = -\rho U\Gamma = 4\pi\rho U^2(1+\mu)\sin\beta, \quad (5.22)$$

for the lift of an airfoil.

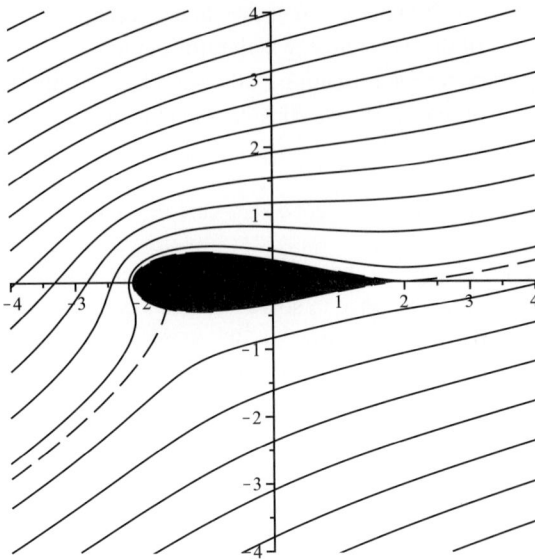

Figure 5.4 The flow around a Joukowsky wing, with an angle of attack $\beta =$ 0.4. The circulation has been chosen according to (5.21), so the flow detaches smoothly from the trailing edge of the wing: the broken line is the dividing streamline that arrives at and leaves the surface of the wing.

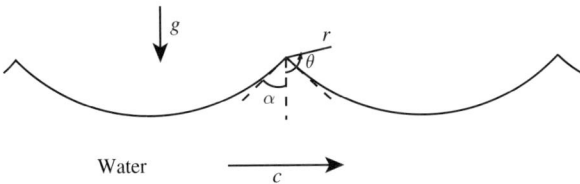

Figure 5.5 Singular traveling waves. The opening angle 2α of the crest is $2\pi/3$ or 120°. The polar coordinates (r, θ) are centered at the crest.

5.3 Stokes waves

A corner singularity can also form on the free surface of a potential flow, provided that surface tension forces can be neglected; these would round off the tip singularity. These so-called Stokes waves can be understood as a limiting case of ordinary water waves, that of the greatest possible height. Their appearance is shown in Fig. 5.5. We will not go into the detail of the theory [131] but simply point out that the local structure of the corner can be understood on the basis of the solutions (5.1) discussed above. In particular, we will calculate the opening angle 2α of the wave crest, which turns out to be $120°$.

The driving force for the formation of the singularity comes from the wave motion at speed c. Accordingly, we consider traveling wave profiles in the form of a wedge of angle 2α:

$$z = f(x - ct) \simeq -(\cot\alpha)\,|x - ct| = -(\cot\alpha)\,|x'|, \qquad (5.23)$$

where c is the wave velocity and (x', z) is a Cartesian system centered at the corner. To find the flow we have to solve Laplace's equation (4.37) below the surface with the Bernoulli equation (4.39) satisfied at the boundary (5.23). Since we are neglecting the effect of surface tension, the pressure p equals the constant pressure p_0 above the traveling wave. In the moving coordinate system, (4.39) becomes

$$\phi_t - c\phi_{x'} + \frac{1}{2}\left|\nabla_{(x',y)}\phi\right|^2 + \frac{p_0}{\rho} - g(\cot\alpha)\,|x'| = 0; \qquad (5.24)$$

the last term on the left is the hydrostatic pressure.

By writing

$$\phi(x', z, t) = \left(\frac{c^2}{2} - \frac{p_0}{\rho}\right)t + \tilde\phi, (x', z), \qquad (5.25)$$

with $\tilde\phi$ harmonic, (5.24) leads to

$$\left|\nabla_{(x',y)}\tilde\phi\right|^2 = 2g(\cot\alpha)\,|x'|. \qquad (5.26)$$

In the reference frame (x', y), the wave is stationary and hence the normal component of the velocity field vanishes:

$$\frac{\partial\tilde\phi}{\partial n} = 0. \qquad (5.27)$$

This means that

$$\left|\nabla_{(x',y)}\tilde\phi\right|^2 = \left(\frac{\partial\tilde\phi}{\partial s}\right)^2,$$

where s is the direction tangential to the interface, and, from (5.26), it follows that

$$\left|\frac{\partial\tilde\phi}{\partial s}\right| = \sqrt{2g\cot\alpha}\,|x'|^{1/2}. \qquad (5.28)$$

The boundary conditions (5.27), (5.28) read, in polar coordinates with origin at the crest (see Fig. 5.5),

$$\frac{\partial \tilde{\phi}}{\partial \theta} = 0, \quad \frac{\partial \tilde{\phi}}{\partial r} = \sqrt{2g \cos \alpha}\, r^{1/2}, \quad \text{at} \quad \theta = \pm \alpha. \tag{5.29}$$

This corresponds to the local solution (5.10) with scaling exponent $\lambda = 3/2$:

$$\tilde{\phi}(r, \theta) = A r^{3/2} \sin \frac{3\theta}{2}. \tag{5.30}$$

We note that the opening angle is 2α instead of α and that θ has been replaced by $\theta - \alpha$ since we have chosen $\theta = 0$ as the line of symmetry. But this means that $\pi/(2\alpha) = 3/2$, and thus $\alpha = \pi/3$. The amplitude A is also determined from the conditions (5.29), giving

$$\alpha = \frac{\pi}{3}, \quad A = \sqrt{\frac{8g \cos \alpha}{9}} = \frac{2}{3}\sqrt{g}. \tag{5.31}$$

On the basis of this local representation for the wave shape and potential close to the crest, it is possible to find higher-order corrections that show, for instance, that the wave's shape is concave locally at both sides of the corner [172, 152].

5.4 Electric fields near tips: Taylor cones

Another interesting case involving local analysis concerns the electric field in the neighborhood of a cone with circular cross-section and opening semiangle α: this provides a simple model for a lightning rod. Owing to the large curvature at the tip of the rod, the electric field is expected to be very large. Let the cone C be a conductor in whose exterior neighborhood an electric field is produced. The electrostatic potential Φ satisfies the equation $\Delta \Phi = 0$ and, on the surface of C, we can assume the potential to vanish relative to a constant potential at infinity. The electric field is then found from

$$\mathbf{E} = -\nabla \Phi. \tag{5.32}$$

Once more, we expect a self-similar solution of the form

$$\Phi = r^{\lambda} f(\theta). \tag{5.33}$$

Using spherical coordinates (r, θ, φ) at the tip of the cone (see Fig. 5.6) and assuming rotational symmetry in the azimuthal angle φ, we have (see (A.22))

$$\Delta \Phi = \frac{1}{r^2} \frac{\partial}{\partial r}\left(r^2 \frac{\partial \Phi}{\partial r}\right) + \frac{1}{r^2 \sin \theta} \frac{\partial}{\partial \theta}\left(\sin \theta \frac{\partial \Phi}{\partial \theta}\right) = 0.$$

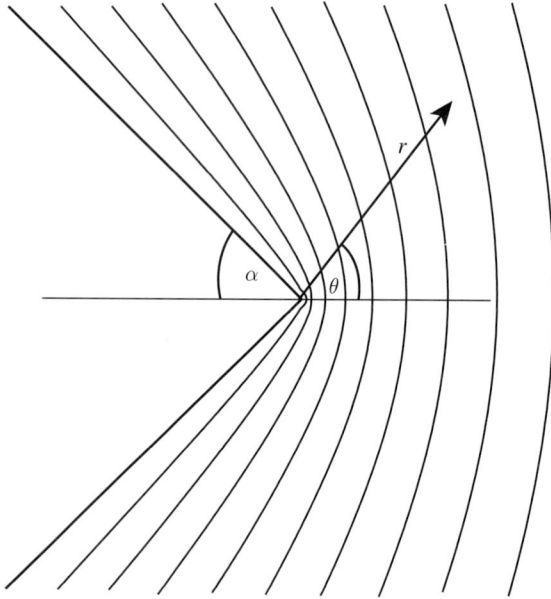

Figure 5.6 Equipotential lines near an axisymmetric cone of opening angle $2\alpha = \pi/2$. From (5.37) the similarity index is $\lambda = 0.463$; the potential Φ is given by (5.36).

This is to be satisfied for $0 \leq r < \infty$ and $0 \leq \theta \leq \pi - \alpha$ with the boundary condition $\Phi = 0$ at $\theta = \alpha$. Using the self-similar solution (5.33), we find

$$\lambda(\lambda + 1)f(\theta) + \frac{1}{\sin\theta}\frac{d}{d\theta}\left(\sin\theta\frac{df(\theta)}{d\theta}\right) = 0, \qquad (5.34)$$

$$f(\pi - \alpha) = 0. \qquad (5.35)$$

The general solution of (5.34) is a linear combination of $P_\lambda(\cos\theta)$ and $Q_\lambda(\cos\theta)$, where P_λ and Q_λ are Legendre functions of the first and second kind, respectively [1]. The function Q_λ contains logarithmic singularities that make $|\mathbf{E}|^2$ non-integrable, so the energy would be infinite. Therefore the asymptotic solution for the electric potential is

$$\Phi = Ar^\lambda P_\lambda(\cos\theta), \qquad (5.36)$$

which is regular for $\theta = 0$ but has a singularity for $\theta = \pi$. From (5.35) we obtain a condition for the index λ:

$$P_\lambda(\cos(\pi - \alpha)) = P_\lambda(-\cos\alpha) = 0. \qquad (5.37)$$

Thus, for a given geometry determined by α, the scaling exponent of the electric field near the tip can be calculated from (5.37), an example of self-similarity of the second kind.

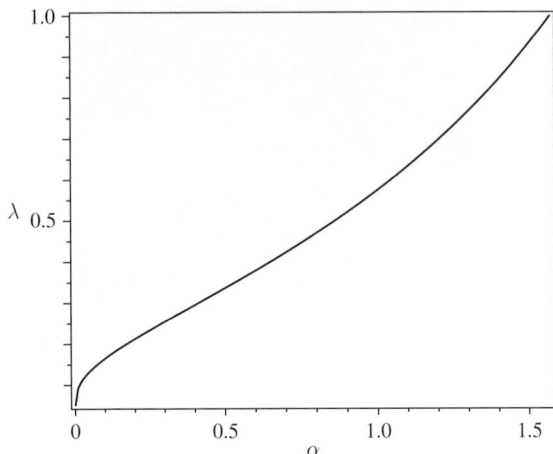

Figure 5.7 The solution of equation (5.37).

Equation (5.37) is transcendental and its solutions have to be found numerically. In Fig. 5.7 we represent the smallest positive solution λ of equation (5.37), as a function of α in the range $\alpha \in [0, \pi/2]$, corresponding to the leading-order behavior near the tip. Notice that $\lambda < 1$ and therefore $|\mathbf{E}|$ is $O(r^{\lambda-1})$ at leading order, with $0 < \lambda < 1$. This implies that conical tips with opening semiangle smaller than $\pi/2$ lead to diverging electric fields. Indeed, the equipotential lines drawn in Fig. 5.6 come closer together near the tip, corresponding to higher field strengths. The divergence becomes stronger with smaller values of α, i.e. when the cone approaches a spike.

A perhaps even more interesting application of the same idea arises if the boundary is not a solid but the free surface of a conducting liquid, as shown in Fig. 5.8. A drop is protruding from the end of a capillary, held in a strong external field. As a result of the electric forces, the drop is deformed into a cone with a very sharp tip. From the tip of the cone a fine jet emerges, the so-called electric jet.

We will focus on the conical part of the solution, whose shape is given by a balance of surface tension and electric forces. The electric field leads to an extra contribution to the normal stress, so the normal force balance (4.24) now reads [134]:

$$p = \gamma \left(\frac{1}{R_1} + \frac{1}{R_2} \right) + \frac{\epsilon_0}{2} E_n^2 = \frac{\gamma}{r \tan \alpha} + \frac{\epsilon_0}{2} E_n^2, \tag{5.38}$$

with p a constant pressure in the static case. From the general form of the potential (5.36) we obtain, for the electric field normal to the surface of the cone,

Figure 5.8 A Taylor cone in the cone-jet mode [164]. The opening half-angle shown is 52°. A large voltage difference is maintained between the capillary on the left and an electrode on the right. The cone-like structure on the right is produced by a cloud of little droplets.

$$E_n = \frac{1}{r}\frac{\partial \Phi}{\partial \theta} = Ar^{\lambda-1}\sin(\pi - \alpha)P'_\lambda(\cos(\pi - \alpha)). \qquad (5.39)$$

Clearly, for (5.38) to hold we need the scaling exponent to be $\lambda = 1/2$; in that case both the electric and surface tension forces go to infinity as $1/r$ near the tip. But this means that (5.37) becomes an equation for the opening angle α, with a given scaling exponent $\lambda = 1/2$. The result $\alpha = 0.86$ can be read off from Fig. 5.7, which corresponds to an angle of about 49.19°, in good agreement with Fig. 5.8.

5.5 Mixed boundary conditions

Singularities also arise if there is an abrupt change of boundary conditions to be satisfied by the potential. Such a change is associated with a vanishing length scale, hence singular behavior is expected near the edge where the two boundary conditions meet. An example taken from electrostatics is shown in Fig. 5.9: a circular disk of radius R has a total charge Q. As a result, the disk is at a potential $\Phi = V$ relative to the potential at infinity. Outside the disk the potential is unknown. However, we know from symmetry that the normal derivative must vanish in the plane of the disk. Thus, in cylindrical coordinates,

$$\Phi(\rho, 0) = V \quad \text{for} \quad 0 \leq \rho \leq R, \qquad (5.40)$$

$$\frac{\partial \Phi}{\partial z}(\rho, 0) = 0 \quad \text{for} \quad \rho > R. \qquad (5.41)$$

We will aim to calculate the behavior of the potential near the edge of the disk, as shown on the right of Fig. 5.9. Locally the boundary can be considered

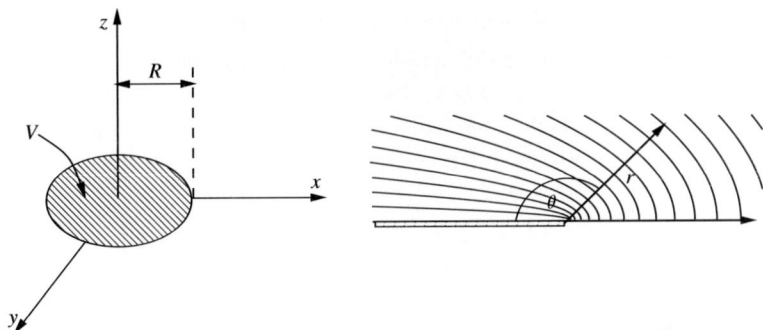

Figure 5.9 On the left, a charged disk; on the right, the potential field at the edge of the disk is shown.

flat, so the problem becomes two dimensional. Introducing polar coordinates based at the edge, we use the local description (5.1):

$$\Phi = r^\lambda f(\theta).$$

Once more we can assume that the potential vanishes on the disk and is $-V$ at infinity; then the boundary conditions are $\Phi = 0$ for $\theta = 0$ and $\partial\Phi/\partial\theta = 0$ for $\theta = \pi$. Since Φ has to verify Laplace's equation, f satisfies the same equation, (5.2), as for the potential flow problem, and the boundary conditions are once more homogeneous:

$$f(0) = 0, \quad f'(\pi) = 0. \tag{5.42}$$

Solutions (5.3) satisfying the first boundary condition (5.42) are given by

$$f(\theta) = A \sin \lambda\theta,$$

while, from the second boundary condition, $\lambda = \lambda_n = n + 1/2$ with $n = 0, \pm1, \pm2, \ldots$, an example of self-similarity of the second kind. The similarity profile becomes

$$f(\theta) = f_n(\theta) = \sin\left(n + \frac{1}{2}\right)\theta.$$

Valid solutions must be such that $|\nabla\Phi|^2$ is bounded locally near $(0, 0)$, which excludes negative values of λ, as for the flow problem. Thus the local expansion of the potential becomes

$$\Phi = a_0 r^{1/2} \sin \frac{\theta}{2} + a_1 r^{3/2} \sin \frac{3\theta}{2} + \cdots, \tag{5.43}$$

with a square root singularity of the electric field $\mathbf{E} = \nabla\Phi$ at the origin.

This result can be compared with the solution of the disk problem in Fig. 5.9, which requires quite sophisticated techniques [120]. Of course, no global solution is available in analytical form if the disk is deformed in any nontrivial way. The local analysis, on the other hand, is not affected by this complication. In cylindrical coordinates, the full potential is [120]

$$\Phi = \frac{2VR}{\pi} \arcsin\left(\frac{2R}{\sqrt{(\rho - R)^2 + z^2} + \sqrt{(\rho + R)^2 + z^2}} \right). \qquad (5.44)$$

Putting $\rho = R - r\cos\theta$ and $z = r\sin\theta$ in this expression and expanding for small r, one finds

$$\Phi \approx \frac{2^{3/2}VR}{\pi} \sqrt{\frac{r}{R}} \sin\frac{\theta}{2},$$

which agrees with (5.43). The benefit of obtaining the full solution is that the prefactor $a_0 = 2^{3/2}V\sqrt{R}/\pi$ is now known.

Example 5.2 (Evaporating drop) Consider an evaporating drop which makes a contact angle of θ_c with the solid surface on which it rests. The vapor diffuses away from the drop, so that in a quasistatic approximation the vapor concentration c in the exterior obeys $\triangle c = 0$. On the solid the boundary condition is that of vanishing flux, $\mathbf{n} \cdot \nabla c = 0$; on the drop surface the vapor concentration is constant, $c = c_s$. Now, the solution (5.3) satisfies $f'(\pi) = 0$ and $f(\theta_c) = 0$ (subtracting the constant c_s), so the simultaneous equations

$$A\cos\lambda\pi - B\sin\lambda\pi = 0, \qquad A\sin\lambda\theta_c + B\cos\lambda\theta_c = 0$$

have to be satisfied. The existence of a nontrivial solution requires that the determinant of the two equations is zero, i.e. that

$$\cos\lambda\pi \cos\lambda\theta_c + \sin\lambda\pi \sin\lambda\theta_c = 0.$$

Using the usual product-to-sum identities, this yields

$$\cos\lambda(\pi - \theta_c) = 0, \qquad (5.45)$$

so that the leading singularity has the scaling exponent

$$\lambda = \frac{\pi}{2\pi - 2\theta_c}, \qquad (5.46)$$

in agreement with the exact solution [61]. The flux of vapor $\mathbf{j}$ from the drop is given by the normal derivative: $\mathbf{j} = -D\nabla c$, where D is the diffusion coefficient. This means that the flux from the drop surface scales as $j \propto r^{\lambda-1}$ as a function of the distance r from the contact line; this *diverges* for $\theta_c < \pi/2$.   □

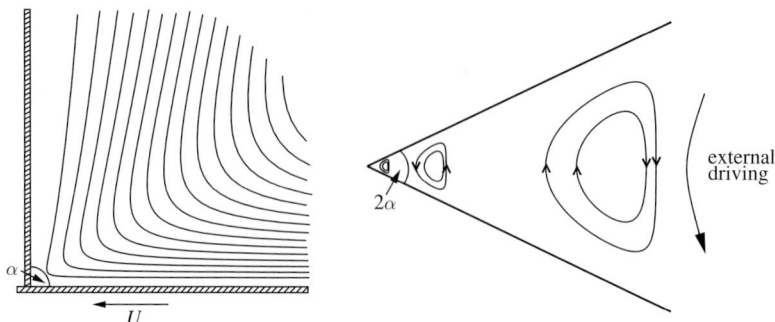

Figure 5.10 Viscous flow in a corner, one side scraping across the other (left), the Taylor scraper. Streamlines are drawn according to the solution (5.51). Corner flow, driven from the exterior (right) produces Moffatt eddies. The streamlines are plotted using (5.58). The angle α on the left is $90°$; on the right $2\alpha = 50°$.

5.6 Viscous flow in corners and Moffatt eddies

The local behavior of a viscous fluid flow sufficiently close to a corner is characterized by a Reynolds number $\text{Re} = rU/\nu$, where r is the distance from the corner and U a typical velocity. Thus the local value of Re is always small and the Stokes equation (4.49) applies. As before, we expect the stream function to be self-similar:

$$\psi = r^\lambda f(\theta). \tag{5.47}$$

Inserting (5.47) into the biharmonic equation (4.49) leads to

$$\lambda^2(\lambda - 2)^2 f(\theta) + [\lambda^2 + (\lambda - 2)^2]f''(\theta) + f^{iv}(\theta) = 0, \tag{5.48}$$

to be satisfied for $0 \leq \theta \leq \alpha$.

We will use this equation to study two very different problems associated with the same geometry. In the first problem, known as the *Taylor scraper* [207], a solid plate is used to scrape a viscous fluid from a solid surface; see Fig. 5.10 (left). In the second problem (see Fig. 5.10 (right)), the flow is driven externally while the solid surfaces remain stationary (as a result, a sequence of eddies, known as *Moffatt eddies*, is produced). Physically, the difference is that in the Taylor scraper problem a fixed velocity U is introduced from the outside, while in the second problem the system is allowed to find its own velocity scale.

In Fig. 5.10 the Taylor scraper is shown in a frame of reference where the point of contact between the two solids is the origin of the coordinate system, and the substrate moves to the right at speed U. Thus the no-slip boundary condition on the walls reads $v_r = -U$, $v_\theta = 0$ for $\theta = 0$ and $v_r = v_\theta = 0$ for

$\theta = \alpha$. Since the problem has only a characteristic speed, the dimensionally correct form of the stream function must be $\psi = Urg(\theta)$, where g is dimensionless. Comparison with (5.47) tells us that therefore the scaling exponent $\lambda = 1$. Indeed, using (5.12), the boundary conditions to be satisfied by ψ are

$$\frac{\partial \psi}{\partial r} = 0, \quad \frac{1}{r}\frac{\partial \psi}{\partial \theta} = -U \quad \text{for} \quad \theta = 0,$$
$$\frac{\partial \psi}{\partial r} = 0, \quad \frac{1}{r}\frac{\partial \psi}{\partial \theta} = 0 \quad \text{for} \quad \theta = \alpha.$$

It follows that $\lambda = 1$ and $f(\theta)$ has to satisfy

$$f(0) = 0, \quad f'(0) = -U, \quad f(\alpha) = 0, \quad f'(\alpha) = 0. \tag{5.49}$$

The Taylor scraper is thus an example of self-similarity of the first kind, as the exponent is fixed by dimensional reasoning.

The general solution of (5.48) in the case $\lambda = 1$ is

$$f(\theta) = A \cos \theta + B \sin \theta + C\theta \cos \theta + D\theta \sin \theta; \tag{5.50}$$

the constants A, B, C, D have to be chosen to satisfy (5.49). The details are left as Exercise 5.2, and the final answer is

$$\psi = \frac{rU}{\alpha^2 - \sin^2 \alpha} \left[-\alpha^2 \sin \theta + \sin^2 \alpha \theta \cos \theta + (\alpha - \cos \alpha \sin \alpha)\theta \sin \theta \right]. \tag{5.51}$$

The equipotential lines of (5.51) are the streamlines of the flow, shown for $\alpha = \pi/2$ in Fig. 5.10 (left). Also in Exercise 5.2, it can be deduced that the normal and shear stresses on the horizontal plate are

$$\sigma_{\theta\theta} = -\frac{2\eta U}{r}\frac{\sin^2 \alpha}{\alpha^2 - \sin^2 \alpha}, \quad \sigma_{r\theta} = \frac{2\eta U}{r}\frac{\alpha - \sin \alpha \cos \alpha}{\alpha^2 - \sin^2 \alpha}, \tag{5.52}$$

respectively. Both expressions diverge as $1/r$ close to the corner. Thus the total forces on both plates, which are integrals over the stresses, are logarithmically infinite. Clearly, they must be resolved on some scale, for example by a tiny gap between the scraper and the wall.

A very different situation arises if no characteristic velocity is imposed on the plates but, rather, the flow is driven from the open end of the corner, as shown experimentally in Fig. 5.11. In this case no value of the exponent λ is imposed; we will show that λ is determined by a solvability condition, as is customary for self-similarity of the second kind. Once more we are looking for a homogeneous eigensolution, while the Taylor scraper corresponds to the inhomogeneous problem.

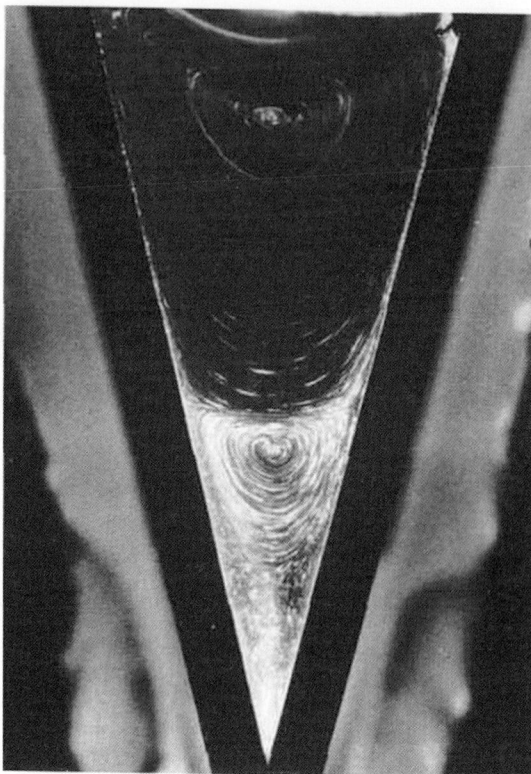

Figure 5.11 Moffatt eddies in a corner of total opening angle 28.5° [204, 212]. The motion is driven by the steady clockwise rotation of a circular cylinder, the base of which is seen just below the free surface, at the top of the photograph. The center of the first eddy is just below the cylinder. A 90 minute exposure shows the next eddy (which appears brighter on account of the lighting), but fails to reveal the third eddy. Indeed, each eddy is predicted by (5.64) to be 400 times weaker than its neighbor above.

Let the fluid occupy the region $-\alpha \leq \theta \leq \alpha$ between the two plates; see Fig. 5.10. The boundary conditions (5.49) turn into

$$f(-\alpha) = 0, \quad f'(-\alpha) = 0, \quad f(\alpha) = 0, \quad f'(\alpha) = 0, \qquad (5.53)$$

and are all homogeneous. The general solution of (5.48) for arbitrary λ is

$$f(\theta) = f_\lambda(\theta) = A \cos \lambda\theta + B \sin \lambda\theta + C \cos(\lambda - 2)\theta + D \sin(\lambda - 2)\theta, \qquad (5.54)$$

excluding the special cases $\lambda = 0, 1, 2$, which are not relevant to this problem.

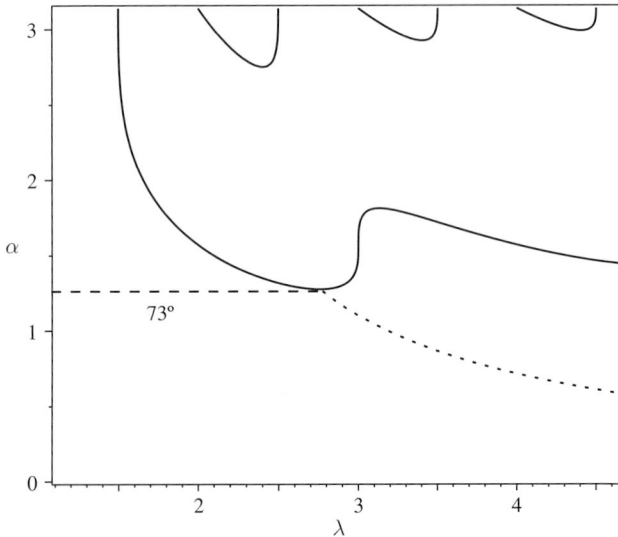

Figure 5.12 Solutions of (5.57). Real solutions are shown as solid lines. Below $\alpha = 73°$, the solutions are complex; here the real part is shown as the dotted line.

If we concentrate on antisymmetric flow about the axis of symmetry (in which case the stream function is symmetric in θ) then $B = D = 0$ and the boundary conditions imply

$$A \cos \lambda\alpha + C \cos(\lambda - 2)\alpha = 0, \tag{5.55}$$

$$A\lambda \sin \lambda\alpha + C(\lambda - 2) \sin(\lambda - 2)\alpha = 0. \tag{5.56}$$

Solutions are only possible if the matrix associated with the system (5.55), (5.56) is singular, that is, if its determinant vanishes:

$$(\lambda - 2) \cos \lambda\alpha \sin(\lambda - 2)\alpha - \lambda \sin \lambda\alpha \cos(\lambda - 2)\alpha$$
$$= -(\lambda - 1) \sin 2\alpha - \sin 2(\lambda - 1)\alpha = 0.$$

Equivalently,

$$\alpha \sin 2(\lambda - 1)\alpha = -(\lambda - 1)\alpha \sin 2\alpha. \tag{5.57}$$

Equation (5.57) is the desired solvability condition for λ. When α is larger than a critical value of approximately 73°, equation (5.57) has real solutions; see Fig. 5.12. As usual, the solution λ_1 with the smallest real part dominates near the corner:

$$\psi \approx Ar^{\lambda_1} [\cos \lambda_1\theta \cos(\lambda_1 - 2)\alpha - \cos(\lambda_1 - 2)\theta \cos \lambda_1\alpha]. \tag{5.58}$$

As α increases from the critical value to 90°, the number of real solutions increases. For acute angles which are smaller than the critical value, no real solutions exist and one has to look for complex roots of (5.57); we then take the real part of ψ. If we write

$$\lambda = 1 + p + iq,$$

the real and imaginary parts of (5.57) are

$$\sin 2p\alpha \cosh 2q\alpha = -p \sin 2\alpha, \qquad (5.59)$$

$$\cos 2p\alpha \sinh 2q\alpha = -q \sin 2\alpha, \qquad (5.60)$$

whose solutions have to be found numerically. In Fig. 5.12 we represent the real roots of equation (5.57) as a solid line and the real part of complex roots of (5.57) as a dotted line. Notice that the real part $\Re\{\lambda\}$ of the complex branch increases with decreasing α. Hence the flow becomes less singular for decreasing values of α.

However, the most interesting feature comes from the fact that λ is complex, as a result of which the solution becomes periodic but on increasingly smaller scales [60, 154]. Namely,

$$r^{\lambda_1} = r^{1+p_1+iq_1} = r^{p_1+1}\left[\cos(q_1 \ln r) + i \sin(q_1 \ln r)\right],$$

and this implies that the real part of the stream function has the following structure:

$$\Re\{\psi\} \approx r^{1+p_1}\left[\cos(q_1 \ln r) f_{\lambda_1}(\theta) + \sin(q_1 \ln r) g_{\lambda_1}(\theta)\right].$$

This means that the stream function obeys the scaling relation

$$\psi\left(e^{-2n\pi/q_1} r_0, \theta\right) = e^{-2n\pi(1+p_1)/q_1} \psi(r_0, \theta), \qquad (5.61)$$

i.e. the same structure repeats itself with period $2\pi/q_1$ in the variable $\ln r$. The discrete symmetry (5.61) is called *discrete self-similarity*; we will come back to this notion in Chapter 12.

As seen in Fig. 5.10 (right), and confirmed experimentally (see Fig. 5.11), the self-similar structure is a sequence of counterrotating eddies. Namely, the transverse flow across the centerline of the corner is

$$v_{\theta=0} = \Re\left\{\frac{\partial \psi}{\partial r}\right\} \propto r^{p_1} \sin(q_1 \ln r + \delta),$$

where δ is a phase factor. Thus the velocity changes sign infinitely often, with zeros at

$$q_1 \ln r + \delta = -n\pi, \quad n = 0, 1, 2, \ldots,$$

so that the center of the nth eddy is at

$$r_n = r_0 e^{-n\pi/q_1}. \tag{5.62}$$

As a result, the size of the vortices falls off in a geometric progression:

$$\frac{r_n}{r_{n+1}} = e^{\pi/q_1} \tag{5.63}$$

and the flow strength $v_n = v_{\theta=0}(r = r_{n+1/2})$ of successive eddies obeys

$$\frac{v_n}{v_{n+1}} = e^{\pi p_1/q_1}. \tag{5.64}$$

The surprise about this solution [154] is that very viscous flow is capable of producing vortical motion, which is usually associated with flow at high Reynolds number. However, in the case of $2\alpha = 28.5°$, shown in Fig. 5.12, we have $p_1 = 8.49$, and $q_1 = 4.43$, and thus successive eddies are smaller by a factor 0.49 and weaker by a factor 2.4×10^{-3}. As a result, the smaller vortices are very difficult to observe [204].

Exercises

5.1   We saw that the hinge solution given in Example 5.1 breaks down for $\alpha = \pi/2$.
   (i) Confirm that for $\alpha = \pi/2$ the leading scaling exponent λ_1 of the homogeneous solution (5.6) has the same value as that of (5.7).
   (ii) Investigate the limit $\alpha - \pi/2 \equiv \epsilon \rightarrow 0$, to show that the hinge solution and the homogeneous solution can be superimposed to cancel the singularity. Show that the next order in ϵ gives the solution

$$\phi = \frac{r^2 \Omega}{\pi} (\theta \sin 2\theta - \ln r \cos 2\theta).$$

   Note the appearance of a logarithmic term.

5.2   In Section 5.6, we considered the Taylor scraper problem shown on the left of Fig. 5.10. It has a similarity solution of the form

$$\psi = r f(\theta).$$

   (i) Calculate the constants A, B, C, D appearing in (5.50).
   (ii) Use the Stokes equation (4.48) in cylindrical coordinates to show that the pressure in the corner is given by

$$p = -\frac{\eta U}{r} \left(f''' + f' \right).$$

Thus calculate the components of the stress tensor for $\theta = 0$ as given in (5.52) and find the total force on the bottom plate. Confirm that a force of the same magnitude but opposite in direction acts on the scraper.

(iii) Show that, in the limit of small angles, the normal stresses behave as

$$\sigma_{\theta\theta} \approx -\frac{6\eta U}{\alpha^2 r}, \qquad \sigma_{r\theta} = \frac{4\eta U}{\alpha r},$$

while, for $\alpha = \pi$,

$$\sigma_{\theta\theta} = 0, \qquad \sigma_{r\theta} = \frac{2\eta U}{\pi r}.$$

Thus a plasterer will hold his trowel at a small angle to maximize the pressure pushing material into small holes. A painter, on the other hand, holds his scraper at a large angle, to minimize the force necessary to apply the paint [207].

5.3 (*) As a model for a viscous fluid drop spreading on a solid substrate, consider the geometry of the Taylor scraper but with the inclined plate replaced by a *free surface* [115]. The angle θ_c between the substrate and the free surface is called the *contact angle*. A spreading drop corresponds to a movement of the bottom plate at speed U to the *right*; see Fig. 5.13.

   (i) Use the similarity solution $\psi = rf(\theta)$ to formulate boundary conditions for f, and thus find f. Use the convention that the free surface is at $\theta = 0$.

   (ii) Show that the flow in the fluid wedge is

$$v_r = U \frac{\theta_c \cos\theta_c \cos\theta - (\cos\theta - \theta\sin\theta)\sin\theta_c}{\theta_c - \cos\theta_c \sin\theta_c},$$

$$v_\theta = U \frac{\theta\cos\theta\sin\theta_c - \theta_c\sin\theta\cos\theta_c}{\theta_c - \cos\theta_c \sin\theta_c}.$$

   (iii) Show that the pressure in the wedge is

$$p = -\frac{\eta U}{r} \frac{2\cos\theta\sin\theta_c}{\theta_c - \cos\theta_c \sin\theta_c}. \tag{5.65}$$

   Use the boundary condition (4.24) to calculate the curvature of the interface. What does this imply for the shape of the free surface?

5.4 In the previous problem the pressure increases in a non-integrable way. This singularity can be weakened by accounting for the *slip* of fluid over the solid surface. Then, in a small region much closer to the contact line than the slip length λ, the fluid velocity is proportional to the shear rate at the wall,

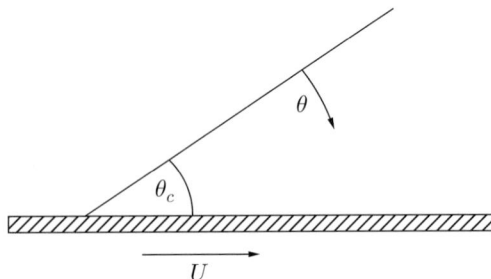

Figure 5.13 The edge of a spreading drop, in a frame of reference in which the contact line is stationary.

$$U = \frac{\lambda}{r} \frac{\partial v_r}{\partial \theta},\qquad(5.66)$$

using the notation of Fig. 5.13. For the precise form of the slip boundary condition, see Exercise 6.2.

(i) Formulate the boundary conditions in terms of the stream function ψ.

(ii) Look for a similarity solution of the form

$$\psi = \frac{Ur^2}{\lambda} f(\theta),$$

and show that the pressure is given by

$$p = -\left(\frac{U\eta}{\lambda} \ln r\right)\left(4f' + f'''\right).$$

(iii) Compute ψ and thus show that

$$p = \frac{U\eta}{\lambda\theta_c} \ln r,$$

which is an *integrable* singularity.

5.5 Consider the geometry of Fig. 5.10 (right), i.e. two plates at an angle 2α, with the plane of symmetry at $\theta = 0$. The plates are hinged and are closing at a constant angular velocity $\dot\alpha = -\Omega$. Consider the viscous limit, so that the stream function $\psi(r, \theta)$ satisfies $\Delta^2\psi = 0$.

(i) Show that problem has a similarity solution

$$\psi = \frac{\Omega}{2} r^2 f(\theta),$$

and write down the boundary conditions for f.

(ii) Show that the solution is [154, 155]

$$f(\theta) = \frac{\sin 2\theta - 2\theta \cos 2\alpha}{\sin 2\alpha - 2\alpha \cos 2\alpha}. \tag{5.67}$$

Note that (5.67) explodes at a critical value α_c, determined by $\tan 2\alpha_c = 2\alpha_c$, which gives $\alpha_c = 2.247$ or $2\alpha_c = 257.45°$.

5.6 In the geometry of Fig. 5.10 (right), assume that fluid is introduced (or sucked out) at a rate Q through a small hole at the apex. This is known as the Jeffrey–Hamel problem [17], for which solutions of the Navier–Stokes equation can be found. Here we consider only the viscous limit $\Delta^2 \psi = 0$. Show that the solution is

$$\psi(r, \theta) = Q \frac{\sin 2\theta - 2\theta \cos 2\alpha}{\sin 2\alpha - 2\alpha \cos 2\alpha},$$

i.e. the angular dependence is the same as in Exercise 5.5.

5.7 For each of Exercises 5.5 and 5.6, the failure of the solution for $\alpha = \alpha_c$ can be understood from the fact that the scaling exponent has the same value as that of the homogeneous solution with boundary conditions (5.53).

(i) Show that, in the case of *symmetric* solutions, the solvability condition (5.57) becomes

$$\alpha \sin 2(\lambda - 1)\alpha = (\lambda - 1)\alpha \sin 2\alpha.$$

(ii) Show that the limits $\lambda \to 2$ (corresponding to hinge flow) and $\lambda \to 0$ (corresponding to source flow) both lead to $\alpha = \alpha_c$.

(iii) As in Exercise 5.1, use a superposition with the corresponding homogeneous solution to construct a solution to the hinge and source problems for $\alpha = \alpha_c$.

5.8 (*) In Section 5.4 we considered conical solutions for a drop of conducting fluid in an external electric field; in that case, there is no electric field inside the fluid. A variation of this problem is that of a dielectric fluid (with dielectric constant ϵ) surrounded by air [140, 181, 201]. Then the boundary conditions are that the electric potential Φ is continuous across the interface and that $E_n = \epsilon \bar{E}_n$, where E_n and $\bar{E}_n$ are the normal components of the electric fields outside and inside the cone of fluid.

(i) Show that the solutions are of the form

$$\Phi = Ar^{1/2} P_{1/2}(\cos \theta), \quad \bar{\Phi} = Br^{1/2} P_{1/2}(-\cos \theta)$$

for the outside and inside of the cone, respectively.

(ii) Applying the boundary conditions, show that, for a given ϵ, the semiangle is determined by the equation

$$P_{1/2}(\cos\alpha)P'_{1/2}(-\cos\alpha) + \epsilon P_{1/2}(-\cos\alpha)P'_{1/2}(\cos\alpha) = 0.$$
$$(5.68)$$

(iii) Plot ϵ as described by (5.68), using e.g. MAPLE. Show that two solutions are found for $\epsilon > \epsilon_c \approx 17.59$. Below this value, no solution exists. Show that on the lower branch $\alpha \to 0$ as $\epsilon \to \infty$ and on the upper branch $\alpha \to \alpha_T \approx 0.86$, which corresponds to Taylor's result for a conducting fluid.

6

Asymptotic expansions of PDEs

Near singularities the structure of the underlying equations of motion often simplifies. For example, in the case of drop formation the fluid neck separating the two pieces of fluid becomes increasingly slender as the singularity is approached. This means that the aspect ratio ϵ of the solution (the ratio of the radial scale and the longitudinal scale) goes to zero as $t' \to 0$. Thus the full axisymmetric Navier–Stokes problem is reduced to an effective one-dimensional equation of motion in this limit. To derive the corresponding equation in a systematic fashion, one has to expand the solution in a small variable ϵ which is a measure of the dimensional reduction.

6.1 Thin film equation

We begin by deriving an equation for a thin film of liquid flowing over a solid surface. An important application is the description of a drop spreading on a solid substrate, which we will study in detail in Chapter 15. We anticipate that, owing to viscous friction near the solid wall, we can use Stokes' equation (4.48) and inertia is unimportant. We begin with the simplest case, in which the flow is uniform in one direction and only depends on the x-coordinate, see Fig. 6.1. The other direction, perpendicular to the substrate, we call z. The corresponding components of the velocity field are denoted by $u(x, z, t)$ and $v(x, z, t)$, respectively. Then the viscous flow equation (4.48) reads, in components,

$$\frac{\partial p}{\partial x} = \eta \left(\frac{\partial^2 u}{\partial x^2} + \frac{\partial^2 u}{\partial z^2} \right), \tag{6.1}$$

$$\frac{\partial p}{\partial z} = \eta \left(\frac{\partial^2 v}{\partial x^2} + \frac{\partial^2 v}{\partial z^2} \right) \tag{6.2}$$

115

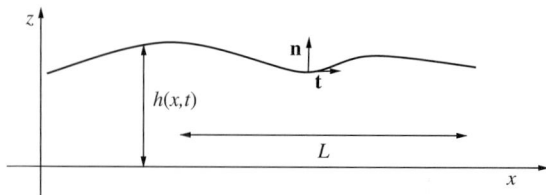

Figure 6.1 Schematic of a flat fluid film.

and is subject to the incompressibility constraint

$$\frac{\partial u}{\partial x} + \frac{\partial v}{\partial z} = 0. \tag{6.3}$$

Let $h(x, t)$ be the thickness of the film. Then the vectors $\mathbf{n}, \mathbf{t}$ normal and tangential to the free surface are

$$\mathbf{n} = \frac{1}{\sqrt{1 + h'^2}} \begin{pmatrix} -h' \\ 1 \end{pmatrix}, \quad \mathbf{t} = \frac{1}{\sqrt{1 + h'^2}} \begin{pmatrix} 1 \\ h' \end{pmatrix}, \tag{6.4}$$

where the prime denotes differentiation with respect to x. Now the boundary conditions (4.24) and (4.25) at the free surface become

$$n_i \sigma_{ij} n_j = \gamma \frac{h''}{(1 + h'^2)^{3/2}}, \quad n_i \sigma_{ij} t_j = 0, \tag{6.5}$$

where σ_{ij} is the stress tensor in the film and the stress from the outer gas atmosphere is neglected. Using the definition (4.12) of the stress tensor, (6.5) becomes

$$-p - 2\eta \frac{1 - h'^2}{1 + h'^2} \frac{\partial u}{\partial x} = \gamma \frac{h''}{(1 + h'^2)^{3/2}}, \tag{6.6}$$

$$\frac{1}{1 + h'^2} \left[-4h' \frac{\partial u}{\partial x} + (1 - h'^2) \left(\frac{\partial u}{\partial z} + \frac{\partial v}{\partial x} \right) \right] = 0, \tag{6.7}$$

where we have used (6.3). On the solid surface the boundary conditions are no-slip:

$$u(x, 0, t) = 0, \quad v(x, 0, t) = 0. \tag{6.8}$$

Finally, the equation of motion for the free surface is given by (4.29):

$$\frac{\partial h}{\partial t} + u(x, h, t)h' = v(x, h, t). \tag{6.9}$$

It is often advantageous to eliminate v from this equation, by noting that

$$\frac{\partial}{\partial x} \int_0^h u(x, z, t) dz = uh' - \int_0^h \frac{\partial v}{\partial z} dz = uh' - v(x, h, t).$$

Here we have used (6.3) and the boundary condition $v(x, 0, t) = 0$. Thus (6.9) is equivalent to

$$\frac{\partial h}{\partial t} + \frac{\partial}{\partial x} \int_0^h u \, dz = 0. \tag{6.10}$$

Equation (6.10) is a statement of volume conservation inside the film: the integral over u is the total volume flux through the film and $(\partial h/\partial t)dx$ is the rate of change of the volume of a slice of the film.

This completes the setup of the problem, which in spite of its apparent simplicity is quite demanding to solve. Inside the fluid one has to solve the partial differential equations (6.1), (6.2), which are coupled via (6.6), (6.7) to the free surface. The free surface can be advanced using (6.9) once the flow field is known. As in any free surface problem, this makes the equations highly nonlinear and tricky to solve even numerically. However, the meat of the problem is contained in a much simpler nonlinear equation for the thickness profile h alone, which we will derive now.

To estimate the size of the different terms in a concrete example, imagine a plate being dipped into a viscous fluid and subsequently withdrawn. Owing to the no-slip condition, a thin fluid film remains on the plate, and is pulled up further onto the plate by viscous forces. Surface tension, however, flattens the film in order to minimize the surface area. In the limit where viscous forces are small compared with the surface tension, in a sense to be made more precise below, the deformation of the surface from a flat state will thus be small. The smallness of the deformation can be characterized by a parameter ϵ defined as the ratio of a typical x-scale h^*/ϵ and a typical z-scale h^*. Let us begin by identifying, in a qualitative fashion, the parameter ϵ.

From the surface tension γ and the viscosity η, one may form a velocity scale

$$v_\eta = \frac{\gamma}{\eta}, \tag{6.11}$$

called the *capillary speed*. To specify the physical problem at hand, we also introduce a characteristic *velocity* scale U, for example the plate speed in the above example. The ratio of the two defines a dimensionless number,

$$\text{Ca} = \frac{U}{v_\eta} \equiv \frac{\eta U}{\gamma}, \tag{6.12}$$

which is known as the *capillary number*. It measures the strength of the external driving relative to the ability of the surface tension to straighten the interface. A small capillary number corresponds to a weakly deformed interface, and thus to a small value of the deformation parameter ϵ.

To identify the relationship between ϵ and Ca, note that the external driving produces surface deformations in x over a length scale h^*/ϵ. This is described by (6.1), in which the second term on the right dominates since a typical z-scale is smaller by a factor ϵ. This means that, taking $\bar{p}$ as a typical pressure scale, we have on the one hand

$$\epsilon \frac{\bar{p}}{h^*} \propto \eta \frac{U}{h^{*2}}.$$

On the other hand (6.6) describes the effect of surface tension, from which we can derive the balance

$$\bar{p} \propto \gamma \frac{\epsilon^2}{h^*}.$$

Eliminating $\bar{p}$ from the two equations we find that

$$\epsilon \propto \mathrm{Ca}^{1/3}; \tag{6.13}$$

thus ϵ does indeed go to zero as Ca becomes small.

We are now in a position to derive the equations more formally, using the smallness of ϵ [163]. According to (6.13) this means that the driving, as described by Ca, is weak. We introduce a transformation

$$\tilde{x} = x\epsilon/h^*, \quad \tilde{z} = z/h^*, \quad \tilde{t} = t\epsilon U/h^* \tag{6.14}$$

of the independent variables and the expansions

$$u(x, z, t) = U\big[\tilde{u}_0(\tilde{x}, \tilde{z}, \tilde{t}) + \epsilon \tilde{u}_1(\tilde{x}, \tilde{z}, \tilde{t}) + \cdots \big], \tag{6.15}$$

$$v(x, z, t) = U\epsilon\big[\tilde{v}_0(\tilde{x}, \tilde{z}, \tilde{t}) + \epsilon \tilde{v}_1(\tilde{x}, \tilde{z}, \tilde{t}) + \cdots \big], \tag{6.16}$$

$$p(x, z, t) = \frac{\eta U}{\epsilon h^*}\big[\tilde{p}_0(\tilde{x}, \tilde{z}, \tilde{t}) + \epsilon \tilde{p}_1(\tilde{x}, \tilde{z}, \tilde{t}) + \cdots \big], \tag{6.17}$$

$$h(x, t) = h^* \tilde{h}(\tilde{x}, \tilde{t}). \tag{6.18}$$

In the following we consider only the leading-order expansion, given by $\tilde{u}_0$, $\tilde{v}_0$, and $\tilde{p}_0$. In our experience, most of the joy is in the leading order! The v-component is of order ϵ, (as implied by (6.16)), so that (6.3) is satisfied at leading order:

$$\frac{\partial u}{\partial x} + \frac{\partial v}{\partial z} = U\epsilon \left(\frac{\partial \tilde{u}_0}{\partial \tilde{x}} + \frac{\partial \tilde{v}_0}{\partial \tilde{z}} \right) = 0.$$

Inserting (6.15)–(6.18) into (6.1), we find

$$\frac{\eta U}{h^{*2}} \frac{\partial \tilde{p}_0}{\partial \tilde{x}} = \frac{\eta U}{h^{*2}} \left(\epsilon^2 \frac{\partial^2 \tilde{u}_0}{\partial \tilde{x}^2} + \frac{\partial^2 \tilde{u}_0}{\partial \tilde{z}^2} \right) = \frac{\eta U}{h^{*2}} \left(\frac{\partial^2 \tilde{u}_0}{\partial \tilde{z}^2} \right) + O(\epsilon^2), \tag{6.19}$$

and (6.2) becomes

$$\frac{\eta U}{\epsilon h^{*2}} \frac{\partial \tilde{p}_0}{\partial \tilde{z}} = \frac{\eta U}{h^{*2}} \left(\epsilon^3 \frac{\partial^2 \tilde{v}_0}{\partial \tilde{x}^2} + \epsilon \frac{\partial^2 \tilde{v}_0}{\partial \tilde{z}^2} \right) = O(\epsilon). \tag{6.20}$$

Thus, by comparing powers in ϵ we have

$$\frac{\partial \tilde{p}_0}{\partial \tilde{x}} = \frac{\partial^2 \tilde{u}_0}{\partial \tilde{z}^2}, \tag{6.21}$$

$$\frac{\partial \tilde{p}_0}{\partial \tilde{z}} = 0. \tag{6.22}$$

In other words, to leading order there are no pressure gradients over the thickness of the film: this is the hallmark of the *lubrication approximation* [17], developed originally by Reynolds [183] to describe the flow between two narrowly spaced moving parts. Only streamwise pressure gradients drive the flow.

The boundary conditions (6.6) and (6.7) can be analyzed in much the same way. The latter gives

$$\left. \frac{\partial \tilde{u}_0}{\partial \tilde{z}} \right|_{\tilde{z}=\tilde{h}} = 0, \tag{6.23}$$

while the former reduces to

$$\tilde{p}_0 = -\frac{\epsilon^3}{\mathrm{Ca}} \frac{\partial^2 \tilde{h}}{\partial \tilde{x}^2}. \tag{6.24}$$

In agreement with our earlier argument, (6.24) implies that ϵ and Ca are related by (6.13). The constant of proportionality is immaterial, and it is simplest to choose $\epsilon = \mathrm{Ca}^{1/3}$.

Now we are in a position to find $\tilde{u}_0$ explicitly. The left-hand side of (6.21) does not depend on $\tilde{z}$, and thus

$$\tilde{u}_0 = A + B\tilde{z} + \frac{\tilde{p}_0'}{2} \tilde{z}^2,$$

where the prime denotes differentiation with respect to $\tilde{x}$. From (6.8) it follows that $\tilde{u}_0(\tilde{x}, 0, \tilde{t}) = 0$ and thus $A = 0$, while from (6.23) we have $B = -\tilde{p}_0'\tilde{h}$. Thus in summary, also using (6.24), we have

$$\tilde{u}_0 = -\frac{\tilde{h}'''}{2} \left(\tilde{z}^2 - 2\tilde{z}\tilde{h} \right), \tag{6.25}$$

an explicit expression for the velocity field. The rescaled form of (6.10) is

$$\frac{\partial \tilde{h}}{\partial \tilde{t}} + \frac{\partial}{\partial \tilde{x}} \int_0^{\tilde{h}} \tilde{u}_0 d\tilde{z} = 0 \tag{6.26}$$

and, inserting (6.25), we finally have

$$\frac{\partial \tilde{h}}{\partial \tilde{t}} + \frac{1}{3}\left(\tilde{h}'''\tilde{h}^3\right)' = 0. \tag{6.27}$$

Note that the velocity field no longer appears; (6.27) is a closed equation for the film thickness alone. It is often more intuitive to write an equation in the original variables, in this case the film thickness $h(x, t)$. Undoing the rescalings (6.14) and (6.18), we find

$$\frac{\partial h}{\partial t} + \frac{v_\eta}{3}\left(h'''h^3\right)' = 0. \tag{6.28}$$

Example 6.1 (Leveling of a thin film) Consider a thin film of thickness h_* with a small sinusoidal perturbation of amplitude δ imposed upon it:

$$h = h_* + \delta \sin kx.$$

Inserting this into (6.28) and linearizing in δ, we obtain

$$\dot{\delta} + \frac{\delta v_\eta h_*^3 k^4}{3} = 0.$$

In other words, the film will relax exponentially to a flat state according to $\delta = \delta_0 \exp(-\lambda t)$, where $\lambda = v_\eta h_*^3 k^4/3$ is a characteristic time scale. □

The one-dimensional description (6.28) can easily be generalized to a thin film of thickness $h(\mathbf{x}, t)$ that covers a surface in two dimensions. The mean curvature of the interface, in the thin film approximation, is now $-\Delta h$, where Δ is the two-dimensional Laplacian. Thus in dimensional form the leading-order contribution to the pressure, replacing the one-dimensional case (6.24), is

$$p = -\gamma \Delta h.$$

The flow produced by pressure differences will be in the direction of the *gradient* of the pressure, so the horizontal flow velocity is now

$$\mathbf{v} = \frac{1}{2\eta}(\nabla p)(z^2 - 2zh),$$

instead of (6.25). The flux is

$$\mathbf{f} = \int_0^h \mathbf{v}\,dz = -\frac{1}{3\eta}(\nabla p)h^3,$$

and thus conservation of mass yields

$$\frac{\partial h}{\partial t} = \frac{1}{3\eta}\nabla \cdot \left(h^3 \nabla p\right) = -\frac{v_\eta}{3}\nabla \cdot \left(h^3 \nabla \Delta h\right). \tag{6.29}$$

Example 6.2 (Gravity-driven thin film flow) On a scale much larger than the capillary length $\ell_c = \sqrt{\gamma/(\rho g)}$, surface tension can be neglected compared with gravity. Gravity must be included in the description, however, so an additional term $-\rho g$ appears on the right-hand side of (6.2). Hence instead of (6.22) we now have

$$\frac{\partial \tilde{p}_0}{\partial \tilde{z}} = -\epsilon \frac{\rho g h^{*2}}{\eta U},$$

with general solution

$$\tilde{p}_0 = A(\tilde{x}) - \epsilon \frac{g h^{*2}}{\nu U} \tilde{z}.$$

Imposing $\tilde{p}_0 = 0$ at the interface $\tilde{z} = \tilde{h}(\tilde{x})$, the pressure becomes

$$\tilde{p}_0 = \epsilon \frac{g h^{*2}}{\nu U} [\tilde{h}(\tilde{x}) - \tilde{z}],$$

which replaces (6.24).

Hence, by introducing

$$\epsilon = \frac{\nu U}{g h^{*2}}, \tag{6.30}$$

we have $\tilde{p}_0' = \tilde{h}'$, but otherwise the nondimensional equations remain unchanged. This means we have to replace $-\tilde{h}'''$ by $\tilde{h}'$ in (6.27), and we find

$$\frac{\partial \tilde{h}}{\partial \tilde{t}} - \frac{1}{3} \left(\tilde{h}' \tilde{h}^3 \right)' = 0.$$

Restoring the dimensions, this leads to the one-dimensional thin film equation,

$$\frac{\partial h}{\partial t} = \frac{\rho g}{3\eta} \left(h^3 h' \right)', \tag{6.31}$$

which is the nonlinear diffusion equation (3.95) with $n = 3$.   □

6.1.1 Hele-Shaw flow

A related problem is that of the flow between two narrowly spaced parallel plates, so-called Hele-Shaw flow [105]. Once more, the idea is that the spacing h (typically a millimeter) is much smaller than any characteristic scale in the direction parallel to the plates. In addition, inertia can often be neglected owing to strong viscous friction at the plates. The details of the derivation are left as an exercise; we will only indicate the general idea. For the velocity field we make the ansatz

$$\mathbf{v} = \mathbf{u}(x, y, t) f(z), \quad f = 6 \left(\frac{z}{h} - \frac{z^2}{h^2} \right), \tag{6.32}$$

where $\mathbf{u} = (u, v)$ is a two-dimensional velocity field in the plane (x, y) of the plates. The velocity field (6.32) satisfies the boundary conditions of no slip at the plates $z = 0$ and $z = h$. The prefactors have been chosen such that $\mathbf{u}$ is the transverse *average* over the velocity field.

Since h is small, the z-derivatives of (6.32) dominate the x- and y-derivatives, and so

$$\triangle \mathbf{v} \approx \mathbf{u} \frac{\partial^2 f}{\partial z^2} = -\frac{12}{h^2} \mathbf{u}.$$

Inserting this into the Stokes equation (4.48), one finds the *Hele-Shaw equation*

$$\nabla p = -\frac{12\eta}{h^2} \mathbf{u}, \tag{6.33}$$

where the pressure $p(x, y, t)$ depends on the transverse coordinates alone. Since $\mathbf{v}$ is incompressible, it follows from (6.32) that $\nabla \cdot \mathbf{u} = 0$. Taking the divergence of (6.33), this implies that the pressure field is harmonic:

$$\triangle_2 p = 0, \tag{6.34}$$

with $\triangle_2$ the two-dimensional Laplace operator.

Example 6.3 (Point sink in Hele-Shaw flow) Fluid is extracted from a Hele-Shaw cell through a hole at the origin at a rate $m < 0$. According to (4.46), the velocity field is

$$\mathbf{u} = \frac{m}{2\pi r} \mathbf{e}_r, \tag{6.35}$$

so that the flux through a circle of radius R is m. Then, from (6.33) we find that the pressure has a logarithmic singularity,

$$p = -\frac{6\eta m}{\pi h^2} \ln r \tag{6.36}$$

at the sink $r = 0$.   □

It also follows from (6.33) that $\mathbf{u}$ is irrotational,

$$\boldsymbol{\omega} = \left(\frac{\partial v}{\partial x} - \frac{\partial u}{\partial y} \right) \mathbf{e}_z = 0.$$

At solid boundaries in the xy-plane the normal component of $\mathbf{u}$ must vanish, but no condition is imposed for the tangential velocity. Thus Hele-Shaw flow satisfies the same equations as two-dimensional inviscid potential flow. This fact was exploited in Fig. 5.2, which shows the flow around an airplane wing squeezed between two flat plates. The only difference between Hele-Shaw flow and inviscid two-dimensional flow is that for Hele-Shaw flow $\mathbf{u}$ can have no circulation, since

$$\oint_C \nabla p \cdot d\mathbf{l} = 0$$

even if C encloses a solid body. This is why the Hele-Shaw flow shown in Fig. 5.2 satisfies $\Gamma = 0$ around the wing.

6.2 Slender jets

Now we turn to an axisymmetric column of fluid, with a view to describing its breakup; see Fig. 1.2. Physically, the situation is quite different from that of a thin film, which is stable in its flat rest state and deforms in response to external driving. By contrast, a fluid jet is inherently unstable and decays spontaneously into drops. Once more we assume the existence of a small aspect ratio ϵ, but this time it will be determined by the intrinsic dynamics of the breakup process. In fact ϵ can be identified as the time distance from the singularity, and thus it goes to zero as the pinch point is approached. As a result the quality of the approximation improves as one gets closer and closer to the pinch point and ultimately the dynamics should capture the singularity exactly.

Since there is no wall to slow down the motion, in drop pinch-off inertia comes into play and so the density ρ enters as a third independent parameter. Using γ, the kinematic viscosity ν, and ρ, we are now able to define length and time scales which are *intrinsic* to the drop dynamics:

$$\ell_\nu = \frac{\nu^2 \rho}{\gamma}, \quad t_\nu = \frac{\nu^3 \rho^2}{\gamma^2}, \tag{6.37}$$

see Exercise 1.1. We are thus able to define all scales relative to those given by (6.37). However, let us first proceed as before by introducing a longitudinal length L which defines the size of a feature along the jet. Thus if a typical length scale characterizing the jet radius is ℓ_r, we have

$$\ell_r = \epsilon L. \tag{6.38}$$

According to Appendix A, the Navier–Stokes equation (4.13) reads, in cylindrical coordinates,

$$\frac{\partial v_r}{\partial t} + v_r \frac{\partial v_r}{\partial r} + v_z \frac{\partial v_r}{\partial z} + \frac{1}{\rho} \frac{\partial p}{\partial r} = \nu \left(\frac{\partial^2 v_r}{\partial r^2} + \frac{\partial^2 v_r}{\partial z^2} + \frac{1}{r} \frac{\partial v_r}{\partial r} - \frac{v_r}{r^2} \right),$$
$$\tag{6.39}$$

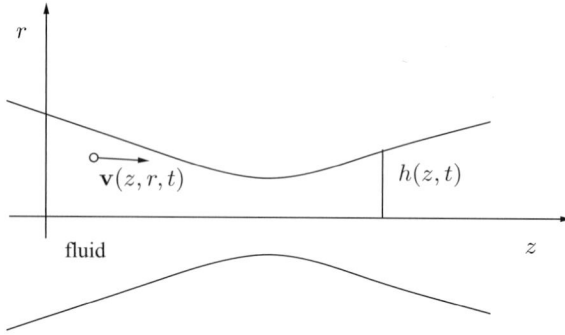

Figure 6.2 The flow geometry of an axisymmetric column of fluid. The velocity field inside the fluid is $\mathbf{v}(z, r) = v_z \mathbf{e}_z + v_r \mathbf{e}_r$.

$$\frac{\partial v_z}{\partial t} + v_r \frac{\partial v_z}{\partial r} + v_z \frac{\partial v_z}{\partial z} + \frac{1}{\rho} \frac{\partial p}{\partial z} = \nu \left(\frac{\partial^2 v_z}{\partial r^2} + \frac{\partial v_z}{\partial z^2} + \frac{1}{r} \frac{\partial v_z}{\partial r} \right), \qquad (6.40)$$

where v_z is the velocity along the axis, v_r is the velocity in the radial direction, and p is the pressure. The continuity equation (4.3) reads

$$\frac{\partial v_r}{\partial r} + \frac{\partial v_z}{\partial z} + \frac{v_r}{r} = 0. \qquad (6.41)$$

Equations (6.39)–(6.41) hold in the fluid domain $0 \le r < h(z, t)$; see Fig. 6.2.

The normal and tangential components of the interfacial stress balance (4.24) are

$$\frac{p}{\rho} - \frac{2\nu}{1 + h'^2} \left[\frac{\partial v_r}{\partial r} + \frac{\partial v_z}{\partial z} h'^2 - \left(\frac{\partial v_z}{\partial r} + \frac{\partial v_r}{\partial z} \right) h' \right] = \frac{\gamma}{\rho} \left(\frac{1}{R_1} + \frac{1}{R_2} \right) \qquad (6.42)$$

and

$$\frac{\nu}{1 + h'^2} \left[2 \frac{\partial v_r}{\partial r} h' + \left(\frac{\partial v_z}{\partial r} + \frac{\partial v_r}{\partial z} \right) (1 - h'^2) - 2 \frac{\partial v_z}{\partial z} h' \right] = 0, \qquad (6.43)$$

respectively, to be evaluated at $r = h$; the prime indicates differentiation with respect to z. Finally, the equation of motion for the free surface is analogous to (6.9):

$$\frac{\partial h}{\partial t} + v_z h' = v_r|_{r=h}. \qquad (6.44)$$

This completes the formulation of the problem, and we return to an analysis of the relevant length and time scales.

In addition to the length scales ℓ_r and L the problem is characterized by a time scale τ. These scales can now be used to estimate the order of magnitude of the different terms in the equations of motion. We begin by noting that the axial component of the velocity field is approximated by $v_z \propto L/\tau$, which may also be viewed as the definition of τ. From (6.41) we have $\partial_z v_z \approx \partial_r v_r$ and hence $v_r \propto \epsilon L/\tau$. The motion is driven by the surface tension, so the main contribution to the pressure comes from the right-hand side of (6.42). Since we assume the radial scale ℓ_r to be small, the mean curvature (4.23) is dominated by the radial curvature $1/h$, and we have $p \propto \gamma/\ell_r$. From the axial component (6.40) of the Navier–Stokes equation we have $\partial_t v_z \propto \partial_z p/\rho$, which means that

$$\gamma/\rho \propto \ell_r L^2/\tau^2 = \epsilon L^3/\tau^2. \tag{6.45}$$

Since the free surface of the jet does not exert a shear force, the velocity gradients in the radial direction are not greater than those in the axial direction, even though ϵ is small. We will demonstrate this statement more precisely below by performing an expansion in the radial direction. For the moment it suffices to say that we can estimate the gradient of v_z by considering its derivative in the axial direction. Hence an estimate of the viscous term in (6.40) is $\nu \partial_z^2 v_z \propto \nu/(L\tau)$. Equating this to $\partial_t v_z$ we find that

$$\nu \approx L^2/\tau. \tag{6.46}$$

For a given problem ν and γ/ρ are of course fixed, so physically speaking the scales ℓ_r, L, and τ of a particular solution adjust themselves to make the various terms in the equations balance. Using (6.45) and (6.46), we can identify

$$L = \epsilon \ell_\nu, \quad \ell_r = \epsilon^2 \ell_\nu, \quad \text{and} \quad \tau = \epsilon^2 t_\nu. \tag{6.47}$$

If the typical scales of a solution behave in this way, the surface tension, inertial, and viscous forces will be balanced.

Having estimated the typical sizes of all terms, we can nondimensionalize accordingly, setting

$$r = \ell_r \tilde{r}, \quad z = L\tilde{z}, \quad t = \tau \tilde{t},$$

$$v_r = \frac{\ell_r}{\tau} \tilde{v}_r, \quad v_z = \frac{L}{\tau} \tilde{v}_z, \quad p = \frac{L^2}{\tau^2} \tilde{p}, \quad h = \ell_r \tilde{h}. \tag{6.48}$$

To make further progress we have to estimate the relative size of the axial and radial gradients in a systematic fashion. We do this by expanding $\tilde{v}_z$ in a power series in ϵr:

$$\tilde{v}_z(\tilde{r}, \tilde{z}, \tilde{t}) = \tilde{v}_0(\tilde{z}, \tilde{t}) + \tilde{v}_2(\tilde{z}, \tilde{t})(\epsilon \tilde{r})^2 + \cdots. \tag{6.49}$$

The key idea is that for small ϵ the higher orders are small and can hopefully be neglected, leading to closure. Inserting (6.49) into the scaled continuity equation

$$\frac{\partial \tilde{v}_r}{\partial \tilde{r}} + \frac{\partial \tilde{v}_z}{\partial \tilde{z}} + \frac{\tilde{v}_r}{\tilde{r}} = 0, \tag{6.50}$$

one finds a corresponding series expansion for $\tilde{v}_r$:

$$\tilde{v}_r(\tilde{r}, \tilde{z}, \tilde{t}) = -\tilde{v}_0'(\tilde{z}, \tilde{t})\frac{\tilde{r}}{2} - \tilde{v}_2'(\tilde{z}, \tilde{t})\frac{\epsilon^2 \tilde{r}^3}{4} - \cdots. \tag{6.51}$$

Here we have used the boundary condition $\tilde{v}_r(\tilde{r} = 0) = 0$, which follows from axial symmetry. Finally, we expand the pressure similarly:

$$\tilde{p}(\tilde{r}, \tilde{z}, \tilde{t}) = \tilde{p}_0(\tilde{z}, \tilde{t}) + \tilde{p}_2(\tilde{z}, \tilde{t})(\epsilon \tilde{r})^2 + \cdots. \tag{6.52}$$

Now we rewrite all equations in terms of the rescaled quantities and insert into them the expansions (6.49), (6.51), and (6.52). Multiplying the axial component of the Navier–Stokes equation (6.40) by τ/L, one finds

$$\frac{\partial \tilde{v}_z}{\partial \tilde{t}} + \tilde{v}_r \frac{\partial \tilde{v}_z}{\partial \tilde{r}} + \tilde{v}_z \frac{\partial \tilde{v}_z}{\partial \tilde{z}} + \frac{\partial \tilde{p}}{\partial \tilde{z}} - \frac{1}{\epsilon^2}\frac{\partial^2 \tilde{v}_z}{\partial \tilde{r}^2} - \frac{\partial^2 \tilde{v}_z}{\partial \tilde{z}^2} - \frac{1}{\epsilon^2}\frac{1}{\tilde{r}}\frac{\partial \tilde{v}_z}{\partial \tilde{r}}$$
$$= \frac{\partial \tilde{v}_0}{\partial \tilde{t}} - \epsilon^2 \tilde{r}^2 \tilde{v}_0' \tilde{v}_2 + \tilde{v}_0 \tilde{v}_0' + \tilde{p}_0' - 2\tilde{v}_2 + \tilde{v}_0'' - 2\tilde{v}_2 + O(\epsilon^2) = 0.$$

In evaluating the right-hand side we have included the leading-order contribution for each term. Thus at order $O(\epsilon^0)$ we find

$$\frac{\partial \tilde{v}_0}{\partial \tilde{t}} + \tilde{v}_0 \tilde{v}_0' = -\tilde{p}_0' + (4\tilde{v}_2 + \tilde{v}_0'') + O(\epsilon^2). \tag{6.53}$$

Performing the same procedure for the radial equation (6.39), multiplying it by τ/ℓ_r we find that the leading contribution comes from the last two terms on the right, giving

$$\epsilon^{-2}\left(\frac{\tilde{v}_0'}{2\tilde{r}} - \frac{\tilde{v}_0'}{2\tilde{r}}\right) + O(\epsilon^0) = 0.$$

This means that the radial equation is satisfied identically at leading order, and we can ignore it. The reason is that, in the absence of a solid wall, only a weak flow is set up in the direction transversal to the main flow direction. (However, if the jet is curved this is no longer the case, and the radial component furnishes a nontrivial equation, in line with the greater number of parameters needed to describe the bending of the jet [149].)

The two boundary conditions (6.42) and (6.43) can be analyzed in the same way, also using the expression (4.23) for the mean curvature. The result of a straightforward calculation, at leading order in ϵ, is

$$\tilde{p}_0 = \frac{1}{\tilde{h}} - \tilde{v}_0' + O(\epsilon^2), \quad -3\tilde{v}_0'\tilde{h}' - \frac{1}{2}\tilde{v}_0''\tilde{h} + 2\tilde{v}_2\tilde{h} + O(\epsilon^2) = 0. \tag{6.54}$$

This can be read as an equation for $\tilde{p}_0$ and $\tilde{v}_2$, which are to be eliminated from (6.53). The result, up to terms of order ϵ^2, is

$$\frac{\partial \tilde{v}_0}{\partial \tilde{t}} + \tilde{v}_0 \tilde{v}_0' = -\left(\frac{1}{\tilde{h}}\right)' + 3\frac{(\tilde{v}_0' \tilde{h}^2)'}{\tilde{h}^2}, \tag{6.55}$$

which is the desired slender jet equation for $\tilde{v}_0$. The kinematic boundary condition (6.44) for $\tilde{h}$ gives immediately

$$\frac{\partial \tilde{h}}{\partial \tilde{t}} + \tilde{v}_0 \tilde{h}' = -\frac{1}{2}\tilde{h}\tilde{v}_0'. \tag{6.56}$$

Equations (6.55), (6.56) constitute a closed system for the one-dimensional velocity field $\tilde{v}_0(\tilde{z}, \tilde{t})$ and the height $\tilde{h}(\tilde{z}, \tilde{t})$. Up to corrections of order $O(\epsilon^2)$, they give a representation of the full Navier–Stokes dynamics with a free surface, which is central to our subsequent discussion. Once again it is useful to reinstate the dimensional variables $h(z, t)$ as well as the axial velocity $v_0(z, t)$; dropping the subscript, we obtain

$$\frac{\partial h}{\partial t} + vh' = -\frac{hv'}{2}, \tag{6.57}$$

$$\frac{\partial v}{\partial t} + vv' = -\frac{\gamma}{\rho}\left(\frac{1}{h}\right)' + 3v\frac{(h^2 v')'}{h^2}. \tag{6.58}$$

Our main objective in deriving the slender jet model (6.57), (6.58) is to describe pinch-off singularities. However, with some small modifications we can expand its usefulness far beyond the neighborhood of the singularity [78].

Namely, in Exercise 6.16 we show that, for (6.57), (6.58), perturbations of arbitrarily small wavelength grow, making the description useless for numerical purposes since errors on the scale of the numerical grid become amplified. The reason is that in the full mean curvature (4.23) only the radial curvature $1/h$ survives from the asymptotics. As shown in Example 6.5 below, the axial contribution to the curvature will ensure that there is no growth at short wavelengths below a limiting wavelength derived by Plateau; see Exercise 6.15. Thus if we replace the pressure term $1/h$ in (6.58) by the expression for the full mean curvature κ, the small-wavelength instability disappears. In the absence of flow (setting v equal to zero), the system of equations includes all equilibrium states of the free surface as solutions. As an additional nicety, one can include the constant acceleration of gravity along the z-axis in the force balance (6.55). In summary, this leads to the velocity equation

$$\frac{\partial v}{\partial t} + vv' = -\frac{\gamma}{\rho}\kappa' + 3v\frac{(h^2 v')'}{h^2} - g, \tag{6.59}$$

with κ given by (4.23).

Example 6.4 (Liquid bridge) A drop of fluid held between two cylindrical plates of radius r_0 (see Fig. 8.1 below) is referred to as a liquid bridge. Often the positions $z_\pm(t)$ of the plates are allowed to move, so that the boundary conditions to be satisfied by (6.57), (6.59) at the plates are

$$h(z_\pm, t) = r_0, \quad v(z_\pm, t) = \dot{z}_\pm. \tag{6.60}$$

The boundary conditions (6.60) allow for a unique solution once the initial conditions for h and v have been prescribed [78]. ☐

Example 6.5 (Rayleigh–Plateau instability) To find how a fluid cylinder of radius r_0 and in the absence of gravity behaves under small perturbations of amplitude ϵ, we insert

$$h(z, t) = r_0 \left(1 + \epsilon e^{\omega t} \cos kz\right), \quad v(z, t) = \epsilon v_0 e^{\omega t} \sin kz$$

into (6.57), (6.59) with $g = 0$. For $\epsilon = 0$ this satisfies both equations identically so, linearizing in ϵ, we find

$$\omega = -\frac{v_0 k}{2} \tag{6.61}$$

and

$$\omega v_0 = -\frac{\gamma}{\rho} \left(\frac{k}{r_0} - r_0 k^3\right) - 3 v v_0 k^2, \tag{6.62}$$

respectively. The two terms proportional to the surface tension come from the radial and the axial contributions to the mean curvature. We can eliminate v_0 between (6.61) and (6.62) to find

$$\omega^2 = \frac{\omega_0^2}{2} \left[(r_0 k)^2 - (r_0 k)^4\right] - 3 v \omega k^2,$$

where $\omega_0^2 = \gamma/(r_0^3 \rho)$. We solve this quadratic equation to obtain

$$\frac{\omega}{\omega_0} = \sqrt{\frac{1}{2}\left(x^2 - x^4\right) + \frac{9}{4}\mathrm{Oh}^2 x^4} - \frac{3}{2}\mathrm{Oh}x^2, \tag{6.63}$$

with $x = k r_0$ and

$$\mathrm{Oh} = \sqrt{\frac{\ell_v}{r_0}} = v\sqrt{\frac{\rho}{r_0 \gamma}}. \tag{6.64}$$

The *Ohnesorge number* Oh measures the relative sizes of the viscous and surface tension forces.

For long wavelengths $x < 1$, equation (6.63) leads to real and positive ω, making the fluid cylinder unstable to small perturbations, the so-called *Rayleigh–Plateau instability*. For $x > 1$, however, the solutions are purely

imaginary and no growth occurs. As shown in Exercise 6.15, the reason is that
in the former case the growth of a sinusoidal perturbation reduces the surface
area, lowering the surface energy, while in the latter case, of short wavelengths,
the surface area is increased. The most amplified, or Rayleigh, wavelength
$\lambda_R = 2\pi r_0/x_R$ is given by the maximum of (6.63) over x:

$$\lambda_R = 2\pi \left(2 + 3\sqrt{2}\mathrm{Oh}\right)^{1/2} r_0. \qquad (6.65)$$

Starting from a broad spectrum of small initial perturbations, the growth of
this wavelength will outrun all others, so the Rayleigh wavelength will be the
one observed macroscopically. Numerically, (6.63) is remarkably close to the
exact result based on linearizing the full Navier–Stokes equation [83]. $\qquad\square$

Example 6.6 (Slender jet equations in conserved form) The slender jet equations for h and v can be seen as conservation laws for mass and momentum,
respectively. Multiplying (6.57) by h we obtain on the one hand

$$\frac{\partial h^2}{\partial t} + \left(h^2 v\right)' = 0, \qquad (6.66)$$

which has the form of a continuity equation for the volume $\pi h^2 dz$ of a slice
of fluid. If on the other hand we multiply (6.59) by h^2 (neglecting gravity) and
use $G' = h^2\kappa$ (see Exercise 6.19), where G is defined as

$$G \equiv -\left(\frac{h}{\sqrt{1 + h'^2}} + \frac{h^2 h''}{(1 + h'^2)^{3/2}}\right), \qquad (6.67)$$

we obtain

$$h^2 \partial_t v + h^2 v v' = -\frac{\gamma}{\rho} G' + 3\nu(a v')'.$$

Combining this with (6.66) yields

$$\partial_t (h^2 v) + (h^2 v^2)' = -\frac{\gamma}{\rho} G' + 3\nu(h^2 v')', \qquad (6.68)$$

which is the continuity equation for the momentum $m = h^2 v$ of a slice of fluid
(omitting a factor of π).

Exercises

6.1 Write down the thin film equation (6.28) including the effects of grav-
     ity and surface tension. Show that small perturbations to a uniform film
     (with thickness h_0), of the form

$$h(x, t) = h_0 + \delta e^{ikx + \omega t}, \quad \delta \ll 1,$$

are such that $\omega \equiv \omega(k) \leq 0$, and discuss the stability of the film. If gravity is reversed, corresponding to the situation where a film hangs from the ceiling, show that $\omega(k)$ may become positive. Compute k_0 such that $\omega(k_0)$ is a maximum and argue that a fingering instability may develop, with a typical separation $2\pi / k_0$ between fingers.

6.2   In deriving (6.28) we assumed that there is no slip at the wall: $u(x, 0) = 0$. Derive the corresponding equation for the case in which the fluid slips over the wall at a speed that is proportional to the shear at the wall,

$$u(x, 0, t) = \lambda \left. \frac{\partial u}{\partial z} \right|_{z=0}, \tag{6.69}$$

where λ is called the *slip length*. Physically, λ is found to be a microscopic length, in the order of nanometers. Show that (6.28) can be generalized to

$$\frac{\partial h}{\partial t} + \frac{v_\eta}{3} \left[h'''(h^3 + 3\lambda h^2) \right]' = 0, \tag{6.70}$$

which includes the case $\lambda = 0$.

6.3   Another version of the thin film flow treated in Section 6.1 is Stokes flow between two closely spaced *solid* surfaces driven by their relative motion. A particular application would be a magnetic head above a rotating hard disk. Let us take the lower surface as occupying the xy-plane and as moving with velocity $\mathbf{U}$. The upper surface $z = h(\mathbf{x})$ is stationary, where $\mathbf{x} = (x, y)$. The crucial property used in the expansion is that $h(\mathbf{x})$ is slowly varying, i.e. $|\nabla h| \ll 1$, with ∇ the gradient operator in the plane. Denote the three-dimensional velocity field as $\mathbf{v} = (\mathbf{u}, w)$, where $\mathbf{u}$ is in the horizontal plane.

Thus if ϵ is a small parameter, we can introduce the expansions

$$\mathbf{u} = \mathbf{u}_0(\mathbf{x}, \tilde{z}) + \epsilon \mathbf{u}_1(\mathbf{x}, \tilde{z}) + \cdots, \tag{6.71}$$

$$p = \epsilon^{-2} p_0(\mathbf{x}, \tilde{z}) + \epsilon^{-1} p_1(\mathbf{x}, \tilde{z}) + \cdots, \tag{6.72}$$

$$\tilde{z} = z/\epsilon, \quad \tilde{h} = h/\epsilon. \tag{6.73}$$

(i)   Show that, to leading order, $p_0 = p_0(\mathbf{x})$ and

$$\nabla p_0 = \eta \frac{\partial^2 \mathbf{u}_0}{\partial \tilde{z}^2}.$$

(ii) Find the velocity profile in the $\tilde{z}$-direction, and calculate the total flux in the xy-plane.

(iii) Use incompressibility to show that

$$\mathbf{U} \cdot \nabla \tilde{h} = \nabla \left(\frac{\nabla p_0}{6\eta} \tilde{h}^3 \right),$$

which is known as *Reynolds' equation*.

6.4 Consider the Taylor scraper problem of the previous chapter (see Fig. 5.10) in the limit of small angles $\alpha \to 0$; then the position of the scraper is described by the function $h(x) = \alpha x$.

(i) Show that in the lubrication limit (in which the pressure is independent of y), the horizontal velocity field is

$$u = \frac{p'}{2\eta}(y^2 - yh) + \frac{U}{h}(y - h),$$

where p' is the horizontal pressure gradient.

(ii) From the condition that the total flux Q over a vertical cross section must vanish, show that

$$p' = -\frac{6U\eta}{h^2}.$$

(iii) Show that the leading-order expressions for the normal and shear stresses, given in Exercise 5.2, are recovered.

6.5 Consider the Hele-Shaw flow between two parallel plates discussed in Section 6.1.1. As a representative model problem, we take a circular disk of radius R in a uniform flow in the x-direction such that the flux per unit length in the spanwise direction is Uh [128]. Any solid obstacle placed in the flow can be approximated locally by a cylinder of corresponding radius of curvature. The assumption we will make is that the plate spacing is small compared with R, so that $\epsilon = h/R$ is a small number. We take the obstacle to be z-independent, so that the flow has no w-component. This suggests the following rescalings:

$$\tilde{z} = z/R\epsilon, \quad \tilde{x} = x/R, \quad \tilde{y} = y/R, \tag{6.74}$$

$$\tilde{\mathbf{u}} = \mathbf{u}/U, \quad \tilde{p} = pR\epsilon^2/(\eta U), \quad \tilde{t} = tU/R. \tag{6.75}$$

(i) Show that, to leading order, as $\epsilon \to 0$ one obtains the equations (suppressing the tildes)

$$\frac{\partial u}{\partial x} + \frac{\partial v}{\partial y} = 0, \quad \frac{\partial p}{\partial x} = \frac{\partial^2 u}{\partial z^2} = 0,$$

$$\frac{\partial p}{\partial y} = \frac{\partial^2 v}{\partial z^2} = 0, \quad \frac{\partial p}{\partial z} = 0.$$

Conclude that the velocity field can be written as

$$\mathbf{v} = -\frac{\nabla p}{12} f(z), \qquad (6.76)$$

where $f(z) = 6(z - z^2)$. The pressure p depends only on x and y and satisfies (6.34).

(ii) Consider the Hele-Shaw flow around a circular cylinder, as described above. Use Exercise 4.1 to show that in cylindrical coordinates the flow field is

$$v_\theta = -\sin\theta \left(1 + \frac{1}{r^2}\right) f(z), \quad v_r = \cos\theta \left(1 - \frac{1}{r^2}\right) f(z).$$
$$(6.77)$$

6.6 (*) In Exercise 6.5 the velocity field satisfies $v_r = 0$ and $v_\theta = -2\sin\theta f(z)$ on the surface of the cylinder $r = 1$. The tangential component does not satisfy the no-slip boundary condition $v_\theta = 0$ on the surface of the cylinder, so there must be a thin region (the boundary layer) which accommodates this discrepancy [128].

(i) Compute the pressure distribution on the cylinder using (6.76), (6.77). Show that the pressure will be almost constant across the thin boundary layer next to the surface of the cylinder; thus show that

$$\frac{\partial^2 v_\theta}{\partial z^2} + \epsilon^2 \frac{\partial^2 v_\theta}{\partial r^2} = 24\sin\theta.$$

(ii) Argue that the boundary layer is of thickness ϵ. Introducing the stretched coordinate $\hat{r} = (r - 1)/\epsilon$, derive an equation for $v_\theta(\hat{r}, \theta, z)$. Show that v_θ satisfies the boundary conditions

$$v_\theta = 0 \quad \text{on} \quad r = 1,$$
$$v_\theta = \to -2(\sin\theta) f(z) \quad \text{for} \quad r \to \infty.$$

(iii) Use a Fourier representation in the z-direction to show that, in the boundary layer,

$$v_\theta = -2\sin\theta f(z) + \sum_{n=0}^{\infty} A_n \cos 2\pi nz \, \sin\theta \, e^{-2\pi n\hat{r}},$$

where $A_0 = -2$ and $A_n = -12/(\pi n)^2$ for $n > 0$.

6.7 (*) The derivation of the thin film equation (6.28) requires that the interface slope is small, $h' \ll 1$. This condition is stronger than needed; it can be relaxed to the requirement that the local slope $\theta(x)$, defined by $\tan\theta(x) = h'$, is *slowly varying* [200]. We confine ourselves to the

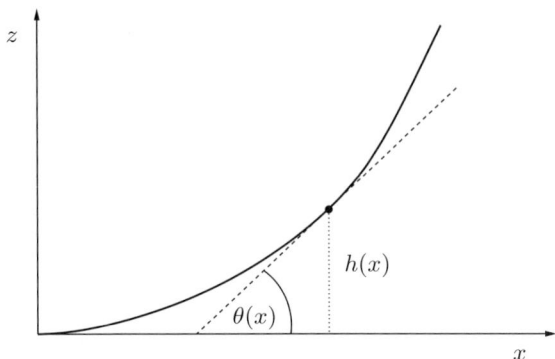

Figure 6.3 A wedge-like profile with slowly varying slope $\theta(x)$.

case in which the total flux q through the film vanishes (as it does if for example there is a stationary contact line at the end of the film) and the solid substrate moves at a speed U to the right. We introduce $\theta(x) = \tilde{\theta}(\tilde{x})$, where $\tilde{x}$ is defined in (6.14) as before, and expand in ϵ. For each x we can define a system of local polar coordinates with origin at $x - h(x)/\tan\theta(x)$, which is the intersection of the x-axis and the tangent line to the interface; see Fig. 6.3.

(i) Show that this local coordinate system is related to the Cartesian system by

$$r(x, z) = \sqrt{\left(\frac{h(x)}{\tan\tilde{\theta}(x)}\right)^2 + z^2},$$

$$\phi(x, z) = \tilde{\theta} - \arctan\left(\frac{z}{h\tan\tilde{\theta}(x)}\right),$$

where we define $\phi = 0$ to be the line tangent to the interface at $(x, h(x))$.

(ii) Expanding the velocity field as

$$\mathbf{v}(x, z) \equiv \tilde{\mathbf{v}}(r(x, z), \phi(x, z)) = \mathbf{v}_0 + \epsilon\mathbf{v}_1 + O(\epsilon^2), \qquad (6.78)$$

show that $\mathbf{v}_0$ is given by the solution to the viscous flow problem in a corner, i.e. Exercise 5.3(ii). Note that (6.78) is different from (6.15), (6.16) in that both components of the velocity scale in the same way.

(iii) Show that the pressure in the wedge is given by (5.65), and thus compute the normal stress $\sigma_{\theta\theta}$ on the free surface. Show that this leads to the equation

$$\frac{\partial \kappa}{\partial x} = -\frac{3Ca}{h^2} F(\theta), \tag{6.79}$$

where

$$F(\theta) = \frac{2}{3} \left(\frac{\tan \theta \sin^2 \theta}{\theta - \cos \theta \sin \theta} \right)$$

and κ is the full two-dimensional curvature. Argue that $\partial_x \kappa$ is of order ϵ^2, and thus (6.79) becomes self-consistent for $Ca^{1/2} \propto \epsilon$, in contrast with (6.13) in the thin film limit.

(iv) Show that in the thin film limit $\theta \to 0$, (6.79) becomes

$$\frac{3Ca}{\tilde{h}^2} = \tilde{h}'''.$$

6.8 Consider a long cylinder of radius R, immersed in a viscous liquid of viscosity η, which is oriented parallel to a solid wall with a narrow gap of size δ between the cylinder and the wall. The cylinder rotates with angular velocity Ω, so the problem can be considered two dimensional. We are interested in the limit $\delta/R \to 0$ [121].

(i) Introduce the scalings

$$X = x/(R\sqrt{\delta}), \quad Y = y/(R\delta),$$

and show that, to leading order, the surface of the cylinder is described by $Y \equiv H = 1 + X^2/2$ and the normal and tangent vectors are

$$\mathbf{n} = \sqrt{\delta} X \mathbf{e}_x - [1 + \delta(1 - H)]\mathbf{e}_y, \quad \mathbf{t} = [1 + \delta(1 - H)]\mathbf{e}_x + \sqrt{\delta} X \mathbf{e}_y.$$

(ii) Scaling the velocities and pressures as

$$u = \Omega R U, \quad v = \Omega R \sqrt{\delta} V, \quad p = \eta \Omega \delta^{-3/2} P,$$

show that the Stokes equation (4.48) gives, to leading order,

$$\frac{\partial P}{\partial Y} = 0, \quad \frac{\partial P}{\partial X} = \frac{\partial^2 U}{\partial Y^2}, \quad \frac{\partial V}{\partial Y} + \frac{\partial U}{\partial X} = 0,$$

if the constraint of incompressibility is included. Formulate the boundary conditions to be satisfied by U and V on the plate and on the cylinder.

(iii) Show that

$$U = \frac{1}{2} P' Y(Y - H) + \frac{Y}{H};$$

thus derive

$$P' = \frac{6}{H^2} - \frac{12Q}{H^3},$$

where

$$q \equiv \int_0^h u\,dy = \Omega R^2 \sqrt{\delta} Q$$

is the flux through the gap between the plate and the cylinder.

(iv) Find $P(X)$ and choose Q such that $P \to 0$ for $X \to \infty$. Use this result to show that the force on the cylinder parallel to the wall vanishes and that the torque is

$$\tau = -\frac{4\pi\eta\Omega R^2}{\sqrt{2\delta}}.$$

6.9 Repeat the analysis of Exercise 6.8 for the case where the cylinder moves parallel to the wall at a speed u_c; it is convenient to consider a frame of reference in which the cylinder is at rest and the plate moves with speed u_c to the left (why?). The scalings are the same as before, with u_c replacing ΩR. Show that the torque is zero while the force parallel to the wall is now

$$f_x = -\frac{4\pi\eta u_c}{\sqrt{2\delta}}.$$

6.10 Consider the spreading of a large drop of viscous liquid on a horizontal surface, so that surface tension can be neglected relative to gravity. Generalizing (6.31) to a two-dimensional surface $h(\mathbf{x}, t)$, the equation of motion becomes

$$\frac{\partial h}{\partial t} = \frac{g\rho}{3\eta}\nabla\cdot\left(h^3\nabla h\right). \tag{6.80}$$

Analogously to Exercise 3.5, the motion of an *axisymmetric* drop is subject to the constraint of mass conservation,

$$V = 2\pi \int_0^R hr\,dr = \text{const.} \tag{6.81}$$

(i) Show that in cylindrical coordinates, and introducing a suitable time scale $\tilde{t}$, (6.80) can be written

$$\frac{\partial h}{\partial \tilde{t}} = \frac{1}{r}(h'h^3r)',$$

where the prime denotes differentiation with respect to r.

(ii) Introduce a similarity solution valid for $R \to \infty$ (dropping the tilde for simplicity),

$$h = t^\alpha f(\xi), \quad \xi = r/t^\beta,$$

and derive the similarity equation. Show that to obtain a solution, and for V to be constant, we must have $\alpha = -1/4$, $\beta = 1/8$. Thus show that the similarity equation becomes

$$\frac{f}{4} + \frac{\xi}{8} f' + \frac{1}{\xi}(f' f^3 \xi)' = 0.$$

(iii) Show that the similarity solution can be solved in the form

$$f^3 = \frac{-3\xi^2}{16} + B.$$

(iv) Using mass conservation, finally show that the radius of the drop increases as

$$R = \left(\frac{2^{10}}{3^4 \pi^3}\right)^{1/8} \left[\left(\frac{\rho g V^3}{3\eta}\right) t\right]^{1/8}.$$

6.11 Non-Newtonian fluids of the power law type are characterized by a shear-dependent viscosity

$$\mu = m \, |D|^{1/\lambda - 1},$$

where D_{ij} is defined by (4.9) and $|D| = \sqrt{D_{ij} D_{ij}}$.

(i) Show that in two dimensions the thin film equation (6.27) for a such a power law is

$$h_t + \frac{1}{\lambda + 2} \left(\frac{\gamma}{m}\right)^\lambda \left(h^{\lambda+2} |h_{xxx}|^{\lambda-1} h_{xxx}\right)_x = 0. \qquad (6.82)$$

(ii) Show that in three dimensions the equation for an axisymmetric film profile $h(r, t)$ becomes

$$h_t - \frac{1}{\lambda + 2} \left(\frac{\gamma}{m}\right)^\lambda \frac{1}{r} \left(r h^{\lambda+2} |\kappa_r|^{\lambda-1} \kappa_r\right)_r = 0, \qquad (6.83)$$

where

$$\kappa = -\frac{1}{r}(r h_r)_r$$

is the curvature.

6.12 Derive the following equations for a thin two-dimensional viscous fluid sheet with surface tension present:

$$h_t + (uh)_x = 0, \qquad (6.84)$$

$$u_t + u u_x = \frac{\gamma}{\rho} h_{xxx} + 4\nu \frac{(u_x h)_x}{h}. \qquad (6.85)$$

Hint: introduce a dimensionless parameter $\epsilon \ll 1$ measuring the ratio of the sheet thickness and the horizontal extension of the sheet, and show that the velocity is horizontal and constant in one coordinate at leading order in ϵ.

6.13 Consider a thin layer of inviscid incompressible fluid, above a flat solid plate; the fluid is subject to a vertical gravitational force. We introduce a dimensionless parameter $\epsilon \ll 1$ measuring the ratio of the fluid layer thickness and the horizontal extension of the layer. Show that for a single horizontal coordinate the velocity is, at leading order in ϵ, purely horizontal and constant in the vertical coordinate. Show that the Euler equation (4.31) leads to the asymptotic model

$$h_t + (uh)_x = 0, \tag{6.86}$$

$$u_t + \left(\frac{u^2}{2} + \frac{p}{\rho}\right)_x = 0, \tag{6.87}$$

where $h(x,t)$ is the free surface elevation with respect to the flat plate, $u(x,t)$ is the x-component of the velocity, and

$$p = \rho g h.$$

This model is called the shallow-water system (or Saint Venant system). Show that for two horizontal dimensions the model reads

$$h_t + (uh)_x + (vh)_y = 0,$$

$$(hu)_t + \left(hu^2 + \frac{1}{2}gh^2\right)_x + (huv)_y = 0,$$

$$(hu)_t + (huv)_x + \left(hv^2 + \frac{1}{2}gh^2\right)_y = 0.$$

6.14 Derive an energy balance for the slender jet equations (6.59), (6.57).
   (i) Use integration by parts to show that

$$\frac{1}{2}\int \partial_t(av^2)dz = \int \frac{(av)'p}{\rho}dz - 3v\int av'^2dz + \text{boundary terms},$$

   where $a \equiv h^2$.
   (ii) Use the variational formulation (4.74) to show that

$$\pi \int (av)'pdz = -\frac{\partial E}{\partial t} + \text{boundary terms},$$

   where

$$E = 2\pi\gamma \int h\sqrt{1 + h_z^2}\,dz.$$

   Thus show that

$$\frac{\pi}{2}\frac{\partial(av^2)}{\partial t} + \frac{\partial E}{\partial t} = -3\pi v\int av'^2dz + \text{boundary terms}, \tag{6.88}$$

   which means that the *total* energy can only decrease.

6.15 Consider the linear stability of a cylindrical fluid column. Following J. A. F. Plateau, instability can occur only if a perturbation is such that it *decreases* the energy of the system. Decomposing the perturbation into sinusoidal modes, the radial profile is

$$h(z) = r_0 + \epsilon \cos kz + O(\epsilon^2),$$

where ϵ is the amplitude of the perturbation. For simplicity, we confine ourselves to axisymmetric perturbations.

(i) Show that in order to conserve mass the perturbation has to be of the form

$$\eta(z) = \epsilon \cos kz - \frac{\epsilon^2}{8r_0}.$$

(ii) Calculate the area of the column (per wavelength) as a function of ϵ. Conclude that ϵ will grow if $kr_0 < 2\pi$.

6.16 Derive the dispersion relation (6.63) for the asymptotic jet equations (6.57), (6.58). Show that they lead to a finite growth rate $\omega/\omega_0 = 1/(6\,\mathrm{Oh})$ in the limit of vanishing wavelength, which will invalidate the model rapidly.

6.17 A particular solution of (6.57), (6.59) with gravity is found by assuming a constant thread radius $h = R(t)$ [106].

(i) Show that the velocity field can be written in the form $v(z,t) = V_0(t) + V_1(t)z$, and derive differential equations for R, V_0, and V_1.

(ii) Solve the differential equation to find $V_1(t)$ and then use this in the other two differential equations to obtain

$$R(t) = R(0)\,[1 + V_1(0)t]^{-1/2},$$

as well as $V_0(t)$.

6.18 Consider a steady stream of liquid falling from a nozzle of radius r_0 under gravity; the exit velocity is v_0. As before, we use r_0 and $\tau = \sqrt{r_0^3 \rho/\gamma}$ as units of length and time, respectively. Show that for a slender stream (6.57), (6.58), the steady solution (with the addition of gravity pointing in the positive z-direction) is described by

$$vv' = -\frac{v'}{2\mathrm{We}^{1/4}v^{1/2}} + 3\,\mathrm{Oh}\left(v'' - \frac{v'^2}{v}\right) + \mathrm{Bo},$$

where $h = \mathrm{We}^{1/4}/v^{1/2}$ and dimensionless constants are defined as in Exercises 1.2 and 1.3.

(i) Show that, if both surface tension and viscosity can be neglected, the radius of the stream is

$$h(z) = \left(\frac{\mathrm{We}}{\mathrm{We} + 2\,\mathrm{Bo}\, z} \right)^{1/4}.$$

This was known to Newton; interpret your result in terms of a freely falling body.

(ii) In the case where *inertia* can be neglected completely, show that an exact solution can be written in the form

$$v(z) = a(z + b)^2,$$

and determine the constants a, b. Also calculate the profile $h(z)$ of the stream.

6.19 Show that

$$G' = -\left(\frac{h}{(1 + h'^2)^{1/2}} + \frac{h^2 h''}{(1 + h'^2)^{3/2}} \right)' = h^2 \kappa' \tag{6.89}$$

where G is defined in (6.67) and the mean curvature κ is defined by (4.23).

PART II

Formation of singularities

7

Drop breakup

7.1 Overview and dimensional analysis

When a drop forms there must be a singularity of the equations of motion, since the separation of one piece of fluid into two is a discontinuous process. Drop pinch-off is driven by surface tension, since a decrease in radius at sufficiently long wavelengths leads to a reduction in surface energy; see Exercise 6.15. Hence the surface tension γ is bound to appear in any expression describing the limiting case where the radius h goes to zero. In the case where the viscosity is irrelevant, the density ρ is the only other parameter. This leads to the scaling law (1.1), which comes from a balance between surface tension and inertia. We now present a more complete discussion, but still based on simple dimensional considerations.

How does the profile look near the breakup point z_0, t_0 as a function of the external parameters ρ (density), ν (kinematic viscosity), γ (surface tension), g (acceleration of gravity), and r_0 (nozzle radius)? We are interested in the limit where $\Delta t = t_0 - t$ and $\Delta z = z - z_0$ are small. Thus, for $\Delta t \to 0$, a typical length scale characteristic for breakup is $\ell_t = \sqrt{\nu \Delta t}$, keeping in mind that

$$[\nu] = \frac{\text{cm}^2}{\text{s}}, \quad [\gamma/\rho] = \frac{\text{cm}^3}{\text{s}^2}, \quad [g] = \frac{\text{cm}}{\text{s}^2}.$$

Using dimensional analysis the representation

$$h = f(\Delta z, \Delta t, \rho, \nu, \gamma, g, r_0)$$

can be reduced to

$$h = \ell_t \bar{f}\left(\frac{\Delta z}{\ell_t}, \frac{\ell_\nu}{\ell_t}, \frac{\ell_c}{\ell_t}, \frac{r_0}{\ell_t}\right), \tag{7.1}$$

where ℓ_ν is the intrinsic viscous scale defined in (6.37) and $\ell_c = \sqrt{\gamma/(\rho g)}$ is the capillary length (typically 1 mm).

143

Table 7.1 *Physical parameters for water, glycerol, mercury, and golden syrup. The values are quoted from [217] and [144]. The internal scales ℓ_ν and t_ν are calculated from the kinematic viscosity ν and from the ratio of surface tension γ and density ρ.*

	Mercury	Water	Glycerol	Golden syrup
ν [m^2/s]	1.2×10^{-7}	10^{-6}	1.18×10^{-3}	0.051
γ/ρ [m^3/s^2]	3.47×10^{-5}	7.29×10^{-5}	5.3×10^{-5}	5.6×10^{-5}
$\ell_\nu = \rho\nu^2/\gamma$ [m]	4.2×10^{-10}	1.38×10^{-8}	0.0279	46
$t_\nu = \nu^3\rho^2/\gamma^2$ [s]	1.4×10^{-12}	1.91×10^{-10}	0.652	4.3×10^4

Of course one might presume that breakup depends on still other, more detailed, characteristics of the initial state of the drop, but in any case r_0 and ℓ_c are sufficient to determine the equilibrium shape and the initial bifurcation of a hanging drop [58]. We would like to argue that, in the limit $\ell_t \to 0$ (the pinch-off limit), external, macroscopic scales such as ℓ_c or r_0 drop out of the problem so $\bar{f}$ should have a *finite* limit for $\ell_c/\ell_t \to \infty$ and $r_0/\ell_t \to \infty$, and one is left with

$$h = \ell_t \bar{f}\left(\frac{\Delta z}{\ell_t}, \frac{\ell_\nu}{\ell_t}\right). \tag{7.2}$$

Thus even if one restricts oneself to *local* quantities in the description of breakup, the similarity form cannot be found completely from dimensional analysis alone. In fact, we now describe three different self-similar regimes, all of which correspond to the breakup of a liquid bridge in which the local radius goes to zero in finite time but with different balances:

(i) *Generic breakup*   Balance between surface tension, viscosity, and inertia
(ii) *Inviscid breakup*   Balance between surface tension and inertia
(iii) *Viscous breakup*   Balance between surface tension and viscosity

In Table 7.1 we show the value of the intrinsic scales ℓ_ν, t_ν for a variety of liquids. Remarkably, the length ℓ_ν varies from a value below the size of an atom for mercury to 46 m for golden syrup. This extreme variation is the reason why different balances are commonly observed.

7.1.1 Surface tension–viscosity–inertia balance

We now assume that all three intrinsic parameters, γ, ρ, and ν, come into play close to breakup. For simplicity we consider the slender jet equations (6.57), (6.58) instead of the full set of Navier–Stokes equations with boundary conditions. We notice that the problem has a special symmetry. If one substitutes $\tilde{h} = h\rho/\gamma$ then ν is the only parameter left in (6.57), (6.58), giving

$$\frac{\partial \tilde{h}^2}{\partial t} + (\tilde{h}^2 v)_z = 0,$$

$$\frac{\partial v}{\partial t} + vv_z = -\left(\frac{1}{\tilde{h}}\right)_z + 3\nu \frac{(v_z \tilde{h}^2)_z}{\tilde{h}^2},$$

and $\ell_t = \sqrt{\nu \Delta t}$ is the only length that can be constructed using the parameter ν. We have argued before that gravity does not come into play near breakup.

Thus if one writes dimensionally correct expressions for $\tilde{h}$, which has dimensions $[\tilde{h}] = \text{s}^2/\text{cm}^2$, (7.2) can in fact be rewritten as

$$\tilde{h} = \frac{(\Delta t)^2}{\ell_t^2} H\left(\frac{\Delta z}{\ell_t}\right). \tag{7.3}$$

This is an example where power law behavior does not follow directly from dimensional analysis; the scaling exponents are fixed only by an additional scaling symmetry [14]. Nevertheless the scaling is determined by local properties alone, and we thus have an example of self-similarity of the first kind. From (7.3) it follows that

$$h_{\min} \propto \Delta t, \quad \ell_z \propto \Delta t^{1/2}, \tag{7.4}$$

where $h_{\min}$ is the minimum radius and ℓ_z is a typical axial length scale. The scaling exponents are $\alpha = 1$ and $\beta = 1/2$, which are rational values as expected for scaling of the first kind.

7.1.2 Surface tension–inertia balance

Next we consider the case where viscosity is small [165], which implies that ℓ_ν is small compared with a reference scale such as r_0; for example, the intrinsic scale of mercury is $\ell_\nu \approx 4 \times 10^{-10}$m, see Table 7.1. Now $\ell_{\text{in}} = (\gamma \Delta t^2/\rho)^{1/3}$ is an inviscid reference length scale, noting that $[\gamma/\rho] = \text{cm}^3/\text{s}^2$. Thus, instead of (7.1) it is more convenient to write

$$h = \ell_{\text{in}} H_{\text{in}}\left(\frac{\Delta z}{\ell_{\text{in}}}, \frac{\ell_\nu}{\ell_{\text{in}}}, \frac{\ell_c}{\ell_{\text{in}}}, \frac{r_0}{\ell_{\text{in}}}\right). \tag{7.5}$$

We are interested in the limit $\ell_{\text{in}} \to 0$, and we assume (as before) that H_{in} has a finite limit for $\ell_c/\ell_{\text{in}} \to \infty$ and $r_0/\ell_{\text{in}} \to \infty$. In addition, we consider a part of the evolution for which $\ell_\nu/\ell_{\text{in}} \ll 1$. Thus, if H_{in} is also regular for $\ell_\nu/\ell_{\text{in}} \to 0$, we have

$$h = \ell_{\text{in}} H_{\text{in}} \left(\frac{\Delta z}{\ell_{\text{in}}} \right).$$ (7.6)

The scalings of the radial and axial length scales are

$$h_{\text{min}} \propto \Delta t^{2/3}, \quad \ell_z \propto \Delta t^{2/3},$$ (7.7)

which corresponds to our earlier result (1.1). Once more, inviscid pinch-off is governed by self-similarity of the first kind, with scaling exponents $\alpha = 2/3$ and $\beta = 2/3$. Note that if ℓ_{in} becomes of the same order as ℓ_ν then one can no longer take the limit $\ell_\nu/\ell_{\text{in}} \to 0$, and one reverts to (7.3) as we will see below.

7.1.3 Surface tension–viscosity balance

Finally, we consider the limiting case where the viscous scale is much *larger* than all other scales in the problem: $\ell_\nu \gg r_0$. Indeed, for many viscous fluids ℓ_ν can become very large; see Table 7.1. In the absence of inertia, the density ρ drops out of the problem and γ and $\eta = \nu\rho$ are the only intrinsic parameters. Now $[\gamma/\eta] = $ cm/s and thus $\ell_{\text{vis}} = v_\eta \Delta t$ ($v_\eta = \gamma/\eta$) is a convenient local length scale. Following the same logic as before, we write

$$h = \ell_{\text{vis}} H_{\text{vis}} \left(\frac{\Delta z}{\ell_{\text{vis}}}, \frac{\ell_\nu}{\ell_{\text{vis}}}, \frac{\ell_c}{\ell_{\text{vis}}}, \frac{r_0}{\ell_{\text{vis}}} \right)$$ (7.8)

and consider the limit $\ell_{\text{vis}} \to 0$. Previously we have assumed that this limit is regular, so that external parameters such as ℓ_c or r_0 also drop out of the problem. However, we now show that this *cannot* be the case, and the limit $r_0/\ell_{\text{vis}} \to \infty$ is singular. This is a sign of self-similarity of the second kind.

In the limit of negligible inertia we can put $\rho = 0$ in (6.58) and obtain

$$0 = \gamma \frac{h_z}{h^2} + 3\eta \frac{(h^2 v_z)_z}{h^2}.$$ (7.9)

Since there is no inertial force, Newton's third law implies that the total force across the cross section of the liquid thread, the tensile force $T(t)$, must be a constant. The existence of such a constant of integration (which can only depend on time) can be used to integrate (7.9) once. Indeed, if we multiply by h^2 to obtain the total force, and integrate over space, we obtain the force balance along the thread:

$$T = \gamma h + 3\eta v_z h^2.$$ (7.10)

Rearranging, one finds that

$$v_z = \frac{v_\eta}{3} \left[\frac{T(t)}{\gamma h^2} - \frac{1}{h} \right],$$ (7.11)

so v is determined directly from h and one has to solve the system (6.57), (7.11). This system remains invariant under the transformation $z = a\tilde{z}$, $v = a\tilde{v}$, that is, the equations do not prescribe any axial length scale. This means that the right-hand side of (7.8) must depend in some way on the initial conditions, as determined by the external parameters r_0 or ℓ_c, to set the axial length scale.

Let us take r_0 as representative of this external length scale. Assuming that the solution remains self-similar but depends on both the combinations $\Delta z/\ell_{vis}$ and r_0/ℓ_{vis}, it follows that

$$h = \ell_{vis} H_{vis} \left(\frac{\Delta z}{\ell_{vis}} \left(\frac{\ell_{vis}}{r_0} \right)^{1-\beta} \right), \tag{7.12}$$

where the scaling exponent β is not determined by dimensional analysis or symmetry arguments. As anticipated, the limit $\ell_{vis}/r_0 \to 0$ is singular: if it existed, h would have to be constant in space and time. Thus (7.12) is an example of self-similarity of the second kind, found previously for (3.1). In both cases the axial length scale of the local problem is determined by the initial conditions; see (3.22). The characteristic length scales are

$$h_{min} \propto \Delta t, \quad \ell_z \propto \Delta t^{\beta}. \tag{7.13}$$

Below we will calculate β using a detailed analysis of the equations of motion (6.57), (7.11).

7.2 Viscous breakup

7.2.1 Lagrangian transformation

The most detailed understanding of breakup exists for the limit in which viscosity is dominant, i.e. for $\ell_v \gg r_0$. To solve (6.57), (7.11) we transform to Lagrangian coordinates, employing a particle marker as the spatial variable, as follows. Let s mark a fluid volume in the form of a slice $h^2(z, t)dz = ds$, so that

$$V = \pi \int_{z_1}^{z_2} h^2(z, t)dz = \pi \int_{s_1}^{s_2} ds = \pi(s_2 - s_1),$$

and let $z(s, t)$ be the position of the volume s at time t. Then one finds that

$$z_s = \frac{1}{h^2}, \quad z_t = v \tag{7.14}$$

(since s marks a material object), which are the transformations needed to pass from Eulerian coordinates to Lagrangian coordinates or vice versa. To transform the spatial derivatives we use $\partial_z = z_s^{-1}\partial_s$.

Example 7.1 (Mass conservation in Lagrangian coordinates) Note that

$$\partial_t h^2(z(s,t),t)\Big|_s = \partial_t h^2\Big|_z + (h^2)_z z_t,$$

so that, using (7.14), the left-hand side of (6.66) becomes in Lagrangian coordinates

$$\partial_t h^2\Big|_z + \left(h^2 v\right)_z = \left(\frac{1}{z_s}\right)_t - \frac{1}{z_s}\left(\frac{1}{z_s}\right)_s z_t + \frac{1}{z_s}\left(\frac{z_t}{z_s}\right)_s = 0,$$

which is consistent with mass conservation.   □

Example 7.2 (Exponentially shrinking filament) Consider a filament of uniform radius h which is shrinking exponentially,

$$h = h_0 e^{-\lambda t}; \tag{7.15}$$

this law is observed in the pinching of viscoelastic fluids [51]. Then, using the first equation in (7.14) and integrating, we have

$$z = \frac{s}{h_0^2} e^{2\lambda t} + z_0(t), \tag{7.16}$$

where we can choose $z_0(t) = 0$. The particle trajectories diverge exponentially from the stagnation point at $z = s = 0$. The velocity field becomes

$$v = z_t = \frac{2\lambda s}{h_0^2} e^{2\lambda t} = 2\lambda z, \tag{7.17}$$

a stagnation point flow.   □

To obtain the equation of motion for a viscous filament we note that differentiating the second equation in (7.14) with respect to z one obtains

$$v_z = \frac{z_{st}}{z_s}. \tag{7.18}$$

Inserting this on the left-hand side of (7.11) yields

$$\frac{z_{st}}{z_s} = \frac{v_\eta}{3}\left(\frac{T(t)}{\gamma} z_s - z_s^{1/2}\right). \tag{7.19}$$

In Lagrangian coordinates the thread radius is

$$H(s,t) = \frac{1}{\sqrt{z_s}}, \tag{7.20}$$

in terms of which (7.19) becomes

$$H_t(s,t) = \frac{v_\eta}{6}\left(1 - \frac{T(t)}{\gamma H(s,t)}\right). \tag{7.21}$$

In particular (7.21) no longer contains a spatial derivative, which shows once again that it cannot contain an intrinsic axial length scale ℓ_z. Note that had the thread tension T been neglected the radius could only *increase*, and no breakup could occur [79]; see Exercise 7.2.

Equation (7.21) appears to be very similar to the simple model (3.71) studied earlier, but the presence of a time-dependent tension $T(t)$ makes the problem much more subtle [165]. To see this, notice that (7.19) can be cast in the form

$$v_s = \frac{v_\eta}{3}\left(\frac{T(t)}{\gamma H^4(s,t)} - \frac{1}{H^3(s,t)}\right). \tag{7.22}$$

Now imagine that a viscous fluid drop is caught between two solid plates, so that the velocity v is zero at either end. Then integrating (7.22) from one plate to the other, the integral over v_s vanishes and one obtains

$$T(t) = \gamma\left(\int_{s_-}^{s_+}\frac{ds}{H^4}\right)^{-1}\int_{s_-}^{s_+}\frac{ds}{H^3}, \tag{7.23}$$

where s_- and s_+ are the boundaries of the domain. In other words the tension $T(t)$ is determined by the profile $H(s,t)$ in a non-local fashion and cannot be prescribed; when solving (7.21) for the profile, (7.23) has to be satisfied simultaneously.

7.2.2 Similarity solutions

In spite of the complication implied by (7.23), the local self-similar problem (7.21) becomes tractable. We introduce the dimensionless variables $t' = v_\eta\Delta t/r_0$, $s' = s/r_0^3$, where r_0 is some axial reference length, and look for the similarity description

$$H = t'r_0\chi(\zeta), \quad T = t'T_0\gamma r_0, \tag{7.24}$$

with $\zeta = s'/t'^\delta$. The relationship to the exponent β defined in (7.12) is $\beta = \delta - 2$, as can be found from passing from Lagrangian to Eulerian variables. Inserting (7.24) into (7.21) and using (3.6) with $\alpha = 1$, we obtain

$$T_0 = \chi + 6\chi^2 - 6\delta\zeta\chi'\chi, \tag{7.25}$$

where T_0 is a constant and the prime denotes differentiation with respect to the argument. Equation (7.25) is an ODE for χ which we can solve, but we have to determine T_0 self-consistently.

As was the case for (3.8), the solution is determined by a regularity condition. Imposing the symmetry and regularity of χ, we expand $\chi(\zeta)$ in a form analogous to (3.19):

$$\chi_i(\zeta) = \chi_m + \zeta^{2i+2} + O(\zeta^{2i+4}), \quad i = 0, 1, 2, \ldots, \tag{7.26}$$

where we have normalized the coefficient of ζ^{2i+2} to unity; this is consistent since any solution of (7.11) is determined only up to a scale factor. The order of the minimum, which is set by the initial conditions, determines which solution is selected, exactly as in the much simpler problem (3.1). Each choice of i corresponds to one member of an infinite sequence of similarity solutions. Again the quadratic minimum ($i = 0$) is expected to be the generic case. Inserting (7.26) into (7.25) we obtain

$$\chi_m + 6\chi_m^2 = T_0, \quad 1 + 12\chi_m - 12\delta(i+1)\chi_m = 0,$$

at orders ζ^0 and ζ^{2i+2}, respectively. Putting $\bar{\delta} = (i+1)\delta$ yields

$$\chi_m = \frac{1}{12(\bar{\delta}-1)}, \quad T_0 = \frac{1}{24}\frac{2\bar{\delta}-1}{(\bar{\delta}-1)^2}. \tag{7.27}$$

Normalizing χ by its minimum, $\chi = \chi_m f$, (7.25) becomes

$$6\delta f \frac{df}{d\ln\zeta} = 6f^2 + \frac{f}{\chi_m} - \frac{T_0}{\chi_m^2}. \tag{7.28}$$

Using $T_0/\chi_m^2 = 6(2\bar{\delta}-1)$, (7.28) can be integrated to give

$$\int \frac{f\,df}{12(\bar{\delta}-1)f + 6f^2 - 6(2\bar{\delta}-1)} = \frac{1}{6\delta}\ln\zeta^{i+1} + C,$$

with C an arbitrary constant. Computing the integral above we obtain

$$\left(f + 2\bar{\delta} - 1\right)^{(2\bar{\delta}-1)/2}(f-1)^{1/2} = \zeta^{i+1}, \tag{7.29}$$

where we find $C = 0$ by comparison with (7.26). Equation (7.29) determines the ith similarity profile, $\chi_i(\zeta) = \chi_m f_i(\zeta)$, implicitly.

To determine the exponent δ we use the constraint (7.23), which in similarity variables becomes

$$T_0 = \chi_m \left(\int_{-\infty}^{\infty} \frac{d\zeta}{f^4}\right)^{-1} \int_{-\infty}^{\infty} \frac{d\zeta}{f^3}; \tag{7.30}$$

it is crucial to note that the integrals are dominated by local contributions near the singularity, so that s_-, s_+ do not appear. To compute the integrals in (7.30) we transform the integration variable from ζ to f, using

$$\int_{-\infty}^{\infty} \frac{d\zeta}{f^i} = 2\int_1^{\infty} \frac{df}{f_\zeta f^i},$$

where from (7.28) f_ζ can be expressed as

$$f_\zeta = \frac{(f + 2\bar{\delta} - 1)(f-1)}{\delta\zeta f}. \tag{7.31}$$

To eliminate ζ from (7.31) we use (7.29), which gives

$$f_\zeta^{-1} = \delta f \left(f + 2\bar{\delta} - 1\right)^{(2\bar{\delta}-1)/[2(i+1)]-1} \left(f - 1\right)^{1/[2(i+1)]-1}. \qquad (7.32)$$

Using (7.27), (7.30) can now be cast in the form

$$\int_1^\infty \left(\frac{2(1-\bar{\delta})}{f^2} + \frac{2\bar{\delta}-1}{f^3}\right)$$
$$\times \left(f + (2\bar{\delta}-1)\right)^{2\bar{\delta}-1/[2(i+1)]-1} \left(f - 1\right)^{1/[2(i+1)]-1} df = 0, \qquad (7.33)$$

which is the equation that determines δ. The integral (7.33) has the same structure as the following integral from [96]:

$$\int_1^\infty x^{-\lambda} (x - 1)^{\mu-1} (x + \beta)^\nu dx$$
$$= \frac{\Gamma(\lambda - \mu - \nu)\Gamma(\mu)}{\Gamma(\lambda - \nu)} F(-\nu, \lambda - \mu - \nu; \lambda - \nu; -\beta), \qquad (7.34)$$

where Γ is the gamma function and F the hypergeometric function, and so (7.33) may be written as

$$K_i(\delta)$$
$$= 2(1-\bar{\delta}) \frac{\Gamma(3-\delta)\,\Gamma\left(\frac{1}{2i+2}\right)}{\Gamma\left(3-\delta+\frac{1}{2i+2}\right)} F\left(1 - \frac{2\bar{\delta}-1}{2i+2}, 3 - \delta; 3 - \frac{2\bar{\delta}-1}{2i+2}; 1 - 2\bar{\delta}\right)$$
$$+ (2\bar{\delta}-1) \frac{\Gamma(4-\delta)\,\Gamma\left(\frac{1}{2i+2}\right)}{\Gamma\left(4-\delta+\frac{1}{2i+2}\right)} F\left(1 - \frac{2\bar{\delta}-1}{2i+2}, 4 - \delta; 4 - \frac{2\bar{\delta}-1}{2i+2}; 1 - 2\bar{\delta}\right)$$
$$= 0. \qquad (7.35)$$

For each i (7.35) is a nonlinear equation for the exponent δ_i which has to be solved numerically; the Eulerian exponent in (7.12) is found from $\beta_i = \delta_i - 2$. The minimum thread radius $\chi_m^{(i)}$, defined by (7.26), can be calculated from (7.27). Solutions for the first few values of i are quoted in Table 7.2.

To summarize, we have found a semi-infinite hierarchy of similarity solutions $i = 0, 1, 2, \ldots$ characterized by exponents $\delta_i = 2 + \beta_i$. However, only the case $i = 0$ is stable, as is argued in more detail in Exercises 7.3 and 7.4. Each exponent δ_i corresponds to a similarity profile $f_i(\zeta)$ to be found from (7.29). To convert to Lagrangian variables we write (7.12) as

$$h(z, t) = r_0 t'^\beta \phi_{\text{vis}}^{(i)}(\xi), \quad \xi = \frac{z'}{t'^\beta}, \qquad (7.36)$$

Table 7.2 *A list of exponents* $\beta_i = \delta_i - 2$, *found from* $K_i(\delta) = 0$ *using MAPLE, with* K_i *given by (7.35). The number* $2i + 2$ *gives the smallest non-vanishing power in a series expansion of the corresponding similarity solution around the origin. Only the solution with* $i = 0$ *is stable. The rescaled minimum radius* χ_m *is found from (7.27).*

i	β_i	χ_m
0	$1.748\,717 \times 10^{-1}$	0.0709
1	$4.538\,527 \times 10^{-2}$	0.0797
2	$1.944\,257 \times 10^{-2}$	0.0817
3	$1.053\,085 \times 10^{-2}$	0.0825
4	$6.530\,260 \times 10^{-3}$	0.0828
5	$4.419\,666 \times 10^{-3}$	0.0832

and identify

$$z' = \frac{\Delta z}{r_0}, \quad t' = \frac{v_\eta \Delta t}{r_0}.$$

Integrating (7.20) we find

$$\Delta z = \int_0^s H^{-2}(\bar{s}, t)d\bar{s}, \tag{7.37}$$

which, in similarity variables (using (7.24) and (7.36)), reads

$$\xi = \frac{1}{\chi_m^2} \int_0^\zeta \frac{d\zeta}{f^2}.$$

Finally, we change variables, using (7.32), to obtain an implicit equation for the similarity profile $\phi_{\text{vis}}^{(i)}$:

$$\frac{\xi}{\bar{\xi}} = \int_1^{\phi_{\text{vis}}^{(i)}/\chi_m^{(i)}} \frac{(f + 2\bar{\delta}_i - 1)^{(2\bar{\delta}_i - 1)/[2(i+1)]-1}}{f(f-1)^{1-1/[2(i+1)]}} df. \tag{7.38}$$

We have introduced a scale factor $\bar{\xi}$ to account for the fact that the axial length scale depends on the initial conditions.

Notice that the axial exponent $\beta_0 \approx 0.175$ of the ground state solution is quite small, so for $t' \to 0$ the typical radial scale t' is much smaller than the axial scale $(t')^{0.175}$; this explains the long neck seen in Fig. 7.1. The profile is symmetric about the minimum at $\xi = 0$, another feature seen clearly in Fig. 7.1. In addition, the slenderness parameter ϵ of the slender jet expansion (see Section 6.2) can be identified as $\ell_r/L = \epsilon \propto (t')^{1-\beta}$, which goes to zero in the limit $t' \to 0$. Thus the leading-order equations, on which our analysis is based, become exact in this limit.

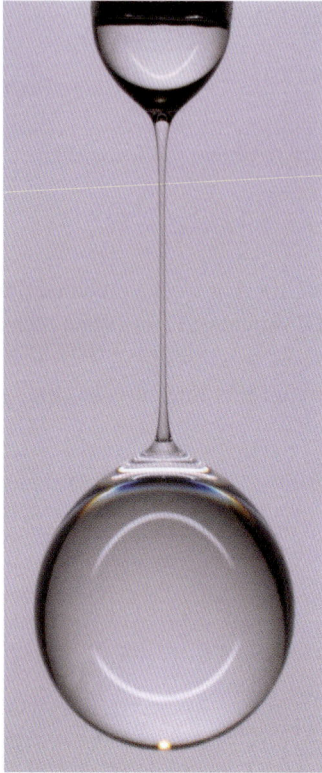

Figure 7.1 A drop of viscous fluid falling from a pipette; note the long neck. Image courtesy of Nick Laan and Daniel Bonn.

As we have argued, the breakup of axisymmetric fluid necks is driven by the reduction in surface energy which comes from the reduction in radius. In the case of a liquid sheet, however, whose equations of motion (6.84), (6.85) were derived in Exercise 6.12, any modulation of the thickness only leads to an increase in surface area. Hence no breakup can occur in the viscously overdamped case, as the following example shows.

Example 7.3 (Viscous liquid sheet with surface tension) In the absence of inertia we can drop the left-hand side of (6.85). Multiplying by h and using the identity

$$ hh_{xxx} = \left(hh_{xx} - \frac{1}{2}h_x^2 \right)_x , $$

which is an analogue of (6.89), one obtains

$$0 = \gamma \left(h h_{xx} - \frac{1}{2} h_x^2 \right)_x + 4\eta \left(u_x h \right)_x .$$

Now we can integrate once to obtain the force balance

$$T(t) = \gamma \left(h h_{xx} - \frac{1}{2} h_x^2 \right) + 4\eta u_x h, \qquad (7.39)$$

which is an analogue of (7.10). Far from a potential pinch-off point, the velocity should go to zero and h should go to a constant, which means that $T(t) = 0$ for a liquid sheet, as opposed to a thread.

If we look at the minimum film thickness $h_{\min}(t)$, we find from (6.84) that on the one hand

$$h_{\min,t} + u_x(x_{\min}, t) h_{\min} = 0, \qquad (7.40)$$

where $x_{\min}$ is the position of $h_{\min}$. On the other hand, using $T(t) = 0$ we have

$$u_x(x_{\min}, t) = -\frac{\gamma}{4\eta} h_{\min} h_{xx}(x_{\min}, t) \leq 0,$$

since $h_{xx}(x_{\min}, t) \geq 0$. Therefore, by (7.40),

$$h_{\min,t} \geq 0$$

so the minimum film thickness cannot decrease and hence breakup can never occur. $\qquad \square$

7.3 Generic breakup

Let us now return to the generic case where surface tension, inertia, and viscosity all come into play and the scaling is determined by (7.4). Then we obtain for the local radius and the velocity

$$h(z, t) = \ell_v t' \phi \left(\frac{z'}{t'^{1/2}} \right), \quad v(z, t) = \frac{\ell_v}{t_v} (t')^{-1/2} \psi \left(\frac{z'}{t'^{1/2}} \right), \qquad (7.41)$$

where $z' = \Delta z / \ell_v$, $t' = \Delta t / t_v$ are dimensionless length and time scales, defined with respect to the intrinsic scales (6.37).

Note that (7.41) could also have been obtained by making the generic similarity assumption (3.47) for the variables h, v:

$$h(z, t) = \ell_v t'^{\alpha_1} \phi(\xi), \quad v(z, t) = \frac{\ell_v}{t_v} t'^{\alpha_2} \psi(\xi), \quad \xi = z' / t'^\beta. \qquad (7.42)$$

Inserting (7.42) into (6.57) and using (3.6) and (3.12) we obtain

$$\frac{t_v}{\ell_v}\left(\frac{\partial h}{\partial t}+vh'+\frac{hv'}{2}\right)$$

$$=(t')^{\alpha_1-1}\left[-\alpha_1\phi+\beta\xi\phi'\right]+(t')^{\alpha_1+\alpha_2-\beta}\left(\psi\phi'+\frac{\phi\psi'}{2}\right)=0;\quad(7.43)$$

so to balance the two powers of t' we must have $\beta=1+\alpha_2$. Similarly, from (6.58) we find

$$\frac{t_v^2}{\ell_v}\left(\frac{\partial v}{\partial t}+vv'+\frac{\gamma}{\rho}\frac{h'}{h^2}+3v\frac{(h^2v')'}{h^2}\right)$$

$$=(t')^{\alpha_2-1}\left(-\alpha_2\psi+\beta\xi\psi'\right)+(t')^{2\alpha_2-\beta}\psi\psi'-(t')^{-\alpha_1-\beta}\frac{\phi'}{\phi^2}$$

$$-(t')^{\alpha_2-2\beta}3\frac{(\phi^2\psi')'}{\phi^2}=0.\qquad(7.44)$$

While the first two powers of t' (describing the fluid acceleration) balance automatically, a balance of inertia and surface tension (the first and third terms) yields $\alpha_2-1=-\alpha_1-\beta$ and thus $\alpha_1=-2\alpha_2$. Balancing inertia and viscosity (the first and fourth terms) gives $\alpha_2-1=\alpha_2-2\beta$, and so $\beta=1/2$. It follows that the other exponents are $\alpha_1=1$ and $\alpha_2=-1/2$, in agreement with (7.41). The *slenderness parameter* ϵ can be identified as $\epsilon\propto t'/t'^{1/2}=t'^{1/2}$, so that (6.57), (6.58) become exact in the limit $t'\to0$.

Note that each term in (7.44) now blows up as $(t')^{-3/2}$; had we kept gravity in the balance, as for example in (6.59), it would clearly have been subdominant in the limit $t'\to0$. Once more the principle of *dominant balance* is at work here, showing that gravity is not important for pinch-off as was implicit in our earlier assumption that ℓ_c drops out from (7.1) in the limit $t'\to0$. Similarly, only the term $\gamma/h\propto(t')^{-1}$ survives from the pressure term in (6.59); all other terms are of lower order, so one recovers equation (6.58) to leading order.

Using the exponents obtained above, (7.43) and (7.44) yield the similarity equations

$$-\phi+\frac{\xi\phi'}{2}=-\psi\phi'-\frac{\psi'\phi}{2},\qquad(7.45)$$

$$\frac{\psi}{2}+\frac{\xi\psi'}{2}=-\psi\psi'+\frac{\phi'}{\phi^2}+3\frac{(\psi'\phi^2)'}{\phi^2},\qquad(7.46)$$

where $\xi=z'/(t')^{1/2}$. This is a system of ordinary differential equations of first order in ϕ and second order in ψ. Three initial conditions are needed to specify a solution locally, corresponding to a large space of possible pinching profiles. Below we will show that the usual matching conditions, which are boundary conditions for the system (7.45), (7.46), select a *discrete* set of similarity solutions, just as in the previous viscous problem.

Returning to our example of the breakup of a fluid sheet, we see that the existence of self-similar solutions is not guaranteed. When trying to balance surface tension with viscosity and inertia, one finds that there are no solutions.

Example 7.4 (Fluid sheet with surface tension, viscosity, and inertia) We look for similarity solutions of (6.84) and (6.85), assuming the scalings $h \propto t'^{\alpha_1}$, $u \propto t'^{\alpha_2}$, and $x \propto t'^{\beta}$. Then the scalings of the two terms of (6.84) with respect to t' are $(t')^{\alpha_1 - 1}$ and $(t')^{\alpha_1 + \alpha_2 - \beta}$. Equating the exponents we obtain $\alpha_1 - 1 = \alpha_1 + \alpha_2 - \beta$, or $\beta = \alpha_2 - 1$. In the same way the exponents appearing in the force balance (6.85) are $\alpha_2 - 1$, $2\alpha_2 - \beta$, $\alpha_1 - 3\beta$, and $\alpha_2 - 2\beta$. Using the value of β obtained from mass conservation, the first two terms, coming from inertia, are automatically of the same order, as in the case of the jet. Comparing the first term (inertia) and the last (viscosity), we find $\alpha_2 - 1 = -\alpha_2 - 2$, and so $\alpha_2 = -\beta = -1/2$, as for the axisymmetric case. However, a balance of the exponent coming from surface tension, $\alpha_1 - 3\beta = \alpha_1 - 3/2$ with $\alpha_2 - 1 = -3/2$ yields $\alpha_1 = 0$. But this means that the sheet thickness would remain asymptotically constant, and no similarity solutions corresponding to pinch-off exist. As we will see below, pinch-off solutions exist only as long as viscosity remains subdominant. □

7.3.1 The universal solution

Behavior at infinity

To find the boundary conditions for $\xi \to \pm\infty$ we have to apply the matching rule (3.50) to the similarity functions ϕ and ψ. This gives

$$\left. \begin{array}{l} \dfrac{\phi(\xi)}{\xi^{\alpha_1/\beta}} = \dfrac{\phi(\xi)}{\xi^2} \to a_0^{\pm} \\[2mm] \dfrac{\psi(\xi)}{\xi^{\alpha_2/\beta}} = \psi(\xi)\xi \to b_0^{\pm} \end{array} \right\} \quad \text{for} \quad \xi \to \pm\infty. \qquad (7.47)$$

In other words, the function ϕ must grow quadratically at infinity but the prefactor will be different on either side of the singularity if the solution is non-symmetric (which turns out to be the case). Similarly, ψ must decay as $1/\xi$ at infinity.

Let us discover which solutions of (7.45), (7.46) have an asymptotic behavior consistent with (7.47). We will try the asymptotic expansions

$$\phi(\xi) = \xi^2 \sum_{i=0}^{\infty} a_i \xi^{-2i}, \quad \psi(\xi) = \frac{1}{\xi} \sum_{i=0}^{\infty} b_i \xi^{-2i}, \qquad (7.48)$$

which have (7.47) as their leading-order behavior. Inserting (7.48) into the equations, and expanding all terms in a powers series in ξ^{-1}, one finds a system of equations for the coefficients a_i, b_i. This is a tedious but ultimately simple calculation, which can be pursued recursively to arbitrarily high order. The main conclusion is that there are only *two* free parameters, and all coefficients are determined in terms of a_0 and b_0 alone.

Example 7.5 (Expansion at infinity) Inserting (7.48) into (7.45), we obtain

$$-\phi + \frac{\xi\phi'}{2} = a1 + O(\xi^{-2}), \quad -\psi\phi' - \frac{\psi'\phi}{2} = 2a_0b_0 - \frac{a_0b_0}{2} + O(\xi^{-2})$$

for the left- and right-hand sides of (7.45), respectively. The leading-order behavior $O(\xi^2)$ of the left-hand side, which comes from the time derivative, vanishes in accordance with the matching condition (7.47). Equating the two expressions we have $a_1 = 3b_0a_0/2$. Similarly, inserting (7.48) into (7.46) we have

$$\frac{\psi}{2} + \frac{\xi\psi'}{2} = \frac{b_1}{2\xi^3} - \frac{3b_1}{2\xi^3} + O(\xi^{-5}),$$

$$-\psi\psi' + \frac{\phi'}{\phi^2} + 3\frac{(\psi'\phi^2)'}{\phi^2} = \frac{b_0^2}{\xi^3} + \frac{2}{a_0\xi^3} - \frac{6b_0}{\xi^3} + O(\xi^{-5}).$$

Thus we obtain the first two coefficients,

$$a_1 = \frac{3b_0a_0}{2}, \quad b_1 = -b_0^2 - \frac{2}{a_0} + 6b_0, \qquad (7.49)$$

in terms of a_0, b_0. This can be pursued to any order; see Exercise 7.5 for the expressions for a_2 and b_2. □

This shows that only a two-dimensional submanifold of solutions of (7.45), (7.46), parameterized by a_0 and b_0, has the correct asymptotics for $\xi \to \infty$ and $\xi \to -\infty$, respectively. To demonstrate that this amounts to the *selection* of a subclass of solutions, one has to consider the stability of the asymptotics (7.48). By perturbing around (7.48) we recover the full three-dimensional space of solutions. If trajectories are driven away from (7.48), a constraint is put on the allowed solution: one parameter must be tuned to exactly the right value in order to be consistent with the matching condition (7.47). However, note that if the manifold (7.48) is attractive (as turns out to be the case for the similarity solution valid *after* pinch-off), all solutions of (7.45), (7.46) automatically have the asymptotic behavior (7.47).

To perturb around (7.48) we write

$$\phi(\xi) = \phi_0(\xi) + \epsilon_1(\xi), \quad \psi(\xi) = \psi_0(\xi) + \epsilon_2(\xi), \qquad (7.50)$$

where ϕ_0, ψ_0 is a solution of the form (7.48), and linearize in ϵ_1 and ϵ_2. The resulting linear system with non-constant coefficients is

$$-\epsilon_1 + \frac{\xi}{2}\epsilon_1' = -\psi_0\epsilon_1' - \phi_0'\epsilon_2 - \frac{\psi_0'}{2}\epsilon_1 - \frac{\phi_0}{2}\epsilon_2', \qquad (7.51)$$

$$\frac{\epsilon_2}{2} + \frac{\xi}{2}\epsilon_2' = -\psi_0\epsilon_2' - \psi_0'\epsilon_2 + \frac{\epsilon_1'}{\phi_0^2} - 2\frac{\phi_0'}{\phi_0^3}\epsilon_1$$

$$+3\epsilon_2'' + 6\frac{\phi_0'}{\phi_0}\epsilon_2' + 6\frac{\psi_0'}{\phi_0}\epsilon_1' - 6\frac{\psi_0'\phi_0'}{\phi_0^2}\epsilon_1.$$

Since the coefficients depend on ξ, the solutions are not simple exponentials. Instead, we need to try a WKB-type analysis [209], in which the exponent contains an arbitrary function $\chi(\xi)$ which needs to be found self-consistently. This means that we write ϵ_1, ϵ_2 in the form

$$\epsilon_1(\xi) = [\exp\chi(\xi)]\sum_{i=0}^{\infty} f_i\xi^{\nu-2i}, \quad \epsilon_2(\xi) = [\exp\chi(\xi)]\sum_{i=0}^{\infty} g_i\xi^{\mu-2i}. \quad (7.52)$$

We anticipate that χ grows faster than linearly at infinity, so the leading-order behavior is generated by the highest derivatives. For example,

$$[\exp\chi(\xi)]' = \chi'\exp\chi(\xi),$$
$$[\exp\chi(\xi)]'' = (\chi'^2 + \chi'')\exp\chi(\xi) \approx \chi'^2\exp\chi(\xi).$$

Inserting this into the second equation in (7.51) one observes that the second term on the left and the fifth on the right grow fastest, leading to the balance

$$\frac{\xi}{2}\chi' = 3\chi'^2,$$

the solution of which is

$$\chi = \frac{\xi^2}{12}, \qquad (7.53)$$

which indeed grows faster than linear. There also is a solution $\chi' = 0$, which corresponds to a change in the parameters of the base solution.

Similarly, from the first equation in (7.51) one can identify the second term on the left and the last on the right as dominant, which leads to

$$f_0\xi^2\xi^\nu = -a_0\xi^3 g_0\xi^\mu.$$

and so $\nu = \mu + 1$ and $f_0 = -a_0 g_0$. The exponent μ is determined from the next order of the second equation in (7.51), which eventually leads to $\mu = b_0/3 - 4$; see Exercise 7.6. A careful analysis to all orders reveals that from the three parameters (a_0, b_0, f_0) all parameters contained in the expansion (7.52) can be determined. This corresponds exactly to the three degrees of freedom expected to exist for the third-order system (7.45), (7.46). Since the sign of (7.53) is positive, the solution will be carried away from the asymptotic behavior (7.47) *unless* $f_0 = 0$. Thus each of the two conditions (7.47) corresponds to a condition on the solution,

$$f_0^{\pm} = 0. \tag{7.54}$$

Behavior near the sonic point

Now we look at the interior of the domain. Just as in the exactly solvable case of a viscous thread there are regularity conditions analogous to (7.26) to be met, which will generate a discrete sequence of solutions. Namely, rewriting (7.45) in the form

$$\phi' = \phi \frac{1 - \psi'/2}{\psi + \xi/2}, \tag{7.55}$$

one sees that there must be a point ξ_0 with

$$\psi(\xi_0) + \xi_0/2 = 0 \tag{7.56}$$

since ξ varies from $-\infty$ to ∞. This indicates that at ξ_0 there is a movable singularity of the equations. Indeed, closer inspection shows that there is a local expansion of the form

$$\phi = \sum_{j=0}^{\infty} \phi_j^{(s)} (\xi - \xi_0)^{(j+1)/2}, \quad \psi = \sum_{j=0}^{\infty} \psi_j^{(s)} (\xi - \xi_0)^{j/2-1}, \tag{7.57}$$

which has a simple pole in ψ as well as algebraic branch points. One can also verify that, apart from the value of ξ_0, $\phi_0^{(s)}$ and $\psi_2^{(s)}$ are freely choosable, which corresponds to the expected three degrees of freedom.

The physical significance of ξ_0 is revealed when considering the equation of motion for a Lagrangian "marker" z', which follows from the second equation in (7.14):

$$\partial_t z'(t) = v(z'(t), t).$$

Rewriting in self-similar coordinates ξ and $\tau = -\ln t'$, one finds

$$\partial_\tau \xi = \xi/2 + \psi(\xi), \tag{7.58}$$

which is the convection equation in similarity variables. Hence, at the point ξ_0, as defined by (7.56), the marker is at rest. Regularity properties on such "stagnation" or "sonic" [145] points often play a similar role in selection.

To avoid the singular solution (7.57), equation (7.55) shows that the condition

$$\psi'(\xi_0) = 2 \tag{7.59}$$

must also be met. In that case one obtains the regular expansions

$$\phi(\xi) = \sum_{i=0}^{\infty} \phi_i (\xi - \xi_0)^i,$$

$$\psi(\xi) = \sum_{i=0}^{\infty} \psi_i (\xi - \xi_0)^i, \tag{7.60}$$

with $\psi_0 = -\xi_0/2$ and $\psi_1 = 2$. Apart from ξ_0, all coefficients in the expansions (7.60) can now be determined in terms of the single parameter $\phi_0 = \phi(\xi_0)$.

Example 7.6 (Expansion at the sonic point) Inserting (7.60) into (7.45), and expanding in $\hat{\xi} = \xi - \xi_0$, one obtains at orders $\hat{\xi}^0$ and $\hat{\xi}^1$, respectively:

$$-\phi_0 = -\frac{\xi_0\phi_1}{2} + \frac{\xi_0\phi_1}{2} - \phi_0, \quad -\phi_1 + \frac{\phi_1}{2} = -\xi_0\phi_2 - 2\phi_1 + \xi_0\phi_2 - \phi_0\psi_2 - \phi_1.$$

The first equation is satisfied identically, and the second yields $\psi_2 = -5\phi_1/(2\phi_0)$. Similarly, from (7.46) one obtains at order $\hat{\xi}^0$

$$-\frac{\xi_0}{4} + \xi_0 = \xi_0 + \frac{\phi_1}{\phi_0^2} + 6\psi_2 + 12\frac{\phi_1}{\phi_0},$$

from which ψ_2 can be eliminated to yield

$$\phi_1 = \frac{\xi_0\phi_0^2}{4(3\phi_0 - 1)}, \quad \psi_2 = -\frac{5\phi_1}{2\phi_0}, \tag{7.61}$$

and so on to arbitrarily high orders.

Selection

Now we put everything together in order to describe the selection mechanism. First, from the local expansion near the sonic point we know that all solutions regular in the interior of the domain can be parameterized in terms of two numbers (ξ_0, ϕ_0), represented as a point in the phase plane of Fig. 7.2. However, in

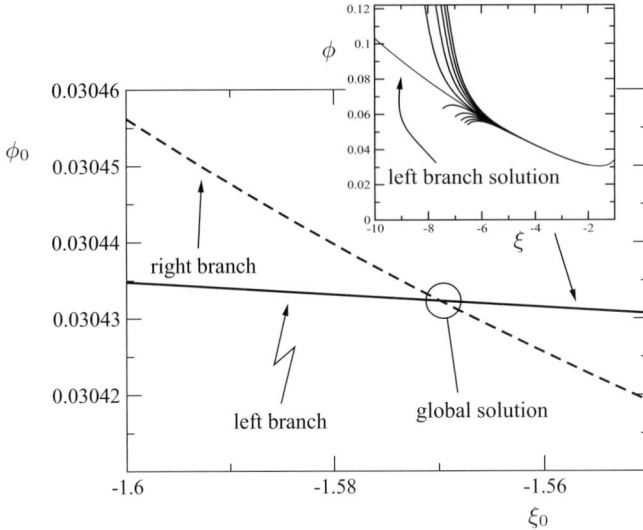

Figure 7.2 The selection mechanism for the universal similarity functions shown in Fig. 7.3; ξ_0 and ϕ_0 are the expansion parameters of (7.60). The solid and broken lines correspond to solutions of (7.45), (7.46) extending to $-\infty$ and ∞, respectively. The crossing point corresponds to a global solution. The inset illustrates the selection at a particular point of the left branch.

general these solutions will not satisfy the boundary conditions (7.47) at infinity, each of which correspond to one constraint on the solution. As a result, solutions which satisfy the condition $f_0^- = 0$ in (7.54), for the left branch, correspond to a line in the $\xi_0\phi_0$-plane (the solid line in Fig. 7.2). The same is true for the right branch, represented as the broken line in Fig. 7.2. The pair of values where the two lines cross evidently corresponds to a solution which satisfies both boundary conditions and thus represents the global solution shown in Fig. 7.3.

To implement this idea numerically one starts with a pair of values (ξ_0, ϕ_0) from which the solution can be integrated outwards to $+\infty$ and $-\infty$. Generically the solution will not be consistent with (7.54); only a one-dimensional submanifold of (ξ_0, ϕ_0) will lead to the correct asymptotic behaviors at $+\infty$ and $-\infty$, respectively. This is illustrated in the inset of Fig. 7.2 for the neighborhood of a particular point on the left branch solution. For a fixed value of ξ_0 the profile ϕ veers off below or above the correct curve if ϕ_0 is chosen too small or too large. By bisection one finds the value $\phi_0(\xi_0)$ belonging to the left branch curve. This procedure of adjusting an initial condition in order to satisfy a boundary condition at another point is known as the *shooting method* [179]. The right branch can be found similarly, by integrating to the right.

Drop breakup

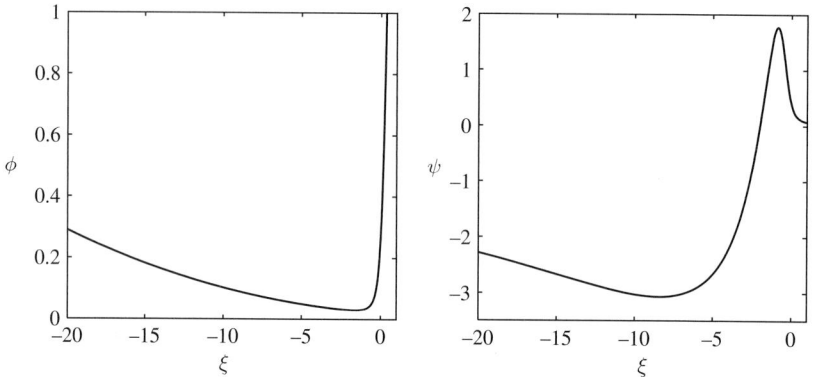

Figure 7.3 The ground state similarity functions ϕ and ψ corresponding to stable self-similar dynamics. Note the strong asymmetry.

A more detailed analysis [34] of the solution curves $\phi_0(\xi_0)$ reveals that the crossing shown in Fig. 7.2 is merely the beginning of a sequence which belongs to an infinite series of solutions. Thus the situation is strikingly similar to other similarity problems we have discussed. In particular, for viscous pinch-off we found the infinite sequence (7.36) of similarity solutions, where the ith similarity profile is given by (7.38). Another related problem is the pinch-off for a thread under surface diffusion, studied in [19], for which also a discrete sequence of solutions was found. All the available numerical evidence [74] suggests that only the first solution in the sequence is linearly stable and thus physically realizable. The same is true for motion by surface diffusion, driven by surface tension. It is tempting to think that there is a structural reason for the fact that there is precisely one stable solution in all those cases, but such an argument remains to be discovered.

The "ground state" solution $\phi(\xi)$ may be compared with an experimental sequence of photographs of a viscous jet near breakup, shown in Fig. 7.4. As the singularity time t_0 is approached, the thread thickness shrinks in proportion to the time distance t' to the singularity. Since the axial scale ℓ_z of the solution only shrinks as $t'^{1/2}$ the fluid neck appears increasingly elongated, leading to a thin thread connecting two drops (of which only one is shown). Another striking feature of the solution is that it is very asymmetric: the constants a_0^+ and a_0^- (reported in Table 7.3) which govern the growth of the profile at infinity differ by about four orders of magnitude. On the steep (+) side (seen on the right in Fig. 7.4) the thread thickens rapidly to merge into a drop. On the other (−) side the thickening is so gradual so as to be almost imperceptible, leading to the appearance of a thread of almost uniform thickness toward the left in Fig. 7.4.

Table 7.3 *Some characteristics of the similarity functions ϕ, ψ before breakup. The first row corresponds to the functions represented in Fig. 7.3, the second row to the most stable of the additional solutions found in [34]. The symbol ξ_0 stands for the position of the stagnation point, where the fluid is at rest in the frame of reference of the interface. The minimum value ϕ is $\phi_{\min}$. The function ψ reaches a maximum (negative) value $\psi_{\max}$. The numbers $a_0^\pm$ stand for the limits $\lim_{\xi \to \pm\infty} \phi(\xi)/\xi^2$ and the numbers $b_0^\pm$ for the limits $\lim_{\xi \to \pm\infty} \psi(\xi)\xi$. All numbers are accurate to the decimal places shown.*

ξ_0	$\phi_{\min}$	$\psi_{\max}$	a_0^+	b_0^+	a_0^-	b_0^-
-1.5699	$0.030\,426$	-3.066	4.635	0.0723	6.047×10^{-4}	57.043
-1.8140	$0.010\,785$	-4.698	52.75			

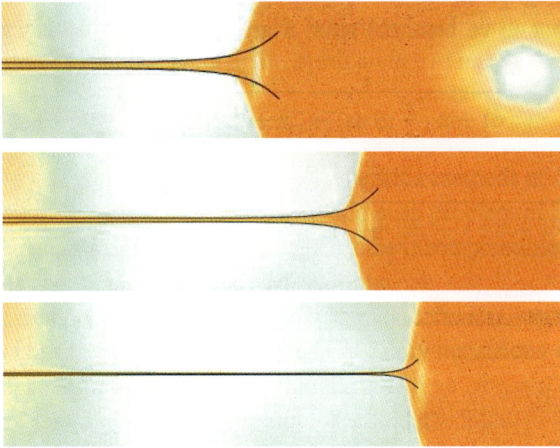

Figure 7.4 A sequence of interface profiles of a jet of glycerol close to the point of breakup [126] (the center of the drop being formed is seen as the bright spot in the top picture). The experimental images correspond to $t_0 - t = 350\,\mu s$, $298\,\mu s$, and $46\,\mu s$ (from top to bottom). The corresponding analytical solutions based on (7.46), (7.45) are superimposed [74]. There is no adjustable parameter in the comparison.

It is important to appreciate that all the parameters characterizing the solution are universal and so are independent of the initial conditions. For example, at a time Δt away from the singularity the minimum radius of a fluid thread is $h_{\min} = 0.0304\gamma\,\Delta t/\eta$, and this is independent of the initial jet radius, a fact that has been tested experimentally [126]. Thus Fig. 7.4 would look much the same for a jet whose initial radius was twice as large. The asymmetry of the solution (as reflected for example by the great disparity between the constants

Figure 7.5 The same experiment as in Fig. 7.4, but with a fluid of higher viscosity, $\ell_\nu = 2$ mm, $t_\nu = 12.8$ ms [126]. The width of the frame corresponds to 2 mm. Just before breakup, the thin thread becomes unstable and produces several "bumps". The moment of breakup shows irregular breakup at several places.

a_0^+ and a_0^-) is also not influenced by the initial condition. Thus the typical configuration, that of a thread (the flat side of the solution) matching on to a drop at the steep side of the solution, cannot be influenced by the choice of initial condition. As a result the appearance of *satellite drops*, which are a remnant of the fluid neck between two main drops, are a universal feature of the nonlinear dynamics of drop breakup.

7.3.2 Stability

A close inspection of the lowest image in Fig. 7.4 in the absence of the added lines would reveal that the interface has become quite wavy and irregular. This becomes even more apparent when the viscosity is higher; see Fig. 7.5. Clearly the similarity solution is prone to the growth of localized perturbations or "bumps". It seems as if this should be expected from the stability analysis of a liquid cylinder, which the solution very much resembles as it becomes thinner. As shown in Exercise 6.15, perturbations can grow as long as the wavelength is greater than 2π times the radius, which is of order t'. Thus one might think that the similarity solution is linearly unstable. The reason why this is not the case is that the *flow* in the thread has a stabilizing effect: the perturbations are both stretched and also convected out of the pinch region, where they no longer grow. Only perturbations which are sufficiently large initially will affect the similarity solution. This leads to a nonlinear instability, which means that perturbations of *finite* amplitude are necessary to destabilize the similarity solution [35, 197, 34].

To investigate the stability around a solution of (7.45), (7.46) we construct the corresponding dynamical system (3.52), using the self-similar variables $\tau = -\ln t'$ and $\xi = z'/t'^{1/2}$:

$$h(z', t') = t'\phi(\xi, \tau), \quad v(z', t') = (t')^{-1/2}\psi(\xi, \tau). \tag{7.62}$$

Inserting (7.62) into (6.58) and (6.57) one obtains

$$\partial_\tau \phi = \phi - \frac{\xi \phi'}{2} - \psi \phi' - \frac{\psi' \phi}{2}, \tag{7.63}$$

$$\partial_\tau \psi = -\frac{\psi}{2} - \frac{\xi \psi'}{2} - \psi \psi' + \frac{\phi'}{\phi^2} + 3\frac{(\psi' \phi^2)'}{\phi^2}, \tag{7.64}$$

where the primes refers to differentiation with respect to ξ. The ground state similarity solution shown in Fig. 7.3 (which we denote by $\bar{\phi}(\xi)$, $\bar{\psi}(\xi)$) is by construction a fixed point of (7.63), (7.64).

To investigate the stability of the similarity solution $\bar{\phi}(\xi)$, $\bar{\psi}(\xi)$ under the dynamics (7.63), (7.64), we consider small perturbations around it. The wavelength of a typical perturbation driven by the Rayleigh–Plateau instability (see (6.65)) is of the order of the radius $h_{\min} \propto t'$, which is much shorter than the axial length scale $t'^{1/2}$ of the similarity solution. Thus we would like to describe a perturbation which oscillates rapidly on the scale of the similarity solution. If $\epsilon \approx t'^{1/2}$ is a typical scale of oscillation in self-similar coordinates, we are led to the following WKB ansatz [34]:

$$\begin{pmatrix} \phi(\xi, \tau) \\ \psi(\xi, \tau) \end{pmatrix} = \begin{pmatrix} \bar{\phi}(\xi) \\ \bar{\psi}(\xi) \end{pmatrix} + \begin{pmatrix} \Phi(\xi, \tau) \\ \Psi(\xi, \tau) \end{pmatrix} \exp\left\{ \frac{iS(\xi, \tau)}{\epsilon} \right\}. \tag{7.65}$$

Here the exponential factor describes a rapid oscillation whose amplitude is given by Φ or Ψ, respectively. Our aim is to derive an equation of motion for the amplitudes, to see how they grow or decay.

If we insert (7.65) into (7.63) and (7.64) and linearize in the amplitudes Φ, Ψ, we obtain

$$\frac{iS_\tau}{\epsilon}\Phi + \Phi_\tau = \Phi - \left(\bar{\psi} + \frac{\xi}{2}\right)\left(\frac{iS'}{\epsilon}\Phi + \Phi'\right) - \bar{\phi}'\Psi$$

$$- \frac{\bar{\psi}}{2}\Phi - \frac{\bar{\phi}}{2}\left(\frac{iS'}{\epsilon}\Psi + \Psi'\right), \tag{7.66}$$

$$\frac{iS_\tau}{\epsilon}\Psi + \Psi_\tau = -\frac{\Psi}{2} - \left(\bar{\psi} + \frac{\xi}{2}\right)\left(\frac{iS'}{\epsilon}\Psi + \Psi'\right) - \bar{\psi}'\Psi$$

$$+ \frac{1}{\bar{\phi}^2}\left(\frac{iS'}{\epsilon}\Phi + \Phi'\right) - \frac{2\bar{\phi}'}{\bar{\phi}^3}\Psi - 3\frac{S'^2}{\epsilon^2}\Psi + \frac{3iS''}{\epsilon}\Psi$$

$$+ 3\Psi'' + 6\frac{\bar{\phi}'}{\bar{\phi}}\left(\frac{iS'}{\epsilon}\Psi + \Psi'\right) + 6\frac{\bar{\psi}'}{\bar{\phi}}\left(\frac{iS'}{\epsilon}\Phi + \Phi'\right)$$

$$- 6\frac{\bar{\psi}'\bar{\phi}'}{\bar{\phi}^2}\Psi. \tag{7.67}$$

We now solve this system using an expansion in the small variable ϵ. An inspection of (7.67) reveals a leading term proportional to Ψ, which is of order ϵ^{-2}, and two leading terms proportional to Φ, which are of order ϵ^{-1}. From a balance between Φ and Ψ one obtains

$$\Psi = \frac{i\epsilon}{3S'} \left(\frac{1}{\bar{\phi}^2} + 6\frac{\bar{\psi}'}{\bar{\phi}} \right) \Phi, \qquad (7.68)$$

which shows that $O(\Psi) = O(\epsilon\Phi)$. With this in mind, the terms of order $O(\epsilon^{-1})$ in (7.66) yield $iS_\tau\Phi + i(\bar{\psi} + \xi/2)S'\Phi = 0$, or

$$S_\tau + \left(\bar{\psi} + \frac{\xi}{2} \right) S' = 0. \qquad (7.69)$$

If we define $\kappa = S'$ then κ/ϵ is the wavenumber of a perturbation. Differentiating (7.69) with respect to ξ we find

$$\kappa_\tau + \left(\bar{\psi} + \frac{\xi}{2} \right) \kappa' = -\left(\bar{\psi}' + \frac{1}{2} \right) \kappa, \qquad (7.70)$$

which describes how the local wavenumber changes in space and time, assuming that its variation is gradual compared with the wavelength itself.

The second term on the left of (7.70) describes the convection of a localized perturbation (or wave packet) at an effective speed $\bar{\psi} + \xi/2$, as described by (7.58). In other words, the equation of motion for the position $\xi_p(\tau)$ of a wave packet in similarity variables is

$$\frac{d\xi_p}{d\tau} = \bar{\psi}(\xi_p) + \frac{\xi_p}{2}. \qquad (7.71)$$

At the stagnation point ξ_0 the packet remains at rest, while if it is placed to the left of the stagnation point, in the neck region, it will be convected further to the left. Our main interest lies in this neck region, as the solution is flat and thin and more vulnerable to the Rayleigh–Plateau instability. The wavenumber of the wave packet is $\kappa(\xi_p(\tau), \tau)$. If we insert this into (7.70) and use (7.71), we find

$$\frac{d\kappa(\xi_p(\tau), \tau)}{d\tau} = -\left(\bar{\psi}'(\xi_p) + \frac{1}{2} \right) \kappa(\xi_p(\tau), \tau). \qquad (7.72)$$

This equation describes how the packet is stretched by the flow, increasing the wavelength. This is the reason why the perturbation becomes eventually ineffectual and no longer grows on the time scale of the similarity solution. The relation (7.71) can be used to transform the temporal change of κ into a spatial variation following the trajectory of a packet. Combining (7.72) and (7.71) we find

$$\frac{1}{\kappa}\frac{d\kappa}{d\xi_p} = -\frac{\bar{\psi}'(\xi_p) + 1/2}{\bar{\psi} + \xi_p/2} = -\frac{1}{d\xi_p}\ln\left|\bar{\psi} + \xi_p/2\right|.$$

Integrating, this results in

$$\frac{\kappa(\xi_p)}{\kappa(\xi_i)} = \frac{\bar{\psi}(\xi_i) + \xi_i/2}{\bar{\psi}(\xi_p) + \xi_p/2}, \tag{7.73}$$

if ξ_i is the initial position of the wave packet.

To calculate the *amplitude* of the wave packet, we insert (7.68) into (7.66) and collect all terms of order ϵ^0, which results in the equation

$$\Phi_\tau + \left(\bar{\psi} + \frac{\xi}{2}\right)\Phi' = \left(1 + \frac{\bar{\psi}'}{2} + \frac{1}{6\bar{\phi}}\right)\Phi \equiv \beta(\xi_p)\Phi. \tag{7.74}$$

Then the amplitude of the wave packet when it is at the position ξ_p follows from

$$\frac{d\Phi(\xi_p(\tau), \tau)}{d\tau} = \beta(\xi_p(\tau))\Phi, \tag{7.75}$$

which can be converted into a spatial evolution using (7.71):

$$\frac{1}{\Phi}\frac{d\Phi}{d\xi_p} = \frac{\beta(\xi_p)}{\bar{\psi}(\xi_p) + \xi_p/2} \equiv \hat{\beta}(\xi_p).$$

Integrating, the amplitude of a packet that has been convected to a position ξ_p, having started at ξ_i, is

$$\frac{\Phi(\xi_p)}{\Phi(\xi_i)} = \exp\left\{\int_{\xi_i}^{\xi_p} \hat{\beta}(\xi')d\xi'\right\}. \tag{7.76}$$

Investigating the argument of the integral, we find that

$$\hat{\beta}(\xi) \approx \begin{cases} \nu/(\xi - \xi_0), & \xi \to \xi_0, \\ 2/\xi, & \xi \to \infty, \end{cases}$$

where

$$\nu = \frac{4 + 1/[3\bar{\phi}(\xi_0)]}{5} \approx 2.991. \tag{7.77}$$

Thus the integral (7.76) is logarithmically divergent as ξ_i approaches ξ_0, so that it seems as if arbitrary growth could be obtained by placing ξ_i arbitrarily close to the stagnation point. Indeed, this corresponds to the perturbation being at rest at the stagnation point, so that the amplitude would have an infinite time to grow. In reality, however, the position of the perturbation is smeared out over the wavelength of the perturbation which scales as $t'^{1/2}$, so the perturbation is pulled apart and rendered ineffectual (and our approximations cease to be valid as well). We can account for this mechanism in a qualitative fashion by placing the initial perturbation at a distance $at'^{1/2}$ to the left of the stagnation point, where a is an empirical factor. This ensures that the dominant contribution is from the lower limit of the integral

$$\int_{\xi_0 - at'^{1/2}}^{\xi_0 - 1} \frac{v}{\xi' - \xi_0} d\xi' = -v \ln at'^{1/2}.$$

As the perturbation is convected to large negative values of ξ_p, once again the integral (7.76) becomes logarithmically divergent and the corresponding contribution from its upper limit is

$$\int_{-1}^{\xi_p} \frac{2}{\xi'} d\xi' = 2 \ln \left| \xi_p \right|.$$

Thus we find that, for large ξ_p and small t', the growth is described by

$$\frac{\Phi(\xi_p)}{\Phi(\xi_i)} \approx e^C \xi_p^2 \left(at'^{1/2} \right)^{-v}, \tag{7.78}$$

where C is a constant. To find C numerically we subtract the asymptotic behavior of $\hat{\beta}$ at both limits, noting that

$$\int_{\xi_0 - at'^{1/2}}^{\xi_p} \left(\frac{v}{\xi' - \xi_0} + \frac{2 - v}{\xi'} \right) d\xi' \approx 2 \ln \left| \xi_p \right| - v \ln \left(at'^{1/2} \right) + (v - 2) \ln |\xi_0| \tag{7.79}$$

for $t' \to 0$ and $\xi_p \to -\infty$. It follows that

$$C = \int_{\xi_0}^{-\infty} \left(\hat{\beta}(\xi) - \frac{v}{\xi' - \xi_0} - \frac{2 - v}{\xi'} \right) d\xi' + (v - 2) \ln |\xi_0| = 2.555. \tag{7.80}$$

The quadratic growth (7.78) with ξ_p of an interfacial perturbation is in step with the quadratic growth of the profile itself, given by (7.48). In terms of the physical thickness $h(z, t)$, given by (7.62), this means that the perturbation becomes time independent and growth ceases, in agreement with the matching condition for $\xi \to \infty$. The size of the perturbation to the velocity field Ψ is determined by (7.68). According to (7.73), $S' = \kappa \propto 1/\xi_p$ for large ξ_p, which implies that

$$\Psi(\xi_p) \propto \frac{\Phi(\xi_p)}{\xi_p^4 \kappa} \propto \frac{1}{\xi_p}.$$

This is once more the same growth as the similarity solution $\bar{\psi}$ of the velocity field; see (7.48).

From these observations it follows that, for a sufficiently small initial perturbation $\Phi(\xi_i)$, a perturbation will be convected away to $\xi = -\infty$, making the ground state similarity solution $\bar{\phi}$, $\bar{\psi}$ *linearly stable*. If, however, the perturbation amplitude is comparable with the size $\bar{\phi} \approx a_0^- \xi^2 = 6 \times 10^{-4} \xi^2$ of the similarity profile itself, we can consider the solution as having been rendered *unstable*. Thus, although we have only considered perturbations to the similarity solution at linear order, we have found that an initial perturbation $\Phi(\xi_i)$ of finite amplitude is necessary to destabilize the pinch-off solution. According to (7.78) this critical amplitude is

$$\Phi_{cr} = a_0^- e^{-C} a^\nu t'^{\nu/2} = 4.7 \times 10^{-5} a^{2.99} (t')^{1.495}. \qquad (7.81)$$

An appropriate value for the constant a must be found in an empirical fashion. By placing perturbations of different wavelengths and amplitudes at the stagnation point and integrating the slender jet equations (6.57), (6.58) numerically, the minimum amplitude necessary to destroy the similarity solution was found in [31]. This occurred at an "optimal" wavelength (in similarity variables) of $\approx 10\phi_0 (t')^{1/2}$, for which the critical amplitude becomes [35]

$$\Phi_{cr} = 6 \times 10^{-7} (t')^{1.495}. \qquad (7.82)$$

In real space, and restoring units, this means that for a thread of radius h_{min} one needs a surface perturbation of size

$$h_c = 3.8 \times 10^{-3} h_{min} \left(\frac{h_{min}}{\ell_\nu} \right)^{1.495} \qquad (7.83)$$

to render the interface unstable. Thus, in the late stages of singularity formation, for which $h_{min} \ll \ell_\nu$, the solution becomes more and more vulnerable to external perturbations, in qualitative agreement with experiment. It is also clear that, the greater the viscosity, the smaller are the perturbations necessary to destabilize a thread of a given thickness, in agreement with the very unstable viscous thread shown in Fig. 7.5. However, to obtain a quantitative estimate of when the pinch-off solution turns unstable we need to calculate the size of the perturbation h_c experienced under typical circumstances. We will now derive an estimate for the amplitude of the perturbations introduced by thermal noise.

7.4 Fluctuating jet equations

To include the effect of thermal noise in the slender jet description, we need to add a noise term to equation (6.58) for the velocity field. We will not revisit here the general theory which is presented in [133] and [158], but merely summarize the results. Essentially the fluid is divided up into elements, each of which is treated similarly to a Brownian particle, i.e. a macroscopic particle subjected to random, uncorrelated, "kicks" [186]. In a one-dimensional description we can define these fluid elements as thin slices of a jet, which in the absence of noise satisfy (6.66), (6.68). Now we add a fluctuating force $s(z, t)$ to the viscous contribution of the slice equation (6.68):

$$3 v h^2 v' + \frac{s}{\pi \rho}. \tag{7.84}$$

Here s is taken from a Gaussian distribution which is uncorrelated in space and time. If we represent s on a discrete spatial and temporal grid z_i, t_i, with constant grid spacings Δz and Δt, respectively, the distribution is then determined completely by its second moments:

$$\langle s(z_{i_1}, t_{j_1}) s(z_{i_2}, t_{j_2}) \rangle = \frac{6 \pi k_B T \eta h^2(z_{i_1})}{\Delta z \Delta t} \delta_{i_1 i_2} \delta_{j_1 j_2}. \tag{7.85}$$

The temperature-dependent prefactor comes from the equipartition theorem, namely the condition that each degree of freedom carries a contribution of $k_B T/2$ to the kinetic energy. Now the slice equation (6.68) with added noise (7.84) becomes

$$\partial_t (h^2 v) + (h^2 v^2)' = -\frac{\gamma}{\rho} G' + 3 v (h^2 v')' + \frac{s'}{\pi \rho}. \tag{7.86}$$

The form (7.86) of this equation of motion is still awkward to treat numerically (see the next chapter for details), since it contains derivatives of the process s which contains jumps from one grid point to the next. This difficulty can be overcome by introducing the integral $P(z) = \int_0^z m(x) dx$ of the momentum $m = h^2 v$, so that (7.86) can be integrated. Using $v = P' h^2$, everything can be expressed in terms of total derivatives of the new variables h and P, as well as s. Nondimensionalizing the equations with the intrinsic scales (6.37), there remains a single quantity [158] $M_l = \ell_T / \ell_v$ characterizing the strength of the noise. Thus we finally arrive at the conserved form of the equations:

$$\partial_t h^2 = -P'', \tag{7.87}$$

$$\partial_t P = -\frac{P'^2}{h^2} - G + 3 h^2 \left(\frac{P'}{h^2} \right)' + D h \xi, \tag{7.88}$$

where

$$\langle \xi(z_{i_1}, t_{j_1}) \xi(z_{i_2}, t_{j_2}) \rangle = \frac{1}{\Delta z \Delta t} \delta_{i_1 i_2} \delta_{j_1 j_2} \tag{7.89}$$

is the noise, normalized to unit variance, and the noise strength is

$$D = M_l \sqrt{\frac{6}{\pi}} = \frac{\ell_T}{\ell_v} \sqrt{\frac{6}{\pi}}. \tag{7.90}$$

An equation of the type (7.88) which includes Gaussian noise is known as a Langevin equation [186].

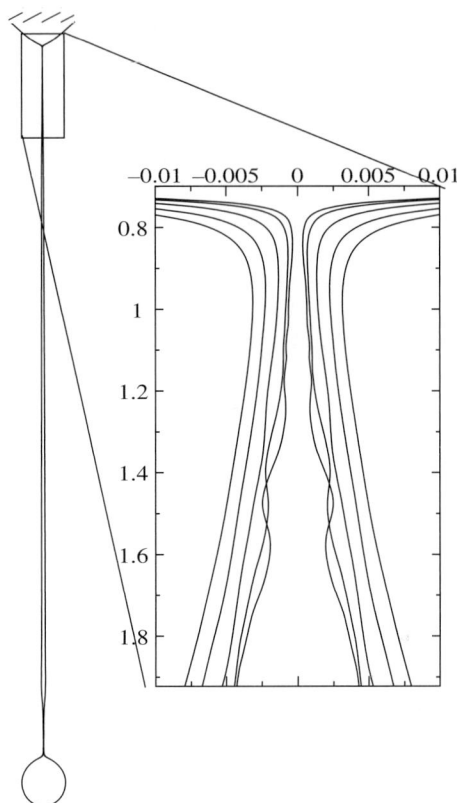

Figure 7.6 A drop of silicicone oil, 100 times more viscous than water, falls from a pipette of radius $r_0 = 0.2$ cm; the lengths shown are in units of r_0. The parameters are Bo $= 1.87$ and Oh $= 0.49$. At room temperature $\ell_T = 4.44 \times 10^{-8}$ cm, and so $D = 7.7 \times 10^{-7}$.

We will simulate (7.87), (7.88) for a drop falling from a pipette of radius $r_0 = 2$ cm, for a fluid 100 times more viscous than water. In this case a long and slender thread is formed, which starts to pinch both at its upper and lower end but the pinch at the upper end comes first. A magnification of the upper pinch region reveals that the interface becomes irregular at a minimum radius of about $h_{min} = 3 \times 10^{-3} r_0$, as seen in the inset of Fig. 7.6. This is in qualitative agreement with the calculation in Section 7.3.2, but for a quantitative comparison we still need an estimate of the size of perturbation h_c produced by thermal noise.

To this end we calculate the additional surface energy E_k necessary to create a sinusoidal perturbation of wavenumber k and amplitude h_c on a cylinder of radius h_{min}, which corresponds to the profile

$$h(z) = h_{\min} + h_c \cos kz.$$

Then the extra area per wavelength is

$$\delta A = 2\pi h_{\min} \int_0^\lambda \left(\sqrt{1 + h_z^2} - 1 \right) dz \approx \pi h_{\min} \int_0^\lambda h_z^2 dz$$

in the limit of small amplitudes, and we find that

$$E_k = \gamma \delta A = \pi^2 h_c^2 k h_{\min}. \tag{7.91}$$

Equating this with $k_B T/2$, according to the equipartition theorem, and assuming an optimum wavenumber of $k = 2\pi/(10h_{\min})$ for which the similarity solution is most sensitive to noise [31], this results in

$$h_c \approx \ell_T \sqrt{\frac{10}{4\pi^2}} \approx 1.26 \times 10^{-8} \text{cm}.$$

Combining this with (7.83), we finally obtain

$$h_{\text{thr}} = 5D^{0.4} \text{Oh}^2 r_0 \approx 4.3 \times 10^{-3} r_0 \tag{7.92}$$

as the threshold radius at which thermal instability sets in, in reasonable agreement with our simulation. Experimentally, however, instability is observed to set in at radii about a factor of 10 greater for comparable fluid viscosities [35], a discrepancy which remains unexplained.

7.5 Inviscid breakup

Finally, we return to the breakup of an inviscid fluid in air, for which the scaling is described by (7.7). In particular, the radial and axial length scales are of the same order, so the lubrication approximation is not applicable. Instead one needs to solve the full inviscid equations (4.32)–(4.39) in Section 4.5.1, which has to be done numerically. The result of a typical simulation is seen in Fig. 7.7, which shows that the fluid neck is close to cone-shaped, and then turns over, so that the neck in fact makes a small indentation into the drop [47, 57]. In particular, a representation of the profile as the graph of a single-valued function $h(z, t)$ cannot describe the solution globally.

The inviscid equations for the free surface $\mathbf{x} = (z, r)$ can be recast in similarity variables using the length $\ell_{\text{in}} = (\gamma \Delta t^2/\rho)^{1/3}$ identified by (7.6):

$$R = \frac{r}{\ell_{\text{in}}}, \quad Z = \frac{z - z_0}{\ell_{\text{in}}}, \quad \mathbf{X} = (Z, R). \tag{7.93}$$

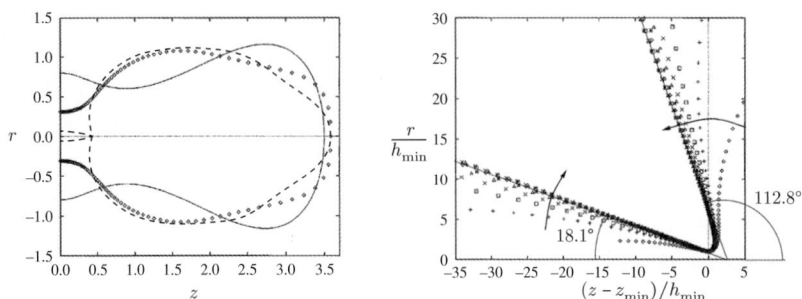

Figure 7.7 Numerical simulation of the pinch-off of an inviscid fluid in a neutral (gas) environment ($D = 0$). Left-hand panel: global view of a typical drop. The solid line with the wide neck is the initial shape. The points show an intermediate shape. The broken line shows the shape just before pinch-off; note the strong asymmetry, with a narrow conical neck on one side and the slightly indented round drop on the other. Right-hand panel: close-up of the numerical profiles near the pinch point, rescaled using the minimum radius $h_{\min}$. The arrows indicate the direction of time. The profiles converge towards a similarity solution which consists of two cones (indicated by the thin solid lines), one inside the other. The cones have opening half angles of $18.1°$ and $112.8°$, respectively. Reprinted with permission from [57]. Copyright 1998 by the American Physical Society.

Since $\ell_{\mathrm{in}}/\Delta t$ is a characteristic velocity scale, the similarity forms of the velocity and of the curvature are

$$\phi(\mathbf{x}, t) = \frac{\ell_{\mathrm{in}}^2}{\Delta t} \Phi(\mathbf{X}), \quad \kappa = \frac{K}{\ell_{\mathrm{in}}}. \tag{7.94}$$

According to (7.93), we have

$$\frac{\partial \mathbf{x}}{\partial t} = \dot{\ell}_{\mathrm{in}} \mathbf{X} = -\frac{2\ell_{\mathrm{in}}}{3\Delta t} \mathbf{X},$$

so the kinematic boundary condition (4.26) becomes

$$\left(\frac{2}{3}\mathbf{X} + \nabla\Phi \right) \cdot \mathbf{N} = 0, \tag{7.95}$$

where $\mathbf{N}$ is the normal to the time-independent (self-similar) interface shape. The normal-stress boundary condition (4.40) becomes

$$-\frac{\Phi}{3} + \frac{2}{3}\mathbf{X} \cdot \nabla\Phi + |\nabla\Phi|^2 + K = 0, \tag{7.96}$$

where the stress exerted by the outside medium has been ignored. Since both z and r are rescaled in the same way, the equation to be satisfied by the potential in the interior of the fluid remains the same:

$$\triangle \Phi = 0. \tag{7.97}$$

Solutions to the self-similar system of equations (7.95)–(7.97) were found in [137]. One can avoid solving Laplace's equation in the interior of the fluid domain by using the so-called boundary integral technique. The idea is to use Green's theorem to reduce the solution of Laplace's equation with the boundary condition (7.96) to an integral over the boundary, that is the interface, alone. We will not discuss the details of this method here but will come back to another version of it in Section 7.7, to solve a Stokes flow problem. Figure 7.7 (right) shows the result of a dynamical calculation, which was rescaled according to (7.93), except that the minimum radius $h_{\min}$ of the fluid neck was used instead of $\ell_{\rm in}$. The profiles are observed to converge onto a similarity solution which asymptotes to a cone on either side. The cones have opening half-angles of 18.1° and 112.8° respectively, and are joined smoothly around the minimum. The minimum radius $h_{\min}$ is given by

$$h_{\min} = 0.7(\gamma \Delta t^2/\rho)^{1/3} \equiv 0.7\ell_{\rm in}. \tag{7.98}$$

The cone angles as well as (7.98) are universal properties of the Euler solution: they are independent of the initial conditions. The scaling law (7.98) has been verified experimentally to great precision, see Fig. 1.2 (right), as well as the value of the smaller angle [43].

The inviscid breakup dynamics may also be considered in the case where the external fluid possesses significant inertia. In that case a more general form of the normal stress balance (4.40) has to be used and (7.97) has to be solved in the exterior as well. A nice physical realization of the case where the inner and outer fluid possess equal densities is the breakup of a soap bubble in air [47], the latter playing the role of the outer fluid. The soap film (whose own density is neglected in this approximation) supplies a tension equivalent to the surface tension between two immiscible liquids.

A study of inviscid breakup for a general density ratio $D = \rho_{\rm out}/\rho_{\rm in}$ was performed in [137]. The solution retains the double-cone structure found for $D = 0$ (as seen in Fig. 7.7), but the value of the angles as well as that of the prefactor governing the minimum radius (7.98) depend on the dimensionless parameter D. However, new phenomena occur if the outer density is significantly greater than that of the inside. Dynamical simulations reveal that, for $D \geq 6.2$, pinch-off solutions are subject to an oscillatory instability, which means that a simple self-similar solution of the type (7.93) is no longer realized. Whether (7.93) is perhaps to be replaced by a more complex form of self-similarity, to be discussed in Chapter 12, remains an open question. In the limit $D \to \infty$, in which the interior medium (typically a gas) no longer plays a role, the scaling changes completely and the radial scaling exponent α is much closer to $1/2$. In addition there is a slow variation of the value of the exponent, which we will examine in Chapter 9.

Rather remarkably, the situation is different for an inviscid sheet in that the self-similarity is now of the second kind [39]. The reason is that pinch-off is controlled by two different length scales, one for the sheet thickness and the other for the horizontal variation.

Example 7.7 (Pinch-off of an inviscid sheet) The self-similar solutions to the inviscid thin film equation are of the form

$$h = r_0 t'^{\alpha_1} H(X), \quad u = \frac{r_0}{\tau} t'^{\alpha_2} U(X), \tag{7.99}$$

where $X = x'/t'^{\beta}$. The spatial and temporal variables are nondimensionalized using a length scale r_0 imposed externally and the time scale $\tau = \sqrt{\rho r_0^3/\gamma}$. From (6.84) and (6.85) one obtains the relations (cf. Example 7.4); see Section 7.3.

$$\alpha_1 - 1 = \alpha_1 + \alpha_2 - \beta,$$
$$\alpha_2 - 1 = 2\alpha_2 - \beta = \alpha_1 - 3\beta,$$

not including the viscous term. Thus two exponents can be expressed in terms of the third:

$$\alpha_1 = 4\beta - 2, \quad \alpha_2 = \beta - 1, \tag{7.100}$$

where β is free. Breakup will happen if $\alpha_1 > 0$, implying $\beta > 1/2$. The breakup will be with a slender geometry if $\alpha_1 > \beta$, which implies that $\beta > 2/3$.

The similarity equations become

$$(2 - 4\beta)H + \beta X H' + (UH)' = 0, \tag{7.101}$$

$$(1 - \beta)U + \beta X U' + UU' = H''' + 4\frac{(U'H)'}{H}; \tag{7.102}$$

they were treated in [39]. Solutions with the correct boundary conditions exist for a particular value of β only, which was found numerically as $\beta = 0.6869 \pm 0.005 > 2/3$. This shows that the breakup solution remains slender, which is consistent with the assumptions underlying (6.84) and (6.85). $\square$

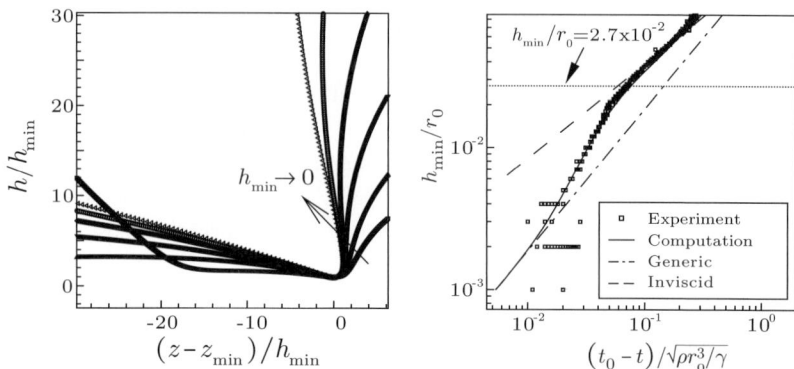

Figure 7.8 On the left, profiles from a Navier–Stokes simulation of the pinch-off of a water drop (Oh $= 1.81 \times 10^{-3}$); the axes have been rescaled by $h_{\min}$. The values of $h_{\min}/r_0$ for the curves are, outermost to innermost: ($\times 10^2$) 7.09, 3.58, 1.75, 0.876, 0.445, 0.221. On the right, $h_{\min}/r_0$ for an 83% glycerol–water mixture (Oh $= 1.63 \times 10^{-1}$), showing the crossover from the inviscid to the generic similarity solution. Reprinted with permission from [44]. Copyright 2002 by the American Physical Society.

7.6 Crossover

We have just described three different similarity solutions, corresponding to different balances in the equation of motion. What is actually observed in a given experiment or simulation depends on the initial conditions, which are characterized by the initial radius r_0 of the liquid column. The size of r_0 relative to the *intrinsic* length scale ℓ_ν is measured by the Ohnesorge number Oh (6.64).

If Oh is small (as is typically the case for water), the viscous term in (4.13) is very small initially and the inviscid solution (7.93) is observed. This is shown in Fig. 7.8 (left), which displays a simulation of the full Navier–Stokes equation – yet the profile converges onto the double-cone shape predicted by the *inviscid* similarity solution. However, as pinch-off is approached the viscous term becomes increasingly important and the inviscid solution fails eventually.

The relative size of the viscous and inertial forces can be estimated by the local Reynolds number [143],

$$\mathrm{Re}_{\mathrm{loc}} = \frac{v\ell_z}{\nu}. \tag{7.103}$$

When this number is of order one, the forces must be of the same order as each other. Using the facts that typical velocities scale as $v \propto \ell_{\mathrm{in}}/\Delta t$ and that $\ell_z \approx \ell_{\mathrm{in}}$ for inviscid pinch-off, we have

$$\mathrm{Re}_{\mathrm{loc}} \approx \frac{\ell_{\mathrm{in}}^2}{\nu \Delta t} = \frac{\Delta t}{t_{\nu}}.$$

Thus crossover is expected to occur when $\Delta t \sim t_{\nu}$, at which point the viscosity can no longer be neglected. At the same time the minimum radius is comparable with the viscous length scale ℓ_{ν}, resulting in the following prediction for the length scale for crossover from inviscid to generic scaling:

$$\frac{h_{\mathrm{min}}^{\mathrm{in}\to\mathrm{gen}}}{r_0} \approx \mathrm{Oh}^2. \tag{7.104}$$

Such a crossover is seen in Fig. 7.8 (right), where the minimum radius first follows the inviscid solution and then starts to cross over, at about $h_{\mathrm{min}}/r_0 = 2.5 \times 10^{-2}$, until the generic solution is finally reached at $h_{\mathrm{min}}/r_0 = 3 \times 10^{-3}$. Thus taking the value $\mathrm{Oh} = 1.63 \times 10^{-1}$ used in the experiment shown in Fig. 7.8 for reference, the observed initial crossover is in good agreement with (7.104), taking $\mathrm{Oh}^2 = 2.6 \times 10^{-2}$.

If $\mathrm{Oh} \gg 1$ then inertia is negligible initially and the viscous similarity solution of Section 7.2 applies. However, this solution also is not valid in the limit of $t' \to 0$. Namely, the axial scale scales as $\ell_z \propto t'^{\beta}$ and thus $\mathrm{Re}_{\mathrm{loc}} \propto \ell_z^2/t' \propto (t')^{2\beta-1}$. This means that the Reynolds number becomes large in the limit $t' \to 0$ if $\beta < 1/2$. Since $\beta \approx 0.17$ (see Table 7.2), inertia eventually becomes relevant leading to a crossover to the asymptotic, highly asymmetric, solution. Both the predictions of symmetric pinching and its transition to the generic solution are confirmed in Fig. 7.9. A more careful analysis of the crossover from viscous to generic scaling may be performed in Exercise 7.8.

7.7 Fluid–fluid breakup

So far we have for the most part neglected the presence of an external fluid. The pinching of a viscous fluid inside another viscous fluid leads to a very different problem, an experimental photograph of which is shown in Fig. 7.10. The reason for the profound changes in the dynamics of breakup is that the outer fluid imposes a stationary reference frame, and prevents the flow velocity from diverging, which is contrary to what is predicted by the generic similarity solution (7.41). Since the size of the pinch region goes to zero, the Reynolds number also goes to zero and the Stokes equation must be applied [143]. The crossover scale predicted by this argument may, however, be quite small and so it usually does not apply to breakup in air, where the crossover will be below the mean free path length and classical hydrodynamics no longer applies.

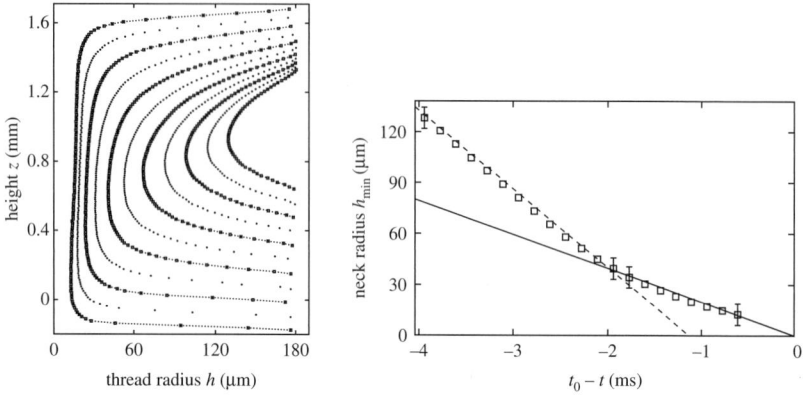

Figure 7.9 On the left, the neck of a falling drop of a glycerol–water mixture [187], Oh = 0.49, Bo = 0.047, showing the transition from symmetric (viscous) to asymmetric (generic) pinching. (Reprinted with permission from [187]. Copyright 2001 by the American Physical Society.) Accordingly the rate of pinching changes at the transition point see experimental measurements of the neck radius on the right (squares). The theoretical predictions for the viscous and generic solutions are shown by the broken and the solid lines, respectively. For the viscous solution, the theoretical prediction is $h_{\min} = 0.709(t_0 - t)v_\eta$ and for the generic solution it is $0.0304(t_0 - t)v_\eta$.

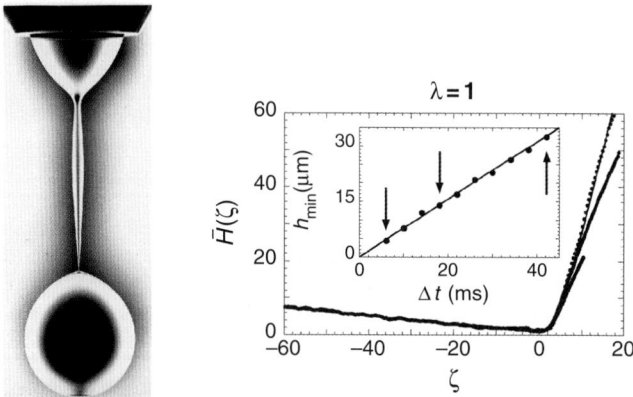

Figure 7.10 On the left, a drop of glycerin dripping through polydimethylsiloxane (PDMS) near pinch-off [52]. The nozzle diameter is 0.48 cm, and the viscosity ratio is $\lambda = 0.95$. On the right, the inset shows the minimum radius $h_{\min}(t)$ as a function of time for the drop shown on the left. The solid line in the inset gives the theoretical prediction. The main figure shows the traveling wave solution $\bar{H}(\zeta)$ as defined by (7.121). The dots are rescaled experimental profiles corresponding to the times indicated by the arrows in the inset. The solid line gives the theory as obtained from a numerical solution of the integro-differential equations (7.122), (7.117).

Two-fluid breakup presents us with a new type of similarity solution, which has the form of a *traveling wave* in the dynamical system description (3.52). For simplicity we consider the case where the viscosities η of the fluid in the drop and of the external fluid are the same; the general case is very similar but computationally more complicated. In fact we have to solve the Stokes equation (4.48) both inside and outside the drop, with boundary conditions (4.24), (4.25) at the interface. This problem can, however, be reduced to an equation for the interface alone by using the so-called boundary integral technique [177].

The idea is that the interface can be viewed as a continuous distribution of point forces, representing the surface tension. Namely, (4.24) and (4.25) imply that there is a net force on a fluid element at a boundary point $\mathbf{x}'$, given by

$$\mathbf{f}(\mathbf{x}) = -\kappa\delta(\mathbf{x} - \mathbf{x}')\mathbf{n} \equiv \mathbf{F}\delta(\mathbf{x} - \mathbf{x}'), \qquad (7.105)$$

where δ is the three-dimensional δ-function, which can be seen as the limit of functions concentrated increasingly in the origin, its total volume being kept constant. This property can be summarized by the formula

$$\int_V \delta(\mathbf{x} - \mathbf{x}')d\mathbf{x}' = \begin{cases} 1, & \mathbf{x} \in V, \\ 0, & \mathbf{x} \notin V. \end{cases} \qquad (7.106)$$

Now we write down the solution of Stokes' equation subject to a point force and integrate over all contributions. This means that we need to solve the Stokes equation

$$\eta\Delta\mathbf{v} - \nabla p + \mathbf{F}\delta = 0, \qquad (7.107)$$

the solution of which (called the Stokeslet) in an infinite domain is

$$v_i(\mathbf{x}) = \frac{1}{\eta}J_{ij}(\mathbf{x} - \mathbf{x}')F_j. \qquad (7.108)$$

It can be verified by explicit calculation (see Exercise 7.9) that the tensor $\mathbf{J}$ is given by

$$J_{ij}(r) = \frac{1}{8\pi}\left(\frac{\delta_{ij}}{r} + \frac{r_i r_j}{r^3}\right), \qquad \mathbf{r} = \mathbf{x} - \mathbf{x}'. \qquad (7.109)$$

Performing the integration over the free surface $\partial\Omega$, we obtain an explicit expression for the velocity:

$$\mathbf{v}(\mathbf{x}) = -v_\eta \int_{\partial\Omega} \kappa\mathbf{J} \cdot \mathbf{n}\, d\sigma'. \qquad (7.110)$$

Equation (7.110) determines the velocity field, given the shape of the boundary. The motion of the surface is determined by the convection equation (6.9), for which we need only the velocity evaluated at points

$$\mathbf{x} = (\cos\theta)h(z)\,\mathbf{e}_x + (\sin\theta)h(z)\,\mathbf{e}_y + z\mathbf{e}_z$$

lying on the boundary. In addition we assume the drop to be axisymmetric, so it is sufficient to calculate the motion at the azimuthal angle $\theta = 0$.

Once more, returning to (7.8) it is tempting to suppose that in the limit $\ell_{\mathrm{vis}} \to 0$ the similarity solution for a pinching thread inside another liquid is

$$h = \ell_{\mathrm{vis}} H_{\mathrm{out}}\left(\frac{\Delta z}{\ell_{\mathrm{vis}}}\right), \qquad \ell_{\mathrm{vis}} = v_\eta \Delta t. \tag{7.111}$$

However, this assumption fails yet again for reasons even more subtle than in the single-fluid case. To understand this note that, according to (3.16), H_{out} must be linear at infinity so the interface has the shape of a cone:

$$H_{\mathrm{out}}(\xi) \to s_{\pm}\xi, \quad \xi \to \pm\infty, \tag{7.112}$$

where $\xi = \Delta z / \ell_{\mathrm{vis}}$. Observe that this asymptotic behavior appears to be borne out by the experimental results shown in Fig. 7.10 and in particular by the rescaled version of the profile shown on the right.

Using the similarity form (7.111) it is a simple matter to transform the integral (7.110) to similarity variables, defining the self-similar velocity as

$$\mathbf{V}(\xi) = \frac{\mathbf{v}(\mathbf{x})}{v_\eta}, \quad \mathbf{x} = h(z)\mathbf{e}_x + z\mathbf{e}_z.$$

First we have to perform the integration over the azimuthal angle θ, which gives integral kernels that depend on the z-coordinate only:

$$\bar{\mathbf{J}}(\xi, \xi') = H(\xi') \int_0^{2\pi} \mathbf{J}(\mathbf{r})d\theta,$$

$$\mathbf{r} = [H(\xi) - H(\xi')\cos\theta]\mathbf{e}_x - H(\xi')\sin\theta\,\mathbf{e}_y + (\xi - \xi')\mathbf{e}_z. \tag{7.113}$$

Explicit expressions for $\bar{\mathbf{J}}$ can be given in terms of elliptic integrals [177]. The self-similar velocity becomes

$$\mathbf{V}(\xi) = -\int_{-\Delta z_b / \ell_{\mathrm{vis}}}^{\Delta z_b / \ell_{\mathrm{vis}}} \kappa(\zeta)\bar{\mathbf{J}}(\xi, \zeta)\mathbf{n}(\zeta)d\zeta, \tag{7.114}$$

where Δz_b marks some boundary point outside the singular region.

In the limit $\ell_{\mathrm{vis}} \to 0$, the limits of the integral (7.114) go to infinity so we must test for convergence. Since only the behavior for large ζ matters, we may assume that H is a cone, $H(\zeta) = s_{\pm}\zeta$, and that ξ lies at its origin, $\xi = 0$. With these assumptions one finds that

$$\mathbf{r} = -s_{\pm}\zeta\left(\cos\theta\,\mathbf{e}_x + \sin\theta\,\mathbf{e}_y\right) - \zeta\mathbf{e}_z, \quad \mathbf{n} = \frac{\cos\theta\,\mathbf{e}_x + \sin\theta\,\mathbf{e}_y - s_{\pm}\mathbf{e}_z}{\sqrt{1 + s_{\pm}^2}}.$$

For symmetry reasons, there can be only a z-component of the velocity field, and (7.109) becomes

$$J_{3j}(\mathbf{r})n_j = -\frac{1}{8\pi} \frac{s_{\pm}}{\sqrt{1 + s_{\pm}^2 \zeta}},$$

so that the angular integration in (7.113) becomes trivial. Noting that the curvature of a cone is $\kappa = 1/H(\zeta)$, one finds that the integral (7.114) *diverges* logarithmically:

$$\mathbf{V}(\xi) \approx -b \ln \frac{\Delta z_b}{\ell_{\mathrm{vis}}}, \quad \ell_{\mathrm{vis}} \to 0, \tag{7.115}$$

with constant

$$b = -\frac{1}{4} \left(\frac{s_+}{1 + (s_+)^2} + \frac{s_-}{1 + (s_-)^2} \right). \tag{7.116}$$

It is evident from the experiment that the slopes $s_{\pm}$ are not equal, and thus $b \neq 0$. But this means that the self-similar description (7.111) is not consistent and results in an *infinite* self-similar velocity $\mathbf{V}$. The reason is that the two cones on either side of the minimum exert a force, resulting in a translation of the pinch point in the direction of the shallower cone. The velocity can be made to converge if one takes this translation into account; this amounts to subtracting the singularity of the integral (7.114) by defining

$$\mathbf{V}^{\mathrm{fin}}(\xi) = \lim_{A \to \infty} \left(-\int_{-A}^{A} \kappa(\zeta) \bar{\mathbf{J}}(\xi, \zeta) \mathbf{n}(\zeta) d\zeta + b(\ln A)\mathbf{e}_z \right), \tag{7.117}$$

which is finite. The original velocity, however, now depends linearly on $\tau = -\ln \Delta t$:

$$\mathbf{V}(\xi, \tau) = \mathbf{V}^{\mathrm{fin}}(\xi) - b\tau \mathbf{e}_z \tag{7.118}$$

up to arbitrary constants. This corresponds to a uniform translation in the coordinate system (ξ, τ) of the "dynamical systems" description.

Now we will also interpret the original expression (7.111) as time dependent and write

$$h = \ell_{\mathrm{vis}} H(\xi, \tau). \tag{7.119}$$

Inserting this into the convection equation (6.9) we obtain

$$H_\tau = H - \left(\xi + V_z^{\mathrm{fin}}\right) H_\xi + V_r^{\mathrm{fin}} + b\tau H_\xi. \tag{7.120}$$

This equation has a solution in the form of a traveling wave:

$$H(\xi, \tau) = \bar{H}(\zeta), \quad \mathbf{V}^{\mathrm{fin}}(\xi, \tau) = \bar{\mathbf{V}}(\zeta), \quad \text{where} \quad \zeta = \xi - b\tau. \tag{7.121}$$

The profiles $\bar{H}$, $\bar{V}$ of the traveling wave obey the equation

$$\bar{H} - \left(\xi + \bar{V}_z^{\text{fin}}\right)\bar{H}_\xi - \bar{V}_r^{\text{fin}} = 0, \tag{7.122}$$

where now everything is finite. Equation (7.122), coupled to (7.117), is an integro-differential equation which needs to be solved numerically [52]. For the case of equal viscosities, considered here, the minimum radius H_{min} and the asymptotic slopes are

$$H_{\text{min}} = 0.0328, \quad s_- = -0.105, \quad s_+ = 4.81. \tag{7.123}$$

As seen in Fig. 7.10 (right), the agreement between theory and experiment is truly remarkable. A detailed study of the case of an arbitrary viscosity ratio $\lambda = \eta_{\text{in}}/\eta_{\text{out}}$ between the inner and the outer fluids was performed in [198].

Exercises

7.1 Show that in the Lagrangian coordinates defined by (7.14) the leading-order slender jet equations (6.57), (6.58) become

$$z_{tt} = -\frac{\gamma}{2\rho}\frac{z_{ss}}{z_s^{3/2}} + 3\nu\left(\frac{z_{ts}}{z_s^2}\right)_s.$$

Demonstrate that, if z_{tt} can be neglected (as in the case of a very viscous fluid), one obtains (7.19) after integration.

7.2 Show that if the full curvature term is taken into account, the viscous flow equation (7.10) is replaced by

$$T = -\gamma G + 3\eta v_z h^2, \tag{7.124}$$

where G is defined by (6.67).

(i) Show that for an isolated viscous drop not subject to gravity (for example a drop in free fall) the tension $T(t)$ must vanish. Hint: consider the force balance at the end of the drop.

(ii) Show that if z is a local minimum of the profile then $h(z)$ can only increase. As a result, an isolated inertialess drop can never break up [79]!

7.3 Consider the stability of viscous pinch-off solutions of the form (7.24).

(i) Show that the dynamical system description for the problem reads

$$\chi_\tau = \frac{1}{6} + \chi - \delta\zeta\chi' - \frac{T_0(\tau)}{6\chi},$$

where the tension is

$$T_0(\tau) = \left(\int_{-\infty}^{\infty} \frac{d\zeta}{\chi^4} \right)^{-1} \int_{-\infty}^{\infty} \frac{d\zeta}{\chi^3}.$$

Evidently the self-similar pinch-off solutions $\bar{\chi}$, $\bar{T}_0$ described in Section 7.2 are fixed points of the dynamical system.

(ii) Consider small perturbations of the form

$$\chi(\zeta, \tau) = \bar{\chi}(\zeta) + \epsilon \chi_m e^{\nu \tau} P(\zeta), \quad T_0 = \bar{T}_0 + \epsilon \bar{T}_0 e^{\nu \tau} \tilde{T}.$$

Show that three eigenvalues follow from invariances of the system, namely $\nu = 1$ with eigenfunction $P = f - \delta \zeta f_\zeta$, $\nu = \delta$ with $P = f_\zeta$, and $\nu = 0$ with $P = \zeta f_\zeta$. For $\nu = 1$ the perturbation to the tension is $\tilde{T} = 1$; in the other cases $\tilde{T} = 0$.

7.4 (*) In the notation of the previous exercise, show that the eigenvalue equation governing perturbations around a similarity solution $\bar{\chi}$ is

$$(\nu - 1)P = -\delta \zeta P_\zeta + \frac{\bar{T}_0 P}{6 \chi_m^2 f^2} - \frac{\bar{T}_0 \tilde{T}}{6 \chi_m^2 f},$$

where $\tilde{T}$ satisfies the constraint

$$\tilde{T} = -3 \left(\int_{-\infty}^{\infty} \frac{d\zeta}{f^3} \right)^{-1} \int_{-\infty}^{\infty} \frac{P d\zeta}{f^4} + 4 \left(\int_{-\infty}^{\infty} \frac{d\zeta}{f^4} \right)^{-1} \int_{-\infty}^{\infty} \frac{P d\zeta}{f^5}.$$

(i) Transform the eigenvalue equation to one for which f is the independent variable.

(ii) In [76] it is shown that the only eigensolution with $\tilde{T} \neq 0$ is the one belonging to $\nu = 1$, derived in the previous exercise. This justifies considering only the case $\tilde{T} = 0$. Show that for $\tilde{T} = 0$ the solution of the eigenvalue equation is

$$P = \frac{1}{f} \left(f + 2\bar{\delta} - 1 \right)^{1-\nu(2\bar{\delta}-1)/(2\bar{\delta})} \left(f - 1 \right)^{1-\nu/(2\bar{\delta})}.$$

(iii) Conclude that the eigenvalues for the ith similarity solution must satisfy

$$\nu_j^{(i)} = \delta_i (2i + 2 - j), \quad j = 0, 1, 2, \ldots$$

(iv) Show that, to satisfy the constraint $\tilde{T} = 0$, only eigenvalues with odd j are allowed. Conclude that the ground state ($i = 0$) solution is stable, while all solutions of higher order are unstable.

7.5   Pursuing the expansion of Example 7.5 (see Section 7.3) to higher order, show that

$$a_2 = (b_1 a_0 - a1 b_0)/4, \quad b_2 = -2b_0 b_1 + 2a_1/a_0^2 - 6b_0 a_1/a_0. \quad (7.125)$$

7.6   Analyze the expansion (7.52) in greater detail. Show that $\mu = b_0/3 - 4$, and use MAPLE to obtain

$$f_1 = -6a_0 g_0 \mu - a_0 g_1 - 6 f_0 \mu - 2b_0 f_0 - a_1 g_0 - 24 a_0 g_0 + 6 f_0;$$

derive a similar equation for g_1.

7.7   The motion of an axisymmetric solid of radius $h(z, t)$ that evolves under the dynamics of surface diffusion, driven by surface tension, is described by the PDE [19]

$$\frac{\partial h}{\partial t} = \frac{1}{h} \left(\frac{h}{\sqrt{1 + h_z^2}} \kappa_z \right)_z, \quad (7.126)$$

where κ is given by (4.23).

(i) Show that there are similarity solutions to (7.126) of the form

$$h(z, t) = t'^\alpha H \left(\frac{z}{t'^\alpha} \right)$$

which describe the breakup of the column; find the exponent α. Show that, to leading order, the far field must behave as $H \approx a\xi$.

(ii) Analyze the behavior of small perturbations ϵ to the far-field behavior using a WKB analysis. Show that there exist three modes,

$$\epsilon_i \propto |\xi|^{-5/3} \exp \left\{ \tfrac{3}{4} \alpha_i \left[(1 + a^2)^2 \xi^4 \right]^{1/3} \right\},$$

with $\alpha_1 = -1, \alpha_{2,3} = (1 \pm i\sqrt{3})/2$. Conclude that there should be at most a discrete set of similarity solutions which satisfy the boundary conditions for the far field.

7.8   Use the form (7.12) of the viscous solution to argue that the transition to the generic solution involving inertia occurs when the minimum radius is [75]

$$\frac{h_{\min}^{\text{vis}\rightarrow\text{gen}}}{r_0} \approx \left(\frac{\text{Re}_{\text{cr}} \ell_v}{r_0} \right)^{1/(2\beta-1)},$$

where Re_{cr} is the critical value of the local Reynolds number (7.103) where crossover occurs. This predicts a strong dependence on viscosity, not found experimentally [187].

7.9   Derive the Stokes flow around a sphere of vanishing size in a fluid of viscosity η.

(i) Write down, using Exercise 4.4, the velocity field of a sphere of radius R drawn through the fluid by a force $\mathbf{F}$.

(ii) Take the limit $R \to 0$ to show that the velocity field generated by a point force $\mathbf{f} = \mathbf{F}\delta(\mathbf{r})$ in the origin is

$$u_i = \frac{1}{\eta} J_{ij} F_j, \quad \text{with} \quad J_{ij} = \frac{1}{8\pi} \left(\frac{\delta_{ij}}{r} + \frac{r_i r_j}{r^3} \right).$$

(iii) Show that the stress distribution σ_{ij} generated by the above point force is

$$\sigma_{ij} = \frac{1}{\eta} K_{ijk} F_k, \quad \text{with} \quad K_{ijk} = -\frac{3}{4\pi} \frac{r_i r_j r_k}{r^5}.$$

7.10 To derive expressions corresponding to J_{ij} and K_{ijk} in two dimensions, one has to integrate their three-dimensional counterparts along z, where $\mathbf{r} = (x, y, z)$, to produce a line force in the z-direction. However, this gives a diverging result for J_{ij}, a fact known as *Stokes' paradox*. To remedy this problem, an equal and opposite force has to be placed at a distance Δ from the original line force. Show that in the limit of large Δ this leads to

$$J_{ij}^{2D} = \frac{1}{4\pi} \left(-\delta_{ij} \ln \frac{r}{\Delta} + \frac{r_i r_j}{r^2} \right)$$

and

$$K_{ijk}^{2D} = -\frac{1}{\pi} \frac{r_i r_j r_k}{r^4}.$$

Any physical problem in two dimensions must have a vanishing total force acting on the fluid, in which case the parameter Δ drops out.

8

A numerical example: drop pinch-off

Numerical methods play an important role in our understanding of singularities. Simulations often provide crucial pointers to the local structure of a singularity, and they reveal which physical effects dominate near a singularity. In addition, the analytical descriptions that we are able to obtain encompass only the local structure of the solution. One often has to rely on numerics in order to capture how local singular solutions are connected on the global scale.

Regrettably, the development of numerical codes is often considered a pursuit best left to the specialist. Our aim is to highlight some of the fundamental ideas that go into the numerical description of singular behavior. As an example, we take the description of a capillary bridge of liquid collapsing under gravity (see Fig. 8.1), which we will describe in some technical detail. Two aspects are of particular importance:

Stability Solutions which are close to singular involve a wide range of time scales. As a result, great demands are placed on the stability of the numerical scheme being used. This issue is addressed by using so-called *implicit* numerical schemes.

Adaptability As the singularity is approached, the solution evolves on smaller and smaller length and time scales. It is crucial that the numerical scheme adapts to these changes, by adjusting the time step and by refining the computational grid in a small region around the singularity.

8.1 Finite-difference scheme

A liquid drop is held between two endplates of radius r_0 which are a distance L apart; cf. Example 6.4 in Section 6.2. We solve equations (6.59), (6.57) on a grid z_i, $i = 1, \ldots, k$, which divides up the computational domain; see Fig. 8.2.

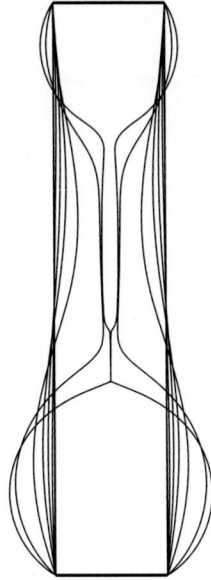

Figure 8.1 Simulation of a drop, between two solid plates, collapsing under gravity. The Ohnesorge number is $\mathrm{Oh} = \sqrt{\ell_\nu/r_0} = 0.22$ and the length of the bridge is ten times the radius r_0. In physical terms this corresponds to a drop of silicone oil with surface tension $\gamma = 20$ dyne/cm collapsing in a bridge of length $L = 0.143$ cm, with a viscosity 37 times that of water.

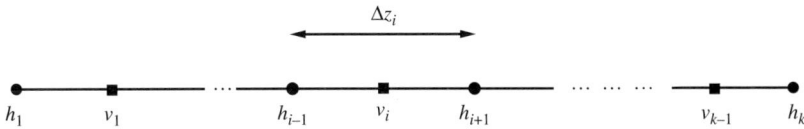

Figure 8.2 A schematic of the grid. The velocities $v_i = v(z_{i+1/2})$ are evaluated at the midpoint between two grid points at which the radius $h_i = h(z_i)$ is evaluated.

Since the total length of the bridge is L, we have $z_1 = 0$ and $z_k = L$. The grid spacing between two points is denoted as $\Delta z_i = z_{i+1} - z_i$; below we discuss in more detail how Δz_i is chosen to represent a given problem accurately. The radius $h(z)$ is represented by its values on this grid. We will see shortly that the scheme becomes more streamlined if the velocity is represented by its values at the midpoints $z_{i+1/2} \equiv (z_i + z_{i+1})/2$. Thus, in summary,

$$h_i = h(z_i), \quad v_i = v(z_{i+1/2}), \tag{8.1}$$

which means that the state of the system is represented by $2k - 1$ real numbers.

To calculate the right-hand side of (6.59) one needs to know the gradient of the pressure p' at the midpoint $z_{i+1/2}$. Let us assume for the moment that p is represented in terms of z_i, $p_i = p(z_i)$, and let us expand p into a Taylor series centered around $z_{i+1/2}$:

$$p(z) = p_{i+1/2} + p'_{i+1/2}(z - z_{i+1/2})$$
$$+ \frac{p''_{i+1/2}}{2}(z - z_{i+1/2})^2 + \frac{p'''_{i+1/2}}{6}(z - z_{i+1/2})^3 + \cdots .$$

If we evaluate this at z_i and z_{i+1} and take differences we obtain

$$p'_{i+1/2} = \frac{p_{i+1} - p_i}{\Delta z_i} - \frac{p'''_{i+1/2}}{24} \Delta z_i^2 + \cdots , \qquad (8.2)$$

where $z_{i+1} - z_{i+1/2} = z_{i+1/2} - z_i = \Delta z_i/2$. Owing to this symmetry, the term linear in Δz_i has dropped out from (8.2). In other words the *centered-finite-difference formula* in (8.2) represents the derivative at the midpoint up to corrections of quadratic order in the grid spacing: one speaks of this as a second-order approximation.

Example 8.1 (Interpolating polynomial) Another way to derive a finite-difference formula for the derivative of the pressure is to construct the unique polynomial which interpolates p at the points z_i, $z_{i+1/2}$, and z_{i+1}. A particularly elegant form of this polynomial is its Newton form:

$$p(z) = p_{i+1/2} + (p_{i+1} - p_{i+1/2})\frac{2(z - z_{i+1/2})}{\Delta z_i}$$
$$+ (p_{i+1} - 2p_{i+1/2} + p_i)\frac{2(z - z_{i+1/2})(z - z_{i+1})}{\Delta z_i^2}, \qquad (8.3)$$

for which there is a simple recursive algorithm [54]. The theory also provides an expression for the error, which we omit here. Differentiating (8.3) at $z_{i+1/2}$ yields the same result as (8.2) to leading order; an analysis of the error gives a quadratic error term, again in consistency with (8.2). Taking the second derivative of (8.3) one finds that

$$p''_{i+1/2} = \frac{4(p_{i+1} - 2p_{i+1/2} + p_i)}{\Delta z_i^2} + O(\Delta z_i^2), \qquad (8.4)$$

which is the centered-finite-difference formula for the second derivative. The order of the error term follows once more from symmetry (if the second derivative at z_i were desired then the error would be of first order, a second-order result requiring four grid points). $\qquad\square$

Our goal is to develop a finite-difference approximation which is of second order throughout. To this end we need a finite-difference expression for p_i which is at least of third order:

$$p_i = p_i^{\text{fin}} + O\left(\Delta z_i^3\right).$$

Inserting this into (8.2), one obtains

$$\frac{p_{i+1}^{\text{fin}} - p_i^{\text{fin}}}{\Delta z_i} = p_{i+1/2}' + O\left(\Delta z_i^2\right).$$

The finite-difference approximation for p must work for arbitrary grid spacings, since our grid will be highly refined close to the pinch point. The highest derivative appearing in the expression for p is h''; let us describe how a finite-difference approximation is obtained which is correct to third order. As in Example 8.1, we construct the polynomial which attains the values h_i at a given number of n grid points z_i. Evidently the order of the polynomial will be $n - 1$, since the number of polynomial coefficients must equal the number of grid points. With this polynomial in hand one can calculate $h(z)$ locally to within an error of $O(\Delta z_i^n)$. Now we can find approximations for the derivatives of h by differentiating the interpolating polynomial. The jth derivative is a polynomial of order $n - j - 1$, hence the error will be $O(\Delta z_i^{n-j})$. In other words, to find the second derivative to third order one needs $n = 5$ grid points.

For symmetry we calculate $h''(z_i)$ by using the five surrounding grid points $h_{i-2}, \ldots, h_{i+2}$; we denote the corresponding finite-difference approximation by h_i''. In view of translational invariance the result can depend only on the grid spacings $\Delta z_{i-2}, \ldots, \Delta z_{i+1}$. Since the derivative is a linear operation the result must also be linear in the h_i, leading to

$$h_i'' = \sum_{j=-2}^{2} a_j(\Delta z_{i-2}, \ldots, \Delta z_{i+1}) h_{i+j}. \tag{8.5}$$

Explicit expressions for a_j are obtained easily, e.g. using MAPLE, by calculating the interpolating polynomial through the required five grid points, differentiating, and then finding the coefficients for each h_{i+j}. For dimensional reasons, each of the five coefficients a_j is of degree -2 in the grid spacings. In the same way an expression for h_i' can be obtained (which is fourth-order accurate).

Example 8.2 (Five-point centered-finite-difference formula for the second derivative) Assuming that all grid spacings are equal, $\Delta z_i \equiv \Delta$, the finite-difference formula for the second derivative at the center of the five-point stencil is fourth-order accurate. The following MAPLE code computes the first coefficient a_{-2} (with $D \equiv \Delta$):

```
p:=interp([-2*D,-D,0,D,2*D],[h1,h2,h3,h4,h5],x):
% calculate interpolating polynomial
hpp:=diff(p,x$2): % compute second derivative
x:=0: % evaluate at center
simplify(coeff(hpp,h1));
% calculate coefficient of h_1 in expression for h''
```

Thus one obtains

$$a_{-2} = a_2 = -\frac{1}{12\Delta^2}, \quad a_{-1} = a_1 = \frac{4}{3\Delta^2}, \quad a_0 = -\frac{5}{2\Delta^2} \qquad (8.6)$$

for the coefficients in (8.5). The general case is no more difficult to code but the formulae are much more cumbersome (cf. Exercise 8.3).   □

Taking into account the full equation for the pressure, we use the approximation

$$p_i^{(\text{fin})} = \gamma \left(\frac{1}{h_i(1 + h_i'^2)^{1/2}} - \frac{h_i''}{(1 + h_i'^2)^{3/2}} \right). \qquad (8.7)$$

The formulae for $v'(z_{i+1/2})$ and $v''(z_{i+1/2})$ are identical to (8.5) if the grid z_i is replaced by the midpoints $z_{i+1/2}$. This means that $a_j(\Delta z_{i-2}, \ldots, \Delta z_{i+1})$ is to be replaced by $a_j(\Delta \bar{z}_{i-2}, \ldots, \Delta \bar{z}_{i+1})$, where $\Delta \bar{z}_i = (\Delta z_i + \Delta z_{i+1})/2$ is the spacing of the midpoints. Now the finite-difference versions of (6.57) (evaluated on z_i) and (6.59) (evaluated on $z_{i+1/2}$) become

$$rh_i \equiv \frac{dh_i}{dt} + \bar{v}_i h_i' + \frac{1}{2} h_i \bar{v}_i' = 0, \qquad (8.8)$$

$$rv_i \equiv \frac{dv_i}{dt} + v_i v_i' + \frac{p_{i+1}^{\text{fin}} - p_i^{\text{fin}}}{\Delta z_i} - 3vv_i'' - 6v\frac{v_i' \bar{h}_i'}{\bar{h}_i} + g = 0. \qquad (8.9)$$

Here $\bar{v}_i$ and $\bar{v}_i'$ denote the approximations for the velocity and its derivative at z_i. To compute them we use the four surrounding values, $v_{i-3/2}, \ldots, v_{i+3/2}$ at the midpoints. Correspondingly, $\bar{h}_i$ and $\bar{h}_i'$ denote approximations for h and its derivative evaluated at the midpoint $z_{i+1/2}$, which are calculated using $h_{i-1}, \ldots, h_{i+2}$.

Finally we implement the boundary conditions. We model a situation where a fluid drop is held between two disks of radius r_0, so that the boundary conditions for the discretization are $h_1 = h_k = r_0$. Since the disks are not moving, we have $v(0) = v(L) = 0$. In our staggered discretization, v_1 is the velocity at $z_{3/2} = \Delta z_1/2$. Strictly speaking, we would have to extrapolate to $z_1 = 0$ to impose the velocity boundary condition.

However, we cannot expect our one-dimensional description to be a faithful description of the three-dimensional flow dynamics near the endplates so it does not seem worthwhile to spend a great effort on a precise implementation of the boundary conditions. Essentially, all the interesting dynamics takes place near pinch-off. Thus we chose to put $v_1 = v_{k-1} = 0$ in the discretization, incurring a first-order error; as a result, there are no equations to be satisfied for rh_1, rv_1, rh_k, or rv_{k-1}. For rh_2, rv_2, rh_{k-1} and rv_{k-2} there is no full five-point stencil available, so we use a simple three-point finite-difference approximation. For example, for rh_2 we use

$$rh_2 \equiv \frac{dh_2}{dt} + \frac{v_1 + v_2}{2} \frac{h_3 - h_1}{2\Delta z_1} + \frac{h_2}{2} \frac{v_3 - v_1}{\Delta z_1} = 0,$$

and analogous expressions for rv_2.

Example 8.3 (Finite-difference scheme for the reaction–diffusion equation) As an example of the finite-difference discretization of a single nonlinear equation, let us take the reaction–diffusion equation (3.10), using a constant grid spacing Δz. Then the finite-difference approximations are

$$u_i' = \frac{u_{i+1} - u_{i-1}}{2\Delta z}, \quad u_i'' = \frac{u_{i+1} - 2u_i + u_{i-1}}{\Delta z^2},$$

which are of second order. Hence the finite-difference approximations for the reaction–diffusion equation are

$$ru_i \equiv \frac{du_i}{dt} - u_i^2 - u_i'' + 2\frac{u_i'^2}{u_i} \tag{8.10}$$

for $2 \le i \le k - 1$. If u satisfies the boundary conditions $u(z_1) = u_l$ and $u(z_k) = u_r$, we can simply put $u_1 = u_l$ and $u_k = u_r$. □

This completes the description of the spatial discretization. In principle the temporal discretization can be derived using similar ideas. However, much more than its accuracy, the usefulness of a time stepping scheme is determined by its stability. We will discuss this crucial issue next.

8.2 Time stepping and stability

Let us assume that the solution is to be found at discrete times, with t_n the latest. We would like to advance the solution to time t_{n+1}, with time step

$\Delta t = t_{n+1} - t_n$. Since (8.8), (8.9) constitutes a set of ordinary differential equations, we begin by discussing the single ODE

$$\dot{x} = f(x). \tag{8.11}$$

Expanding into a Taylor series about t_n, one finds

$$x_{n+1} = x_n + f_n \Delta t + O(\Delta t^2),$$

where $x_n \equiv x(t_n)$ and $f_n \equiv f(x_n)$. This suggests that the simplest possible way of advancing the solution is according to the algorithm

$$x_{n+1}^{(\text{fE})} = x_n + f_n \Delta t, \tag{8.12}$$

known as the forward Euler scheme. Since the time derivative is represented up to corrections of first order in Δt this is a first-order scheme.

Example 8.4 (Forward Euler scheme for hyperbolic growth) The forward Euler scheme for the growth model (2.7) reads

$$u_{n+1} - u_n = K \Delta t u_n^2 \equiv \tau u_n^2. \tag{8.13}$$

For small τ this should approximate the exact solution (2.9), whose discrete version reads

$$u_n = \frac{u_0}{1 - u_0 n \tau}. \tag{8.14}$$

Inserting (8.14) into (8.13), we find that the difference between the left- and the right-hand sides is

$$\frac{u_0}{1 - u_0(n+1)\tau} - \frac{u_0}{1 - u_0 n \tau} - \frac{u_0^2 \tau}{(1 - u_0 n \tau)^2}$$

$$= -\tau^2 \frac{u_0^3}{(1 - n \tau u_0)^2 (1 - u_0(n+1)\tau)},$$

which is indeed quadratically small for small τ.

For larger values of n, (8.14) eventually blows up while the iteration remains bounded for any finite n. It does however grow very rapidly, since for large values of u_n and u_{n+1} we have

$$u_{n+1} \approx \tau u_n^2.$$

Making an ansatz of the form

$$u_n = \frac{r^{\alpha(n)}}{\tau}$$

we obtain $\alpha(n + 1) = 2\alpha(n)$, with solution $\alpha(n) = C2^n$. Thus we find superexponential growth:

$$u_n = \frac{A^{2^n}}{\tau} \tag{8.15}$$

with $A = r^C$ a free parameter (that depends on the initial data).

We conclude that the solutions to the discretization, although growing very fast, will never blow up. As the singularity of the continuous equation is approached, its characteristic time scale becomes shorter and shorter and can no longer be resolved if the numerical time step is held constant. This is an important remark in connection with the description of singularities by numerical discretizations. We can identify a singularity not by the fact that the discretized solution blows up but by looking for similarity characteristics of the discrete problem that indicate the presence of self-similar solutions approaching blowup in the continuous problem. $\qquad\square$

Before we discuss ways to improve the accuracy of the forward Euler scheme (8.12), we have to discuss its *stability*. The question to address is whether tiny perturbations to the time evolution, which would decay away under the exact dynamics (8.11), might grow when using the *discrete* version (8.12). This would of course render the numerical solution useless. Since we are considering small perturbations, on the one hand it is sufficient to discuss a linearized version of (8.11),

$$\dot{x} = -ax \tag{8.16}$$

with $a > 0$. Thus the exact solution of (8.16) decays exponentially:

$$x = x_0 e^{-at},$$

with a^{-1} the characteristic time scale of decay.

On the other hand, the discretized version of (8.16) using the forward Euler scheme leads to the iteration

$$x_{n+1} = x_n - a\Delta t x_n = (1 - a\Delta t)x_n.$$

The discrete solution will decay only if $|(1 - a\Delta t)| < 1$; otherwise it will grow exponentially, although the real solution is known to decay rapidly! In other words, the forward Euler scheme is stable only if

$$\Delta t < \frac{2}{a}. \tag{8.17}$$

If we imagine the system (8.8), (8.9) decomposed into independent eigenmodes (see below), each mode will obey an equation of the form (8.16). The stability criterion (8.17) implies that the time step that one is able to choose is limited by the fastest decaying mode in the problem, which is typically associated with the shortest available wavelength. Thus numerical instability often manifests itself as oscillations on the scale of the grid. Let us find the shortest characteristic time scale (corresponding to the largest a) that is relevant for our system (8.8), (8.9).

To that end we consider a small perturbation on the current solution h_0, v_0:

$$h(z) = h_0 + \epsilon_1 e^{ikz}, \quad v(z) = v_0 + \epsilon_2 e^{ikz}. \tag{8.18}$$

We assume that the wavelength $2\pi/k$ of the perturbation is much smaller than the scale on which h_0 and v_0 are varying, so that the base solution can be taken as being constant. Inserting (8.18) into (6.57), (6.59) and linearizing in $\epsilon_{1/2}$ we find

$$\begin{pmatrix} \dot{\epsilon}_1 \\ \dot{\epsilon}_2 \end{pmatrix} = \begin{pmatrix} -iv_0 k & -ih_0 k/2 \\ -i\gamma k^3/\rho & -3\nu k^2 \end{pmatrix} \begin{pmatrix} \epsilon_1 \\ \epsilon_2 \end{pmatrix}.$$

Here we have taken into account only the largest terms as $k \to \infty$. The eigenvalues of this system for large k are

$$e_{1/2} = \tfrac{1}{2}\left(-3\nu \pm \sqrt{9\nu^2 - 2\gamma h_0/\rho}\right) k^2 + O(k). \tag{8.19}$$

Thus if the local thread radius satisfies $h_0 \ll \ell_\nu$, the most negative eigenvalue is $-3\nu k^2$. Since the largest wavenumber in the system is essentially $k \approx \pi/\Delta z$, the largest a-value we have to deal with is

$$a_{\max} \approx \frac{3\pi^2 \nu}{\Delta z^2}. \tag{8.20}$$

A more accurate way of calculating the eigenvalues is to represent (8.18) on the discrete grid z_i and to evaluate derivatives using the finite-difference formulae (8.8), (8.9). This is known as von Neumann stability analysis [178] and leads to essentially the same result as (8.20) but with somewhat modified prefactors.

The typical size of a structure to be resolved in a simulation has to be considerably larger than the grid spacing Δz, which means that we want to choose Δt much larger than $a_{\max}^{-1}$. As a result it is crucial for the time stepping to be stable even if $\Delta t a_{\max} \gg 1$. The simplest way to find a stable scheme is to replace the point x_n at which the right-hand side of (8.11) is evaluated in the forward Euler scheme by x_{n+1}; this is known as the backwards Euler scheme:

$$x_{n+1}^{(\mathrm{bE})} = x_n + f_{n+1}\Delta t. \tag{8.21}$$

Note that the right-hand side of (8.21) is not expressed in terms of the old time step x_n. Instead, a (generally nonlinear) equation for x_{n+1} has to be solved at each time step. Schemes where x_{n+1} appears on the right-hand side are called *implicit* schemes. Solving a nonlinear equation is a major inconvenience, but we will see that it can be well worth the effort!

Clearly (8.21) is still first-order accurate, but now the iteration for solving (8.16) reads

$$x_{n+1} = \frac{x_n}{1 + a\Delta t}.$$

For any positive value of $a\Delta t$ this iteration will decay to zero, as expected. In other words there is no condition on Δt for decaying modes to be stable; one says that the backwards Euler scheme is *unconditionally stable*. We now describe a simple way to use (8.21) to find a time stepping procedure which is *second-order accurate* yet retains all the stability properties of (8.21). This is done using the very general idea of Richardson extrapolation [179], which permits us to extend *any* numerical scheme to one of higher order. (However, it is unfortunately not necessarily true that the higher-order scheme inherits the same stability properties!)

We first take a step Δt using (8.21) and then repeat the same step using two steps of size $\Delta t/2$. Next, we combine the two results in such a way that the leading-order error cancels. Namely, comparing the result x_{n+1} from (8.21) with the true solution x_{n+1}^{true}, we have on the one hand

$$x_{n+1} = x_{n+1}^{\text{true}} + c\Delta t^2 + O(\Delta t^3)$$

(the coefficient c is easy to compute but its value is immaterial). On the other hand, the result of two half steps is

$$x_{n+1}^{(2)} = x_{n+1}^{\text{true}} + 2c\Delta t^2/4 + O(\Delta t^3),$$

since the error accumulates over two steps. Thus if one combines the two results so as to cancel the error, one obtains

$$x_{n+1}^{\text{new}} \equiv 2x_{n+1}^{(2)} - x_{n+1} = x_{n+1}^{\text{true}} + O(\Delta t^3), \tag{8.22}$$

a second-order accurate result!

Example 8.5 (Stability of Richardson scheme) To test the stability of (8.22) we apply the composite scheme to our model equation (8.16). Putting $\omega = a\Delta t$ for brevity, we find that

$$x_{n+1} = \frac{x_n}{1 + \omega}$$

and

$$x_{n+1}^{(2)} = \frac{x_{n+1/2}}{1 + \omega/2} = \frac{x_n}{(1 + \omega/2)^2}.$$

Hence

$$x_{n+1}^{\text{new}} = 2x_{n+1}^{(2)} - x_{n+1} = \left(\frac{2}{(1+\omega/2)^2} - \frac{1}{1+\omega}\right)x_n \equiv f x_n, \qquad (8.23)$$

with

$$f = \frac{4 + 4\omega - \omega^2}{(2+\omega)^2(1+\omega)}. \qquad (8.24)$$

For positive ω we have

$$|f| < \frac{|4 + 4\omega - \omega^2|}{|4 + 4\omega + \omega^2|} < 1,$$

so the scheme is stable. In fact this statement extends to any complex ω with a positive real part; see Exercise 8.6. $\qquad\square$

As an additional benefit the principle of step halving provides us with a convenient error estimate, if we compute

$$\text{error} = x_{n+1}^{(2)} - x_{n+1}. \qquad (8.25)$$

We adapt the time step Δt self-consistently in such a way that the error remains below certain bounds. Adaptive step refinement is crucial for the description of singular phenomena, as the characteristic time scale, and thus the necessary step size, is expected to vary over many orders of magnitude. That said, there is no reason *ever* to integrate an equation without adaptive step-size control, be it singular or not! Specifically, if the error is above a certain threshold, $\epsilon_t = 10^{-2}$, say, we reject the step and cut Δt in half. If the error has been below this threshold for 10 steps and is currently below $\epsilon_t/2$, we increase Δt by a factor of $2^{1/4}$. The above procedure is a robust way for Δt to be adjusted to this time scale automatically; it was adopted for the simulation shown in Fig. 8.1.

All the above generalizes easily to the system (8.8), (8.9). Let $(h_i^{(n)}, v_i^{(n)})$ be the old state of the system. To simplify the notation we represent this state by a single vector g_i. For k grid points, its dimension is $2k - 5$, taking into account that the boundary values are fixed. To make a time step we want to find increments $dg_i^{(n)}$ of the dependent variables such that

$$g_i^{(n+1)} = g_i^{(n)} + dg_i^{(n)}. \qquad (8.26)$$

In the backwards Euler step we replace all variables by $g_i^{(n+1)}$ and make the approximation

$$\frac{dg_i}{dt} = \frac{dg_i^{(n)}}{\Delta t}. \qquad (8.27)$$

We then have to solve the nonlinear system of equations (8.8), (8.9) for the increments dh_i, dv_i. Let us write the system (8.8), (8.9) symbolically as

$$rg_i = 0, \tag{8.28}$$

which is to be solved for $dg_i^{(n)}$.

We solve (8.28) using Newton's method [179], which is very efficient if one starts from a good guess. As an initial guess we use the value extrapolated from the previous time step Δt_{prev}:

$$dg_i^{(n+1)} = \frac{\Delta t}{\Delta t_{\text{prev}}} (g_i^{(n)} - g_i^{(n-1)}). \tag{8.29}$$

The error control guarantees that the initial guess is good, and we find that two Newton iterations suffice to reduce the maximum value $\max_i\{rg_i\}$ of all the residuals by a factor 10^{-6}. The update of the Newton iteration is

$$dg_i^{\text{up}} = dg_i - s_i,$$

where s_i is the solution of the linear problem

$$J_{ij}s_j = rg_i(dg_i) \tag{8.30}$$

and J_{ij} is the Jacobi matrix

$$J_{ij} = \frac{\partial(rg_i)}{\partial(dg_j)}. \tag{8.31}$$

In one spatial dimension $\mathbf{J}$ is a band-diagonal matrix, so (8.30) can be solved with an effort proportional to the size k of the problem [179]. Thus the computational effort required is (up to a substantial prefactor) equivalent to that of an explicit scheme where the results are given directly in terms of the old time step.

Example 8.6 (Jacobi matrix for reaction–diffusion equation) The Jacobi matrix $\mathbf{J}$ for the nonlinear diffusion equation (3.10) can be calculated from (8.31) using the spatial discretization (8.10) derived earlier. When calculating the derivatives (with $g_i^{(n)} \equiv u_i^{(n)}$), we evaluate the right-hand side at the new time step using (8.26): $u_i = u_i^{(n)} + du_i^{(n)}$. The time derivative is calculated from (8.27). Thus we obtain the matrix

$$
\begin{pmatrix}
1 & 0 & & 0 & & & & & \\
0 & dt^{-1}+b_2 & c_2 & & & & & & \\
0 & a_3 & dt^{-1}+b_3 & c_3 & & \mathbf{0} & & & \\
& & & \cdot & & & & & \\
& & \mathbf{0} & & \cdot & & & & \\
& & & & & \cdot & & & \\
& & & & a_{k-2} & dt^{-1}+b_{k-2} & c_{k-2} & 0 \\
& & & & & a_{k-1} & dt^{-1}+b_{k-1} & 0 \\
& & & & & 0 & 0 & 1
\end{pmatrix},
$$

where

$$
a_i = -\frac{1}{\Delta z^2} - 4\frac{u_i'}{u_i}, \quad b_i = -2u_i + \frac{2}{\Delta z^2} - 2\frac{u_i'^2}{u_i^2}, \quad c_i = -\frac{1}{\Delta z^2} + 4\frac{u_i'}{u_i}.
$$

We prefer to write the problem formally as a k-dimensional set of equations, so that $\mathbf{J}$ is a $k \times k$ matrix. However, since u_1 and u_k do not couple to any other variable they will remain at the values set by the boundary conditions.   □

Now we are set to advance the solution. We use the backwards Euler scheme to compute $g_i^{(n+1)}$. Next we take two successive steps of size $\Delta t/2$ to find $g_i^{(n+1,2)}$. The solution is finally updated to

$$
g_i^{\text{new}} = 2g_i^{(n+1,2)} - g_i^{(n+1)}. \tag{8.32}
$$

As the error we simply take the maximum norm over all the coefficients:

$$
\text{error} = \max_i \left\{ \left| g_i^{(n+1,2)} - g_i^{(n+1)} \right| \right\}. \tag{8.33}
$$

8.3 Grid refinement

Adapting the grid to a singular phenomenon is just as important as adapting the time step. In the simplest case a singularity is a small-scale feature in a solution which is varying smoothly at a finite distance from the singular point. Thus only the neighborhood of the singularity needs to have a fine grid, and this leads to huge savings in the number of grid points required; the grid is refined in response to the characteristics of the solution. In higher dimensions adaptive grid refinement is a very subtle issue [175] but in one dimension the situation can be kept simple, as we explain now.

Let us assume, as is the case in our simulation, that there is only one point where the solution comes close to pinch-off. Let the minimum radius be h_{min}

and the position of the minimum z_{min}. We measure the typical size of the singularity by the half-width of this minimum: let $z_{l/r}^{(1/2)}$ be the points to the left and right of the minimum where the radius has doubled: $h(z_{l/r}^{(1/2)}) = 2h_{min}$. Then the width to the left and right of the minimum is defined as

$$w_{l/r} = |z_{min} - z_{l/r}^{(1/2)}|. \tag{8.34}$$

We choose the minimum grid spacing Δz_{min} on the basis of the width of the singularity:

$$\Delta z_{min} = \min\{0.05w_l, 0.05w_r, \Delta z_{max}\}. \tag{8.35}$$

Here Δz_{max} is the grid spacing away from the singularity; it was chosen as 5×10^{-3} in the simulation. Near the center of the singularity Δz_i is kept constant, so as to perturb the solution as little as possible, and to achieve the highest accuracy. Everywhere in the interval $[z_{min} - 4w_l, z_{min} + 4w_r]$ we choose $\Delta z_i = \Delta z_{min}$.

Outside this interval we allow Δz_i to increase by a factor $a = 1.02$ at each step until the maximum value Δz_{max} is reached. Thus, from the right-hand edge of the central interval the iteration reads

$$\Delta z_{i+1} = \max\{a\Delta z_i, \Delta z_{max}\}, \tag{8.36}$$

which means that the grid size increases exponentially with the index i and that the number of grid points k_{tr} to achieve the transition between the central and the outer regions is

$$k_{tr} = \log_a\left(\frac{\Delta z_{max}}{\Delta z_{min}}\right). \tag{8.37}$$

Thus the number of points needed to resolve the singularity is essentially a constant for the central region plus a contribution that varies logarithmically with the resolution. A typical sequence of grids is shown in Fig. 8.3, with particular emphasis on the last stages of the evolution. The minimum radius reached is $h_{min} = 10^{-5}$, at which point the minimum grid spacing is $\Delta z_{min} = 1.2 \times 10^{-5}$. The final number of grid points $k = 2409$, while a constant grid spacing would have required $k \approx 10^5$.

To update the grid to the current solution we use a simple procedure. Whenever h_{min} has decreased by 5%, a new grid is created according to the current solution. With the new grid in hand we interpolate the old solution to the new grid. Namely, around each point z_j^{new} on the new grid, we choose four old grid points to the left and four to the right. We then compute the interpolating polynomial through these $2 \times 4 = 8$ nodes using the old profile $h_i^{old} = h(z_i^{old})$. The value of the polynomial at z_j^{new} gives the new value $h_j^{new} = h(z_j^{new})$. Note that

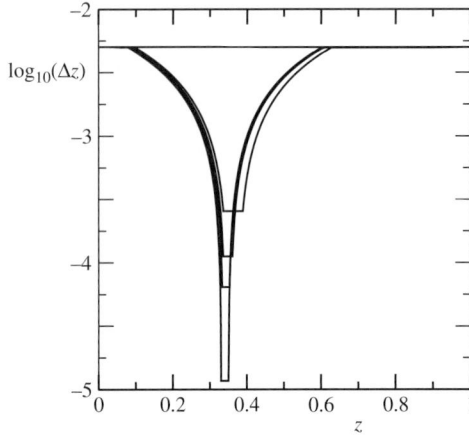

Figure 8.3 The logarithm of the grid spacing as function of position. The neighborhood of the position x_{min} of the minimum radius h_{min} is increasingly refined by the procedure.

near the boundaries the grid is not changing, so no interpolation is necessary. The same procedure is used for v but at the midpoints $z_{i+1/2}$.

Using the method described here we were able to follow the solution down to about $h_{min} = 10^{-5}$. It is remarkably difficult to go much further because the solution becomes quite susceptible to perturbations, as shown in Section 3.2. The wavelength of the most dangerous perturbations are of the order of h_{min}, which scales as t', while the total width ℓ_z of the singular region scales as $t'^{1/2}$. Thus, to ensure stability the grid spacing is set (at least for small radii) by h_{min}, not the much larger axial scale ℓ_z. This means that during the last stages the number of grid points increases as $(t')^{-1/2}$, which becomes prohibitively large.

8.4 Analysis of pinching

We now analyze the neighborhood of the pinch-off singularity in more detail, taking a particular set of parameters (see Fig. 8.1 for the large-scale picture). In the simulation we chose the distance L between the plates as the length scale, so z runs between 0 and 1. The plate radius was chosen as $r_0 = 0.1$, and the initial profiles were $h(z, 0) = r_0$ and $v(z, 0) = 0$. In the equations (6.57), (6.59) we put $\gamma/\rho = 1$, which is equivalent to measuring the time in units

$$\tau = \left(\frac{\rho L^3}{\gamma} \right)^{1/2}.$$

The viscosity was chosen as $\nu = 0.07$, which in our system of units amounts to a viscous length scale $\ell_\nu = \nu^2 = 0.0049$. Thus, for the initial radius r_0, viscous effects are relatively small but the generic pinching regime, characterized by a balance of surface tension, viscosity, and inertia, should begin to become relevant when $h_{\min} \approx \ell_\nu$. At the end of the simulation the time step $\Delta t = 1.9 \times 10^{-8}$ and the smallest grid spacing $\Delta z_{\min} = 1.2 \times 10^{-5}$.

We begin the analysis of the singularity by finding the singularity time t_0 and the position of the pinch point z_0, using the time dependence of the minimum radius $h_{\min}$ and its position $x_{\min}$. The extrapolation to values z_0, t_0 is of course aided by the fact that we know in advance which scaling behavior to expect. In particular $h_{\min}$ approaches zero linearly; see (7.4). Thus we can determine the singularity time t_0 to about 10 decimal places by plotting $h_{\min}$ as a function of t and fitting the last few data points with a straight line, giving $t_0 \approx 0.6213$.

The minimum should behave asymptotically as

$$h_{\min} = \frac{\phi_{\min}}{\nu}(t - t_0),$$

which gives, using the value for $\phi_{\min}$ from Table 7.3,

$$\log_{10} h_{\min} = -0.3619 + \log_{10}(t - t_0).$$

As seen on the left of Fig. 8.4, the value unity for the slope agrees well with the numerics. A linear fit over the last decade gives -0.3621 for the constant coefficient, in excellent agreement with theory.

The right-hand side of Fig. 8.4 tests a more subtle aspect of the singularity; the motion of the minimum at $z_{\min}$. According to the definition of the similarity variable, we have

$$z_{\min} - z_0 = \xi_{\min} \nu^{1/2}(t_0 - t)^{1/2},$$

where $\xi_{\min}$ is the position of the minimum of the similarity function ϕ (see Fig. 7.3): $\xi_{\min} \approx -1.6024$ [73]. Taking the logarithm we obtain

$$\log_{10}(z_{\min}) - z_0 = -0.37268 + \frac{1}{2}\log_{10}(t_0 - t).$$

We first determine the spatial pinch point z_0 by plotting $z_{\min}$ as a function of $\sqrt{t - t_0}$ and interpolating linearly, obtaining $z_0 \approx 0.3389$. Next we plot $\log_{10}(z_{\min} - z_0)$ against $\log_{10}(t - t_0)$, see Fig. 8.4 (right), and find the expected slope, $1/2$. A linear fit gives -0.3759 for the constant, again in very good agreement with theory.

Having found t_0 and z_0, we can now describe the entire profile without adjustable parameters, using (7.41). We rescale both $h(z, t)$ and $v(z, t)$ to find the similarity profiles $\phi(\xi)$ and $\psi(\xi)$; see Fig. 8.5. To illustrate the convergence

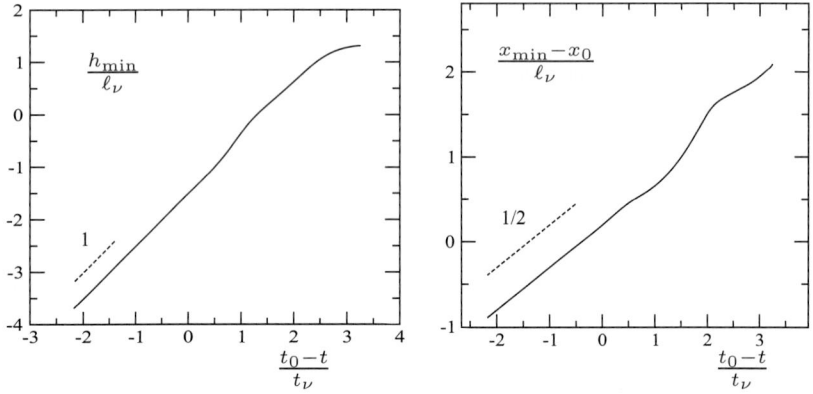

Figure 8.4 The minimum radius and its position for the simulation shown in Fig. 8.1. All lengths and times are normalized by their intrinsic values.

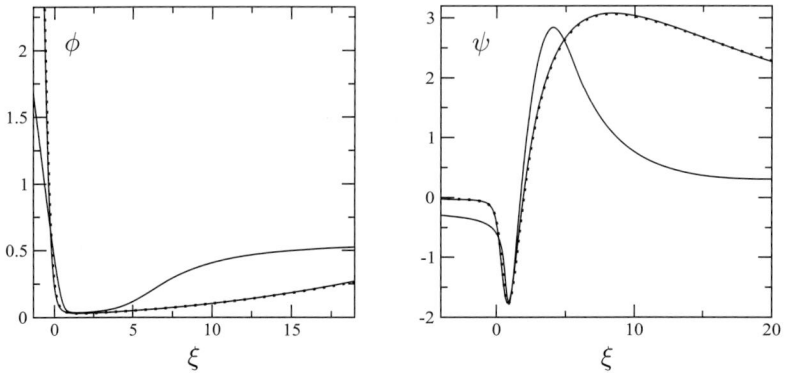

Figure 8.5 The same similarity profiles as shown in Fig. 7.3, with, for comparison, the same numerical simulation as that seen in Fig. 8.1. Shown is the neighborhood of the pinch point x_0 for two snapshots, taken at $h_{min}/r_0 = 10^{-2}$ and $h_{min}/r_0 = 10^{-5}$. The profiles of h and v have been rescaled according to (7.41) to yield the similarity profiles ϕ and ψ. The dots are the theoretical predictions for the profiles.

onto the theoretical solution, we consider the profile when $h_{min} = 10^{-2}$ and the last profile when $h_{min} = 10^{-5}$. In the similarity variable ξ the region over which there is agreement expands in time. For the profile at $h_{min} = 10^{-2}$, reasonable agreement near the center of the pinch region is found. For the last profile, almost perfect agreement has been reached over the entire range of ξ shown in Fig. 8.5. This is consistent with a roughly *constant* region of convergence in real space, which implies that this region expands as $1/\sqrt{t - t_0}$ in the similarity variable.

Exercises

8.1 Write down the second-order polynomial which interpolates a function h at grid points z_{i-1}, z_i, and z_{i+1}. Thus derive for the first derivative a three-point finite-difference formula which is second-order accurate for arbitrary grid spacing.

8.2 Using the MAPLE code given in Example 8.2, calculate the coefficients a_j of the finite-difference formula (8.5) for h' and h''', assuming a constant grid spacing $\Delta z_i \equiv \Delta$. In the case of the second derivative, the result is (8.6); for the first derivative the answer is given by

$$a_{-2} = -a_2 = \frac{1}{12\Delta^2}, \quad a_{-1} = -a_1 = -\frac{2}{3\Delta^2}, \quad a_0 = 0.$$

What is the accuracy of the respective formulae?

8.3 Use a variant of the MAPLE code of Example 8.2 to calculate the finite-difference formulae of the previous problem in the case of arbitrary grid spacing. For example, the coefficient a_{-2} of the second derivative is

$$a_{-2} = -\frac{2}{D}(\Delta z_{-1}\Delta z_1 + 2\Delta z_{-1}\Delta z_0 - \Delta z_0^2 - \Delta z_0\Delta z_1),$$

where

$$D = (\Delta z_0 + \Delta z_1 + \Delta z_{-1} + \Delta z_{-2})$$
$$\times(\Delta z_0 + \Delta z_{-1} + \Delta z_{-2})\Delta z_{-2}(\Delta z_{-1} + \Delta z_{-2}).$$

8.4 In the numerical scheme described in this chapter we imposed boundary conditions for the velocity not at the endpoints, but at $z_{3/2}$ and $z_{k-1/2}$, respectively. This is accurate to first order only. Using polynomial interpolation, describe a numerical scheme that is at least second-order accurate and imposes an arbitrary velocity $v(0) \equiv v_-(t)$ at the left endpoint.

8.5 Consider the following implicit version of Example 8.4:

$$u_{n+1} - u_n = \tau u_{n+1}^2. \tag{8.38}$$

Show that the discrete solution (8.14) solves (8.38) to leading order in τ. From (8.38) find u_{n+1} as a function of u_n and show that the iteration becomes indefinite at some finite value of n. Show that the maximum possible value of u_n is $1/(2\tau)$.

8.6 Use MAPLE to plot the curve $|f| = 1$, where f is the amplification factor of the Richardson scheme defined by (8.24). Convince yourself that $|f| < 1$ for all ω with positive real part.

8.7 An alternative to the Richardson scheme (8.22) is the so-called θ-weighted Crank–Nicolson algorithm,

$$\frac{x_{n+1} - x_n}{\Delta t} = \theta f_{n+1} + (1 - \theta) f_n. \tag{8.39}$$

(i) Use the model equation (8.16) to show that (8.39) is second-order accurate if $\theta = 1/2$ (this is the Crank–Nicolson scheme).

(ii) Show that, for $\theta > 1/2$, (8.39) is unconditionally stable.

(iii) Show that, for $\theta = 1/2$ and $a\Delta t \to \infty$, oscillations are not damped.

8.8 An implicit time-stepping scheme requires the solution of a nonlinear equation; this is costly, particularly in higher dimensions. An ingenious way around this problem was proposed by Douglas and Dupont [64, 67]. To illustrate their method, consider the (nonlinear) model equation (8.11). The idea is to write (8.11) in the identical form

$$\dot{x} = f(x) + bx - bx,$$

but to treat the first two terms on the right explicitly and the last term implicitly. Show that this results in the scheme

$$x_{n+1} = \frac{x_n(1 + b\Delta t) + \Delta t f_n}{1 + b\Delta t}, \tag{8.40}$$

which does *not* require the solution of an implicit equation.

(i) Show that (8.40) is unconditionally stable if $b > |f'(x)|/2$ for all x; to that end, approximate $f(x)$ locally as $f \approx -ax$.

(ii) What is the order of the scheme? Compute the error and conclude that b should be as small as possible, consistent with the stability requirement.

8.9 Use Richardson extrapolation to convert the scheme of Exercise 8.8 to a second-order scheme. Using the model equation (8.16) show that it is still unconditionally stable as long as $b > a/2$.

8.10 A model of Hele-Shaw flow with surface tension is [53]

$$\dot{h} + (hv)' = 0, \quad v = h'''. \tag{8.41}$$

Using centered differences, develop a spatial discretization scheme which is second-order accurate for constant grid spacing.

8.11 Write down a centered finite-difference scheme for (6.66) and (6.68) such that discrete versions of mass and momentum are conserved *exactly*.

8.12 If one uses the standard expression (4.23) to discretize the mean curvature, it will fail near the endpoints $z = \pm R$ of a circle

$$h(z) = \sqrt{R^2 - z^2},$$

although its mean curvature is exactly $2/R$. The reason is that the derivatives h_z and h_{zz} diverge near the endpoints, where the profile behaves as $h \approx \sqrt{2R}(z + R)^{1/2}$ (for the left endpoint).

(i) Show that the expression (4.75) for κ in terms of $a = h^2$ does not suffer from this problem.

(ii) Devise a three-point centered-difference scheme based on (4.75).

(iii) Find a finite-difference expression for the endpoint $z = -R$ which is second-order accurate.

8.13 Apply Newton's method to the one-dimensional problem $f(x) = 0$. Starting from an initial guess x_0, construct the tangent line to f at x_0. Use this to construct an approximate solution to $f(x) = 0$.

(i) Show that this leads to the iteration

$$x_{n+1} = x_n - \frac{f(x_n)}{f'(x_n)}.$$

(ii) Show that Newton's method exhibits quadratic convergence near a root x^*.

8.14 Consider the one-dimensional linear diffusion equation

$$\partial_t u(x, t) = u'',$$

with boundary conditions $u(0, t) = u_-$ and $u(1, t) = u_+$.

(i) Write down a backwards Euler scheme with three-point centered-difference discretization in the spatial domain.

(ii) Write down the Jacobian matrix J_{ij} needed for the solution of the implicit step. How is J_{ij} modified in the case of *periodic* boundary conditions?

(iii) Explain how Richardson extrapolation can be used to make the scheme second order in time.

8.15 Consider a tridiagonal matrix J_{ij}, as obtained in the previous problem. The general structure is $J_{ii} = b_i$ for $i = 1, \ldots, N$, $J_{ii-1} = a_i$ for $i = 2, \ldots, N$, and $J_{ii+1} = c_i$ for $i = 1, \ldots, N-1$. All other entries are zero. Construct an algorithm for the solution of the linear problem (8.30) using Gaussian elimination [178]. This means that all entries below the diagonal are eliminated using row operations. Subsequently, the solution can be found iteratively by starting from the last row, and working one's way upwards.

8.16 In the case of periodic boundary conditions, additional entries appear in the lower left-hand and upper right-hand corners of the matrix, equal to

$J_{1N} = a_1$ and $J_{N1} = c_N$, so that the matrix is no longer tridiagonal. Show that the resulting matrix can be written in the form

$$J_{ij} = J_{ij}^{\text{trid}} + u_i v_j,$$

where J_{ij}^{trid} is a tridiagonal matrix, $u_1 = \gamma$, $u_N = c_N$, $v_1 = 1$, and $v_N = a_1/\gamma$, where γ is an arbitrary parameter.

(i) To avoid round-off errors in the scheme below, a good choice is $\gamma = -b_1$. How does J_{ij}^{trid} differ from the original tridiagonal matrix?

(ii) For the implicit treatment we have to solve the matrix equation

$$J_{ij}s_j = r_i.$$

Show that the solution is

$$s_i = y_i - \left(\frac{v_i y_i}{1 + v_i z_i} \right) z_i,$$

where $\mathbf{y}$ and $\mathbf{z}$ are solutions of $J_{ij}^{\text{trid}} y_j = b_i$ and $J_{ij}^{\text{trid}} z_j = u_i$. This is known as the Sherman–Morrison formula [179]; in other words, one is only required to solve tridiagonal problems. The method can be generalized to additional entries in the form of block matrices.

9

Slow convergence

9.1 Mean curvature flow

In previous chapters we considered the convergence toward a particular similarity solution by a local analysis of the *dynamical system* close to its fixed point. We found the "ground state" similarity solution to be stable, with all the relevant eigenvalues negative. Now we consider a particular, but frequently observed, case: one eigenvalues is *zero* (excluding the zero eigenvalues which result from the existence of a continuous family of similarity solutions, as in Table 3.1). This means that to a linear approximation the system is in a neutral equilibrium; only a *nonlinear* analysis will reveal whether the fixed point is stable. This implies that the approach to the similarity solution is very slow at best, since the nonlinear terms are small.

A first example of such slow dynamics is that of pinching by mean curvature flow. As we can see from (4.26), only the projection of the speed of the interface in the direction normal to itself matters for its motion. In mean curvature flow this speed v_n of normal motion is proportional to the *mean curvature* of the interface. This can be considered in any dimension, but we will look at the physically most important case, that of a two-dimensional surface in three-dimensional space. For simplicity, we assume that the surface is axisymmetric.

A beautiful physical realization of mean curvature flow is furnished by the shape changes of a crystal as it melts or freezes in response to gradients of the surface or gravitational energy [161], as shown in Fig. 9.1. On a scale that is sufficiently small (i.e. below the capillary length $\ell_c = \sqrt{\gamma/(\rho g)}$, which is 1 mm for a typical fluid–gas interface), one finds that $v_n = -\Lambda \kappa$ where κ is (twice) the mean curvature and Λ is a constant [119]. This is a valid description only because in the system under study the latent heat of melting is extremely small, so the melting is not limited by temperature gradients.

207

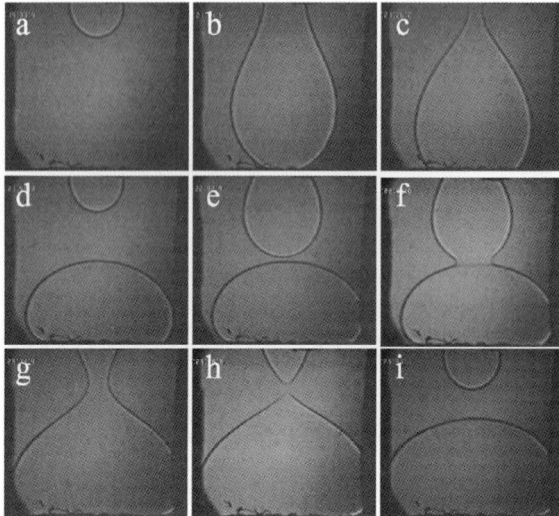

Figure 9.1 Nine images (of width 3.5 mm) showing how a ^{3}He crystal "flows" down from the upper part of a cryogenic cell into the lower part. (Reprinted with permission from [118]. Copyright 2004 by the American Physical Society.) The recording takes a few minutes. The temperature is just above the temperature $T_{min} = 0.32$ K where the latent heat vanishes. The crystal first "drips" down, so that a crystalline "drop" forms at the bottom, (a)–(c); then a second drop appears (d) and comes into contact with the first (e); coalescence is observed (f) and subsequently breakup occurs (h).

In Fig. 9.1 one may observe two breakup events, close to frames (c) and (h), at which the local radius goes to zero. If the radius of the axisymmetric crystal is $h(z, t)$, its shape is an isosurface of $C(r, z, t) = h(z, t) - r$ and the free surface motion (4.27) becomes

$$h_t = \mathbf{v} \cdot \mathbf{n} \sqrt{1 + h'^2}. \tag{9.1}$$

Using the mean curvature (4.23) and $\mathbf{v} \cdot \mathbf{n} = v_n = -\Lambda\kappa$, the equation of motion becomes

$$h_t = \Lambda \left(-\frac{1}{h} + \frac{h''}{1 + h'^2} \right). \tag{9.2}$$

To establish a connection with the examples considered in Chapter 3, we note that the transformation $h = 1/u$ leads to $h_t = -u_t/u^2$. This means that, in the u-variable, (9.2) has the form

$$u_t = u^3 + u'' + \text{terms in } u', \tag{9.3}$$

which is similar to (3.9) but with a cubic nonlinearity and additional nonlinear terms involving u'. In the present case, using the results to be developed below one can show that near the singularity these additional terms are small compared with u''. As a result, the singularities of axisymmetric mean curvature flow and those of reaction–diffusion equations with cubic nonlinearity share the same structure; see Exercise 9.4.

9.2 Center-manifold analysis

We begin by constructing a similarity solution of (9.2) as before. We nondimensionalize lengths, using for example the radius $r_0 = 0.5$ mm of the capillary out of which the crystal grows, as in Fig. 9.1. Using the constant Λ in (9.2) we can also construct a time scale r_0^2/Λ. Thus if t_0 is the time of the singularity and z_0 its location along the z-axis then

$$t' = \frac{\Lambda(t_0 - t)}{r_0^2}, \quad z' = \frac{z - z_0}{r_0} \tag{9.4}$$

are dimensionless measures of the temporal and spatial distances from the singularity. With these conventions the similarity solution becomes

$$h(z, t) = r_0 t'^\alpha \phi\left(\frac{z'}{t'^\beta}\right). \tag{9.5}$$

Inserting this into (9.2) we find

$$t'^{\alpha-1}\left(-\alpha\phi + \beta\xi\phi'\right) = t'^{-\alpha}\frac{1}{\phi} + t'^{\alpha-2\beta}\left(\frac{\phi''}{1 + t'^{2(\alpha-\beta)}\phi'^2}\right),$$

so to obtain a balance we require $\alpha - 1 = -\alpha$ and $\alpha - 1 = \alpha - 2\beta$. This leads to $\alpha = \beta = 1/2$ and the similarity equation

$$-\frac{\phi}{2} + \xi\frac{\phi'}{2} = \frac{\phi''}{1 + \phi'^2} - \frac{1}{\phi}, \quad \text{with} \quad \xi = \frac{z'}{t'^{1/2}}. \tag{9.6}$$

This equation has a constant solution,

$$\bar{\phi}(\xi) = \sqrt{2}, \tag{9.7}$$

which corresponds to a cylinder. However, (9.7) has to be dismissed since it does not agree with the matching condition (3.16) (the crystal would have to shrink uniformly, i.e. cylindrically). Instead, (3.16) requires the similarity function ϕ to be *linear* at infinity:

$$\phi(\xi) = a_\pm\xi, \quad \xi \to \infty.$$

The normal course of action would be to look for other solutions of the similarity equation (9.6) which exhibit linear growth at infinity, as required by the matching condition. However, mean curvature flow is special: a highly nontrivial result shows that regular solutions of (9.6) with such boundary conditions do not exist [5, 116]! The consequence of this observation is that we must grudgingly return to the fixed point solution (9.7), but the perturbations around it must be time dependent [6]. To study this time dependence we put

$$h(z, t) = r_0 t'^{1/2} \phi(\xi, \tau), \quad \tau = -\ln t', \tag{9.8}$$

so that the *dynamical system* becomes

$$\phi_\tau = \frac{\phi}{2} - \xi \frac{\phi'}{2} + \frac{\phi''}{1 + \phi'^2} - \frac{1}{\phi}. \tag{9.9}$$

We expand about the constant solution (9.7) by putting $\phi = \sqrt{2} + g$. It is easy to verify the identity

$$\frac{1}{\phi} = \frac{1}{\sqrt{2} + g} = \frac{1}{\sqrt{2}} - \frac{g}{2} + \frac{g^2}{2^{3/2} + 2g}$$

and thus

$$g_\tau = g - \frac{\xi g'}{2} + \frac{g''}{1 + g'^2} - \frac{g^2}{2^{3/2} + 2g}, \tag{9.10}$$

where $g = 0$ is the fixed point. As before we linearize the right-hand side of (9.10), and this leads to the eigenvalue problem

$$\nu g = \mathcal{L} g \equiv g - \xi g'/2 + g'', \tag{9.11}$$

which is the same as (3.44), studied earlier. For simplicity, we consider only even perturbations and, according to Example 3.3, the even eigenfunctions are

$$g_i(\xi) = H_{2i}(\xi/2),$$

with eigenvalues

$$\nu_i = 1 - i, \quad i = 0, 1, \dots \tag{9.12}$$

As usual the eigenfunctions are orthogonal with respect to a suitably defined inner product:

$$\langle g_i g_j \rangle \equiv \int_{-\infty}^{\infty} g_i(\xi) g_j(\xi) e^{-\xi^2/4} d\xi = 2\sqrt{\pi} \delta_{ij} 2^{2i} (2i)!. \tag{9.13}$$

Example 9.1 (Even eigenfunctions) Using Table 3.2 the first few eigenfunctions g_i are $g_0 = 1$, $g_1 = \xi^2 - 2$, and $g_2 = \xi^4 - 12\xi^2 + 12$. Note that the asymptotic behavior of these eigenfunctions is *not* consistent with

$g_i(\xi) \propto \xi^{(\alpha-\nu_i)/\beta} = \xi^{2i-1}$, as given by (3.59), and thus, like the base profile (9.7), does not correspond to a time-independent outer solution. □

Example 9.2 (Time translation) According to (9.12) the eigenvalue for $i = 0$ is $\nu_0 = 1$, the eigenvalue predicted by time translation invariance; see Table 3.1. The corresponding eigenfunction $\alpha\bar{\phi} - \beta\xi\bar{\phi}_\xi$ is a constant for the constant solution (9.7), in agreement with $g_0 = 1$. The eigenvalue $\nu = \beta = 1/2$ predicted by spatial invariance does not appear, since we are considering only symmetric solutions, so the minimum is fixed at $z' = 0$. □

Now we investigate the time dependence of g by expanding into a series of eigenfunctions. If we substitute

$$g(\xi, \tau) = \sum_{j=1}^{\infty} a_j(\tau) g_j(\xi) \tag{9.14}$$

into $g_\tau = \mathcal{L}$, we get

$$\sum_{j=1}^{\infty} \frac{da_j}{d\tau} g_j = \sum_{j=1}^{\infty} a_j \mathcal{L} g_j = \sum_{j=1}^{\infty} \nu_j a_j g_j. \tag{9.15}$$

Taking the inner product of (9.15) with g_i and using (9.13), we find

$$\frac{da_i}{d\tau} = \nu_i a_i = (1 - i)a_i. \tag{9.16}$$

Thus, as expected, all perturbations with $i > 1$ decay since $\nu_i < 0$. However, for $i = 1$ there is a *vanishing* eigenvalue. This eigenvalue cannot be explained away by the existence of a continuous family of solutions, as listed in Table 3.1. Instead, $\nu_1 = 0$ implies that in the direction of the eigenvector g_1 the linear problem does not contract toward the fixed point. Instead, we have to look at the *nonlinear* behavior to determine the dynamics in the g_1-direction. For our purposes it is enough to consider the leading-order (quadratic) nonlinearity:

$$g_\tau = \mathcal{L}g + NL(g) = \mathcal{L}g - 2^{-3/2}g^2 + O(g^3). \tag{9.17}$$

How do we cope with this nonlinear problem? The idea is that the system is "slaved" by the slow dynamics in the g_1-direction, as indicated schematically in Fig. 9.2. Far from the fixed point the a_1-direction extends to an *invariant manifold* along which the dynamics are described by a one-dimensional equation. In the other directions the dynamics contract rapidly onto the invariant manifold, so in the long-time limit, all a_j with $j \geq 2$ will be small compared

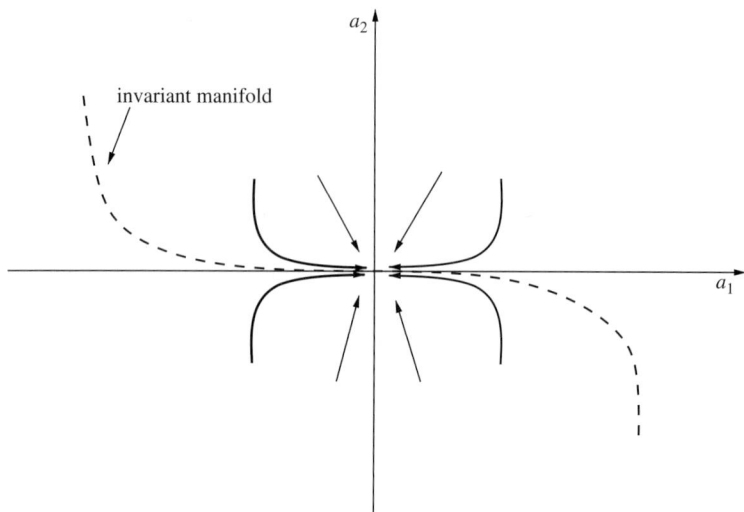

Figure 9.2 Schematic of the neighborhood of the fixed point (at the origin); only the first two directions of the expansion (9.14) are represented. The slow dynamics takes place along the invariant manifold; all other directions converge quickly onto it. Near the fixed point the invariant manifold points in the a_1-direction.

with a_1, as we will confirm below. Using (9.14) we can thus write, for the nonlinear term of (9.17),

$$- 2^{-3/2}g^2 = -2^{-3/2}\left(\sum_{j=1}^{\infty} a_j g_j\right)^2 = -2^{-3/2}a_1^2 g_1^2 + O(a_1 a_j). \quad (9.18)$$

By taking the inner product of (9.17) with g_1, we obtain

$$\frac{da_1}{d\tau}\langle g_1^2 \rangle = -2^{-3/2}a_1^2\langle g_1^3 \rangle; \quad (9.19)$$

the linear term vanishes. Multiplying (9.74) (see Exercise 9.1) by g_1 and using (9.13) we find that $\langle g_1^3 \rangle = 8\langle g_1^2 \rangle$, so that

$$\frac{da_1}{d\tau} = -2^{3/2}a_1^2. \quad (9.20)$$

The minus sign means that the fixed point $a_1 = 0$ is still being approached, but at a much slower rate than in the linear case. Indeed, the solution of (9.20) is

$$a_1 = 1/(2^{3/2}\tau), \quad (9.21)$$

which vanishes according to a power law as $\tau \to \infty$ rather than the usual exponential convergence. In "real time" this means that convergence onto a cylindricer is *logarithmic*, since $\tau = -\ln t'$. Indeed, the leading-order behavior is given by

$$h(z,t) = r_0 t'^{1/2} \left(\sqrt{2} + a_1(\tau) g_1(\xi) \right), \tag{9.22}$$

where $g_1 = \xi^2 - 2$. Hence

$$h(z,t) = r_0 t'^{1/2} \left(\sqrt{2} - \frac{2-\xi^2}{2^{3/2}\tau} + \cdots \right), \tag{9.23}$$

where the next term involves the eigenfunction $g_2(\xi)$. Note that (9.23) breaks down for large ξ since the asymptotic behavior does not correspond to a time-independent outer solution, as we have seen.

Example 9.3 (Blowup solution in self-similar form) By introducing a new self-similar variable $\eta = \xi/\sqrt{\tau}$, (9.23) can be written as

$$h(z,t) = r_0 t'^{1/2} \left(\sqrt{2} - \frac{1}{\sqrt{2}\tau} + \frac{\eta^2}{2^{3/2}} + \cdots \right).$$

In the limit $\tau \to \infty$ this tends to a solution of the form

$$h(z,t) = r_0 t'^{1/2} P(\eta), \tag{9.24}$$

where

$$P(\eta) = \sqrt{2} \left(1 + \frac{\eta^2}{4} \right) + O(\eta^4). \tag{9.25}$$

In Exercises 9.2 and 9.3 $P(\eta)$ is calculated in more detail. The fact that the profile converges to a similarity form in the new variable η, which contains a logarithmic correction, is the essence of *type-II self-similarity* [6]. ☐

Now we confirm that the amplitude a_2 corresponding to the next lowest eigenvalue is indeed slaved by the slow dynamics of a_1. To this end we compute the inner product of (9.17) with g_2, using (9.18):

$$\frac{da_2}{d\tau} \langle g_2^2 \rangle = (1-2)a_2 \langle g_2^2 \rangle - 2^{3/2} a_1^2 \langle g_1^2 g_2 \rangle. \tag{9.26}$$

Multiplying (9.74) by g_2 and once more invoking orthogonality, we find that $\langle g_1^2 g_2 \rangle = \langle g_2^2 \rangle$ and obtain

$$\frac{da_2}{d\tau} = -a_2 - 2^{-3/2} a_1^2 + \ldots = -a_2 - \frac{\sqrt{2}}{32\tau^2} + \ldots \tag{9.27}$$

The convergence of a_2 is not exponential, either, owing to the inhomogeneities in (9.27) which decay algebraically only. As a result, the time derivative of a_2 on the left of (9.27) is of lower order than a_2 itself. Instead, the decay of a_2 is determined by a_1: the balance is between a_2 and the inhomogeneity coming from a_1. The result is

$$a_2 \approx -\frac{\sqrt{2}}{32\tau^2}, \tag{9.28}$$

which is indeed a slower decay than that of a_1, justifying our previous assumptions.

At higher order, terms with $\ln \tau$ appear as well. A more detailed analysis which picks up all the logarithmic terms shows that the minimum h_m of the neck behaves as [80]

$$h_m = r_0(2t')^{1/2}\left(1 - \frac{1}{2\tau} - \frac{3 + 17\ln\tau}{8\tau^2}\right), \tag{9.29}$$

omitting terms of order τ^{-3}. A useful way to interpret this result is to write it as the evolution of an *effective* scaling exponent $\alpha(\tau)$. Namely, define a *local* scaling exponent by

$$\alpha(\tau) = -\frac{d\ln(h_m/r_0)}{d\tau}. \tag{9.30}$$

If $h_m(t')/r_0$ were a power law, this would give the value of the exponent. Now

$$\ln\frac{h_m}{r_0} \approx \frac{\ln 2}{2} - \frac{\tau}{2} + \ln\left(1 - \frac{1}{2\tau}\right) \approx \frac{\ln 2}{2} - \frac{\tau}{2} - \frac{1}{2\tau}$$

and thus

$$\alpha(\tau) = \frac{1}{2} - \frac{1}{2\tau^2} < \frac{1}{2}. \tag{9.31}$$

The local scaling exponent varies over a logarithmic time scale. This means that if an exponent were determined by fitting a power law over a limited scaling range, then its value could be significantly *smaller* than the asymptotic value and thus might suggest a spurious new scaling law [119].

9.3 Bubbles

9.3.1 Basics

Our next example of slow dynamics is the pinch-off of a bubble inside water; see Fig. 9.3. Just as in the "inverse" case of a drop of water in air, finite-time pinch-off occurs but a bubble is formed instead of a drop. Down to a neck radius of a few microns the effect of the air inside the bubble is negligible, and the water can be considered inviscid and irrotational. Thus at first one

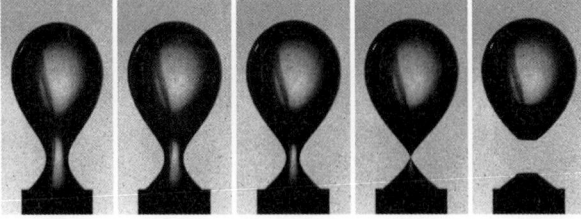

Figure 9.3 The pinch-off of an air bubble in water. (Reprinted with permission from [208]. Copyright 2007, AIP Publishing LLC.) The air is released through a pipette submerged under water, and begins to rise. A bubble separates at a finite time t_0.

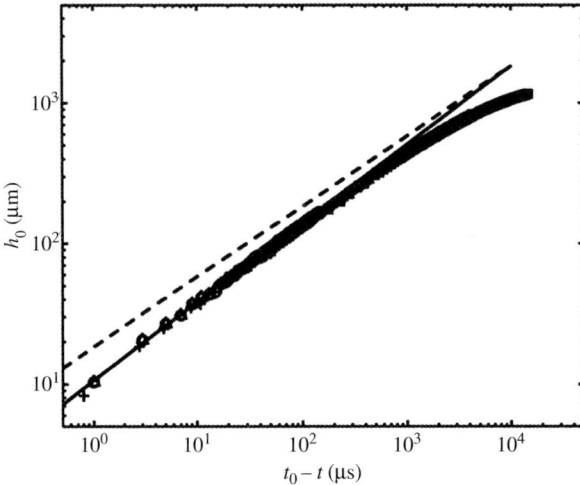

Figure 9.4 A measurement of the minimum radius h_0 of the neck of a bubble, as function of the time to pinch-off. (Reprinted with permission from [208]. Copyright 2007, AIP Publishing LLC.) The slope of the fit to the experimental data yields a scaling exponent $\alpha = 0.56$, significantly larger than 1/2 (broken line).

might think that the pinch-off dynamics is very similar to fluid-drop pinch-off in the inviscid limit, as described by (1.1). However, even a superficial examination of Fig. 9.3 shows that this is not the case. In the case of a bubble the profile around the pinch-point looks almost perfectly symmetric, while the drop profile is very asymmetric and even overturns. Turning to the scaling of the minimum radius h_0 of a bubble, an exponent $\alpha \approx 0.56$ is observed (see Fig. 9.4), much smaller than the exponent 2/3 seen for drop pinch-off. Thus the speed $\dot{h}_0$ of pinch-off diverges more quickly than would be expected on the basis of a surface-tension–inertia balance, which means that asymptotically surface tension must be subdominant.

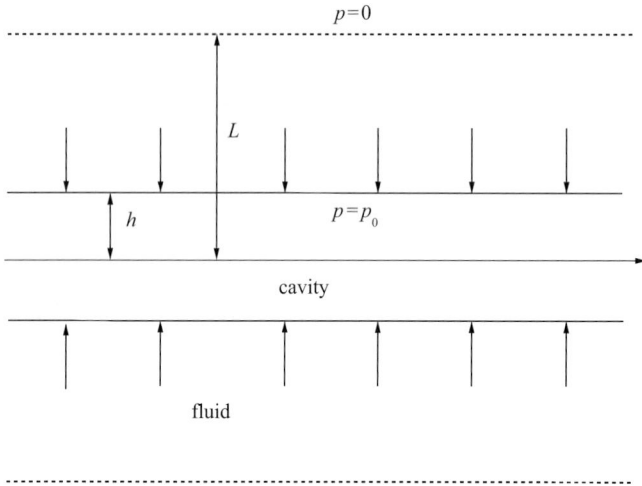

Figure 9.5 Schematic of the collapse of a cylindrical cavity of radius $h(t)$. The flow in the exterior is purely radial, and is set in motion because the cavity pressure $p_0 < 0$ is smaller than the pressure $p = 0$ at a distance L from the center.

Let us try to understand the rapid collapse of a bubble using a highly simplified model: we describe the bubble by a *cylindrical* cavity of radius $h(t)$; see Fig. 9.5. The velocity potential in the exterior is

$$\phi = F(t)\ln\frac{r}{L}, \tag{9.32}$$

chosen to vanish at a distance L. This we can think of as the (constant) radius of a fluid cylinder on whose surface the pressure vanishes. The potential (9.32) leads to the purely radial velocity field

$$\mathbf{v} = \nabla\phi = \frac{F(t)}{r}\mathbf{e}_r. \tag{9.33}$$

Thus, viewed from the exterior, the collapsing bubble or cavity acts as a two-dimensional source of fluid. The fluid velocity on the surface of the cavity equals the radial speed of the cavity itself: $F/h = \dot{h}$.

Now we consider the Bernoulli equation (4.39), neglecting the effect of gravity. At $r = L$ the velocity is small, and the choice (9.32) ensures that the pressure is held at $p(r) \approx 0$. If, however, the cavity is held at a constant pressure p_0 (and surface tension is neglected), evaluating (4.39) for $r = h$ gives

$$\dot{F}\ln\frac{h}{L} + \frac{F^2}{2h^2} + \frac{p_0}{\rho} = 0$$

or

$$\frac{d(h\dot{h})}{dt}\ln\frac{h}{L}+\frac{\dot{h}^2}{2}=-\frac{p_0}{\rho}. \tag{9.34}$$

The initial driving force for the collapse might be that the cavity pressure p_0 is negative but, as the flow accelerates, the right-hand side of (9.34) becomes subdominant. For a leading-order balance we have to equate the two terms on the left.

Since the equation of motion contains logarithms, we try the generalized power law

$$h\propto t'^{\alpha}\left(-\ln t'\right)^{\beta}\equiv t'^{\alpha}\tau^{\beta}, \tag{9.35}$$

where $t'=t_0-t$. Time is written in units of $L\rho^{1/2}/p_0^{1/2}$. Then $\ln h\approx-\alpha\tau+\beta\ln\tau\approx-\alpha\tau$, and so

$$\frac{d(h\dot{h})}{dt}\ln h\approx-\alpha^2(2\alpha-1)(t')^{2\alpha-2}\tau^{2\beta+1},\qquad\frac{\dot{h}^2}{2}\approx\frac{\alpha^2}{2}(t')^{2\alpha-2}\tau^{2\beta}.$$

Evidently this does not balance unless we have $\alpha=1/2$, in which case the term on the left cancels and we have to consider the next-to-leading order terms. If we retry (9.35) with $\alpha=1/2$, we find

$$\frac{d(h\dot{h})}{dt}\ln h\approx\frac{\beta\tau^{2\beta}}{2t'},\qquad\frac{\dot{h}^2}{2}\approx\frac{\tau^{2\beta}}{8t'}.$$

To achieve a balance we must have $\beta=-1/4$, and thus

$$h\propto\frac{t'^{1/2}}{\tau^{1/4}}. \tag{9.36}$$

This confirms that simply by focusing the inertia of the surrounding fluid into a line singularity one achieves extremely rapid collapse, much faster than surface tension could provide. Indeed, spherically symmetric collapse is even more effective: a classical solution by Rayleigh [133] implies that the radius $R(t)$ of a *spherical* cavity collapses as $t'^{2/5}$, faster still! The scaling (9.36) predicts logarithmic corrections to the scaling exponent $\alpha=1/2$, but they are too small to explain the data shown in Fig. 9.4. What is missing is a proper description of the spatial dependence of the profile, which clearly is far from cylindrical (see Fig. 9.3). Once the collapse is no longer uniform along the axis focusing becomes less effective; this is consistent with the larger scaling exponent $\alpha>1/2$ observed experimentally.

Example 9.4 (Cavity in a viscous fluid) Before we continue, we will return to the situation illustrated in Fig. 3.6; the breakup of a fluid (or gas) in a viscous environment in such a way that the inner viscosity can be neglected. We now demonstrate that the radius collapses at a constant rate, as implied by (3.71). We have shown previously that the drop profile is locally slender (see (3.81)), since the axial length scales as $t'^{1/2}$, which is much larger than the thread thickness, which is proportional to t'. Then, to leading order, the drop can be considered as a hollow cylinder of radius $h(t)$ surrounded by a viscous fluid of viscosity η. Using the same reasoning as in the inviscid case, from the incompressibility of the flow and the kinematic condition on the surface of the cylinder the velocity field is

$$\mathbf{v} = \frac{\dot{h}(t)h(t)}{r}\mathbf{e}_r. \tag{9.37}$$

For a radial flow, the Laplacian of the vector $\mathbf{v}$ is (see Appendix A)

$$\Delta\mathbf{v} = \left(\frac{\partial^2 v_r}{\partial r^2} + \frac{1}{r}\frac{\partial v_r}{\partial r} - \frac{v_r}{r^2}\right)\mathbf{e}_r = 0,$$

so (9.37) solves the Stokes equation with a constant pressure distribution. At the boundary the normal stress jump across the interface must balance the Laplace pressure γ/h:

$$\frac{\gamma}{h} = \sigma_{rr} + p_{\text{in}} = -\Delta p(t) - 2\eta\frac{\dot{h}}{h}. \tag{9.38}$$

For $h \to 0$ the pressure jump Δp becomes negligible, so then (9.38) is given simply by

$$\frac{\partial h}{\partial t} = -\frac{\gamma}{2\eta} \equiv -\frac{v_\eta}{2}. \tag{9.39}$$

Thus the profile indeed collapses everywhere at a constant rate $K = v_\eta/2$, which has been normalized to unity in (3.71). $\qquad\qquad\square$

9.3.2 Slender body theory

We aim to describe the inviscid motion in the unbounded fluid domain surrounding an axisymmetric cavity, allowing for a slow variation of the radius along the z-axis. Instead of a two-dimensional sink flow, which is uniform in the z-direction, we will try a superposition of three-dimensional sinks. This approach is known as slender body theory [12] and has a history going back to the beginning of the twentieth century. If the sink is located at the origin, the potential is

$$\phi = \frac{C}{|\mathbf{r}|} = \frac{C}{\sqrt{z^2 + r^2}}. \tag{9.40}$$

A continuous distribution of such sinks along the axis thus has potential

$$\phi = \int_{-L}^{L} \frac{C(x)dx}{\sqrt{(z-x)^2 + r^2}}, \tag{9.41}$$

where $C(x)$ remains to be determined. Note that (9.41) is so far an exact solution of Laplace's equation (4.37).

Making use of the fact that the bubble has a *slender* shape, we assume that the velocity is mostly radial: $v_r \gg v_z$. Taking the derivative of (9.41) in the r-direction, the value of v_r at the surface of the bubble $r = h(z)$ is given by

$$v_r(z) = -\int_{-L}^{L} \frac{C(x)h(z)dx}{\left[(z-x)^2 + h^2(z)\right]^{3/2}}. \tag{9.42}$$

Here and in the following, we suppress the time dependence of all profiles. The integral (9.42) is dominated by local contributions, as can be seen by considering the substitution $\eta = (z - x)/h(z)$, which gives

$$v_r(z) = -\frac{1}{h(z)} \int_{(z-L)/h(z)}^{(z+L)/h(z)} \frac{C(z - \eta h(z))}{(1 + \eta^2)^{3/2}} d\eta.$$

In a slender geometry we have $h \ll L$ and, taking the limit $h \to 0$, we obtain

$$v_r(z) \approx -\frac{C(z)}{h(z)} \int_{-\infty}^{\infty} \frac{1}{(1 + \eta^2)^{3/2}} d\eta = -\frac{2C(z)}{h(z)}. \tag{9.43}$$

Since $v_z \ll v_r$, the equation of motion (6.44) for a collapsing cavity of radius $h(z)$ becomes $\dot{h} \approx v_r$, and using (9.43) we find the distribution of sinks to be

$$C(z) \approx -\frac{\dot{a}(z)}{4},$$

where $a \equiv h^2$. We are now in a position to use Bernoulli's equation (4.39) once more, evaluated at the edge of the cavity. Since $v_z \ll v_r$ it has the form

$$\dot{\phi} + \frac{1}{2}v_r^2 + \frac{p_0}{\rho} = 0,$$

where p_0 is the gas pressure in the cavity and gravity has been neglected. Using (9.41) and (9.43) we arrive at

$$\int_{-L}^{L} \frac{\ddot{a}(x, t)dx}{\sqrt{(z-x)^2 + a(z, t)}} = \frac{\dot{a}^2(z, t)}{2a(z, t)} + \frac{4p_0}{\rho}, \tag{9.44}$$

restoring the explicit dependence on all the arguments, as the equation of motion for a slender bubble or cavity in the form of an integral equation.

9.3.3 Cavity dynamics

In the idealized geometry represented in Fig. 9.5, the collapse and therefore the acceleration $\ddot{a}$ of the interface are z-independent. The situation is thus analogous to that of the constant solution (9.7) in the case of mean curvature flow, which we found to be inconsistent with the matching condition (3.16). In terms of (9.44), a constant value of $\ddot{a}$ implies that the integral is determined by the boundary conditions since it diverges logarithmically for large values of x. By contrast, if $\ddot{a}$ were localized around the pinch point (which we assume to lie at $z = 0$) the integral converges locally. However, we have to keep track of not only the minimum cross section a_0 but also the width Δ of the pinch region. For definiteness we define Δ using the curvature of the interface at the pinch point, so we have

$$a_0(t) = a(0, t), \quad \Delta(t) = \left(\frac{2a_0}{a_0''}\right)^{1/2}, \tag{9.45}$$

where $a_0''(t) \equiv (\partial_{zz}a)(0, t)$. In accordance with the experimental profiles in Fig. 9.3, we assume that the profile $a(z, t)$ is symmetric with respect to the pinch point.

To obtain an equation of motion for $a_0(t)$, we evaluate (9.44) at the pinch point $z = 0$ and use the fact that the constant pressure p_0 is subdominant:

$$\int_{-L}^{L} \frac{\ddot{a}(x, t)dx}{\sqrt{x^2 + a_0}} = \frac{\dot{a}_0^2}{2a_0}. \tag{9.46}$$

To calculate the integral we assume that the acceleration has the similarity form

$$\ddot{a}(z, t) = \ddot{a}_0\phi(\eta), \quad \eta = \frac{z}{\Delta}, \tag{9.47}$$

where $\phi(\eta)$ is time independent to leading order, as we will show later. Indeed, in Exercise 9.6 the full similarity function may be calculated explicitly. We take the profile $\phi(\eta)$ (an even function) to be normalized in such a way that $\phi(0) = 1$. Then, in the limit $\Delta \to 0$, (9.46) becomes

$$\ddot{a}_0 F(\epsilon) = \frac{\dot{a}_0^2}{2a_0}, \tag{9.48}$$

where

$$F(\epsilon) = \int_{-\infty}^{\infty} \frac{\phi(\eta)d\eta}{\sqrt{\eta^2 + \epsilon^2}} \tag{9.49}$$

and the *aspect ratio* is

$$\epsilon = \frac{\sqrt{a_0}}{\Delta} \equiv \sqrt{\frac{a_0''}{2}}. \tag{9.50}$$

Our calculation will be based on the assumption that ϵ becomes small for $t' \to 0$, i.e. that the profile approaches a cylindrical shape asymptotically. We will confirm this assumption once we have calculated the relevant quantities.

We assume that on the one hand $\phi(\eta)$ vanishes at infinity, so the integral (9.49) converges at its upper limit. On the other hand, (9.49) has a logarithmic divergence at its lower limit if we set $\epsilon = 0$, so $F(\epsilon)$ will behave as $\ln \epsilon$ for small ϵ. To show this more formally we write

$$F = 2 \int_1^\infty \frac{\phi(\eta) d\eta}{\sqrt{\eta^2 + \epsilon^2}} + 2 \int_0^1 \frac{[\phi(\eta) - 1] d\eta}{\sqrt{\eta^2 + \epsilon^2}} + 2 \int_0^1 \frac{d\eta}{\sqrt{\eta^2 + \epsilon^2}}.$$

The first two integrals have finite values which depend on the precise form of $\phi(\eta)$ (note that $\phi(\eta) - 1 = O(\eta^2)$). However, the last integral diverges logarithmically as $\epsilon \to 0$ since

$$\int_0^1 \frac{d\eta}{\sqrt{\eta^2 + \epsilon^2}} = \int_0^{\epsilon^{-1}} \frac{d\zeta}{\sqrt{1 + \zeta^2}} = \ln \epsilon - \ln \left(\sqrt{1 + \epsilon^2} - 1 \right)$$

$$= -\ln \epsilon + O(\epsilon^0).$$

Thus, to leading order, $F(\epsilon) \approx -2 \ln \epsilon$; note that the shape of the similarity function $\phi(\eta)$ does not enter the numerical value of the prefactor. As a result (9.48) becomes

$$-2\ddot{a}_0 \ln \epsilon \equiv -\ddot{a}_0 \ln \frac{a_0}{\Delta^2} = \frac{\dot{a}_0^2}{2a_0}. \tag{9.51}$$

This equation is the same as (9.34), which was derived from the collapse of a cylindrical cavity, except for the fact that the argument of the logarithm is now the aspect ratio of the cavity. To close the system we also need an equation for Δ.

To that end we evaluate the second derivative of (9.44) at $z = 0$, to obtain

$$\int_{-L}^{L} \ddot{a}(x, t) \left(\frac{2x^2 - a_0}{(x^2 + a_0)^{5/2}} - \frac{a_0''}{2(x^2 + a_0)^{3/2}} \right) dx = \frac{\dot{a}_0 \dot{a}_0''}{a_0} - \frac{\dot{a}_0^2 a_0''}{2a_0^2}, \tag{9.52}$$

where we have used the fact that $\partial_z a = 0$ at the minimum. Employing the similarity form (9.47), equation (9.52) can be written as

$$\frac{\ddot{a}_0}{\Delta^2} G_1(\epsilon) - 2 \frac{\ddot{a}_0 \epsilon^2}{\Delta^2} G_2(\epsilon) = \frac{\dot{a}_0 \dot{a}_0''}{a_0} - \frac{\dot{a}_0^2 a_0''}{2a_0^2}, \tag{9.53}$$

with

$$G_1(\epsilon) = \int_{-\infty}^{\infty} \phi(\eta) \frac{2\eta^2 - \epsilon^2}{(\eta^2 + \epsilon^2)^{5/2}} d\eta, \quad G_2(\epsilon) = \int_{-\infty}^{\infty} \frac{\phi(\eta)}{2(\eta^2 + \epsilon^2)^{3/2}} d\eta. \tag{9.54}$$

For $\epsilon = 0$, the integrand of $G_2(0)$ diverges as $1/\eta^3$ at its lower limit. To calculate the leading divergence of $G_2(\epsilon)$ for small ϵ, we write it in the equivalent form

$$G_2(\epsilon) = \int_0^\infty \frac{\phi(\eta) - 1}{(\eta^2 + \epsilon^2)^{3/2}} d\eta + \int_0^\infty \frac{d\eta}{(\eta^2 + \epsilon^2)^{3/2}},$$

adding and subtracting the leading divergence. Now the integrand of the first integral behaves as $1/\eta$ for $\epsilon = 0$ and $\eta \to 0$, giving a divergence as $\ln \epsilon$. For the second integral, we find in the limit $\epsilon \to 0$

$$\int_0^\infty \frac{d\eta}{(\eta^2 + \epsilon^2)^{3/2}} = \frac{1}{\epsilon^2} \int_0^\infty \frac{d\zeta}{(1 + \zeta^2)^{3/2}} = \frac{1}{\epsilon^2},$$

and thus

$$G_2(\epsilon) = \frac{1}{\epsilon^2} + O(\ln \epsilon). \tag{9.55}$$

Turning to $G_1(\epsilon)$, we have

$$G_1(\epsilon) = 2 \int_0^\infty \frac{[\phi(\eta) - 1](2\eta^2 - \epsilon^2)}{(\eta^2 + \epsilon^2)^{5/2}} d\eta + 2 \int_0^\infty \frac{2\eta^2 - \epsilon^2}{(\eta^2 + \epsilon^2)^{5/2}} d\eta. \tag{9.56}$$

The second integral in (9.56) would normally diverge as $1/\epsilon^2$ but fortuitously it vanishes:

$$\frac{2}{\epsilon^2} \int_0^\infty \frac{2\zeta^2 - 1}{(1 + \zeta^2)^{5/2}} d\zeta = 0.$$

Hence we are left with another logarithmically diverging integral, since $\phi(\eta) - 1 = O(\eta^2)$, and we can calculate the leading order without knowing $\phi(\eta)$ explicitly. Adding and subtracting the remaining divergence in (9.56), we have

$$G_1(\epsilon) = 2 \int_0^1 \frac{[\phi(\eta) - 1 - \phi''(0)\eta^2/2](2\eta^2 - \epsilon^2)}{(\eta^2 + \epsilon^2)^{5/2}} d\eta$$

$$+ \phi''(0) \int_0^1 \frac{\eta^2(2\eta^2 - \epsilon^2)}{(\eta^2 + \epsilon^2)^{5/2}} d\eta + 2 \int_1^\infty \frac{[\phi(\eta) - 1](2\eta^2 - \epsilon^2)}{(\eta^2 + \epsilon^2)^{5/2}} d\eta.$$

The only diverging integral as $\epsilon \to 0$ is the second on the right, which behaves as follows:

$$\int_0^1 \frac{\eta^2(2\eta^2 - \epsilon^2)}{(\eta^2 + \epsilon^2)^{5/2}} d\eta = \int_0^{\epsilon^{-1}} \frac{\zeta^2(2\zeta^2 - 1)}{(1 + \zeta^2)^{5/2}} d\zeta = -2 \ln \epsilon + O(\epsilon^0),$$

and so

$$G_1(\epsilon) = -2\phi''(0) \ln \epsilon + O(\epsilon^0). \tag{9.57}$$

Inserting (9.55) and (9.57) into (9.53), we can see that the logarithmically diverging contribution from $G_1(\epsilon)$ dominates and so we obtain

$$-2\ddot{a}_0'' \ln \epsilon \equiv -\ddot{a}_0'' \ln \frac{a_0}{\Delta^2} = \frac{\dot{a}_0 \dot{a}_0''}{a_0} - \frac{\dot{a}_0^2 a_0''}{2a_0^2}. \tag{9.58}$$

Here we have used the fact that differentiating (9.47) twice gives $\ddot{a}_0'' = (\ddot{a}_0/\Delta^2)\phi''(0)$. Equations (9.51) and (9.58) form a coupled system of ODEs for a_0 and Δ, which we will now solve.

9.3.4 Approach to the fixed point

As in our treatment of mean curvature flow, we would like to describe the *approach* to the fixed point. This can be done by allowing for a slow time dependence of the exponents, which in analogy to (9.30) we will define as

$$2\alpha \equiv -\frac{1}{a_0}\frac{da_0}{d\tau}, \quad 2\delta \equiv -\frac{1}{a_0''}\frac{da_0''}{d\tau}, \tag{9.59}$$

with $\tau = -\ln t' \equiv -\ln(t_0 - t)$. As a unit of time one can choose, for example, $\sqrt{\rho L^3/\gamma}$, where L is the characteristic width of the bubble. Using the relations

$$\frac{\dot{a}_0}{a_0} = -\frac{2}{t'}\alpha, \quad \frac{\ddot{a}_0}{a_0} = -\frac{2}{t'^2}\left(\alpha + \alpha_\tau - 2\alpha^2\right) \tag{9.60}$$

and

$$\frac{\dot{a}_0''}{a_0''} = -\frac{2}{t'}\delta, \quad \frac{\ddot{a}_0''}{a_0''} = -\frac{2}{t'^2}\left(\delta + \delta_\tau - 2\delta^2\right), \tag{9.61}$$

we transform to new variables α, δ. Dividing (9.51) and (9.58) by a_0 and a_0'', respectively, and using (9.60) and (9.61), one finds

$$\left(\alpha_\tau + \alpha - 2\alpha^2\right)\ln\epsilon^2 = \alpha^2, \tag{9.62}$$

$$\left(\delta_\tau + \delta - 2\delta^2\right)\ln\epsilon^2 = 2\alpha\delta - \alpha^2, \tag{9.63}$$

where the subscript denotes the τ-derivative. The time dependence of $\ln\epsilon$ is found from integrating

$$(\ln\epsilon)_\tau = \frac{\epsilon_\tau}{\epsilon} = \frac{\dot{\epsilon}}{\epsilon}t' = \frac{1}{2}\frac{\dot{a}_0''}{a_0''}t' = -\delta. \tag{9.64}$$

Now we analyze the system (9.62)–(9.64) in the neighborhood of its fixed point, which corresponds to cylindrical collapse. In addition to α and δ we introduce a third variable $v(\tau) = 1/\ln \epsilon^2$, so that the equations of motion become

$$\alpha_\tau + \alpha - 2\alpha^2 = \alpha^2 v, \tag{9.65}$$

$$\delta_\tau + \delta - 2\delta^2 = (2\alpha\delta - \alpha^2)v, \tag{9.66}$$

$$v_\tau = 2\delta v^2. \tag{9.67}$$

A fixed point is found at $(\alpha, \delta, v) = (1/2, 0, 0)$, consistently with the cylindrical analysis. Putting $\alpha(\tau) = 1/2 + u(\tau)$, the leading-order behavior of (9.65)–(9.67) becomes

$$\begin{pmatrix} u \\ \delta \\ v \end{pmatrix}_\tau = \begin{pmatrix} 1 & 0 & 1/4 \\ 0 & -1 & -1/4 \\ 0 & 0 & 0 \end{pmatrix} \begin{pmatrix} u \\ \delta \\ v \end{pmatrix} + \begin{pmatrix} 0 \\ 0 \\ 2\delta v^2 \end{pmatrix}. \tag{9.68}$$

Clearly the linearization around the fixed point has eigenvalues $1, 0$, and -1. The positive eigenvalue corresponds to time translation; see Table 3.1. However, the vanishing eigenvalue implies that we have to retain the nonlinearities on the right of (9.68) in order to describe the slow approach to the fixed point.

We can simplify (9.68) further by noting that the derivatives u_τ and δ_τ are of lower order in the first two equations corresponding to (9.68); this is analogous to our treatment of (9.27) in the mean curvature problem. Thus to leading order the first two equations become $u + v/4 = 0$ and $-\delta - v/4 = 0$, respectively. As a result we have $u = \delta$ and $v = -4\delta$, which means that the variables u and v are "slaved" by the slow variable δ, which obeys the nonlinear equation

$$\delta_\tau = -8\delta^3. \tag{9.69}$$

While in the mean curvature flow problem we were left with a quadratic nonlinearity, here the nonlinearity is of *third* order, corresponding to an even slower approach. In the limit $\tau \to \infty$ the cubic equation (9.69) is solved by $\delta = 1/(4\sqrt{\tau})$, so it follows that

$$\alpha = \frac{1}{2} + \frac{1}{4\sqrt{\tau}} + O(\tau^{-1}), \quad \delta = \frac{1}{4\sqrt{\tau}} + O(\tau^{-1}). \tag{9.70}$$

To calculate the corrections of order $1/\tau$ the explicit form of $\phi(\eta)$ is needed; see Exercise 9.7. Note that, using (9.70),

$$\phi''(0) = \frac{\ddot{a}_0''}{\ddot{a}_0/\Delta^2} = 2\frac{\ddot{a}_0''/a_0}{\ddot{a}_0/a_0} = 2\frac{\delta_\tau + \delta - 2\delta^2}{\alpha_\tau + \alpha - 2\alpha^2} = -2 + O\left(\frac{1}{\sqrt{\tau}}\right), \tag{9.71}$$

which is consistent with $\phi(\eta)$ being time independent to leading order.

In Fig. 9.6 we present a full numerical simulation of inviscid potential flow in the exterior of an elongated bubble with surface tension. This means that

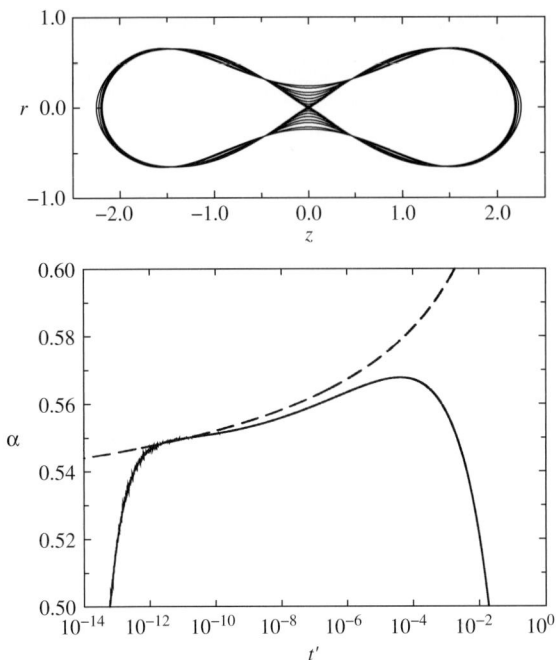

Figure 9.6 Upper panel: a simulation of the collapse of an elongated cavity using the full irrotational flow dynamics to model the exterior; see [84]. Lower panel: the exponent α as calculated from the numerical simulation (solid line) and the leading-order asymptotic theory (broken line). After an initial transient, theory and numerical simulation agree, except at the very end of the simulation, where t' is extremely small. The sudden drop in the exponent as determined from the simulation comes from the numerical uncertainty in determining the singularity time t_0.

$\Delta\phi = 0$ is solved in the exterior subject to the boundary condition (4.40) on the surface of the bubble. The inertia of the gas inside the bubble is neglected, so we have $\phi^{(in)} = 0$. Initially the bubble pinches, owing to surface tension, but the late stages are dominated by inertia. Time is measured in units of $\sqrt{\rho L^3/\gamma}$ and all lengths have been normalized by L. The minimum radius h_0 is used to calculate

$$\alpha = (t'/h_0)dh_0/dt',$$

and the pinch-off time t_0 is estimated directly from the numerical data (the solid line). The apparently rapid drop of α at around $t' = 10^{-11}$ merely illustrates the accuracy to which t_0 can be determined. The broken curve is the leading-order prediction, given by equation (9.70), and agrees well with the numerics. Notice that, between $t' = 10^{-4}$ and $t' = 10^{-12}$, α changes only by about 0.02 or 4%. This illustrates why in a typical experiment (see Fig. 9.4) the minimum radius is well described by a power law fit over a range of three decades. The value $\alpha = 0.56$, estimated by this fit, is well in line with the values seen in Fig. 9.6 over a comparable range of t' values.

Finally, we return to the scaling behavior of the minimum h_0 and to the question of slenderness. Integrating $(\ln a_0)_\tau = -2\alpha \approx -1 - 1/(2\sqrt{\tau})$, we find that

$$\ln a_0 = -\tau - \sqrt{\tau} + \text{const},$$

and thus

$$h_0 \propto t'^{1/2} e^{-\sqrt{\tau}/2}. \tag{9.72}$$

This is different from our original answer (9.36), which came from the assumption of a cylindrical cavity and which predicted a much smaller correction to power law scaling. Integrating (9.64), with $\delta = 1/(4\sqrt{\tau})$, we find that the aspect ratio behaves as follows:

$$\epsilon \propto e^{-\sqrt{\tau}/2}. \tag{9.73}$$

This is good news since the smallness parameter of our theory decays to zero rapidly, making the calculation self-consistent.

Exercises

9.1   Use Example 9.1 to show that

$$g_1^2 = g_2 + 8g_1 + 8g_0. \tag{9.74}$$

9.2   Use (9.28) to show that at the next order the similarity function introduced by (9.24) and (9.25), valid in the limit $\tau \to \infty$, is

$$P(\eta) = \sqrt{2}\left(1 + \frac{\eta^2}{4} - \frac{\eta^4}{32} + \dots\right).$$

9.3   The similarity function $P(\eta)$ (see (9.25)) can in fact be calculated exactly [6].
   (i) Using the transformation $\phi(\xi, \tau) = P(\xi/\sqrt{\tau}, \tau)$, derive the dynamical system for $P(\eta, \tau)$.
   (ii) Show that, for $\tau \to \infty$, the dynamical system becomes

$$P_\tau = \frac{P}{2} - \frac{\eta}{2}P_\eta - \frac{1}{P}. \tag{9.75}$$

(iii) Assuming that $P_\tau \to 0$ for $\tau \to \infty$, solve (9.75). Show that the only solution consistent with (9.25) is

$$P(\eta) = \sqrt{2 + \eta^2},\qquad(9.76)$$

and confirm that the result agrees with Exercise 9.2.

9.4 Consider blowup for solutions of the reaction–diffusion equation

$$u_t = u^p + u_{xx}\qquad(9.77)$$

with $p > 1$, which is a generalized form of (3.9). Show that the constant profile

$$u = t'^{-1/(p-1)} (p-1)^{1/(p-1)}$$

is a solution of (9.77); you can use the fact that no other self-similar solution exists. Repeat the center-manifold analysis of Section 9.2 to describe the blowup of u in finite time.

9.5 Analyze the pinching of axisymmetric hypersurfaces of $n + 1$ dimensions, evolving according to *Ricci flow* [80]. If the radius of the neck is ψ and s is the arc length of the generatrix, the equation becomes

$$\psi_t = \psi_{ss} - \frac{(n-1)(1-\psi_s^2)}{\psi}.\qquad(9.78)$$

Find a similarity description of $\psi(s, t)$ similar to that for mean curvature flow.

9.6 (*) As shown in detail in [87], when (9.44) is evaluated at an arbitrary position z one obtains an equation analogous to (9.51) but where ϵ can be treated as being z-independent to leading order:

$$-2\ddot{a}(z, t) \ln \epsilon(\tau) = \frac{\dot{a}^2(z, t)}{2a(z, t)}.\qquad(9.79)$$

(i) Introduce the new variable $d = a/a_\tau$ and use (9.73) to show that

$$d_\tau - d = 1 - \frac{1}{2\sqrt{\tau}}.\qquad(9.80)$$

(ii) Solve (9.80) to show that

$$\ln a(z, t) = C_2(z) - \int_{\tau_0}^\tau \frac{d\tau'}{C_1(z)e^{\tau'} + 1 - 1/(2\sqrt{\tau})},$$

where $C_1(z)$, $C_2(z)$ are constants of integration; conclude that C_1 must have the expansion $C_1(z) = Bz^2 + B_4z^4 + \dots$.

(iii) Show that to leading order, as $\tau \to \infty$,

$$a(z,t) = a_0(t)\left(1 + \eta^2\right).$$

(iv) Finally use (9.79) to show that

$$\phi(\eta) = \frac{1}{1 + \eta^2} \tag{9.81}$$

to leading order. Confirm that this result is consistent with (9.71).

9.7   Use the result of the previous exercise to show that, at the next order, (9.51) and (9.58) are replaced by

$$-\ddot{a}_0 \ln \frac{\epsilon^2}{4} = \frac{\dot{a}_0^2}{2a_0}$$

and

$$-\ddot{a}_0'' \ln \frac{\epsilon^2 e^2}{4} = \frac{\dot{a}_0 \dot{a}_0''}{a_0} - \frac{\dot{a}_0^2 a_0''}{2a_0^2},$$

respectively. Analyze this system in the limit $\tau \to \infty$ to show that

$$\alpha = \frac{1}{2} + \frac{1}{4\sqrt{\tau}} - \frac{1}{2\tau} + O(\tau^{-3/2}), \quad \delta = \frac{1}{4\sqrt{\tau}} + O(\tau^{-3/2}),$$

contain the next-order corrections to (9.70).

9.8   (*) The nonlinear Schrödinger equation, with a critical nonlinearity and in one space dimension, reads

$$i\varphi_t + \varphi_{xx} + |\varphi|^4\varphi = 0.$$

(i) Show that there exists a solution of the form (with $\tau = -\ln t'$ and $\xi = x'/t'^{1/2}$)

$$\varphi(x,t) = \frac{e^{i\mu_0\tau}e^{-i\xi^2/8}}{t'^{1/4}}\varphi_0(\xi),$$

where

$$\varphi_0(\xi) = \frac{(3\mu_0)^{1/4}}{\sqrt{\cosh(2\sqrt{\mu_0}\xi)}}.$$

(ii) Look for more general solutions, of the form

$$\varphi(x,t) = e^{i\mu(t) - i\beta(t)z^2/4}\lambda^{1/2}(t)\varphi_a(z),$$

by writing a system of ODEs to be satisfied by $\mu(t)$, $\beta(t)$, and $\lambda(t)$, with $z = \lambda(t)x$; here $\varphi_a(z)$ satisfies the equation

$$-\varphi_{a,\xi\xi} + \varphi_a - \frac{1}{4}az^2\varphi_a - |\varphi_a|^4\varphi_a = 0.$$

(iii) If one allows a to vary slowly in time, a semiclassical analysis yields the following extra condition to be satisfied by $h(t) \equiv \sqrt{a(t)}$:

$$h_t = -c\lambda^2 \frac{e^{-S_0/h}}{h},$$

where S_0 and c are positive constants. By introducing

$$\alpha = -\frac{\lambda_\tau}{\lambda}, \quad \gamma = -\frac{\beta_\tau}{\beta}, \quad \delta = -\frac{h_\tau}{h},$$

show that solutions close to $(\alpha, \gamma, \delta) = (1/2, 0, 0)$ are such that

$$\alpha(\tau) \sim \frac{1}{2} - \frac{1}{2\tau \ln \tau}, \quad \gamma(\tau) \sim \frac{1}{\tau \ln \tau}, \quad \delta(\tau) \sim \frac{1}{\tau \ln \tau}.$$

Notice that the presence of factors $\ln \tau = \ln(-\ln t')$ implies corrections which vary extremely slowly in time. Hint: use the results of Exercise 2.10.

9.9 (*) The Keller–Segel model for chemotaxis describes the aggregation of microorganisms driven by chemotactic stimuli [107, 108]. In two space dimensions the problem has biological meaning. If we describe the density of individuals by $u(x, t)$ and the concentration of the chemotactic agent by $v(x, t)$ then the Keller–Segel system reads

$$u_t = \Delta u - \chi \nabla \cdot (u \nabla v),$$
$$\Gamma v_t = \Delta v + (u - 1),$$

where Γ and χ are non-negative constants. We consider the particular case $\Gamma = 0$.

(i) Show that there exists an explicit solution $u(x, t) = a(t)$ such that $a(t) \sim (t')^{-1}$ as $t \to t_0$.

(ii) By performing center-manifold analysis, show that there exists a radial solution such that, at leading order ($\tau = -\ln t'$),

$$u(r, t) \sim \frac{1}{R^2(t)} U\left(\frac{r}{R(t)}\right)$$

with

$$R(t) = C \exp\left(-\frac{\tau}{2} - \frac{\sqrt{\tau}}{\sqrt{2}} - \frac{\ln \tau}{4} + \frac{\ln \tau}{4\sqrt{\tau}}\right) [1 + o(1)]$$

and

$$U(\xi) = \frac{8}{\chi(1 + \xi^2)^2}.$$

This implies that individuals aggregate into a Dirac delta of total mass $8\pi/\chi$ in finite time.

10

Continuation

An important issue is what happens *after* a singularity has occurred. A very simple example was considered in Section 3.5; now we will turn to some physical examples in more detail. First, we consider the breakup of a liquid drop, which marks the transition from one piece to two separate pieces of liquid. To know how the two pieces evolve after breakup one needs to *continue* across the singularity. In particular, is the continuation *unique* or are there several possible ways for the solution to evolve after the singularity? As in Section 3.5 we will consider two very different approaches to continuation. In the first approach we construct a similarity solution that applies directly after the singularity, by solving two separate problems corresponding to the two pieces resulting from breakup. We show that they are determined uniquely by the pre-breakup dynamics. In the second approach we regularize the equations on a small scale, so that a true singularity never occurs.

10.1 Post-breakup solution: viscous thread

In Chapter 7 we saw that the asymptotics of drop breakup is described by the similarity solution (7.41). We described in some detail how a unique, stable, similarity solution is selected; this is shown in Fig. 7.3. Now we show how the knowledge of the pre-breakup solution permits us to construct a unique post-breakup solution. The strategy has already been laid out and hinges on the matching conditions (3.84), which transfer information about the behavior of the pre-breakup solution *far from the pinch point* to the post-breakup solution.

However, there remains a technical challenge, illustrated in Fig. 10.1. In the immediate neighborhood of the tip, where the slope h' of the profile diverges, the solution can no longer be considered slender. As a result the derivation of

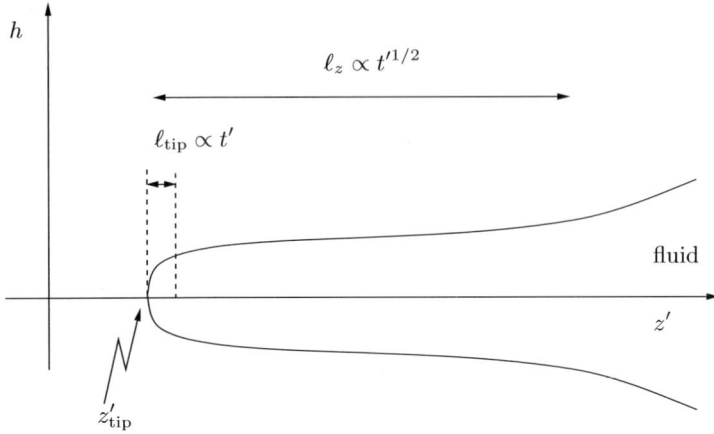

Figure 10.1 A schematic of a receding thread after breakup. The size of the tip region ℓ_{tip} is the same as the typical radius of the fluid thread, much smaller than the axial extension of the thread.

the slender jet equations (6.57), (6.58) is not valid; the equations break down at the tip. Instead, we must consider a separate tip region, whose width ℓ_{tip} is comparable with its radial extension. In the language of matched asymptotics the tip is the inner region, while the receding thread, over which the slender jet approximation applies, is the outer region. At the matching point between the two, both radial scales must coincide and so we must have $\ell_{\text{tip}} \approx \ell_\nu |t'|$, using the scaling of the outer solution. This means that in the scaling limit $t' \to 0$ the size of the tip region goes to zero *in similarity variables*; see Fig. 10.1. By matching the inner (finger) solution to the outer (thread) solution, we can determine the correct boundary condition for the outer similarity solution at some matching point ξ_m.

We begin with the similarity description after breakup, which is

$$h(z,t) = \ell_\nu |t'| \phi_a\left(\frac{z'}{|t'|^{1/2}}\right), \quad v(z,t) = v_\eta |t'|^{-1/2} \psi_a\left(\frac{z'}{|t'|^{1/2}}\right). \quad (10.1)$$

Since after breakup $t > t_0$ we have $|t'| = (t - t_0)/t_\nu$, which corresponds to a change in sign of the time variable relative to the pre-breakup solution (7.41). As a result, all the terms in (7.45), (7.46) which come from the time derivative change sign, and the similarity equations valid after breakup are

$$\phi_a - \frac{\xi \phi_a'}{2} + \psi_a \phi_a' = -\frac{\psi_a' \phi_a}{2}, \quad (10.2)$$

$$-\frac{\psi_a}{2} - \frac{\xi \psi_a'}{2} + \psi_a \psi_a' = \frac{\phi_a'}{\phi_a^2} + 3\frac{(\psi_a' \phi_a^2)'}{\phi_a^2}. \quad (10.3)$$

The asymptotic structure of ϕ_a and ψ_a for large $|\xi|$ is still described by expansions of the form (7.48), which are determined completely by the two leading-order amplitudes a_0 and b_0 (the expressions for the coefficients in terms of a_0, b_0 being different, of course).

The matching condition (3.84) for h reads, in similarity variables,

$$\lim_{t' \to 0} |t'| \phi_b \left(\frac{\pm \Delta z}{|t'|^{1/2}} \right) = \lim_{t' \to 0} |t'| \phi_a \left(\frac{\pm \Delta z}{|t'|^{1/2}} \right) ;$$

so, using (7.48), we obtain $a_0^\pm|_b = a_0^\pm|_a$. A similar argument holds for the velocity profile v, and the matching conditions connecting the pre- and post-breakup solutions become

$$a_0^\pm|_a = a_0^\pm|_b , \quad b_0^\pm|_a = b_0^\pm|_b . \tag{10.4}$$

Recall that the key to finding a unique solution before breakup, and thus to finding unique values of $a_0^\pm|_b$ and $b_0^\pm|_b$, was the fact that the asymptotics (7.47) is unstable as the similarity equations are integrated toward $\xi \to \pm\infty$. Solutions are driven away from the asymptotic behavior (7.48), the corrections growing as $\exp(\xi^2/12)$. After breakup the second term on the left of (10.3) changes sign, which turns the factor into $\exp(-\xi^2/12)$. This means that *all* solutions of (10.2), (10.3) are attracted to the correct behavior (7.47) at infinity. Since according to (10.4) the amplitudes $a_0^\pm$ and $b_0^\pm$ are the same before and after breakup, (10.2), (10.3) have to be solved subject to the condition (7.47), with $a_0^\pm$ and $b_0^\pm$ as reported in Table 7.3.

To find the similarity solution after breakup we integrate (10.2), (10.3) from the tip and apply the shooting method to satisfy (10.4). However, we *cannot* apply the "natural" boundary condition $\phi_a = 0$ at the tip since the pressure term ϕ_a'/ϕ_a^2 would then go to infinity. Instead the correct boundary condition must be determined from matching to an inner solution (the finger solution), which we describe now.

Finger solution

Since the radial and axial length scales are comparable near the tip we must solve the full three-dimensional axisymmetric equations. However, the local Reynolds number on the scale of the tip size ℓ_{tip} is $v_\eta |t'|^{-1/2} \ell_{\text{tip}}/\nu \propto |t'|^{1/2}$, which goes to zero. Thus inertia drops out and we only have to consider the Stokes limit. To make the presentation more transparent, we will calculate the tip solution on the basis of the slender jet equations with full curvature, although this is strictly speaking not valid near the tip. The true solution, based on the full Stokes equation, is given in [77] and empirically is found to be very close to the simplified solution presented here.

We start from (7.124) with the tension T in the thread set equal to zero, since there is no force acting on the tip. Since the thread solution (10.1) becomes very long and slender in the limit $|t'| \rightarrow 0$, we expect the finger solution to asymptote to a constant radius far from the tip. The expectation that radial and axial scales are equal motivates the similarity solution

$$h = \ell_v |t'| H(Z), \quad v = v_\eta V(Z), \tag{10.5}$$

where $Z = z/(|t'|\ell_v)$; we choose the tip of the finger to be at $Z = 0$, so that $H(0) = 0$. Inserting (10.5) into (6.66) and into (7.124) with $T = 0$, we obtain the similarity equations

$$2H^2 - Z(H^2)' + (H^2 V)' = 0, \tag{10.6}$$

$$-3V' = \frac{1}{H(1 + H'^2)^{1/2}} + \frac{H''}{(1 + H'^2)^{3/2}}. \tag{10.7}$$

We are looking for solutions which approach a constant value $H = H_\infty$ far from the tip. Equation (10.6) implies that $V' = -2$, so (10.7) determines the finger radius H_∞ to be $1/6$; see Fig. 10.2. Solutions to (10.6), (10.7) are found by adjusting the curvature at the tip until $H(Z)$ conforms with the expected value, $1/6$, for large Z. The behavior of V for large arguments is

$$V(Z) \approx 3Z_m - 2Z, \tag{10.8}$$

where $Z_m \approx 0.087\,55$ is a constant determined numerically. To understand the meaning of Z_m, note that (10.6) can be rearranged as

$$1 - 36H^2 = \left[Z - 12H^2(Z - V) \right]',$$

which we integrate from 0 to a (large) value Z:

$$\int_0^Z (1 - 36H^2)dZ = Z - 12H^2 \left(Z - 3Z_m + 2Z \right),$$

using (10.8). Inserting the asymptotic behavior $H = 1/6$ for large Z, this becomes

$$Z_m = \int_0^\infty \left(1 - 36H^2 \right) dZ. \tag{10.9}$$

However, (10.9) can be written in the form

$$\int_0^Z H^2 dZ = \frac{1}{36} \int_{Z_m}^Z dZ,$$

which implies that Z_m is the tip position of a cylinder having the same volume as the finger (see Fig. 10.2).

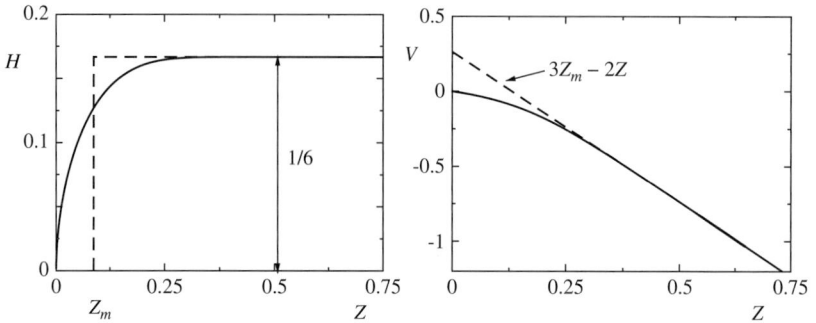

Figure 10.2 The similarity profiles H, V found from integrating (10.6), (10.7), with the boundary condition $H(\infty) = 1/6$. They describe a finger retracting at constant speed, in a frame of reference where the tip is at rest. The value Z_m is the tip position of a cylinder having the same volume as the finger.

Matching

Matching between the inner and outer solutions implies that the radius h has to agree between the outer limit $Z \to \infty$ of the inner solution and the inner limit of the outer solution $\xi \to \xi_m$: $h_m = \ell_v |t'| \phi_a(\xi_m) = \ell_v |t'|/6$ or

$$\phi_a(\xi_m) = \frac{1}{6}. \tag{10.10}$$

We now show that if the similarity equations (10.2), (10.3) are solved with the boundary condition (10.10) then the velocity field near ξ_m agrees with (10.8) and inertia drops out, as anticipated. To this end we transform to the frame of reference of the matching point $z'_m = \xi_m |t'|^{1/2}$, which moves at speed $v_m/v_\eta = \xi_m |t'|^{-1/2}/2$. Therefore, if we introduce the variable $\bar{\xi} = |t'|^{-1/2}(\xi - \xi_m)$, which is $O(1)$ on the scale of the inner solution, the transformation to inner variables moving with the tip becomes

$$\bar{\phi}(\bar{\xi}) = \phi_a(\bar{\xi} + \xi_m), \quad \bar{\psi}(\bar{\xi}) = |t'|^{-1/2} \left[\psi_a(\bar{\xi} + \xi_m) - \xi_m/2 \right]. \tag{10.11}$$

The boundary conditions to be satisfied are $\bar{\phi}(0) = 1/6$ and $\bar{\psi}(0) = 0$.

Inserting (10.11) into the similarity equations (10.2), (10.3), one obtains

$$\bar{\phi} - \frac{\bar{\xi}\bar{\phi}'}{2} + \bar{\psi}\bar{\phi}' = -\frac{\bar{\psi}'\bar{\phi}}{2} \tag{10.12}$$

$$|t'|\bar{\phi}^2 \left(-\frac{\bar{\psi}}{2} - \frac{\bar{\xi}\bar{\psi}'}{2} + \bar{\psi}\bar{\psi}' - \frac{\xi_m}{4|t'|^{1/2}} \right) = \left(\bar{\phi} + 3\bar{\psi}'\bar{\phi}^2 \right)', \tag{10.13}$$

respectively. Using $\bar{\psi}(0) = 0$, (10.12) yields $\bar{\phi} = -\bar{\psi}'\bar{\phi}/2$ as $\bar{\xi} \to 0$, so it follows that

$$\bar{\psi} = -2\bar{\xi} + \text{higher-order terms}. \tag{10.14}$$

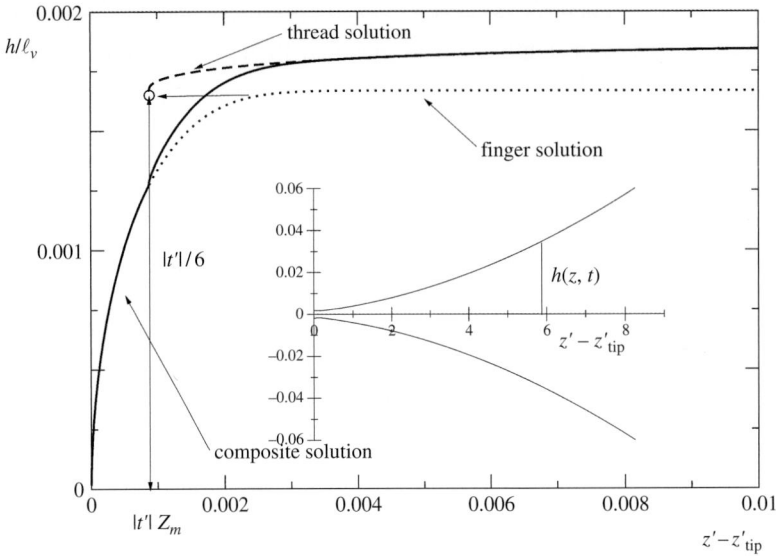

Figure 10.3 A detail of the tip region of a receding thread at $|t'| = 0.01$, showing the inner, finger, solution (broken line), the outer, thread, solution (broken line), and a composite solution constructed from the two (solid line). All lengths are in units of ℓ_v, and the origin is at the tip. The inset shows the composite solution on a larger scale.

Expressed using ψ_a, this becomes

$$\psi_a = \frac{\xi_m}{2} - 2(\xi - \xi_m) + \text{higher-order terms}, \qquad (10.15)$$

which matches the large-Z limit (10.8) of the inner solution. In the limit $t' \to 0$ all inertial terms drop out from the left-hand side of (10.13) and, integrating once, we obtain

$$\bar{\phi} + 3\bar{\psi}'\bar{\phi}^2 = \bar{T}, \qquad (10.16)$$

where $\bar{T}$ is a constant of integration. Inserting $\bar{\phi}(0) = 1/6$ and $\bar{\psi}'(0) = -2$, one obtains $\bar{T} = 0$, which corresponds to our earlier condition that the tension T in (7.124) vanishes.

The matching between the inner and outer solutions is illustrated in Fig. 10.3. In particular, the radius at the end of the thread solution is set by the radius of the inner (finger) solution. There remains, however, an unknown shift between the two solutions. This shift can be determined from volume conservation: this requires that the fluid volume in the tip region must equal the volume inside the outer solution, which asymptotically is a cylinder of constant radius. But this means that $Z = Z_m$ in the finger solution corresponds to $\xi = \xi_m$ in the outer solution; as shown in Fig. 10.3, it follows that the actual tip position is shifted by $|t'|Z_m$ to the left:

$$z'_{\text{tip}} = \xi_m |t'|^{1/2} - t' Z_m \equiv z'_m - t' Z_m, \qquad (10.17)$$

and the velocity of the tip is given by

$$\frac{v_{\text{tip}}}{v_\eta} = \frac{v_m}{v_\eta} - Z_m. \qquad (10.18)$$

Writing $v - v_m$ in terms of the velocity field $V = \left(v - v_{\text{tip}}\right)/v_\eta$ of the inner solution, we have

$$\frac{v - v_m}{v_\eta} = V - Z_m = 2Z_m - 2\frac{z' - z'_{\text{tip}}}{|t'|} = -2\frac{z' - z'_m}{|t'|},$$

using (10.18), (10.8), and (10.17). But this agrees with the inner asymptotics (10.15) of the outer solution since

$$\frac{v - v_m}{v_\eta} = |t'|^{-1/2}\left(\psi_a - \frac{\xi_m}{2}\right) = -2|t'|^{-1/2}(\xi - \xi_m) = -2\frac{z' - z'_m}{|t'|}.$$

This confirms the matching between the two velocity fields.

Full solution

Now we can solve the similarity equations (10.2), (10.3) with the boundary conditions (10.10) and

$$\psi_a(\xi_m) = \frac{\xi_m}{2}, \quad \psi'_a(\xi_m) = -2, \qquad (10.19)$$

which follow from (10.15). The boundary condition for ϕ is unusual in that it starts with a *jump* at the tip. As a result of the jump, the expansions of ϕ_a and ψ_a near the tip are themselves singular. Anticipating this fact, we try the ansatz

$$\phi_a = \frac{1}{6} + \phi_1 (\xi - \xi_m)^{\sigma_1} + \dots,$$

$$\psi_a = \frac{\xi_m}{2} - 2(\xi - \xi_m) + e_0 (\xi - \xi_m)^{\sigma_2} + \dots \qquad (10.20)$$

First, we insert (10.20) into (10.2), which yields, to the leading non-vanishing order,

$$\frac{-5\sigma_1\phi_1}{2}(\xi - \xi_m)^{\sigma_1} + \frac{\sigma_2 e_0}{6}(\xi - \xi_m)^{\sigma_2 - 1} = 0.$$

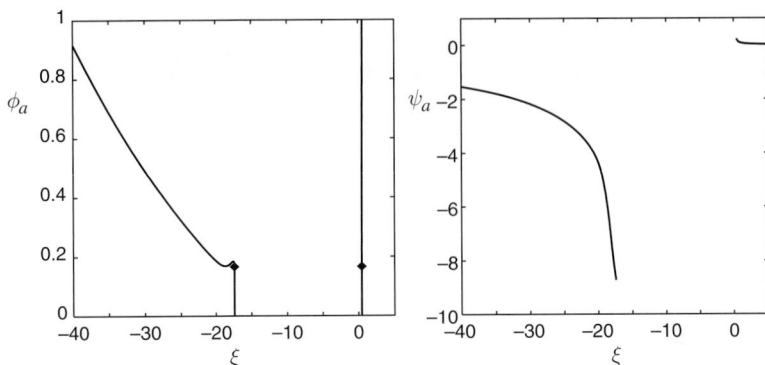

Figure 10.4 The similarity functions ϕ_a and ψ_a. The asymptotic behavior for $\xi \to \pm\infty$ is the same as before breakup. On the left, the rapidly receding "neck" part of the solution; on the night, the drop part of the solution. The points ξ_{neck} and ξ_{drop}, from where the interface is plane, are marked by diamonds.

We conclude that $\sigma_2 = \sigma_1 + 1$ and $\sigma_2 e_0 = 30\sigma_1\phi_1$. Similarly, from (10.3) one finds the leading-order balance

$$-36\sigma_1\phi_1(\xi - \xi_m)^{\sigma_1 - 1} + 3\sigma_2(\sigma_2 - 1)e_0(\xi - \xi_m)^{\sigma_2 - 2} = 0.$$

Equating the prefactors yields $36\sigma_1\phi_1 = 3\sigma_2(\sigma_2 - 1)e_0 = 90\sigma\phi_1\sigma_1$ (the exponents are equal identically), and hence $\sigma_1 = 36/90 = 2/5$. This means that the expansions (10.20) start with the exponents $\sigma_1 = 2/5$ and $\sigma_2 = 7/5$, and the general expansion becomes

$$\phi_a = \frac{1}{6} + \phi_1(\xi - \xi_m)^{2/5} + \phi_2(\xi - \xi_m)^{3/5} + \cdots,$$

$$\psi_a = \frac{\xi_m}{2} - 2(\xi - \xi_m) + e_0(\xi - \xi_m)^{7/5} + e_1(\xi - \xi_m)^{8/5} + \cdots. \tag{10.21}$$

An explicit calculation shows that all the coefficients can be calculated from the two parameters ϕ_1 and ξ_m; see Exercise 10.1. Given a pair (ϕ_1, ξ_m), one can use the expansions (10.21) to find an initial condition for the similarity equations (10.2), (10.3) valid close to the tip (this is necessary because the equations are singular for $\xi = \xi_m$). The equations are integrated from a positive value of ξ_m to $\xi = +\infty$, to find the solutions near the drop, and from a negative value of ξ_m to $\xi = -\infty$ to describe the neck; see Fig. 10.4. As argued above, the solution converges exponentially to the asymptotic behavior (7.47), which is determined by two free constants, a_0, b_0. Solving the post-breakup problem consists in adjusting ϕ_1 and ξ_m until a_0, b_0 match their corresponding values *before* breakup, as reported in Table 7.3.

Table 10.1 *Characteristics of the similarity functions ϕ_a, ψ_a after breakup.*
The tip position of the left-hand, or neck, side is ξ_{neck} and the expansion
coefficient ϕ_1, see (10.21), is ϕ_{neck}. Correspondingly, ξ_{drop} and ϕ_{drop} uniquely
determine the "drop" side of ϕ_a and ψ_a. The values of $a_0^{\pm}$ and $b_0^{\pm}$ are the
same as before breakup; see Table 7.3.

ξ_{neck}	ϕ_{neck}	ξ_{drop}	ϕ_{drop}
-17.452	0.06183	0.4476	0.6180

The pair (a_0^+, b_0^+) corresponds to the "drop" side of the problem, and
(a_0^-, b_0^-) to the neck side. We denote the pair (ξ_m, ϕ_1) by $(\xi_{drop}, \phi_{drop})$ in
the case of the drop (the right-hand part of the solution in Fig. 10.4), and
$(\xi_{neck}, \phi_{neck})$ in the case of the neck (the left-hand part of the solution). The
numerical values of the parameters are reported in Table 10.1. The left-hand
part of the solution represents a receding thread; the right-hand part is a small
perturbation to the drop. Although obtained by the same procedure, they look
markedly different, as the constants a_0 and b_0 are quite different on the left and
right of the singularity.

The similarity solution for the receding thread (the outer solution) is shown
as the broken line in Fig. 10.3 (but oriented the other way around). It breaks
down at the tip, while the finger solution (the inner solution), shown as the dotted line, fails away from the tip. However, by combining the inner and outer
solutions it is possible to obtain a *composite solution* (the solid line) which is
valid everywhere. The idea is to add the two solutions together and then subtract the solution in the overlap region; in our case, this is the cylinder solution
$h = \ell_v |t'|/6$, which is both the outer limit of the inner solution and the inner
limit of the outer solution. The two solutions are glued together at the matching
point Z_m, ξ_m. Then the result for the composite solution, shown in Fig. 10.3, is

$$h_{comp} = \ell_v |t'| R \left(\frac{z' - z'_{tip}}{|t'|} \right)$$

$$+ \ell_v |t'| \begin{cases} 0, & z' - z'_{tip} \leq Z_m |t'|, \\ \phi_a \left(\dfrac{z' - z'_{tip}}{|t'|^{1/2}} - |t'|^{1/2} Z_m \right) - \dfrac{1}{6}, & z' - z'_{tip} > Z_m |t'| . \end{cases}$$

$$(10.22)$$

It follows from the construction that the composite solution approaches the finger solution as one moves toward the tip and approaches the thread solution far
from the tip, with a continuous transition between the two. The time evolution
after breakup, as described by the composite solution (10.22), is illustrated in

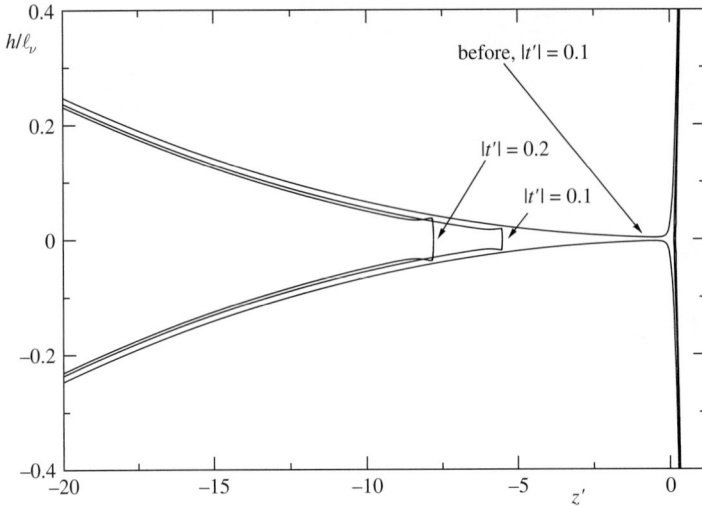

Figure 10.5 The profile $h(z', t')$ before and after the singularity. The solutions approach one another far from the pinch point. Note the strong asymmetry between the thread and the drop.

Fig. 10.5. Note that the tip is rounded on a small scale which corresponds to the radial dimension. After breakup, one part of the solution consists of a rapidly receding thread; the other part is visible as a small protuberance to the drop, which heals off to a smooth interface.

A direct comparison with experiment is much more difficult for the post-breakup solution than for the pre-breakup solution. Owing to the high speed of the receding thread, the air drag is significant, and this slows down the thread [74]. In addition, breakup tends to become irregular in the final moments before pinch-off, as we confirmed in Section 7.3.2. This affects the receding thread, which breaks at several places almost simultaneously; see Fig. 7.5. A possible way to compare with the theory is to perform new experiments for smaller viscosity, where the instability is less pronounced. Obviously, this places greater demands on the temporal and spatial resolution of the experiment.

10.2 Regularization: thread formation for viscoelastic materials

Now we will describe the second approach to continuing the breakup of a liquid thread across the singularity. As in (3.60), we add another term to the equation of motion so that a singularity is avoided. As a result a thin thread

Figure 10.6 High-speed video image of a jet of dilute (0.01 wt%) aqueous poly-acrylamide solution (nozzle radius $r_0 = 0.3$ mm) undergoing capillary thinning [51]. The polymer relaxation time is $\lambda = 0.012$ s, and the elasto-capillary length scale (10.40) is $\ell_e = 0.051$ mm.

forms, which joins the two pieces which were formerly separate. We now describe a more physical version of this type of regularization. Figure 10.6 shows a jet of water to which a small percentage of a long, flexible, polymer has been added. The jet no longer breaks up but instead the individual "drops" remain connected through very thin threads. Many biological fluids, such as human saliva, contain such macromolecules.

10.2.1 Dilute polymer solutions

In equilibrium a polymer forms a densely packed coil. In a breaking thread, however, the polymer is subjected to an extensional flow which stretches the molecule. In a simple approximation [27] the polymers can be described as a set of elastic springs submerged in the fluid that are extended along with the flow. This produces a tension in the spring, which resists further stretching. As a result, the thinning of the fluid filament is slowed down and a thread is formed. The polymers in solution introduce a new time scale into the problem: the *relaxation time* of the polymer or *Zimm's constant*, λ. As described above, the polymers support an extra normal stress $\sigma_{zz} - \sigma_{rr}$ that adds to the solvent's viscous stress. Here σ_{zz} and σ_{rr} are components of the polymeric stress tensor σ in cylindrical coordinates. However, stretching occurs only in the axial direction while the flow compresses in the radial direction. Therefore we can write $\sigma_p = \sigma_{zz}$ for the polymeric stress, neglecting σ_{rr}, which is very small. In conclusion, (6.59) is replaced by

$$\frac{\partial v}{\partial t} + vv' = -\frac{\gamma}{\rho}\kappa' + \frac{1}{h^2}\left[h^2\left(3v_s v' + \sigma_p\right)\right]',\qquad(10.23)$$

where $v_s =$ is the solvent's kinematic viscosity and a prime denotes the derivative with respect to z as usual. Also, we are neglecting the effects of gravity, which is not important for the local behavior we are investigating.

The difficult part is to derive an equation for σ_p. If, however, the polymers can be modeled as linear noninteracting springs, which are carried along perfectly with the fluid, then the task is relatively simple and leads to the so-called Oldroyd-B model [27]. Omitting the details of the derivation, the equation for σ_p becomes

$$\frac{\partial \sigma_p}{\partial t} + v \sigma_p' = 2 v' \sigma_p + 2 \frac{v_p}{\lambda} v' - \frac{\sigma_p}{\lambda}. \tag{10.24}$$

The first term on the right comes from the stretching of the polymers along with the flow, while the second describes the back reaction on the flow and the stress created by it. Finally, the polymer strands relax at rate λ, leading to a reduction in the stress described by the third term. More realistic models contain a term which limits the stretching on account of the polymers' finite length. If this is taken into account, self-similar breakup can occur [182]. We will now consider a particular limit of (10.24) in which λ goes to infinity. In this case the stresses *never* relax, and the polymer contribution behaves like a perfectly elastic (rubber) material [86].

To consider this limit we use the Lagrangian transformation (7.14), where the subscript s marks a volume of fluid. Introducing the notation $u \equiv h^{-2} = z_s$, (7.18) can be written as

$$v' = \frac{u_t}{u}. \tag{10.25}$$

Performing essentially the same calculation as in Example 7.1 we transform (10.24) to Lagrangian coordinates, obtaining

$$\sigma_p + \lambda \left(\left. \frac{\partial \sigma_p}{\partial t} \right|_s - 2 \frac{u_t}{u} \sigma_p \right) = 2 v_p \frac{u_t}{u}. \tag{10.26}$$

Note that σ_p is now understood to be a function of s and t. Equation (10.26) can be rewritten as

$$e^{-t/\lambda} \frac{\partial}{\partial t} \left(\frac{1}{u^2} e^{t/\lambda} \sigma_p \right) = 2 \frac{E}{\rho} \frac{u_t}{u^3} = -\frac{E}{\rho} \left(\frac{1}{u^2} \right)_t, \tag{10.27}$$

where

$$E = \frac{\eta_p}{\lambda} \tag{10.28}$$

is an elastic modulus [7].

Now (10.27) can be integrated to give

$$\sigma_p = u^2 e^{-t/\lambda} \left[-\frac{E}{\rho} \int_0^t e^{\tau/\lambda} \left(\frac{1}{u^2} \right)_\tau d\tau + \frac{\sigma_0}{u_0^2} \right], \tag{10.29}$$

where σ_0 is the initial stress and $u_0 = h^{-2}(t = 0)$ is the initial jet shape. Integration by parts yields

$$\sigma_p = -\frac{E}{\rho} + \frac{u^2}{u_0^2} \left(\frac{E}{\rho} + \sigma_0 \right) e^{-t/\lambda} + \frac{E}{\rho \lambda} u^2 e^{-t/\lambda} \int_0^t e^{\tau/\lambda} \frac{1}{u^2} d\tau, \tag{10.30}$$

where the integrals are taken along Lagrangian paths, i.e. paths with a constant particle label s. However, for $\lambda \to \infty$ the last term can be neglected and we obtain

$$\sigma_p = -\frac{E}{\rho} + \frac{u^2}{u_0^2}\left(\frac{E}{\rho} + \sigma_0\right),\tag{10.31}$$

which is a local relationship between the stress and the thread radius in Lagrangian coordinates.

Example 10.1 (Johnson–Segalman model) The Johnson–Segalman constitutive relation [195] is a generalization of (10.24):

$$\frac{\partial \sigma_p}{\partial t} + v\sigma_p' = 2av'\sigma_p + 2\frac{v_p}{\lambda}v' - \frac{\sigma_p}{\lambda},\tag{10.32}$$

where the parameter a, $0 < a \leq 1$, describes the relative "slip" between the flow and polymeric deformation. Instead of (10.26), in Lagrangian coordinates we now have

$$\sigma_p + \lambda\left(\frac{\partial \sigma_p}{\partial t} - 2a\frac{u_t}{u}\sigma_p\right) = 2v_p\frac{u_t}{u}$$

or, equivalently,

$$e^{-t/\lambda}\left(\frac{1}{u^{2a}}e^{t/\lambda}\sigma_p\right)_t = -\frac{E}{\rho a}\left(\frac{1}{u^{2a}}\right)_t.$$

Integrating in time, it follows that

$$\sigma_p = -\frac{E}{\rho a} + \frac{u^{2a}}{u_0^{2a}}\left(\frac{E}{\rho a} + \sigma_0\right)e^{-t/\lambda} + \frac{E}{\rho\lambda a}u^{2a}e^{-t/\lambda}\int_0^t e^{\tau/\lambda}\frac{1}{u^{2a}}d\tau;$$

so, taking the limit $\lambda \to \infty$, we arrive at

$$\sigma_p = -\frac{E}{\rho} + \frac{u^{2a}}{u_0^{2a}}\left(\frac{E}{\rho} + \sigma_0\right).\tag{10.33}$$

As expected, for $a < 1$ there is a weaker relationship between stress and deformation than that in (10.31). $\square$

For the case where the initial jet has an almost constant radius r_0, it follows that $u_0^{-1} = r_0^2$. If the initial stress is small as well ($\sigma_0 \approx 0$), (10.31) contains only Lagrangian labels at the same time t, and so the Lagrangian quantities can be replaced by their Eulerian counterparts. We thus obtain $\sigma_p \propto 1/h^4$ for

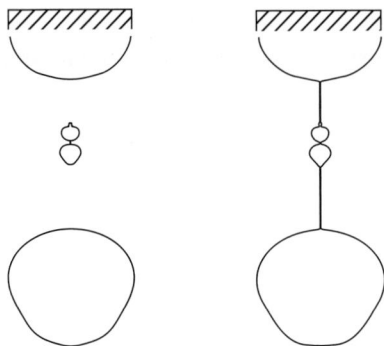

Figure 10.7 A drop falling from a faucet, forming satellites. The parameters are Bo $= 1$ and Oh $= 6.7 \times 10^{-3}$. On the left, the solution of (6.57), (6.59) is continued through the singularity, using "surgery". On the right, we use (10.34) to regularize the solution; the elastic length scale (10.40) is $\ell_e/r_0 = 6.3 \times 10^{-3}$. Thin threads are formed between drops.

small h, which is a classical result for rubber elasticity [210]. Now (10.23) becomes, rewriting κ' with the help of (6.89),

$$\rho \left(\frac{\partial v}{\partial t} + v v' \right) = \frac{1}{h^2} \left(-\gamma G + 3\eta_s h^2 v' + \frac{E r_0^4}{h^2} \right)' . \qquad (10.34)$$

As usual, (10.34) has to be solved together with the continuity equation (6.66). It is apparent from (10.34) that the elastic stresses will continue to grow as h becomes small. Since there is no mechanism for relaxing the stresses, at some point the surface tension and elastic forces will balance and h reaches a minimum. This behavior is illustrated in Fig. 10.7 (right), where the parameter E has been set to a small value such that pinching is arrested only when a thin thread has been reached. On the left, the simulation has been continued by jumping directly from a pre-breakup to a post-breakup solution. Note that the two profiles resemble each other very closely.

10.2.2 The beads-on-a-string configuration

Now we will describe the long-time limit of the system (6.66), (10.34) analytically [85]. Since the motion of the jet is slowed down by the polymer, eventually inertia becomes unimportant, and the left-hand side of (10.34) can be ignored. Then, in analogy to the case of a viscous thread (see (7.10)), (10.34) can be integrated and we obtain the force balance

$$-\gamma G + 3\eta_s h^2 v' + \frac{E r_0^4}{h^2} = T(t). \qquad (10.35)$$

Once more $T(t)$ is the tensile force in the thread. As suggested by Fig. 10.6 and the simulation diagram shown on the right of Fig. 10.7, the solution we are seeking consists of a thin filament connected to almost spherical drops.

The first step consists in the analysis of the evolution of the filament. As h thins, the fluid is squeezed out from within the filament, leading to a linear velocity profile (extensional flow). To be precise, near the stagnation point $z = 0$ we can approximate the profiles to leading order as follows:

$$h(z, t) = h_0(t) + O(z), \quad v(z, t) = v_1(t)z + O(z^2). \tag{10.36}$$

Inserting (10.36) into (6.66), (10.35), one finds

$$\dot{h}_0 + \frac{1}{2}h_0 v_1 = 0, \tag{10.37}$$

$$\gamma h_0 + 3\eta_s h_0^2 v_1 + \frac{Er_0^4}{h_0^2} = T(t). \tag{10.38}$$

Note that a balance occurs at leading order, which shows that a filament of uniform thickness is a local solution.

From (10.37) we get

$$v_1 = -\frac{2\dot{h}_0}{h_0},$$

which, when substituted into (10.38), leads to the equation

$$\gamma h_0 - 6\eta_s h_0 \dot{h}_0 + \frac{Er_0^4}{h_0^2} = T(t). \tag{10.39}$$

For $E = 0$ we recover the purely viscous case, which admits the similarity solution (7.24), representing breakup in finite time. When $E > 0$, however, the elastic term Er_0^4/h_0^2 becomes relevant for sufficiently small values of h_0, and breakup is arrested. The left-hand side of (10.39) contains positive terms only since for a thinning filament $\dot{h}_0 < 0$, and so it is larger than $3(Er_0^4\gamma^2/4)^{1/3}$, which is the minimum of

$$\gamma h_0 + \frac{Er_0^4}{h_0^2}.$$

In particular there is always tension in the string, and this has to be included as part of the solution. The minimum is reached for $h_0 = (2Er_0^4/\gamma)^{1/3}$, which suggests introducing the elasto-capillary length scale

$$\ell_e = \left(\frac{Er_0^4}{\gamma}\right)^{1/3}. \tag{10.40}$$

When the filament thickness h is of order ℓ_e, the whole jet evolves towards a stationary beads-on-a-string configuration, which we compute by using matched asymptotics. The unknown tension, which must necessarily be positive, will be part of this solution.

Example 10.2 (Breakup or non-breakup in the Johnson–Segalman model) Before we continue, we will discuss under what conditions a filament is formed as the slip parameter a in (10.32) changes. From (10.33), and keeping in mind that $u \propto h^{-2}$, we find, for $a > 0$,

$$\sigma_p \propto h^{-4a}$$

as $h \to 0$. The equation equivalent to (10.39) that results in this case is

$$\gamma h_0 - 6\eta_s h_0 \dot{h}_0 + \frac{E r_0^{4a}}{a h_0^{4a-2}} = T(t). \tag{10.41}$$

A filament can only form if a minimum of

$$F(h_0) \equiv \gamma h_0 + \frac{E r_0^{4a}}{a h_0^{4a-2}}$$

exists, since the tension can only relax; it cannot increase spontaneously. Since

$$F'(h_0) = \gamma - \frac{4a - 2}{a} \frac{E r_0^{4a}}{h_0^{4a-1}}$$

has a root only if $a > 1/2$, we conclude that breakup will happen if $a < 1/2$ while filaments will form for $a > 1/2$. If the polymer stretching does not follow the flow sufficiently well ($a < 1/2$) then the elastic stresses will not build up sufficiently to overcome surface tension and breakup will occur. If, however, the polymer follows the flow faithfully ($a > 1/2$) then a filament forms and breakup is arrested. □

The matching of the asymptotic regions assumes that the limiting filament radius is much smaller than the drop radius, which is of the order of the initial jet radius r_0. In other words, we are assuming that the limit

$$\frac{\ell_e}{r_0} = \left(\frac{E r_0}{\gamma} \right)^{1/3} \ll 1 \tag{10.42}$$

is satisfied. It corresponds to the limit in which the relaxation time λ is large, when the polymers are very long and flexible. In that case there are two well-distinguished regions: the filaments and the drops, as indicated in Fig. 10.8.

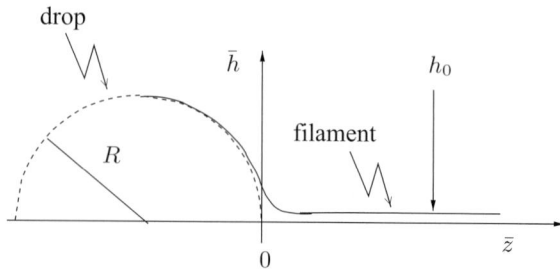

Figure 10.8 A schematic of a filament of radius h_0 connected to a drop of radius R. The origin $\bar{z} = 0$ is placed at the junction of the spherical drop and the filament. The filament and drop are connected by an intermediate solution described by (10.55).

The regions are connected by an inner or boundary layer solution, which we describe now.

In the long-time limit we are looking for time-independent profiles h, v which have a similarity form:

$$h(z, t) = \ell_e \bar{h}\left(\frac{z}{\ell_e}\right), \quad T(t) = \gamma \ell_e \bar{T}, \tag{10.43}$$

$$v(z, t) = v_\eta \bar{v}\left(\frac{z}{\ell_e}\right). \tag{10.44}$$

The drop radius and the tensile force have been made dimensionless using the elasto-capillary length scale ℓ_e. As for the velocity, we notice that the time derivative drops out, and (6.66) can be integrated to give

$$\bar{v}\bar{h}^2 = v_0 h_0^2. \tag{10.45}$$

Assuming that there is no imposed flow far from the pinch region, we have $v_0 = 0$, so $\bar{v} = 0$ everywhere. Thus the viscous term also drops out, and the balance (10.35) is now between surface tension and elastic forces alone:

$$\frac{\bar{h}}{\sqrt{1 + \bar{h}'^2}} + \frac{\bar{h}^2 \bar{h}''}{(1 + \bar{h}'^2)^{3/2}} + \frac{1}{\bar{h}^2} = \bar{T}. \tag{10.46}$$

We want to find an asymptotic solution of (10.46) that has the structure shown in Fig. 10.8; this consists of a filament of constant thickness attached to an almost spherical drop (the outer regions). The two outer regions are connected by a transition (inner) region located around $\bar{z} = 0$. The profile has been renormalized according to (10.43) to make the thread thickness of order one. In the same variables, the drop radius R is of order r_0/ℓ_e and thus very large: $R \gg 1$.

Inside the filament the profile corresponds to the filament radius h_0, and thus $\bar{h} \approx h_0$ for $\bar{z} \gg 1$. For constant $\bar{h}$, (10.46) yields the condition

$$h_0 + \frac{1}{h_0^2} = \bar{T}. \tag{10.47}$$

If we multiply (10.46) by the factor $-\bar{h}'/\bar{h}^3$, the equation can be integrated to give

$$\frac{1}{\bar{h}\sqrt{1 + \bar{h}'^2}} + \frac{1}{4\bar{h}^4} = \frac{\bar{T}}{2\bar{h}^2} + C. \tag{10.48}$$

If we evaluate (10.48) at the top of the drop, where $\bar{h} = R$ and $\bar{h}' = 0$, we find that the constant of integration is $C \approx 1/R + O(R^{-2})$.

Evaluating (10.48) once more inside the filament, and using $R \gg h_0$, we find that to leading order

$$2h_0 + \frac{1}{2h_0^2} = \bar{T}. \tag{10.49}$$

Thus we can use solve (10.47) and (10.49) to give

$$h_0 = \frac{1}{2^{1/3}}, \quad \bar{T} = \frac{3}{2^{1/3}}, \tag{10.50}$$

so we have already performed the main task, that of calculating the thread radius.

From (10.48) the profile can be written in the form

$$\bar{h}' = -\left(\frac{4\bar{h}^2}{\left[3/2^{1/3} + 2\bar{h}^2/R - 1/(2\bar{h}^2)\right]^2} - 1\right)^{1/2}. \tag{10.51}$$

We will now confirm that this solution matches to a drop on one side and to a filament of uniform radius on the other side. Namely, the equation for a spherical drop is

$$\bar{h} = \sqrt{R^2 - (\bar{z} + R)^2},$$

which, for $|\bar{z}| \ll R$, i.e. toward the inner region, becomes

$$\bar{h} \approx \sqrt{2R}\sqrt{-\bar{z}}. \tag{10.52}$$

For $R^{1/2} \ll \bar{h} \ll R$, equation (10.51) simplifies to

$$\bar{h}' \approx -\frac{R}{\bar{h}},$$

the solution of which is (10.52), so the profile does indeed approach a circle. Toward the filament, for $\bar{h} \ll R^{1/2}$, equation (10.51) simplifies to

$$\bar{h}' \approx -\left(\frac{4\bar{h}^2}{[\bar{T}-1/(2\bar{h}^2)]^2}-1\right)^{1/2}, \tag{10.53}$$

where condition (10.49) guarantees that the right-hand side of (10.53) vanishes inside the filament, i.e. for $\bar{h}=h_0$. If we linearize (10.53) around $\bar{h}=h_0$ by putting

$$\bar{h}=h_0(1+\delta),$$

we find that

$$\delta' \approx -\sqrt{3}\delta + O(\delta^2), \tag{10.54}$$

so $\bar{h}$ converges *exponentially* toward h_0. Integrating (10.53), the "universal" (i.e. R-independent) part of the profile can be written (up to quadrature) as

$$\int_{2^{-1/3}}^{\bar{h}} \left(\frac{4g^2}{[3/2^{1/3}-1/(2g^2)]^2}-1\right)^{-1/2} dg = \bar{z}. \tag{10.55}$$

The final filament radius h_f, in dimensional form, is found from (10.50):

$$h_f = \left(\frac{Er_0^4}{2\gamma}\right)^{1/3}. \tag{10.56}$$

Now consider a drop attached to a thread on either side. If there are variations of the initial radius r_0, this translates into slightly different radii h_f^R and h_f^L on either side of the drop, and so the tensions in the two filaments are not the same. This means that there is a net force on the drop, which according to (10.47) is given by

$$F = \left(\gamma h_f^R + \frac{Er_0^4}{(h_f^R)^2}\right) - \left(\gamma h_f^L + \frac{Er_0^4}{(h_f^L)^2}\right) \approx \gamma k(h_f^R - h_f^L),$$

with

$$k = \left(1 - \frac{2Er_0^4}{\gamma((h_f^R+h_f^L)/2)^3}\right).$$

This leads to the subsequent dynamics of the drops, which happens on a time scale much larger than the time scale for the jet to relax to the beads-on-a-string shape. Drops can oscillate, translate, and coalesce [141].

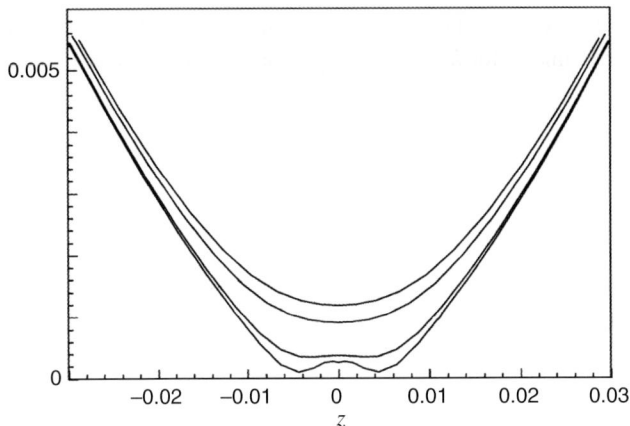

Figure 10.9 Numerical simulation of the evolution of bubble shapes close to pinch-off, showing the formation of a satellite bubble at a density ratio $\Lambda = \rho_g/\rho = 0.005$ [95].

10.3 Crossover: bubbles and satellites

We now return to the problem of bubble pinch-off described in Chapter 9, and we describe how the presence of air regularizes the singularity. The expression for α provided in (9.70) predicts the evolution of a cavity very well but only for an intermediate range of times. However, as the singularity is approached, gas inertia comes into play no matter how small the inner-to-outer density ratio is, and the rate of pinching is slowed down at the center $z = 0$. As a result, the first pinch of the bubble takes place at $z \neq 0$ in spite of the initial mirror symmetry of the problem, and a small satellite drop is formed in the middle; see Fig. 10.9. Satellite bubbles have also been reported experimentally [125, 208]. We can account for this phenomenon by including the effect of gas inertia in the equations of motion (9.62)–(9.64).

We will consider the breakup of a bubble immersed in a fluid such that the inner-to-outer density ratio $\Lambda = \rho_g/\rho$ is small. The scaling exponent α must drop to zero for sufficiently small values of h_0 as a consequence of the increasingly larger pressures at $z = 0$ needed to accelerate the gas longitudinally. We will develop a simple model which, while retaining the dominant physical mechanisms described above, permits us to both reproduce the time evolution of the exponent α and scale the size of the satellite bubbles obtained as a function of Λ [95].

On the one hand, under the slenderness approximation the gas velocity can be calculated using the continuity equation

$$\frac{\partial a}{\partial t} + (av_g)' = 0,$$

where a/π is the cross-sectional area as before. Approximating the cavity by a cylindrical shape $a(z,t) \simeq a_0(t)$, this leads to

$$v_g(z,t) = -\frac{\dot{a}_0}{a_0}z. \tag{10.57}$$

On the other hand, modeling the gas motion as inviscid in one dimension along the z-axis the Euler equation (4.31) simplifies to

$$\rho_g \left(\frac{\partial v_g}{\partial t} + v_g v_g' \right) = -p_g'. \tag{10.58}$$

Thus in our original cavity dynamics (9.44) the constant cavity pressure has to be replaced by the z-dependent gas pressure: $p_0 = p_g(z)$.

To derive a new equation of motion for a_0 and Δ, we need $p_g(0)$ and $p_g''(0)$. Taking the derivative of (10.58) and using (10.57), we find

$$p_g'' = -\rho_g \left(2\frac{\dot{a}_0^2}{a_0^2} - \frac{\ddot{a}_0}{a_0} \right). \tag{10.59}$$

To evaluate the gas pressure at the origin we integrate (10.58) over the typical longitudinal distance Δ, as given by (9.45):

$$p_g(0) = p_g(\Delta) + \rho_g \left(2\frac{\dot{a}_0^2}{a_0^2} - \frac{\ddot{a}_0}{a_0} \right) \frac{\Delta^2}{2}. \tag{10.60}$$

The pressure $p_g(\Delta)$ can be taken as the constant value of the pressure inside the main bubble, which we neglect.

According to (9.44) the gas pressure is accounted for by adding $4p_g(0)/\rho$ on the right-hand side of the first cavity equation, (9.51), which leads to

$$-\frac{\ddot{a}_0}{a_0} \ln \epsilon^2 = \frac{\dot{a}_0^2}{2a_0^2} + \frac{2\Lambda}{\epsilon^2} \left(2\frac{\dot{a}_0^2}{a_0^2} - \frac{\ddot{a}_0}{a_0} \right). \tag{10.61}$$

To obtain a second equation we add $4p_g''/\rho$ to the right-hand side of (9.58), which gives

$$-\frac{\ddot{a}_0''}{a_0''} \ln \epsilon^2 = \frac{\dot{a}_0 \dot{a}_0''}{a_0 a_0''} - \frac{\dot{a}_0^2}{2a_0^2} - \frac{2\Lambda}{\epsilon^2} \left(2\frac{\dot{a}_0^2}{a_0^2} - \frac{\ddot{a}_0}{a_0} \right). \tag{10.62}$$

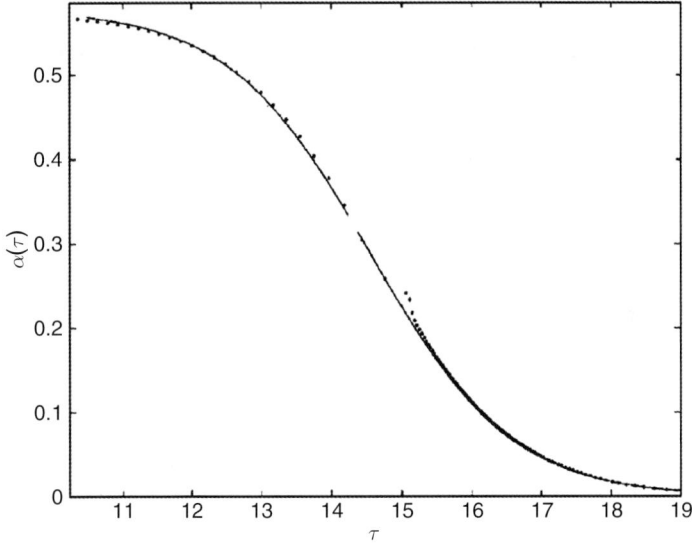

Figure 10.10 The scaling exponent $\alpha(\tau)$ as calculated from a full numerical simulation of the Euler equations (dots) and from a solution of (10.63)–(10.65). Note that for $\tau > 15$ the numerical results are affected by an instability triggered by the large gas velocities inside the bubble neck.

Thus, using the transformations (9.60), (9.61) as well as $v(\tau) = 1/\ln \epsilon^2$, the equations of motion (9.65)–(9.67) can be generalized to

$$\left(\alpha_\tau + \alpha - 2\alpha^2 \right) = \alpha^2 v + \frac{2\Lambda v}{\epsilon^2} \left(2\alpha^2 + \alpha + \alpha_\tau \right), \qquad (10.63)$$

$$\left(\delta_\tau + \delta - 2\delta^2 \right) = (2\alpha\delta - \alpha^2)v - \frac{2\Lambda v}{\epsilon^2} \left(2\alpha^2 + \alpha + \alpha_\tau \right), \quad (10.64)$$

$$v_\tau = 2\delta v^2. \qquad (10.65)$$

Of course, if $\Lambda = 0$ then one retrieves the previous case, for which the pressure inside the bubble need not be considered and the asymptotic behavior of the pinching solution is described by (9.70). However, for finite Λ the gas pressure inside the bubble neck rises and the pinching slows down. This situation is reproduced by a numerical solution of (10.63)–(10.65), shown in Fig. 10.10. The scaling exponent $\alpha(\tau)$ goes to zero close to the pinch point, a behavior very different from the dynamics without gas shown in Fig. 9.6. Comparison with a full numerical simulation of the Euler equation demonstrates [95] that the system (10.63)–(10.65) indeed captures the effect of the gas pressure faithfully.

In order to estimate the typical diameter d_b of the gas bubbles formed, we calculate the characteristic value τ_c for which the ratio of the liquid and the gas momentum becomes unity, then the two terms on the right-hand side of (10.63) are of the same order. This occurs when $\epsilon^2 \approx 8\Lambda$, where we have used the fixed point values $\alpha = 1/2$ and $\delta = 0$. Using (9.73), the critical time is found to be given by

$$-\sqrt{\tau_c} \approx \ln 8\Lambda. \tag{10.66}$$

Finally, inserting τ_c into (9.72) we find that the critical neck radius at which the gas pressure becomes important is

$$h_c \approx e^{-\tau_c/2 - \sqrt{\tau_c}/2} \approx \sqrt{8\Lambda}e^{-(\ln 8\Lambda)^2/2}. \tag{10.67}$$

The total length $2\Delta_c \equiv 2\Delta(\tau_c)$ over which the profile is deformed (see Fig. 10.9), can be estimated as

$$\Delta_c = \frac{h_c}{\epsilon} \approx \frac{h_c}{\sqrt{8\Gamma}} \approx e^{-(\ln 8\Lambda)^2/2}. \tag{10.68}$$

Thus using volume conservation, $2\pi h_c^2 \Delta_c = \pi d_b^3/6$, as well as (10.67) and (10.68) we finally obtain

$$d_b \approx \left(12h_c^3 e^{\sqrt{\tau_c}/2}\right)^{1/3} \approx (96\Lambda)^{1/3} e^{-(\ln 8\Lambda)^2/2}. \tag{10.69}$$

With an air–water density ratio $\Lambda = 0.0012$, (10.69) gives $d_b \simeq 10^{-5}$. If the air is replaced by SF$_6$, the density ratio is increased to $\Lambda = 0.0062$ and $d_b \simeq 9 \times 10^{-3}$. Since the typical length scale is set by the size of the bubble, the high-Reynolds-number breakup of a 1 cm bubble will give $d_b \simeq 0.1\mu$m in water, and $d_b \simeq 90\,\mu$m in SF$_6$. This explains why satellite bubbles are not readily observable in air but become visible if the bubble is made up of a much heavier gas, as was observed in [208].

Exercises

10.1 Use the expansions (10.21) to show that

$$e_0 = \frac{60}{7}\phi_1, \quad e_1 = 0, \quad \text{and} \quad e_2 = -\frac{120}{7}\phi_1^2. \tag{10.70}$$

Optionally also show that

$$\phi_2 = 0, \quad \phi_3 = \frac{24}{7}\phi_1^2, \quad \text{and} \quad e_3 = -\frac{5}{72}\xi_m;$$

thus the coefficients of the expansion are indeed determined in terms of the two parameters ϕ_1 and ξ_m.

10.2 Consider the post-breakup solution of the viscous fluid thread (assuming that inertial effects can be neglected) whose pinching was described in Section 7.2.

(i) Show that before breakup the (ground state) similarity solution behaves as

$$\chi_b \approx \chi_m \zeta^{1/\delta}, \quad \zeta \to \infty.$$

(ii) Use (7.124) to show that in the absence of inertia the tension $T(t)$ vanishes after breakup.

(iii) Show that in Lagrangian coordinates the post-breakup similarity solution (the outer solution) is

$$\chi_a = \chi_m \zeta^{1/\delta} + \frac{1}{6},$$

which agrees with (10.10) at $\zeta = 0$.

(iv) (*) Conclude that that tip recedes at a speed

$$v_e = \frac{\pi}{\chi_m^\delta} 2^{2-\delta} 3^{2-\delta} \delta (2 - \delta)(1 - \delta) \,(\operatorname{cosec} \pi \delta)\, t^{\delta-3}.$$

10.3 In Example 3.2 we showed that the pinch-off problem

$$\frac{\partial h}{\partial t} = -1 + h''$$

has similarity solutions of the form $h(x, t) = t'g(\xi)$, with $\xi = x/t'^{1/2}$, where

$$g_a(\xi) = 1 + a(\xi^2 - 2), \quad 0 < a < 1/2.$$

(i) Show that the post-breakup similarity solution is

$$g_a = -1 + (2 + \xi^2)\left(a - \sqrt{\pi}C\right)$$
$$+ C\left[\sqrt{\pi}(\xi^2 + 2)\operatorname{erf}\left(\frac{\xi}{2}\right) + 2\xi e^{-\xi^2/4}\right],$$

where the error function is defined in (4.73).

(ii) Conclude that for the continuation to be unique one must impose another condition at the tip, for example

$$\frac{\partial h}{\partial x}\left(x_{\text{tip}}\right) = 0.$$

10.4 Owing to the overturning of the profile, the full potential flow equations have to be used to describe inviscid pinch-off; see Section 7.5. However, the inviscid slender jet equations can be used to describe the retracting filament after breakup [33]. Show that inserting

$$h(z,t) = \ell_{in} H\left(\frac{z}{\ell_{in}}\right), \quad v(z,t) = \frac{\ell_{in}}{\Delta t} V\left(\frac{z}{\ell_{in}}\right)$$

into the slender jet equations (6.58), (6.57) with $\nu = 0$ leads to the following similarity equations *after* breakup:

$$\frac{2}{3}H - \frac{2\xi}{3}H' = -VH' - \frac{1}{2}V'H,$$

$$\frac{V}{3} + \frac{2\xi}{3}V' = VV' + \left(\frac{1}{H(1+H'^2)^{1/2}} - \frac{H''}{(1+H'^2)^{3/2}}\right)'.$$

(i) Expand around the tip at $\xi = \xi_{tip}$ to show that solutions are parameterized by a_0 and ξ_{tip}, where

$$H \approx a_0(\xi - \xi_{tip})^{1/2}, \quad V \approx \frac{2\xi_{tip}}{3}.$$

(ii) Show that the far-field behavior of the profile is determined by

$$H(\xi) = \xi \sum_{i=0}^{\infty} a_i \xi^{-3i/2}, \quad V(\xi) = \frac{1}{\xi^{1/2}} \sum_{i=0}^{\infty} b_i \xi^{-2i}.$$

The coefficients a_0, b_0 follow from the pre-breakup dynamics and all other coefficients can be computed from it.

(iii) Use a WKB analysis to show that perturbations around the mean profile are governed by $\epsilon \propto \exp \chi(\xi)$, where

$$\chi = \pm\frac{2i}{3}\left(\frac{9}{8}\frac{a_0}{(1+a_0^2)^{3/2}}\right)^{1/2}\xi^{3/2}.$$

Conclude that there is a unique solution for each pair (a_0, b_0); comment on the behavior of the capillary waves growing on the receding thread.

10.5 In Section 10.2.1 we analyzed the limit in which the polymer relaxation time λ goes to infinity. If one keeps the exponential factor $e^{-t/\lambda}$ in the step from (10.30) to (10.31) then

$$\sigma_p = -\frac{E}{\rho} + \frac{u^2}{u_0^2}\left(\frac{E}{\rho} + \sigma_0\right)e^{-t/\lambda}.$$

(i) Show that, in this case, the radius h_0 of the filament thins according to

$$h_0(t) \propto e^{-t/(3\lambda)}.$$

(ii) Find the profiles of the velocity and normal stress along the filament at leading order.

(iii) (*) Repeat the matching procedure of Section 10.2.2 to obtain the solution for the boundary layer between the drop and the filament.

10.6 In the case of breakup for the Johnson–Segalman model, when $a < 1/2$ (cf. Example 10.2), use Lagrangian coordinates to find the following equation, which is analogous to (7.19):

$$\frac{z_{st}}{z_s} = \frac{v_\eta}{3}\left(\frac{T(t)}{\gamma}z_s - z_s^{1/2} - kz_s^{2a}\right),$$

where k is a function of s that depends on the initial stress distribution. What is the rate as a function of a at which filaments thin towards breakup? Note: treat the cases $a \le 1/4$ and $a > 1/4$ separately.

10.7 (*) Repeat the same analysis as in Exercise 10.5 but using the Eulerian formulation (10.23), (10.24) directly.

10.8 Suppose that, instead of regularizing bubble collapse by means of finite gas density effects, we invoke a physical mechanism leading to the system

$$\alpha_\tau + \alpha - 2\alpha^2 = \alpha^2 v + \frac{\Lambda v^2}{\epsilon^p}, \tag{10.71}$$

$$\delta_\tau + \delta - 2\delta^2 = (2\alpha\delta - \alpha^2)v - \frac{\Lambda v^2}{\epsilon^p}, \tag{10.72}$$

$$v_\tau = 2\delta v^2, \tag{10.73}$$

where $0 < \Lambda \ll 1$ and $p > 1$. All unknowns in the equations have the same meaning as in Section 10.3. Discuss the physical implications of this regularization and whether it leads to the formation of satellites. If satellites are formed, what is their size?

PART III

Persistent singularities: propagation

11

Shock waves

11.1 Burgers' equation

Shock waves were the first singularities of PDEs to be studied in detail [188]. As seen in the experimental results in Fig. 1.6, a shock consists of a wave which steepens gradually to produce a jump (e.g. in density) in finite time. This discontinuity then propagates through the gas. In the first few sections of this chapter we treat the simplest models of shock formation and propagation, which have a single variable only. We will then move on to more realistic descriptions using several variables, in particular considering compressible gas flow.

As a simple model for gas motion down a tube, let us consider the nonlinear wave equation [222]

$$\rho_t + c(\rho)\rho_x = 0. \tag{11.1}$$

Here ρ is the density, x the distance down the tube in which the wave is propagating, and c the wave speed, which is allowed to depend on the density. The simplest nonlinear case, in which c depends linearly on the density, leads to Burgers' equation,

$$u_t + uu_x = 0. \tag{11.2}$$

Here and in the following we denote the dependent variable, which in this case is the wave speed, by u.

Equation (11.2) can be solved exactly using the method of characteristics [133, 50]. This method relies on the fact that the speed u remains constant along the family of characteristic curves (or *characteristics*, for short) defined by

$$x(s, t) = u_0(s)t + s. \tag{11.3}$$

259

Each characteristic is labeled by its initial position s, and $u_0(s) = u(s, 0)$ is the initial condition for $u(s, t)$. Thus

$$u(x(s, t), t) = u_0(s) \tag{11.4}$$

is a solution of (11.2), given implicitly in terms of the parameter s: indeed,

$$0 = \left.\frac{du(x(s, t), t)}{dt}\right|_s = u_t + x_t u_x = u_t + u u_x, \tag{11.5}$$

so (11.2) is satisfied. In addition (11.3), (11.4) ensure that the initial condition is verified for $t = 0$.

When two characteristics cross, a jump in u, known as a *shock*, is produced, as illustrated by the following:

Example 11.1 (Burgers shock) The characteristics of Burgers' equation are shown in the lower panel of Fig. 11.1 for a simple initial condition given by

$$u_0(x) = \begin{cases} 2 & \text{for } x \leq 0, \\ 1 + \cos x & \text{for } 0 < x < \pi, \\ 0 & \text{for } x \geq \pi, \end{cases} \tag{11.6}$$

shown in the upper panel. Whenever $u_0(x)$ has a negative slope the slope of the characteristic curve (11.3) increases with increasing x when t is plotted as a function of x; see Fig. 11.1. This means that characteristics cross, so, according to (11.4), the speed becomes multivalued at some finite time t_0. This event is illustrated in Fig. 11.2, which shows the evolution of the profile constructed from (11.3), (11.4). □

For given parameter s, a crossing of characteristics first occurs when

$$\frac{\partial x}{\partial s} = u_0'(s)t + 1 = 0,$$

i.e. at a time $t = -1/u_0'(s)$. The earliest of these times gives the singularity time:

$$t_0 = \min_s \left\{ -\frac{1}{u_0'(s)} \right\}; \tag{11.7}$$

see Fig. 11.1. Denoting by s_m the s value for which the minimum is assumed, according to (11.3) a singularity will first form at

$$x_0 = s_m - \frac{u_0(s_m)}{u_0'(s_m)}. \tag{11.8}$$

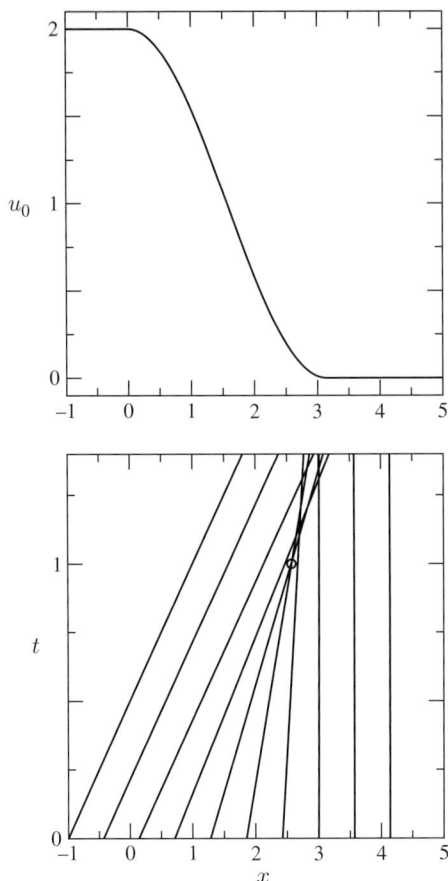

Figure 11.1 The lower panel shows the characteristics (11.3) of Burgers' equation (11.2), with time t plotted against position x. According to (11.3) the smaller u_0, the greater the slope. The circle marks the point where the characteristics cross for the first time. The upper panel gives the initial conditions u_0 (x).

In Example 11.1 the most negative value of u_0' (which is -1) occured at $s_m = \pi/2$. Thus t_0 was 1 and x_0 equaled $1 + \pi/2$; this is where the profile acquired a vertical tangent and a shock first formed.

For times greater than t_0 the solution (11.3), (11.4) turns over; see Fig. 11.2. Since physically u must be a single-valued function of x (for example the density in a shock tube must be a single-valued function of position), we are seeking solutions which experience a jump (the shock) at a certain position x_s. However, it is not clear at which x_s to insert the shock, which connects the speed values u_ℓ and u_r to the left and right of the shock, respectively. In other

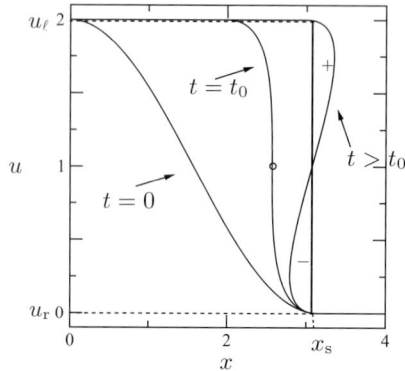

Figure 11.2 The evolution of the solution (11.3), (11.4) of (11.2) with initial condition (11.6). A circle marks the point where the profile first assumes a vertical tangent. For $t > t_0$ a shock can be inserted at x_s, between the values u_ℓ and u_r, to make the solution single-valued. The condition (11.10) requires that the two regions marked $+$ and $-$ have equal areas.

words, the solution can be continued in an infinity of ways to times $t > t_0$, and Burgers' equation (11.2) is not sufficient to determine which solution is the one that is selected physically.

The key to selecting the physically correct solution is to rewrite (11.2) in the *conservation* form

$$u_t + (f(u))_x = 0, \tag{11.9}$$

where $f(u) = u^2/2$ is the flux. It is clear that (11.9) is equivalent to (11.2) as long as there is no singularity, but, as we will see, (11.9) ensures unique continuation. We will study the rationale behind (11.9) in more detail below but the idea is that (11.9) requires the flux into the shock to be the same as the flux out, each evaluated in the frame of reference of the shock, which moves at an unknown speed u_s. The flux into the shock from the left is $f_\ell = (u_\ell - u_s)^2/2$, while the flux out of the shock is $f_r = (u_r - u_s)^2/2$. Equating f_ℓ and f_r, we arrive at

$$u_s = \frac{u_\ell + u_r}{2}, \tag{11.10}$$

which is a jump condition to be satisfied across the shock. In the context of compressible gas dynamics, (11.10) is known as the *Rankine–Hugoniot condition* of shock dynamics. On propagating the shock position according to $\dot{x}_s = u_s$, with initial condition $x_s(t_0) = x_0$, a unique continuation is found.

Example 11.2 (Propagation of a Burgers' shock) In the case of the initial condition (11.6), illustrated in Fig. 11.2, the shock continues to propagate at the constant speed $u_s = 1$ on account of the symmetry of the two lobes marked + and −. To show this, consider the characteristic curve

$$x = u_0(s)t + s = (1 + \cos s)t + s \tag{11.11}$$

as a function of s. For $t > t_0 = 1$, (11.11) can have three solutions: s_ℓ, s_r, and s_s. Anticipating that s_s, corresponding to the initial shock, is given by $s_s = s_m = \pi/2$, the shock position is given by

$$x_s = t + \frac{\pi}{2}, \tag{11.12}$$

and the shock speed is $u_s = 1$.

To verify that (11.10) is satisfied, we set $x = x_s$; then (11.11) becomes

$$0 = (\cos s)t + s - \frac{\pi}{2}.$$

Substituting $s' = s - \pi/2$, this can be written as

$$s' - (\sin s')t = 0. \tag{11.13}$$

Apart from $s' = 0$, for $t > 1$ this has solutions $s'_{\ell/r} = \mp\Delta s$, where the value of Δs is immaterial. The corresponding velocities are

$$u_{\ell/r} = 1 + \cos s_{\ell/r} = 1 - \sin s'_{\ell/r} = 1 \pm \sin \Delta s,$$

so that clearly

$$1 = u_s = \frac{u_\ell + u_r}{2}$$

is satisfied. □

There is also an elegant and constructive way to find x_s at any time $t > t_0$, given the (overturned) profile. The method is to insert the shock in such a way as to make the areas of the two lobes created in the profile (the regions marked + and − in Fig. 11.2) equal. This is equivalent to saying that the area

$$A = \int_{u_\ell}^{u_r} (x - x_s)du \tag{11.14}$$

must vanish (so that each lobe is counted with a different sign). To demonstrate this fact, we show that A is constant,

$$\frac{d}{dt}A = \int_{u_\ell}^{u_r} \left(u - \frac{u_\ell + u_r}{2} \right) du = 0,$$

using $\dot{x} = u$ and $\dot{x}_s = (u_\ell + u_r)/2$. In the first step we use the fact that the integrand always vanishes for $u = u_\ell$ and $u = u_r$, so the time derivatives with respect to the limits do not contribute. But $A(t_0) = 0$ (the loop lobes are vanishingly small as the shock first forms), and so the statement follows.

11.2 Similarity description

We now analyze the neighborhood of the shock singularity from the perspective of similarity methods. This will permit us to understand the universal properties of the shock close to t_0, both before and after the singularity, independently of any particular initial condition. Since (11.2) is invariant under a shift in speed, we can assume without loss of generality that $u_0(s_m) = 0$, and thus that $x_0 = s_m$ according to (11.8). This means that the speed is zero at the singularity.

The local behavior of (11.2) near t_0 can be obtained using the scaling (setting $t' = t_0 - t$)

$$u(x,t) = |t'|^\alpha U\left(\frac{x'}{|t'|^{\alpha+1}}\right), \tag{11.15}$$

which solves (11.2). The similarity equation becomes

$$\pm\left[-\alpha U + (1+\alpha)\xi U_\xi\right] + U U_\xi = 0, \tag{11.16}$$

where the plus sign refers to times before shock formation and the minus sign to times after shock formation. We begin with an analysis of the time before the singularity and describe the steepening of the wave profile.

The similarity equation (11.16) is homogeneous and thus can be solved by the separation of variables [18]. For $\alpha \neq 0$ the implicit solution before the singularity is

$$\xi = -U - CU^{1+1/\alpha}, \tag{11.17}$$

where C is a constant of integration. The special case $\alpha = 0$ has the solution $U = -\xi$. However, this is inconsistent with the matching condition (3.16), which would require U to be constant at infinity, so this solution has to be discarded. The exponent α is undetermined and we have an example of self-similarity of the second kind. As before, a discretely infinite sequence of exponents α_n remains, selected by the requirement that (11.17) must define a smooth function for all ξ. Namely, for the solution (11.17) to be differentiable in the origin, $1 + 1/\alpha$ must be an integer. In addition, for (11.17) to be defined on the whole ξ-axis the right-hand side must be odd, so $1 + 1/\alpha$ must be an odd integer. This leads to

$$\alpha_i = \frac{1}{2i+2}, \quad i = 0, 1, 2, \ldots \tag{11.18}$$

The constant C in (11.17) must be positive but is otherwise arbitrary. It is set by the initial conditions, another characteristic of self-similarity of the second kind. However, C can be normalized to unity by rescaling x and U.

The dynamical system corresponding to the self-similar solution (11.15) is

$$U_\tau - \alpha U + (1+\alpha)\xi U_\xi + U U_\xi = 0, \tag{11.19}$$

and so the eigenvalue equation for a perturbation P around the base profile $\bar{U}_i$ becomes

$$(\alpha_i - \nu)P - (1+\alpha_i)\xi P_\xi - P(\bar{U}_i)_\xi - P_\xi \bar{U}_i = 0, \quad i = 0, 1, \ldots \tag{11.20}$$

Here $\bar{U}_i$ is the ith similarity function defined by (11.17) for the exponent α_i given by (11.18).

The eigenvalue equation (11.20) is solved by transforming from the variable ξ to the variable $\bar{U}_i$, using (11.17):

$$P\left[(\alpha_i - \nu)(1 + (2i+3)\bar{U}_i^{2i+2}) + 1\right] = \frac{\partial P}{\partial \bar{U}}\left[\alpha_i \bar{U}_i + (1+\alpha_i)\bar{U}_i^{2i+3}\right]. \tag{11.21}$$

Again the solution

$$P = \frac{\bar{U}_i^{3+2i-2\nu(i+1)}}{1 + (2i+3)\bar{U}_i^{2i+2}} \tag{11.22}$$

is found by the separation of variables. The exponent $3 + 2i - 2\nu(i+1)$ must be an integer for (11.22) to be regular at the origin, so the eigenvalues are

$$\nu_j = \frac{2i+3-j}{2i+2}, \quad j = 0, 1, \ldots \tag{11.23}$$

It is straightforward to confirm that (11.22) satisfies the growth condition (3.59). As usual the eigensolutions alternate between even and odd. However, we are interested in the *first* instance, given by (11.7), at which a shock forms. This implies that the second derivative of the profile must vanish at the location of the shock and thus that the amplitude of the $j = 2$ perturbation must be exactly zero.

Thus for $i = 0$ the remaining eigenvalues are $\nu = 3/2, 1, 0, -1/2, \ldots$; the first two are the eigenvalues $\nu = \beta = 1+\alpha$ and $\nu = 1$, coming from spatial and temporal invariance; see Table 3.1. The vanishing eigenvalue occurs because there is a continuous family of solutions parameterized by the coefficient C in (11.17). All the other eigenvalues are negative, which shows that the similarity solution given in (11.24) below is stable. In the same vein, for $\alpha_1 = 1/4$

there are two more positive exponents, since now $\nu = 5/4, 1, 1/2, 1/4, \ldots$ so the solution must be unstable. The same is of course true for all higher-order solutions. In conclusion the ground state solution $\bar{U}_0$ given by (11.24) is the generic form of shock formation.

Thus, in close analogy to Section 3.2, only the lowest-order solution is selected, and so the similarity solution before the singularity is

$$u(x, t) = t'^{1/2} U_b \left(\frac{x'}{t'^{3/2}} \right), \tag{11.24}$$

where U_b is defined implicitly by the third-order curve

$$\xi + U_b + C U_b^3 = 0. \tag{11.25}$$

Example 11.3 (Local description of a shock) Let us verify the cubic form of the solution (11.25) for Example 11.1, using the exact solution (11.4). To obtain a local description we expand the profile (11.6) about $s_m = \pi/2$, which yields

$$u_0(s) = 1 - s' + s'^3/6 + \ldots, \tag{11.26}$$

where $s' = s - s_m$ as before. Note that the quadratic term vanishes, so we have continued the expansion to third order. This is a general feature, as seen in Exercise 11.2.

We want to write the solution in a coordinate system moving with the shock speed $u_0(s_m) = 1$, so we introduce $x' = x - x_0 + u_0(s_m)t'$, where $t' = t_0 - t$. Then from (11.3) we obtain

$$x' = u_0(s)t + s - x_0 + t' \approx (s' - s'^3/6)t' + s'^3/6,$$

using (11.26). Clearly, x' is of order s'^3, so t' is of order s'^2 and we can neglect $s'^3 t'$, to write

$$x' = s't' + s'^3/6 \tag{11.27}$$

up to terms of higher order. The velocity $\Delta u = u_0 - u_0(s_m)$ relative to the shock speed is $\Delta u = -s' + O(s'^3)$ according to (11.26). Inserting this into (11.27), we finally obtain

$$x' + \Delta u\, t' + \Delta u^3/6 = 0, \tag{11.28}$$

which is the same as (11.25) when put into the similarity form (11.24). For our particular case we can identify the constant C as $1/6$. $\square$

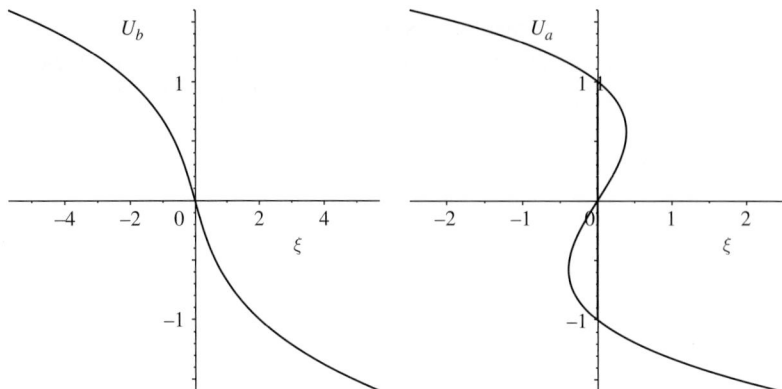

Figure 11.3 Similarity solutions before and after the shock. On the left, the solution of (11.25); on the right, the solution of (11.31). The curve is non-unique in the center, where the position of the jump (the vertical line between 1 and −1) is determined by an energy argument.

Now we proceed to the corresponding continuation. The similarity equation for $t > t_0$ is (11.16) with a minus sign, and the solution in this case is

$$\xi - U_a + \bar{C}U_a^{1+1/\alpha} = 0. \tag{11.29}$$

The matching condition

$$\lim_{t' \to 0} u_b(\Delta x, t) = \lim_{t' \to 0} u_a(\Delta x, t) \tag{11.30}$$

requires that

$$CU^3 = \bar{C}U^{1+1/\alpha},$$

which implies that $\bar{C} = C$ and $\alpha = 1/2$. The corresponding solution

$$\xi - U_a + CU_a^3 = 0 \tag{11.31}$$

is plotted in Fig. 11.3. Thus, although the matching condition selects a unique curve, the latter is not single-valued and there must be a jump in the solution at some ξ_s such that $-2/\sqrt{27} \le \xi_s \le 2/\sqrt{27}$ in the region where there are three possible values of U.

As discussed above, the jump condition (11.10) fixes the value of ξ_s. In the outer variable the position of the shock is $x_s = \xi_s |t'|^{3/2}$, and so

$$u_s = \dot{x}_s = 3\xi_s |t'|^{1/2}/2. \tag{11.32}$$

Thus the jump condition (11.10) becomes

$$3\xi_s = U_\ell + U_r. \tag{11.33}$$

From symmetry it is clear that this condition is verified for $\xi_s = 0$. The shock solution is described by the curve shown in Fig. 11.3, where the two branches of the solution are connected by a vertical line through the origin. In contrast to drop breakup the singularity (shock) is now a persistent structure, which will continue to propagate.

Example 11.4 (Local description after shock formation) The local form (11.28) of the solution is valid after shock formation as well, except that now $t' = t - t_0$, leading to the opposite sign in front of the second term. Thus we again obtain a cubic form of the type (11.31) that is valid after shock formation. □

11.3 Conservation laws: shocks and unique continuation

We have seen that by writing Burgers' equation (11.2) in the conserved form (11.9) and demanding conservation of the flux f across the shock, unique continuation is guaranteed. We will now investigate more systematically the conditions under which unique continuation is obtained, using a generalized form of Burgers' equation:

$$u_t + (f(u))_x = 0. \tag{11.34}$$

We will see below that f is required to be convex ($f''(u) > 0$ for all u) for (11.34) to be consistent.

Analogously to the case for Burgers' equation (see Exercise 11.1), u is constant along the characteristic curves

$$x(s,t) = f'(u_0(s))t + s,$$

where $u_0(s) = u(s,0)$. The characteristics will cross when $\partial x/\partial s = 0$, so a singularity will develop at

$$t_0 = \min_s \left\{ -\frac{1}{f''(u_0(s))u_0'(s)} \right\}.$$

If the conserved form of the equation is used, the shock propagates with a uniquely defined speed. To show this, we integrate (11.34) from $x = a$ to $x = b$, obtaining

$$\frac{d}{dt}\int_a^b u(x,t)dx + f(u(b)) - f(u(a)) = 0. \tag{11.35}$$

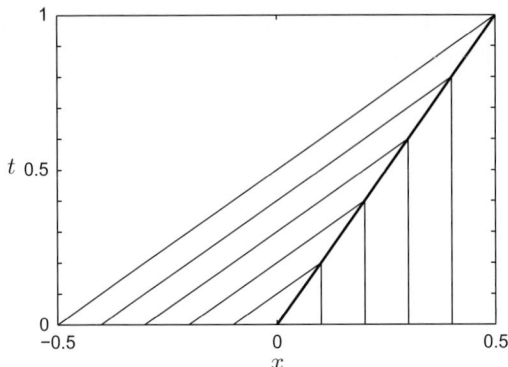

Figure 11.4 Characteristics in the neighborhood of a Burgers' shock (the bold line) obeying the entropy condition (11.38). All characteristics emanating from the shock can be traced back to the initial condition at $t = 0$.

If there is a shock located at $x_s(t)$ (with $a < x_s(t) < b$) one finds for the first term in (11.35)

$$\frac{d}{dt}\int_a^b u(x,t)dx = \frac{d}{dt}\int_a^{x_s(t)} u(x,t)dx + \frac{d}{dt}\int_{x_s(t)}^b u(x,t)dx$$

$$= u_\ell\frac{dx_s}{dt} - u_r\frac{dx_s}{dt} + \int_a^{x_s(t)}\frac{\partial}{\partial t}u(x,t)dx + \int_{x_s(t)}^b\frac{\partial}{\partial t}u(x,t)dx$$

$$= (u_\ell - u_r)\frac{dx_s}{dt} - [f(u_\ell) - f(u(a))] - [f(u(b)) - f(u_r)]. \quad (11.36)$$

In the last step we have used that $u_\ell = u(x_s^-(t), t)$ is the value of u on the left of the shock and $u_r = u(x_s^+(t), t)$ the value of u on the right of the shock. Inserting (11.36) into (11.35) one obtains the following jump condition:

$$u_s \equiv \frac{dx_s}{dt} = \frac{f(u_\ell) - f(u_r)}{u_\ell - u_r} \equiv \frac{[f]}{[u]}. \quad (11.37)$$

For the flux $f = u^2/2$ one arrives at our earlier condition (11.10).

Figure 11.4 shows the neighborhood of a shock, with time plotted as function of position. Thus the slope of the bold line represents the inverse of the shock speed u_s. Causality, i.e. the condition that events at a given time are determined only by earlier events and not by future events, demands that each point on the shock curve can be connected by a characteristic to the initial condition, i.e. the line $t = 0$. It is clear that this is possible only if the slope of the characteristics to the left of the shock is smaller than that of the shock and the slope of the characteristics to the right of the shock is greater than that of the shock. This condition is due to Lax and is called the *entropy condition*, for reasons that we will discuss shortly; it implies that

$$f'(u_\ell) > u_s > f'(u_r). \tag{11.38}$$

First, note that if f is convex then the condition on u_s implied by (11.38) is satisfied by u_s as given by (11.37); see Exercise 11.3. Further, it follows that

$$u_\ell > u_r. \tag{11.39}$$

Thus if f is convex (Burgers' equation, illustrated in Fig. 11.2, is an example of this), any shock formation from smooth initial data will satisfy (11.39).

Nevertheless, we may ask if there are other possibilities for continuing a shock which has been imposed as a discontinuous initial condition. Let us consider two different singular initial conditions, one of which satisfies the entropy condition (11.39) for convex f and the other of which does not. The first condition is a backward facing step,

$$u_0(x) = \begin{cases} 1 & \text{for } x < 0, \\ 0 & \text{for } x \geq 0; \end{cases} \tag{11.40}$$

the other is a forward facing step,

$$u_0(x) = \begin{cases} 0 & \text{for } x < 0, \\ 1 & \text{for } x \geq 0. \end{cases} \tag{11.41}$$

The initial condition (11.40) clearly satisfies (11.39), so it can be continued as a shock wave. The characteristics are shown in Fig. 11.4 and are consistent with causality, as expected.

In the case of (11.41), however, we have a shock with $u_\ell < u_r$ and (11.39) is violated. Points immediately to the right of the discontinuity move with velocity $u_r = 1$ and points to the left with velocity $u_\ell = 0 < u_r$. Hence we are left with a region $u_\ell t < x < u_r t$ where the solution is undefined. If we try to construct a global solution to (11.34), there are two possibilities. The first consists in continuing (11.41) as a shock which propagates at a velocity u_s given by (11.37). The undefined region is filled with characteristics having a slope $1/u_\ell$ to the left of the shock and a slope $1/u_r$ to the right of the shock:

$$u = u_\ell \quad \text{for } (u_\ell - c)t < x - ct < 0,$$
$$u = u_r \quad \text{for } 0 < x - ct < (u_r - c)t;$$

see Fig. 11.5 (left). However, this solution *violates* the entropy condition (11.38). Clearly, characteristics which intersect the shock are *not* traceable to the initial condition.

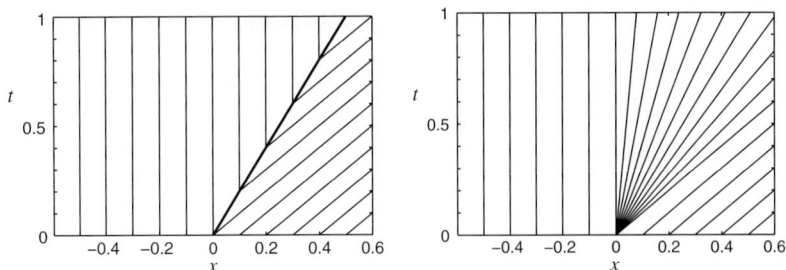

Figure 11.5 Two different continuations of Burgers' equation with initial condition (11.41). On the left, a rarefaction shock; on the right, a rarefaction wave.

A second possibility is to continue the solution by inserting a so-called *rarefaction wave*; see Fig. 11.5 (right). This is a self-similar solution of the form

$$u(x, t) = r(\xi), \quad \xi = x/t, \tag{11.42}$$

where we assume that the shock is centered at $x = 0$. The boundary conditions are $r(\xi) \to u_\ell$ as $\xi \to \xi_\ell$ and $r(\xi) \to u_r$ as $\xi \to \xi_r$. Inserting (11.42) into (11.34), we obtain

$$-\xi r_\xi + (f(r))_\xi = 0,$$

with implicit solution

$$f'(r(\xi)) = \xi.$$

Therefore we can construct a rarefaction wave in the form

$$r(\xi) = \begin{cases} u_\ell & \text{for } \xi \leq \xi_\ell, \\ (f')^{-1}(\xi) & \text{for } \xi_\ell < \xi < \xi_r, \\ u_r & \text{for } \xi \geq \xi_r. \end{cases} \tag{11.43}$$

The rarefaction wave interpolates between the speeds of propagation to the left and right of the undefined region, as seen in Fig. 11.5 (right).

In principle, both a rarefaction shock propagating with a velocity given by (11.37) and the rarefaction wave defined by (11.43) are valid ways of continuing the singular initial condition (11.41). However, only the rarefaction wave satisfies the entropy (causality) condition. It is also apparent that all characteristics on the right of Fig. 11.5 are connected to the line $t = 0$, while this is not the case on the left. This is another expression of the fact that the rarefaction wave (11.43) obeys causality, while the rarefaction shock does not.

11.4 Viscosity solutions

The root of the non-uniqueness described above lies in the fact that the solutions are singular. As long as the solution is smooth, the evolution remains well defined. In order to avoid the problem of non-uniqueness one can introduce a viscous regularization into (11.34) and thus obtain solutions which remain smooth. Upon taking the limit of vanishing viscosity the solutions become singular once more, but a unique solution can be selected which obeys the conditions derived previously.

Therefore we will take equation (11.34) and introduce a viscous term νu_{xx} on the right-hand side, to get

$$u_t + (f(u))_x = \nu u_{xx}. \tag{11.44}$$

We seek traveling wave solutions of the form

$$u(x, t) = U(\xi), \quad \xi = \frac{x - u_s t}{\nu}, \tag{11.45}$$

which corresponds to a smooth profile of thickness ν; this replaces the discontinuities of the previous sections. Inserting (11.45) into (11.44), the shape of the profile is described by

$$-u_s U' + (f(U))' = U''. \tag{11.46}$$

In analogy to (3.15), the inner problem (11.46) has to satisfy matching conditions, in the limit $\nu \to 0$, which ensure that the inner solution agrees with the outer solution, described by (11.34). The velocity immediately to the left of the jump is u_ℓ and to the right of the jump is u_r. Thus in the limit $\nu \to 0$ it follows that $U(\xi)$ has to satisfy the boundary conditions

$$U(\xi) \to u_\ell \quad \text{as} \quad \xi \to -\infty, \tag{11.47}$$

$$U(\xi) \to u_r \quad \text{as} \quad \xi \to +\infty. \tag{11.48}$$

By integrating (11.46) from $-\infty$ and using the fact that U' vanishes at $-\infty$, one obtains the first-order equation

$$U' = -u_s(U - u_\ell) + f(U) - f(u_\ell). \tag{11.49}$$

Evaluating this at $+\infty$ yields

$$u_s = \frac{[f(U)]_{-\infty}^{+\infty}}{[U]_{-\infty}^{+\infty}}, \tag{11.50}$$

which is the jump condition (11.37) obtained earlier.

If $f(u)$ is strictly convex, as in Burgers' equation, then a solution to (11.49) with (11.47), (11.48), and (11.50) exists if and only if the entropy condition

(11.39) is satisfied. The reason is that the right-hand side of (11.49) has zeros at $U = u_r$ and u_ℓ. Since $f(u)$ is convex, however, it follows that $U' < 0$ for u between u_r and u_ℓ. Hence $U(\xi)$ is a smooth decreasing function, and in the limit $\nu \to 0$ the traveling wave (11.45) tends to a discontinuity satisfying (11.39). If one starts from a discontinuous initial condition, with $u_\ell < u_r$ for $t = 0$, such a connection by means of U does not exist and a rarefaction shock cannot be the limit of viscous solutions as $\nu \to 0$. Instead, viscosity solutions (which are smooth for $t > 0$) converge to a rarefaction wave rather than a shock in the limit of vanishing viscosity.

Example 11.5 (Energy dissipation) In the case where $f(u) = u^2/2$, (11.44) becomes

$$u_t + u u_x = \nu u_{xx}, \tag{11.51}$$

which is called the viscous Burgers' equation (this is the equation introduced by Burgers originally [37]). Thus the similarity equation (11.46) can be written as

$$2\frac{dU}{d\xi} = (U - u_s)^2 - \Delta^2, \tag{11.52}$$

where $\Delta = (u_\ell - u_r)/2$ measures the strength of the shock and we have used $u_s = (u_\ell + u_r)/2$. Now (11.52) can be integrated to give an explicit form for the shock profile,

$$U = u_s - \Delta \tanh \frac{\Delta \xi}{2}. \tag{11.53}$$

In other words the total width of the regularized shock is ν/Δ, i.e. the viscosity divided by the strength of the shock.

Multiplying (11.51) by u, the equation can be written in the form

$$\frac{\partial}{\partial t}\frac{u^2}{2} = -\left(\frac{u^3}{3}\right)_x + \nu u u_{xx} = -\left(\frac{u^3}{3} - \nu u u_x\right)_x - \nu u_x^2,$$

which can be interpreted as a one-dimensional version of the energy balance (4.17). Integrating over space and disregarding boundary terms (which describe the energy flowing into and out of a control volume), we obtain

$$\dot{E} \equiv \frac{\partial}{\partial t} \int \frac{u^2}{2} dx = -\nu \int u_x^2 dx. \tag{11.54}$$

In the limit $\nu \to 0$ the right-hand side goes to zero and energy is conserved, as long as u remains smooth. However, this is no longer the case in the presence of shocks. To compute the rate of energy dissipation inside a shock, we insert the traveling wave solution (11.45) into (11.54):

$$\dot{E} \equiv -\nu \int u_x^2 dx = -\int_{-\infty}^{\infty} U'^2 d\xi = -\int_{u_\ell}^{u_r} U' dU,$$

using $d\xi = (d\xi/dU)dU$ in the last step. Now we substitute (11.52) to find

$$\dot{E} = -\frac{1}{2} \int_{u_\ell}^{u_r} \left[(U - u_s)^2 - \Delta^2 \right] dU = -\frac{\Delta^3}{2} \int_1^{-1} \left(V^2 - 1 \right) dV = -\frac{2\Delta^3}{3}.$$

$$(11.55)$$

Thus even in the limit $\nu \to 0$ there is a finite rate of energy dissipation inside the shock, which from its definition (11.54) is manifestly negative. Thus it follows once more that $\Delta > 0$, which is equivalent to the entropy condition (11.39). We could also have obtained (11.55) by considering the energy fluxes into and out of the shock as given by the conservative dynamics (11.2). Namely, in a frame of reference moving at the speed u_s of the shock, the influx is $(u_\ell - u_s)^3/3$ while the outflux is $(u_r - u_s)^3/3$. The difference between the two gives

$$\dot{E} = -(\text{influx} - \text{outflux}) = -\frac{2\Delta^3}{3},$$

in agreement with (11.55). Energy dissipation leads to an increase in entropy so (11.55) expresses the fact that the entropy increases, as required by thermodynamics. □

A generalization of the above approach leads to another characterization of entropy solutions, introducing an arbitrary convex function $\eta(u)$ called the convex entropy density. Unfortunately, the sign convention for this function in the context of hyperbolic systems is such that it always decreases in time. Multiplying (11.44) by the derivative of η with respect to its argument, one finds

$$\eta'(u)u_t + \eta'(u)(f(u))_x = \nu\eta'(u)u_{xx}.$$

Introducing the auxiliary function

$$G'(u) = \eta'(u)f'(u),$$

this can be written as

$$\frac{\partial}{\partial t}\eta(u) + (G(u))_x = \nu\eta'(u)u_{xx}. \qquad (11.56)$$

Integrating over x, and disregarding boundary terms, yields

$$\frac{d}{dt} \int \eta(u)dx = \nu \int \eta'(u)u_{xx}dx = -\nu \int \eta''(u)(u_x)^2 dx \leq 0. \qquad (11.57)$$

The relation (11.57) is called a global entropy inequality. In the limit $\nu \to 0$, (11.57) still holds so for a solution of (11.34) with propagating shocks we find that

$$\frac{d}{dt} \int \eta(u)dx \leq 0 \qquad (11.58)$$

for any convex entropy function $\eta(u)$. Equation (11.58) corresponds to $\dot{E} \leq 0$ in Example 11.5, with $\eta = u^2/2$.

What is not so evident is that the reverse also holds: a solution u of (11.34) such that (11.58) is satisfied for any entropy function is unique. This fact was proved by Kruzkov in 1969 [127]. Hence, for scalar conservation laws, entropy solutions coincide with solutions of (11.44) in the limit of vanishing viscosity and are unique.

11.5 Compressible gas flow

Now we turn to a more complete description of shocks in an inviscid, compressible gas, described by mass conservation (4.1) on the one hand and by the Euler equation (4.31) on the other. We will still confine ourselves to motion in one dimension, i.e. to planar shocks, or those in a shock tube. We begin by deriving the jump conditions which connect two states of a system described by several variables. We then move on to the problem of continuation for systems of equations, which we have treated previously for the case of a single variable alone. Denoting the first component of the velocity $\mathbf{v}$ by u, the equations of compressible gas dynamics become

$$\rho_t + (\rho u)_x = 0, \qquad (11.59)$$

$$u_t + uu_x = -\frac{p_x}{\rho}. \qquad (11.60)$$

This system has to be complemented by an equation of state $p(\rho)$ which follows from the thermodynamics of the gas. Since we are considering ideal fluids for which dissipative processes (such as viscous friction or heat conduction) are negligible, this is an *adiabatic* equation of state (for a system at constant entropy). For an ideal gas the entropy per unit mass is

$$s = c_v \ln \frac{p}{\rho^\gamma}, \qquad (11.61)$$

where γ is called the adiabatic exponent. For a monatomic gas $\gamma = 5/3$, while for a diatomic gas such as air $\gamma = 7/5$. Thus the adiabatic equation is

$$p = A\rho^{\gamma}, \tag{11.62}$$

where A depends on the value of s.

In principle, (11.59)–(11.62) should provide a complete description of gas motion. However, on account of the convective nonlinearity on the left-hand side of (11.60), the system is susceptible to the formation of shocks by the same mechanism as that present in Burgers' equation. As demonstrated by (11.55), energy is dissipated inside a shock, which means (as we will demonstrate explicitly) that s is not constant *across* the shock; in particular the constant A will be different to the left and to the right of the shock. To deal with the presence of discontinuities, the dynamics need to be reformulated in terms of conservation laws for the mass, momentum, and energy, since these remain conserved across the shock. First, with the help of (11.59), formula (11.60) can be written as a continuity equation for the momentum ρu:

$$(\rho u)_t + \left(\rho u^2 + p\right)_x = 0. \tag{11.63}$$

To replace (11.62) we use the conservation of energy, which in the *incompressible* case and in the absence of viscosity reads (cf. (4.17))

$$\left(\frac{\rho u^2}{2}\right)_t + \left(u\left(\frac{\rho u^2}{2} + p\right)\right)_x = 0.$$

In the case where the fluid is compressible we need to account for changes in the internal energy ϵ of the gas as well, so the total energy becomes

$$e = \rho\frac{u^2}{2} + \rho\epsilon \equiv \rho\frac{u^2}{2} + \frac{p}{\gamma - 1}. \tag{11.64}$$

Here we have used the formula for the internal energy of an ideal gas [133]. Now the conservation of energy becomes

$$e_t + (u(e + p))_x = 0. \tag{11.65}$$

Expressing p in terms of ρ using (11.62), it is a simple yet tedious task to confirm that the conserved equations (11.59), (11.63), and (11.65), together with (11.64), are indeed equivalent to the original formulation (11.59)–(11.62) as long as the solutions are smooth (see Exercise 11.4).

Now let us assume that a shock develops at a finite time t_0. Performing operations very similar to (11.36) for the three equations (11.59), (11.63), and

(11.65), we conclude that the jump in the quantities ρ, u, and p across the shock (denoted by square brackets) must satisfy the equations

$$u_s[\rho] = [\rho u], \tag{11.66}$$

$$u_s[\rho u] = \left[\rho u^2 + p\right], \tag{11.67}$$

$$u_s[\rho e] = \left[(\rho e + p) u\right], \tag{11.68}$$

where u_s is the velocity of propagation of the shock. Equations (11.66)–(11.68) are the analogues of (11.37) and are known as the Rankine–Hugoniot conditions. Since the equations of fluid mechanics are Galilean invariant we can fix our coordinate system at the shock, and hence its velocity in this frame will be zero. Therefore, the left-hand sides of (11.66)–(11.68) are zero and we obtain the following jump relations across the shock:

$$\rho_r u_r = \rho_\ell u_\ell, \tag{11.69}$$

$$\rho_r u_r^2 + p_r = \rho_\ell u_\ell^2 + p_\ell, \tag{11.70}$$

$$(\rho_r e_r + p_r)u_r = (\rho_\ell e_\ell + p_\ell)u_\ell. \tag{11.71}$$

Let us find out what this implies for the relation between the thermodynamic quantities on the left and on the right of the shock. Inserting (11.69) into (11.70) and writing $M = \rho_r u_r = \rho_\ell u_\ell$, it follows that

$$u_r - u_\ell = \frac{p_\ell - p_r}{M}, \quad u_r + u_\ell = M\frac{\rho_\ell + \rho_r}{\rho_\ell \rho_r}. \tag{11.72}$$

Now the energy equation (11.71) can be written as

$$0 = e_r - e_\ell + \frac{p_r}{\rho_r} - \frac{p_\ell}{\rho_\ell} = \frac{1}{2}(v_r - v_\ell)(v_r + v_\ell) + \frac{p_r\rho_\ell - p_\ell\rho_r}{\rho_r\rho_\ell} + \epsilon_r - \epsilon_\ell.$$

Using (11.72) this can be rewritten as

$$\frac{1}{2}(p_r + p_\ell)\left(\rho_r^{-1} - \rho_\ell^{-1}\right) + \epsilon_r - \epsilon_\ell = 0, \tag{11.73}$$

which is known as the *Hugoniot equation* for the shock.

Since

$$\epsilon = \frac{p}{\rho(\gamma - 1)}, \tag{11.74}$$

(11.73) permits us to calculate (ρ_ℓ, p_ℓ) to the left of the shock, given (ρ_r, p_r) to the right of the shock. In other words (11.73) determines the state of the gas after the shock has passed through, in terms of the initial state of the gas.

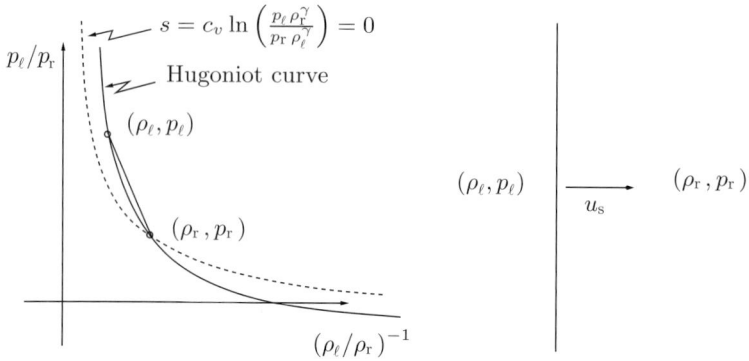

Figure 11.6 On the left, a schematic of the Hugoniot curve of shock dynamics. On the right, the shock passes through the system and the state (ρ_ℓ, p_ℓ) of the gas is converted into the state (ρ_r, p_r). The broken line on the left gives the adiabatic equation of state (11.62), corresponding to constant entropy.

This is illustrated schematically in Fig. 11.6, where we plot the ratio p_ℓ/p_r as function of $(\rho_\ell/\rho_r)^{-1}$. From (11.73) we find that

$$\frac{p_\ell}{p_r} = \frac{\rho_\ell/\rho_r - \mu^2}{1 - \mu^2 \rho_\ell/\rho_r}, \qquad (11.75)$$

where

$$\mu^2 = \frac{\gamma - 1}{\gamma + 1}.$$

The initial state of the system thus corresponds to the point $(1, 1)$ in the diagram; the final state is some other point along the Hugoniot curve.

Example 11.6 (Strong shock) The relations simplify in the limit of a *strong shock*, defined by $p_\ell \gg p_r$. From (11.75) it follows that in this case $\rho_\ell/\rho_r = \mu^{-2}$, and so

$$\frac{\rho_\ell}{\rho_r} = \frac{\gamma - 1}{\gamma + 1}. \qquad (11.76)$$

Inserting this into (11.72) and solving for u_ℓ and u_r, one deduces that

$$u_r = \sqrt{\frac{(\gamma + 1)p_\ell}{2\rho_r}}, \qquad u_\ell = \sqrt{\frac{(\gamma - 1)^2 p_\ell}{2(\gamma + 1)\rho_r}}. \qquad (11.77)$$

$\square$

Coming back to (11.75), note that the adiabatic gas law (11.62) corresponds to

$$\frac{p_\ell}{p_r} = \left(\frac{\rho_\ell}{\rho_r}\right)^\gamma, \tag{11.78}$$

shown as the broken line in Fig. 11.6. Clearly if the final state lies on the Hugoniot curve (11.75) then it *cannot* satisfy (11.78), which explains why the system (11.59)–(11.62) does not describe shocks correctly. To appreciate the meaning of this observation, note that (11.78) is the adiabatic gas law, and hence the entropy is constant along the broken line shown in Fig. 11.6. However, we know from fundamental thermodynamic principles that the entropy can only increase. In fact, we leave it as Exercise 11.5 to show that energy is dissipated inside a shock even in the limit of vanishing viscosity, so the entropy must strictly increase as the shock converts the gas from the state on the right to that on the left:

$$s_\ell > s_r. \tag{11.79}$$

We can see that, to the left of the initial state, the broken line always lies below the Hugoniot curve and on the right it lies above it. This means that, for a given ρ_ℓ, the full curve corresponds to a higher pressure p_ℓ than the pressure corresponding to constant entropy. In other words, all final states to the left satisfy (11.79), whereas states to the right violate it and are excluded on physical grounds. We thus have

$$p_\ell > p_r, \quad u_\ell > u_r \tag{11.80}$$

on account of (11.72). This motivates the terminology "entropy condition" for (11.38).

One can also show that (11.80) can be derived from the requirement of causality, i.e. that characteristics crossing the shock must be traceable to the initial condition [50]. Indeed, (11.79) itself is an expression of causality since the entropy for any physical system with dissipation can only increase in the course of time; this is known as the second law of thermodynamics. The first law of thermodynamics is embodied by the conservation of energy (11.71) across the shock.

11.5.1 Unique continuation for systems

Let us generalize the analysis of Section 11.3 by considering an initial condition that consists of the discontinuous transition between two states $\mathbf{u}_\ell$ and $\mathbf{u}_r$ of a system of n conservation laws of the form

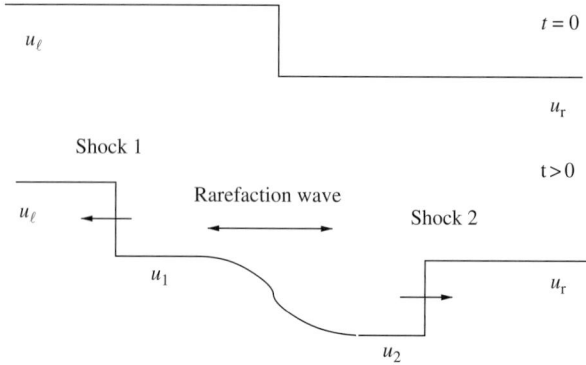

Figure 11.7 The continuation after a discontinuity (at $t = 0$) for one of the components of a system of three conservation laws. Initially there is a jump from u_ℓ to u_r. For $t > 0$ the solution consists of two shocks, connecting u_ℓ with u_1 and u_2 with u_r, together with a rarefaction wave connecting u_1 with u_2.

$$\frac{\partial \mathbf{u}}{\partial t} + \frac{\partial \mathbf{f}(\mathbf{u})}{\partial x} = 0. \tag{11.81}$$

In the case of compressible gas dynamics, treated in the previous section, this would be the system (11.59), (11.63), and (11.65) with $\mathbf{u} = (\rho, \rho u, e)$ and $n = 3$. In general it will not be possible to satisfy the Rankine–Hugoniot conditions (11.66)–(11.68) with a single shock propagating at some velocity u_s, since the system is overdetermined. Instead, we have to introduce $n - 1$ intermediate states to satisfy all the conditions. In addition we have to make sure that a phenomenological Lax causality condition, analogous to (11.38), is satisfied for each shock. If this is not possible then the shock has to be replaced by a rarefaction wave, as illustrated in Fig. 11.7. However, for each state $\mathbf{u}$ there now exist n different possible characteristic speeds, corresponding to the eigenvalues $\lambda_k(\mathbf{u})$, $k = 1, 2, \ldots, n$, of the Jacobian matrix of $\mathbf{f}(\mathbf{u})$. For definiteness let us assume that for each state the eigenvalues are numbered in such a way that $\lambda_1(\mathbf{u}) \leq \lambda_2(\mathbf{u}) \leq \cdots \leq \lambda_n(\mathbf{u})$.

To explain this construction, let us discuss the case $n = 2$ and consider the discontinuity between $\mathbf{u}_\ell = (u_\ell^{(1)}, u_\ell^{(2)})$ and $\mathbf{u}_r = (u_r^{(1)}, u_r^{(2)})$. In this case there are two eigenvalues, $\lambda_1(\mathbf{u})$, $\lambda_2(\mathbf{u})$, with $\lambda_1(\mathbf{u}) \leq \lambda_2(\mathbf{u})$ for any $\mathbf{u}$. In general the jump from $\mathbf{u}_\ell$ to $\mathbf{u}_r$ will not satisfy the Rankine–Hugoniot condition, since

$$\frac{f_1(\mathbf{u}_r) - f_1(\mathbf{u}_\ell)}{u_r^{(1)} - u_\ell^{(1)}} \neq \frac{f_2(\mathbf{u}_r) - f_2(\mathbf{u}_\ell)}{u_r^{(2)} - u_\ell^{(2)}}.$$

To overcome this we introduce an intermediate state $\mathbf{u}_1 = (u_1^{(1)}, u_1^{(2)})$ such that the jumps from $\mathbf{u}_\ell$ to $\mathbf{u}_1$ and from $\mathbf{u}_1$ to $\mathbf{u}_r$ satisfy the corresponding Rankine–Hugoniot conditions:

$$s_1 \equiv \frac{f_1(\mathbf{u}_1) - f_1(\mathbf{u}_\ell)}{u_1^{(1)} - u_\ell^{(1)}} = \frac{f_2(\mathbf{u}_1) - f_2(\mathbf{u}_\ell)}{u_1^{(2)} - u_\ell^{(2)}} \tag{11.82}$$

$$s_2 \equiv \frac{f_1(\mathbf{u}_r) - f_1(\mathbf{u}_1)}{u_r^{(1)} - u_1^{(1)}} = \frac{f_2(\mathbf{u}_r) - f_2(\mathbf{u}_1)}{u_r^{(2)} - u_1^{(2)}}, \tag{11.83}$$

which constitute a system of two equations with two unknowns, $u_1^{(1)}$ and $u_1^{(2)}$, from which we can expect a discrete set of solutions.

Next we have to decide whether the connections between $\mathbf{u}_\ell$ and $\mathbf{u}_1$ and between $\mathbf{u}_1$ and $\mathbf{u}_r$ are shocks, moving with velocities s_1, s_2 given by (11.82) and (11.83) respectively, or rarefaction waves. In order to have a shock between $\mathbf{u}_\ell$ and $\mathbf{u}_1$, the Lax causality condition requires

$$\lambda_2(\mathbf{u}_\ell) > \lambda_1(\mathbf{u}_\ell) > s_1 > \lambda_1(\mathbf{u}_1), \tag{11.84}$$

which guarantees that the shock is traceable to the initial condition via the characteristics associated with the states $\mathbf{u}_\ell$ and $\mathbf{u}_1$. Analogously, a shock between $\mathbf{u}_1$ and $\mathbf{u}_r$ will satisfy the causality condition if

$$\lambda_2(\mathbf{u}_1) > s_2 > \lambda_2(\mathbf{u}_r) > \lambda_1(\mathbf{u}_r). \tag{11.85}$$

When (11.84) is violated, the connection must be made with a rarefaction wave, and analogously if (11.85) is violated. However, connections with rarefaction waves between $\mathbf{u}_\ell$ and $\mathbf{u}_1$ are possible only if $\lambda_1(\mathbf{u}_\ell) < \lambda_1(\mathbf{u}_1)$, and correspondingly between $\mathbf{u}_1$ and $\mathbf{u}_r$; see Exercise 11.7.

Example 11.7 (Shallow water system) To illustrate the calculation explicitly let us consider a system of two conservation laws, describing inviscid fluid motion in a rectangular open channel. The height of the water surface with respect to the base of the channel is denoted by $h(x, t)$ and the velocity in the direction of the channel is $u(x, t)$. The equations (11.81) become in this case (see Exercise 6.13)

$$h_t + (uh)_x = 0, \tag{11.86}$$

$$(hu)_t + \left(u^2 h + g \frac{h^2}{2} \right)_x = 0, \tag{11.87}$$

corresponding to mass and momentum conservation, respectively; the last term in (11.87) comes from the hydrostatic pressure $p = \rho g h$: The system (11.86), (11.87) is formally equivalent to (11.59), (11.62), and (11.63) with $\gamma = 2$ and $A = g/2$; $h(x, t)$ takes the role of the density. However, we will not be requiring energy conservation, so in Example 11.8 below we impose only the first two jump conditions, (11.66) and (11.67). Indeed, shock solutions (which are known as hydraulic jumps in this context) do not conserve energy; see Exercise 11.6.

The system (11.86), (11.87) can be written in the matrix form

$$\begin{pmatrix} h \\ hu \end{pmatrix}_t + A \begin{pmatrix} h \\ hu \end{pmatrix}_x = \begin{pmatrix} 0 \\ 0 \end{pmatrix},$$

with

$$A = \begin{pmatrix} 0 & 1 \\ gh - u^2 & 2u \end{pmatrix}. \qquad (11.88)$$

The eigenvalues of A are

$$\lambda_{1,2}(h, u) = u \pm \sqrt{gh}, \qquad (11.89)$$

so that the characteristics satisfy the equations

$$\frac{dx_1}{dt} = u + \sqrt{gh},$$

$$\frac{dx_2}{dt} = u - \sqrt{gh}.$$

$\square$

Example 11.8 (Discontinuous initial conditions) Now we consider the specific case of the discontinuity between the states

$$(h_\ell, u_\ell) = (1, \gamma), \quad (h_r, u_r) = (1, -\gamma), \qquad (11.90)$$

illustrated schematically in Fig. 11.8, where γ is a parameter measuring the strength of the shock. The space–time diagram is shown in Fig. 11.9. We can choose units such that $g = 1$. Now we have to find an intermediate state (h_1, u_1) such that the Rankine–Hugoniot conditions with respect to both (h_ℓ, u_ℓ) and (h_r, u_r) are satisfied:

$$\frac{u_1 h_1 - u_\ell h_\ell}{h_1 - h_\ell} = \frac{u_1^2 h_1 + h_1^2/2 - u_\ell^2 h_\ell - h_\ell^2/2}{u_1 h_1 - u_\ell h_\ell} = s_1, \qquad (11.91)$$

$$\frac{u_r h_r - u_1 h_1}{h_r - h_1} = \frac{u_r^2 h_r + h_r^2/2 - u_1^2 - h_1^2/2}{u_r h_r - u_1 h_1} = s_2. \qquad (11.92)$$

These equations yield the system

$$\frac{u_1 h_1 - \gamma}{h_1 - 1} = \frac{u_1^2 h_1 + h_1^2/2 - \gamma^2 - \gamma}{u_1 h_1 - \gamma},$$

$$\frac{-\gamma - u_1 h_1}{1 - h_1} = \frac{\gamma^2 + \gamma - u_1^2 h_1 - h_1^2/2}{-\gamma - u_1 h_1},$$

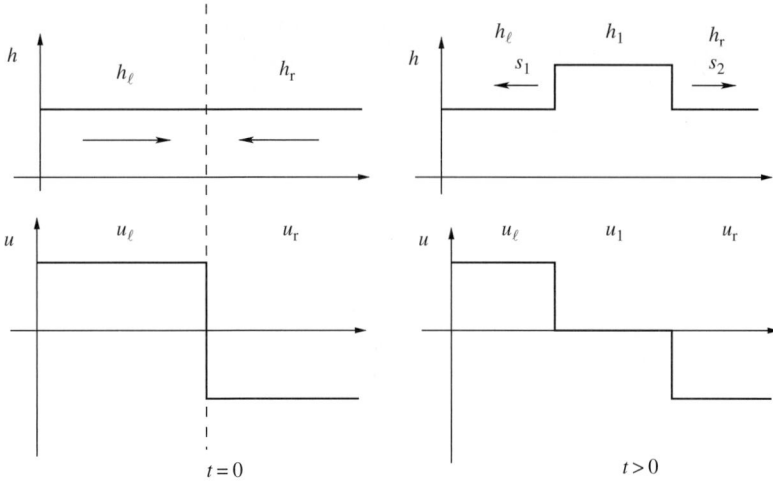

Figure 11.8 The continuation for the shallow water system (11.86), (11.87) with initial conditions (11.90).

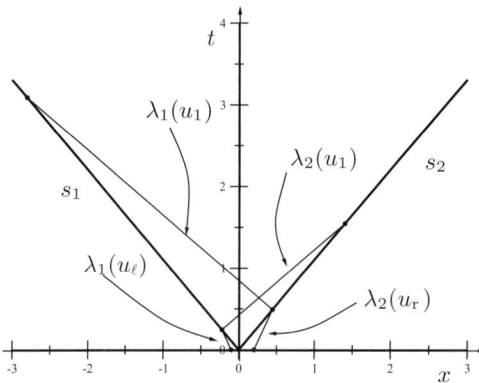

Figure 11.9 The two shocks seen in Fig. 11.8 have velocities $s_{1/2}$. Both shocks can be traced through characteristics back to the initial conditions.

with solution $u_1 = 0$; here h_1 is a root of

$$\frac{1}{2}h_1^3 - \frac{1}{2}h_1^2 - (\gamma + \gamma^2)h_1 + \gamma = 0. \tag{11.93}$$

For $\gamma > 0$, (11.93) has three roots: one negative root, which is unphysical, and two positive roots.

For definiteness, let us take $\gamma = 1/2$; the limit of small γ will be considered in Exercise 11.8. Then, the two possible values of h_1 are 1.5514 and 0.5732. The first leads to a solution with two shocks, which we will derive now. The other value leads to a solution that admits no shocks, and whether connections via rarefaction waves are possible is left as Exercise 11.7. For $h_1 = 1.5514$ the velocities for possible shocks are $s_1 = -0.90678$ and $s_2 = 0.90678$, according to (11.91), (11.92). Since s_1 lies between the values

$$\lambda_1(h_\ell, u_\ell) = u_\ell - \sqrt{gh_\ell} = -\frac{1}{2}$$

and

$$\lambda_1(h_1, u_1) = u_1 - \sqrt{gh_1} = -1.2456,$$

a shock propagating at s_1 satisfies the Lax condition (11.84); since s_2 lies between the values

$$\lambda_2(h_1, u_1) = u_1 + \sqrt{gh_1} = 1.2456$$

and

$$\lambda_2(h_r, u_r) = u_r + \sqrt{gh_r} = -1/2 + 1 = 1/2,$$

a shock propagating with s_2 satisfies (11.85). Thus the solution consists of two shocks traveling with velocities ± 0.90678 (see Fig. 11.9). Between the shocks, $(h, v) = (1.5514, 0)$, and outside the shocks the initial condition remains unaltered; see Fig. 11.8.   □

   The above situation can be generalized to systems of n conservation laws. As above, we can always order the eigenvalues in the form $\lambda_1(\mathbf{u}) \leq \lambda_2(\mathbf{u}) \leq \cdots \leq \lambda_n(\mathbf{u})$. Then continuation after a discontinuity between $\mathbf{u}_\ell$ and $\mathbf{u}_r$ is performed as follows. First, by imposing Rankine–Hugoniot conditions one can find a one-parameter family of states $\mathbf{u}_1$ that can be connected to $\mathbf{u}_\ell = \mathbf{u}_0$ and also a one-parameter family of states $\mathbf{u}_2$ that can be connected to $\mathbf{u}_1$ and, in general, a one-parameter family of states $\mathbf{u}_k$ that can be connected to $\mathbf{u}_{k-1}$. The condition that $\mathbf{u}_{n-1}$ connects with $\mathbf{u}_n = \mathbf{u}_r$ enables one to find a discrete set of states $\mathbf{u}_1, \ldots, \mathbf{u}_{n-1}$. Then each discontinuity between $\mathbf{u}_k$ and $\mathbf{u}_{k+1}$ can be continued either in the form of a shock propagating with the velocity s_k given by the Rankine–Hugoniot condition or as a rarefaction wave. The Lax condition requires that only if s_k lies between $\lambda_k(\mathbf{u}_k)$ and $\lambda_k(\mathbf{u}_{k+1})$ can continuation be performed in the form of a shock. Otherwise, rarefaction waves must be used.

   Whether Lax's causality condition always guarantees a unique solution for (11.81) remains an open problem. An important question is the relation to viscosity solutions. For systems in one spatial dimension, of the form

$$\frac{\partial \mathbf{u}}{\partial t} + \frac{\partial \mathbf{f}(\mathbf{u})}{\partial x} = \nu \frac{\partial^2 \mathbf{u}}{\partial x^2},$$

Bianchini and Bressan [26] showed that, for smooth initial data, the limit $\nu \to 0^+$ is well defined and provides a unique solution to the inviscid system. For higher space dimensions, a complete theory of the formation and interaction of shocks or other types of discontinuities is still lacking. Nevertheless, in situations where a high degree of symmetry is involved one can obtain solutions representing shocks. Next we will describe a special type of solution found by Guderley [100], in which a cylindrical or spherical shock implodes in finite time into a single line or point, respectively.

11.6 Imploding spherical shocks

We assume that a strong spherical shock wave has been created initially and travels inwards to focus on a point; see Fig. 11.10. This creates a strong divergence of the pressure at the origin, which might be useful to achieve, for example, nuclear fusion. Inside the shock sphere $r < R$ we take the gas to be at rest and the density to have a constant value ρ_i; the pressure inside the sphere is much smaller than the pressure behind the shock. For $r > R$ we solve the Euler equation (4.31) and the continuity equation (4.1), which read in spherical symmetry

$$\rho_t + \frac{1}{r^2} \left(r^2 \rho v \right)_r = 0, \tag{11.94}$$

$$v_t + v v_r + \frac{p_r}{\rho} = 0, \tag{11.95}$$

where v is the radial velocity. To close the system it is not possible to use the adiabatic equation of state (11.62), which is valid at constant entropy only. The reason is that entropy is produced inside the shock (see (11.79)) and is then transported through the system. Thus we have to use a transport equation for the entropy (11.61), which yields

$$(\partial_t + v \partial_r) \left(\ln \frac{p}{\rho^\gamma} \right) = 0. \tag{11.96}$$

Now we impose boundary conditions at $r = R(t)$, where the shock is located, using the strong-shock conditions (11.76) and (11.77). In these equations the speeds u_ℓ and u_r are measured relative to the speed u_s of the shock; u_ℓ is the speed just outside the sphere and u_r is the speed inside. Since the gas in front of the shock is at rest in a stationary coordinate system ($u_r = 0$), (11.77) yields

$$-u_s = \sqrt{\frac{(\gamma + 1) p_\ell}{2 \rho_r}}, \quad u_\ell - u_s = \sqrt{\frac{(\gamma - 1)^2 p_\ell}{2(\gamma + 1) \rho_r}}. \tag{11.97}$$

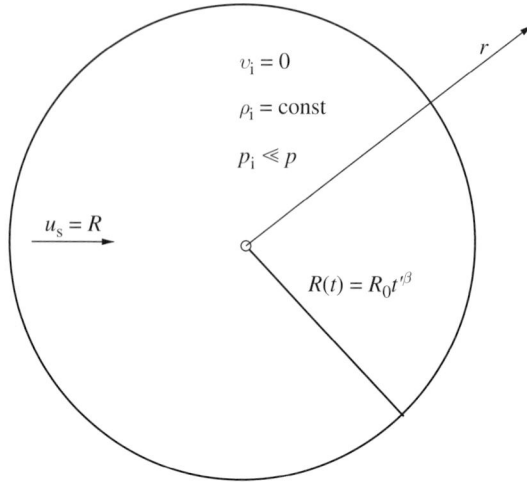

Figure 11.10 A collapsing spherical or cylindrical shock whose radius shrinks to zero at $t' = 0$.

Using (11.97) and the strong-shock condition (11.76) as well as making the identifications $\rho_r = \rho_i$, $p_\ell = p(R, t)$, $u_\ell = v(R, t)$, and $\rho_\ell = \rho(R, t)$, we obtain the boundary conditions

$$v(R, t) = \frac{2u_s}{\gamma + 1}, \quad \rho(R, t) = \rho_i \frac{\gamma + 1}{\gamma - 1}, \quad p(R, t) = \frac{2\rho_i u_s^2}{\gamma + 1}. \quad (11.98)$$

The system (11.94)–(11.96) with the boundary conditions (11.98) is to be solved for $r > R$. It is somewhat more convenient to replace the pressure by the square of the sound speed, $c^2 = \gamma p / \rho$, throughout. We introduce the similarity variable $\xi = r / R(t)$, where the radius obeys the power law

$$R(t) = R_0 t'^\beta. \quad (11.99)$$

Here $t' = (t_0 - t)/T_0$ and R_0 and T_0 are arbitrary length and time scales. In contrast with the case of an explosion driving an expanding shock front (see Example 1.1), the imploding-shock problem does not have an intrinsic scale which would fix a length scale R_0 given T_0. According to (11.99), the shock is located at $\xi = 1$ and (11.94)–(11.96) are solved by the similarity ansatz

$$v(r, t) = \dot{R}\xi V(\xi), \quad \rho(r, t) = \rho_0 \Omega(\xi), \quad c^2(r, t) = \dot{R}^2 \xi^2 Z(\xi). \quad (11.100)$$

The exponent β cannot be determined by dimensional considerations; this appears to have been the first example studied in which the idea of self-similarity of the second kind was appreciated [14].

The system to be satisfied by the similarity functions $V(\xi)$, $\Omega(\xi)$, $Z(\xi)$ is

$$\frac{dV}{d\ln\xi} - (1-V)\frac{d\ln\Omega}{d\ln\xi} = -3V, \tag{11.101}$$

$$(1-V)\frac{dV}{d\ln\xi} - \frac{Z}{\gamma}\frac{d\ln\Omega}{d\ln\xi} - \frac{1}{\gamma}\frac{dZ}{d\ln\xi} = \frac{2Z}{\gamma} - V(\beta^{-1}-V), \tag{11.102}$$

$$(\gamma-1)Z\frac{d\ln\Omega}{d\ln\xi} - \frac{dZ}{d\ln\xi} = \frac{2Z(\beta^{-1}-V)}{1-V}; \tag{11.103}$$

the logarithms of the dependent and independent variables have been introduced for convenience. The boundary conditions (11.98) transform into

$$V(1) = \frac{2}{\gamma+1}, \quad \Omega(1) = \frac{\gamma+1}{\gamma-1}, \quad Z(1) = \frac{2\gamma(\gamma-1)}{(\gamma+1)^2}. \tag{11.104}$$

The matching condition (3.16) implies that

$$V(\xi) \propto \xi^{-1/\beta}, \quad \Omega(\xi) \propto \xi^0, \quad Z(\xi) \propto \xi^{-2/\beta}, \quad \text{for} \quad \xi \to \infty. \tag{11.105}$$

Details of the solution of (11.101)–(11.104) are given in [100, 133, 194, 48]. Since the system (11.101)–(11.104) is autonomous, the solution can be found in the phase plane spanned by the variables V, Z. As usual there are singular points in the equations, so the requirement of a regular solution in the interval $\xi \in (1, \infty)$ leads to additional conditions on the solution. Regular solutions can only be found for specific values of the scaling exponent β, which have to be computed numerically. The results are quoted in Table 11.1 for two values of the adiabatic exponent γ, as well as for both spherical and cylindrical collapse. In the cylindrical case the continuity equation (11.94) is to be replaced by

$$\rho_t + \frac{1}{r}(r\rho v)_r = 0; \tag{11.106}$$

otherwise the problem is the same. Table 11.1 also contains values of the scaling exponents calculated from a simplified geometrical theory, which we will discuss in the following section.

To summarize the results, according to (11.98) and (11.100) the scalings of the shock speed and of the pressure are given by

$$u_s = \dot{R} \propto (t')^{\beta-1}, \quad p \propto (t')^{2\beta-2}, \tag{11.107}$$

as long as the density remains bounded. Since the spatial similarity exponent β is smaller than unity, the shock speed and the pressure diverge as the shock focuses to a point. The divergence is of course stronger (i.e. the value of β is smaller) for the spherical geometry, as there is more efficient focusing. Examples of the density and pressure profiles are shown in Fig. 11.11, for various

Table 11.1 *The scaling exponent β of the shock radius, see (11.99), for adiabatic exponents corresponding to monatomic and diatomic gases. We give the result of the full similarity theory, in the limit of strong shocks, for cylindrical and spherical symmetry (the second and third columns) as well as that for the approximate geometrical shock theory discussed below (the third and fourth columns).*

γ	cylindrical	spherical	cylindrical, geom.	spherical, geom.
5/3	0.815625	0.688377	0.816044	0.689251
7/5	0.835217	0.717173	0.835373	0.717287

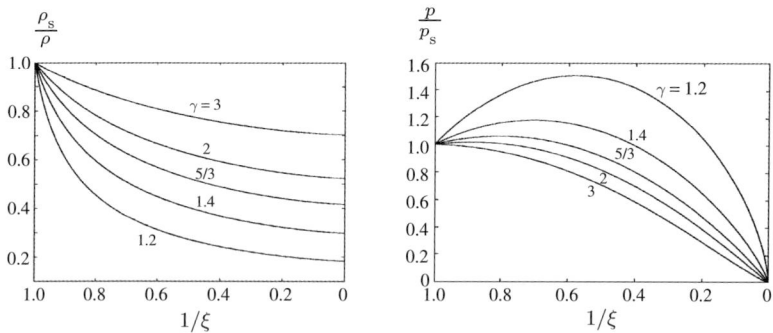

Figure 11.11 The density and pressure behind a collapsing spherical shock front, normalized by their values at the shock position [48]. Here $\xi = r/R(t)$ measures the distance from the focus.

values of γ. The density increases monotonically, reaching a finite value far from the shock. For the most relevant values of γ, the pressure exhibits a maximum behind the shock and decays as $\xi^{2-2/\beta}$ at infinity, as required by (11.100) and (11.105).

The considerable complexity involved in solving the similarity equations (11.101)–(11.104) has motivated a number of studies attempting to simplify the problem; these are summarized in [48]. Notice that for γ approaching 1 the density drops rapidly behind the shock, indicating that the shock motion and the flow become decoupled. This can be shown systematically by demonstrating that in the limit $\gamma \to 1$ the region from which a disturbance to the flow reaches the shock shrinks to zero [48]. This means that the motion of the front becomes a geometrical problem, which we will study now. The only physical ingredient needed is a local relationship between the area of the shock front and its speed. In the case of the Guderley problem this leads to the approximation (11.117) for β, given below; see also Table 11.1.

11.6.1 Geometrical shock dynamics

If the shape of the shock front no longer has a simple symmetry (for example if it is flat or spherical), the problem of shock propagation becomes very complicated. In order to be able to make analytical progress, Whitham [221] developed a simplified description of shock dynamics based on the ideas of geometrical optics. We will confine ourselves to two-dimensional problems, as illustrated in Fig. 11.12.

We introduce an orthogonal curvilinear coordinate system (λ, σ), defined by the advancing shock front (lines of constant λ). Another set of curves are the "rays" (lines of constant σ), which are by definition orthogonal to the shock fronts. The variable σ labels the position along a shock front; since σ is constant along a ray this defines the value of σ along each front. The position along a ray is labeled by the time taken by the shock front to reach a certain position. We normalize time by the vacuum sound speed c_0 ahead of the shock and define $\lambda = c_0 t$. The problem is written in terms of two dependent variables, $M(\lambda, \sigma)$ and $A(\lambda, \sigma)$. The first is the Mach number $M = u_s/c_0$, so that $M(\lambda, \sigma)d\lambda$ is the spatial distance along a ray between λ and $\lambda + d\lambda$. The second variable is defined such that $A(\lambda, \sigma)d\sigma$ is the spatial distance between the rays at σ and $\sigma + d\sigma$. Thus $A(\lambda, \sigma)$ measures how the shock front expands and contracts.

Now an equation can be derived from purely geometrical considerations. If θ measures the direction of a ray with respect to the horizontal, it can be seen, by considering a line parallel to PQ that passes through S in Fig. 11.12, that the change in direction of a ray is

$$\delta\theta = \frac{QR - PS}{PQ}.$$

Next, since $PS = A(\lambda, \sigma)d\sigma$, $QR = A(\lambda + d\lambda, \sigma)d\sigma$, and $PQ = M(\lambda, \sigma)d\lambda$, we obtain

$$\frac{\partial\theta}{\partial\sigma} = \frac{1}{M}\frac{\partial A}{\partial\lambda}. \tag{11.108}$$

Similarly, considering increments of θ along PS one finds

$$\frac{\partial\theta}{\partial\lambda} = -\frac{1}{A}\frac{\partial M}{\partial\sigma}. \tag{11.109}$$

Eliminating θ between (11.108) and (11.109), we obtain the nonlinear wave equation

$$\frac{\partial}{\partial\lambda}\left(\frac{1}{M}\frac{\partial A}{\partial\lambda}\right) + \frac{\partial}{\partial\sigma}\left(\frac{1}{A}\frac{\partial M}{\partial\sigma}\right) = 0. \tag{11.110}$$

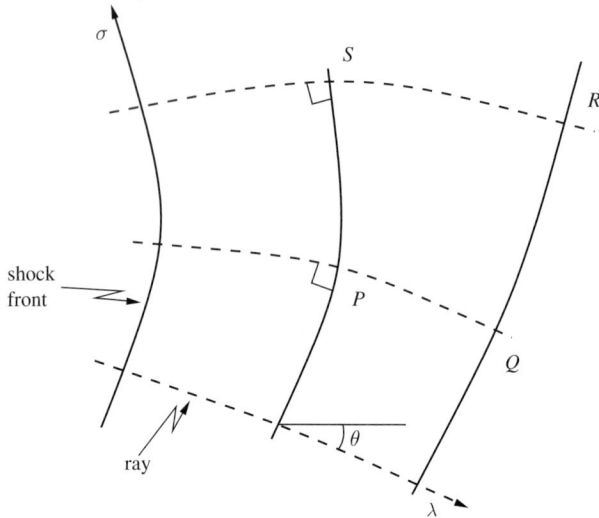

Figure 11.12 The "geometrical optics" approximation of shock dynamics. The shock fronts (solid lines) and rays (broken lines) form an orthogonal plane coordinate system (λ, σ); $\lambda = ct_0$ is the time variable and σ labels the position along a front and is defined to be constant along a ray.

To introduce the physics of shock waves into the geometrical description (11.108)–(11.109), we postulate a *local* relationship of the form

$$A = f(M). \tag{11.111}$$

Once f is known the wave motion (11.108)–(11.109) is defined uniquely. Whitham found a local relationship (11.111) of the form

$$A = \chi M^{-\nu}, \tag{11.112}$$

by considering the propagation of shock waves along tubes with slowly varying cross section [221], where χ is a constant. This relationship is not rigorously valid, since in particular it equates particle paths (which are parallel to the walls of the tube) with rays. For true shock propagation the two are not the same. In practice, however, it has been found that (11.111) yields good results. In the limit of strong shocks, $M \to \infty$, one finds

$$\nu = 1 + \frac{2}{\gamma} + \sqrt{\frac{2\gamma}{\gamma - 1}}, \tag{11.113}$$

where γ is the adiabatic exponent defined by (11.62).

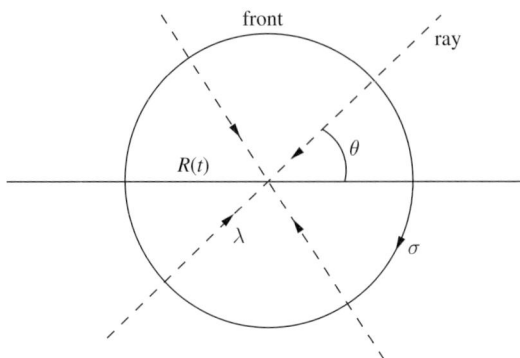

Figure 11.13 A collapsing cylindrical shock front. The variable $\lambda = c_0 t$ measures time and σ is the position along the shock front; we can choose $\sigma = -\theta$, so σ varies from -2π to 0.

Example 11.9 (Collapsing cylindrical shock) As a first application, let us consider the problem of an imploding cylindrical shock wave (see Fig. 11.13), as described by geometrical shock dynamics. The rays point radially toward the origin and can be parameterized by the angle θ, so we can choose

$$\sigma = -\theta, \tag{11.114}$$

bringing the sign conventions in line with Fig. 11.12. On account of symmetry the variables M and A are constant along the circle, and so the circumference is

$$2\pi R = \int_{-2\pi}^{0} A \, d\sigma = 2\pi A;$$

thus $A = R$.

Since the shock speed M is independent of σ, (11.109) is satisfied identically. Inserting (11.114) into (11.108) and using (11.112) one finds

$$-1 = -\nu \chi M^{-\nu-2} \frac{\partial M}{\partial \lambda},$$

which leads to

$$M^{-\nu-1} = \frac{\nu+1}{\nu \chi} (\lambda_0 - \lambda),$$

where λ_0 is a constant of integration. Remembering that λ is proportional to time, M goes to infinity at $\lambda = \lambda_0$. Putting $\lambda' = \lambda_0 - \lambda$ we obtain

$$M = \left(\frac{\nu+1}{\nu \chi} \lambda' \right)^{-1/(\nu+1)}, \tag{11.115}$$

and with (11.112),

$$R = A = \chi \left(\frac{v+1}{v\chi} \lambda' \right)^{v/(v+1)} \propto t'^{v/(v+1)}. \tag{11.116}$$

Thus the scaling exponent β of the imploding shock wave problem is found to be $\beta = v/(1+v)$ in the case of a cylindrical wave. In D spatial dimensions we have $A \propto R^{D-1}(t)$, so this result generalizes to

$$\beta = \frac{v}{D-1+v}. \tag{11.117}$$

Since v can be calculated from (11.113) this provides a very simple estimate of β, which can be compared with the exact value, as we do in Table 11.1. The predictions of the geometrical shock dynamics are remarkably good: the deviation is smaller than 2% for both monatomic and diatomic gases. □

Exercises

11.1 Use characteristics to find the general solution of the nonlinear kinematic equation (11.1).

11.2 Perform the local analysis of Example 11.3 for the case of a general initial condition $u_0(s)$.

11.3 Show both graphically and analytically that if $f''(u) > 0$ for all u then one finds the entropy condition, cf. (11.38),

$$f'(u_\ell) > \frac{f(u_\ell) - f(u_r)}{u_\ell - u_r} > f'(u_r).$$

11.4 Use the conservation of mass equation (11.59) and the Euler equation (11.60), together with the equation of state (11.62), to demonstrate momentum conservation, (11.63), and energy conservation, (11.64). Assume that the solutions are smooth.

11.5 Show that in going from (ρ_r, p_r) to (ρ_ℓ, p_ℓ) along the Hugoniot curve the entropy must increase (cf. (11.79)). Also show that (11.80) holds.

11.6 Consider a stationary hydraulic jump, described by (11.86), (11.87), of a fluid with thickness h_ℓ and speed u_ℓ upstream from the jump to a thickness h_r and speed u_r downstream from the jump. Such a structure, which is stationary in the laboratory frame, forms if for example a stream of water falls into a sink.

(i) Show using mass and momentum conservation that

$$u_\ell^2 \frac{h_\ell}{h_{\rm r}} = \frac{g}{2}(h_{\rm r} + h_\ell)$$

across the jump.

(ii) Show that if $\Delta = h_{\rm r}/h_\ell$ and $\mathrm{Fr} = \sqrt{u_\ell^2/(gh_\ell)}$ is the Froude number then

$$\Delta = -\frac{1 + \sqrt{1 + 8\,\mathrm{Fr}^2}}{2}.$$

In other words, if $\mathrm{Fr} > 1$ (i.e. the flow is faster than the wave speed $\sqrt{gh}$), then the fluid thickness *increases* across the jump.

(iii) Show, using (11.64), that the rate at which energy is lost inside the jump is

$$\dot{\epsilon} = \frac{gu_\ell}{4h_{\rm r}}(h_{\rm r} - h_\ell)^3 .$$

Since this must be positive, we conclude that $\Delta > 1$.

11.7 Show that for $\gamma = 1/2$ the solution $h_1 = 0.5732$ of (11.93) does not allow a rarefaction wave solution. Thus a solution consisting of two shocks is the only solution for the initial condition (11.90) that is physically realizable.

11.8 Discuss the limiting case of a weak initial jump, $\gamma \to 0$, for the initial condition (11.90), both for $\gamma > 0$ (the shock solution) and $\gamma < 0$ (the rarefaction wave solution).

11.9 Consider the initial condition $\gamma = -1/2 < 0$ under the evolution (11.86), (11.87).

(i) Show that this amounts to finding the solution to the problem shown in Fig. 11.8 for $t < 0$.

(ii) Show that a shock solution is not possible; instead, two rarefaction waves must be introduced.

(iii) Look for a similarity solution of (11.86), (11.87) in the form

$$h = H(\xi), \quad u = U(\xi),$$

with $\xi = x/t$, and write down the similarity equations.

(iv) Use the symmetry of the problem to show that the left-propagating rarefaction wave is bounded by $-3/2 \le \xi \le -\sqrt{h_{\rm c}}$ in similarity variables, where $h_{\rm c}$ is the fluid height in the center, to be determined. The second rarefaction wave propagates to the right and is the mirror image of the first. Setting $\xi_1 = -3/2$ and $\xi_2 = -\sqrt{h_{\rm c}}$, the boundary conditions to be satisfied by the similarity equations are $H(\xi_1) = 1$, $H(\xi_2) = h_{\rm c}$, $U(\xi_1) = -1/2$, and $U(\xi_2) = 0$.

11.10 (*) To find h_c in the previous exercise, we need to solve the similarity equations for the rarefaction wave.

    (i) Show that the similarity equations can be written as an eigenvalue equation

$$\xi \mathbf{U}_\xi = A(\mathbf{U})\mathbf{U}_\xi,$$

where $\mathbf{U}(\xi) = (H(\xi), U(\xi)H(\xi))$ is the solution vector and A is the matrix (11.88). Thus show that $\xi = \lambda_p(\mathbf{U})$, $p = 1, 2$, where the λ_p are the eigenvalues.

    (ii) Compute the eigenvalues of A, and thus show that $\xi = U - \sqrt{H}$ across the left-propagating rarefaction wave. Solve the similarity equations of the previous exercise to show that

$$H(\xi) = (C - \xi/3)^2,$$

where C is a constant. Use the boundary conditions derived in the previous exercise to show that $C = 1/2$ and thus $h_c = 9/16$.

11.11 (*) Consider the problem of a strong explosion, treated using dimensional analysis in Example 1.1. The full theoretical treatment is superficially similar to the converging shock wave problem; however, the constraint of constant energy selects a unique exponent, as shown by (1.2). Introduce similarity variables (11.100) as in the collapse problem, where $\xi = r/R(t)$.

    (i) Show that the system to be satisfied by V, Ω, Z is (11.101)–(11.103) with $\beta = 2/5$; the boundary conditions are those in (11.104).

    (ii) Consider an expanding radius $r_0(t) = \xi_0 R(t)$, where $\xi_0 = $ const. Show that

$$E = \int_0^{r_0} r^2 e\, dr,$$

where e, the energy density (11.64), is an integral of motion for all $0 < \xi_0 < 1$. Thus show, using (11.65), that

$$Z = \frac{\gamma(\gamma - 1)(1 - V)V^2}{2(\gamma V - 1)}. \tag{11.118}$$

    (iii) Armed with (11.118), use (11.101), (11.103) to show that

$$\frac{(\gamma^2 + \gamma)V^2 - 2(1 + \gamma)V + 2}{V(\gamma V - 1)\left[5 + (1 - 3\gamma)V\right]} = \frac{d\ln\xi}{dV},$$

which can be integrated to give

$$\xi^5 = \left(\frac{\gamma+1}{2}V\right)^{-2}\left\{\frac{\gamma+1}{7-\gamma}[5-(3\gamma-1)V]\right\}^{\nu_1}$$
$$\times \left(\frac{\gamma+1}{\gamma-1}(\gamma V-1)\right)^{\nu_2} \equiv f(V),$$

and

$$\Omega = \frac{\gamma+1}{\gamma-1}\left(\frac{\gamma+1}{\gamma-1}(\gamma V-1)\right)^{\nu_3}$$
$$\times \left\{\frac{\gamma+1}{7-\gamma}[5-(3\gamma-1)V]\right\}^{\nu_4}\left(\frac{\gamma+1}{\gamma-1}(1-V)\right)^{\nu_5},$$

where

$$\nu_1 = -\frac{13\gamma^2-7\gamma+12}{(3\gamma-1)(2\gamma+1)}, \quad \nu_2 = \frac{5(\gamma-1)}{2\gamma+1}, \quad \nu_3 = \frac{3}{2\gamma+1},$$
$$\nu_4 = -\frac{\nu_1}{2-\gamma}, \quad \nu_5 = -\frac{2}{2-\gamma}.$$

(iv) Finally, show that

$$\Gamma^{-5} = \frac{16\pi}{125}\int_{1/\gamma}^{2/(1+\gamma)} G(V)\left(\frac{V^2}{2}+\frac{Z(V)}{\gamma(\gamma+1)}\right)f'(V)dV,$$

where Γ is the prefactor defined by (1.2). This integral has to be performed numerically and gives $\Gamma = 1.032\,777$ for $\gamma = 7/5$; thus plot the density profile.

11.12 Deduce the Rankine–Hugoniot conditions for the shallow water equations (11.86), (11.87) and from them obtain the relation

$$u_\pm(h) = u_* \pm \sqrt{\frac{g}{2}\left(\frac{h_*}{h}-\frac{h}{h_*}\right)(h_*-h)}$$

for the possible states $(h, u_\pm)$ that can be connected with a shock to (h_*, u_*). Of the two states (h, u_+) and (h, u_-), show that only one satisfies the entropy conditions. Show that this is the one for which fluid particles experience an increase in depth as they pass through the shock.

11.13 In the so-called dam-break problem, fluid is at rest initially but is released suddenly with a jump in height. This corresponds to the following initial conditions for the shallow-water equations in one dimension:

$$h = h_\ell, \quad x < 0,$$
$$h = h_r, \quad x > 0,$$
$$u_\ell = u_r = 0,$$

where $h_\ell > h_r$. Solve the problem for $t > 0$ and show that the solution consists of a shock propagating for $x > 0$ and a rarefaction wave for $x < 0$.

11.14 When a fluid drop impacts a thin liquid layer a hydraulic jump is observed [130], which expands radially. A very simple model for this consists of the following shallow water equations:

$$\frac{\partial h}{\partial t} + \frac{1}{r}\frac{\partial}{\partial r}(rvh) = 0,$$

$$\frac{\partial v}{\partial t} + v\frac{\partial v}{\partial r} = 0,$$

where $v(r, t)$ is the radial velocity and $h(r, t)$ is the thin film free boundary. Show that for $r \geq ct^{1/2}$ the system has self-similar solutions of the form

$$v = t^{-1/2}f(\xi), \quad h = h_\infty g(\xi),$$

where $\xi = r/t^{1/2}$. Show that for such solutions

$$g|_{\xi \to c^+} \sim \frac{A}{(\xi - c)^{1/2}}, \quad g|_{\xi \to \infty} \sim e^{c^4/\xi^4}.$$

(Note: the singularity in $g(\xi)$ at $\xi = c$ represents, in this very simple model, an Edgerton crown [71] that has formed after impact.)

11.15 (*) Strong electric fields in a gas may produce a plasma by ionization of the gas molecules. Thus, the ionization process results in the production of free electrons, which are accelerated by the electric field and impact with new molecules, ionizing them in turn. This results in an avalanche process and the growth of the region occupied by the plasma. If we call ρ and σ the densities of the ions, considered non-movable, and the free electrons respectively, the simplest system modeling the process (the so-called minimal model) [11] reads as follows:

$$\frac{\partial \sigma}{\partial t} + \nabla \cdot (\sigma \mathbf{E}) = \sigma\,|\mathbf{E}|\,e^{-E_0/|\mathbf{E}|}, \tag{11.119}$$

$$\frac{\partial \rho}{\partial t} = \sigma\,|\mathbf{E}|\,e^{-E_0/|\mathbf{E}|}, \tag{11.120}$$

$$\nabla \cdot \mathbf{E} = \sigma - \rho, \tag{11.121}$$

where $\mathbf{E}$ is the electric field and the right-hand sides of equations (11.119), (11.120) take into account the production of ion–electron

pairs by electron impact ionization. Equation (11.121) is Poisson's
equation, in dimensionless form, for the electric field.

(i) For very strong electric fields, one can make the approximation
$e^{-E_0/|\mathbf{E}|} \simeq 1$. For this case, find traveling wave solutions $\sigma(x-ct)$,
$\rho(x-ct)$ such that the charge density contains a shock disconti-
nuity separating the ionized regions from the non-ionized regions,
which propagate with velocity c.

(ii) By subtracting equations (11.119) and (11.120) and differentiating
(11.121) with respect to t, deduce the following equation for the
electric field:

$$\nabla \cdot (\mathbf{E}_t + \sigma \mathbf{E}) = 0.$$

Neglecting magnetic fields this yields $\mathbf{E}_t + \sigma \mathbf{E} = 0$, so that

$$\mathbf{E}(\mathbf{x}, t) = \mathbf{E}(\mathbf{x}, 0) \exp \left(- \int_0^t \sigma(\mathbf{x}, \tau) d\tau \right).$$

Putting

$$u(\mathbf{x}, t) = \exp \left(- \int_0^t \sigma(\mathbf{x}, \tau) d\tau \right),$$

deduce Burgers' equation for $u(\mathbf{x}, t)$ in the cases of cylindrical and
spherical symmetry. Find laws for the growth of the ionized region.

11.16 Consider the linear stability of a planar shock front, using geometrical
shock theory.

(i) Use (11.110) and (11.112) to write down a wave equation for
$M(\sigma, \lambda)$.

(ii) Linearize the wave equation and derive the dispersion relation.
What is the wave speed for a front of constant initial Mach number
$M(\sigma, 0) = M_0$?

(iii) Consider an initial front of the form $x = \epsilon \sin ky$, with constant
Mach number M_0. Show that this translates into an initial condition
for the derivative M_λ.

12

The dynamical system

12.1 Overview

Let us return to the dynamical system (3.52) from a more general perspective. The simplest case, observed for example in drop pinch-off, occurs when the dynamical system possesses a fixed point which is stable: all nontrivial eigenvalues are negative. In that case convergence to the fixed point is exponential or, equivalently, *algebraic* in the original time variable t'. Soon the solution effectively reaches the fixed point and there is very little further change in the self-similar behavior. If, however, one or several eigenvalues around the fixed point vanish, the approach to the fixed point is slow and the behavior is described by a dynamical system whose dimension corresponds to the number of vanishing eigenvalues. The same holds true if the attractor has few dimensions (as does a limit cycle or a low-dimensional chaotic attractor).

Thus, although singular behavior is in principle a problem to be solved in infinite dimensions, in practice the dynamical-system description often reduces to a dynamical problem of few dimensions. In addition one also gains a fruitful means of *classifying*, or at least *characterizing*, singularities.

In this chapter we will explore cases with somewhat more complex dynamics, which go beyond the analysis of a fixed point.

For reference, we will consider a one-dimensional PDE of the form (3.46),

$$u_t = F[u].$$

If (3.46) permits self-similar solutions, then the transformation (3.51),

$$u(x, t) = t'^\alpha f(\xi, \tau),$$

with $\xi = x'/t'^\beta$, transforms the original PDE into the dynamical system form (3.52),

$$f_\tau = G[f] \equiv \alpha f - \beta \xi f_\xi + F[f].$$

We will now give a list of observed behaviors, including examples studied earlier.

298

(I) *Stable fixed points*   The dynamical system possesses a fixed point and all the nontrivial eigenvalues are negative. The eigenvalue with the smallest absolute value determines the rate of convergence toward self-similarity. The fixed point is approached exponentially in the logarithmic variable τ, so the dynamics is described by the self-similar law (3.47). This pure power law behavior is also known as *type-I self-similarity* [6]. For example, the various modes of drop breakup in an inert medium, described in Chapter 7, all have this behavior.

A subclassification divides this case into two subcases, corresponding to self-similarity of the first or second kind. In the former subcase the scaling exponents are defined by an intrinsic (local) property of the equation, and their values are rational. In the latter subcase at least one exponent is determined by a regularity condition, and its value may be irrational.

(II) *Center manifold*   The dynamical system possesses a fixed point, but one or more of the eigenvalues around the fixed point is or are zero. As a result the approach to the fixed point is only algebraic, leading to logarithmic corrections to the scaling. This is called *type-II self-similarity* [6]; it characterizes cases where the blowup rate is different from what would be expected on the basis of a solution of the type (3.47). Pinchoff in axisymmetric mean curvature flow, described in Chapter 9, is an example.

(III) *Traveling waves*   The solutions of (3.46) converge to $u = t'^{\alpha} \phi(\xi + c\tau)$, which is a traveling wave solution of (3.52) with propagation velocity c. This occurred in fluid–fluid breakup, described in Chapter 7.

(IV) *Periodic orbits*   The solutions have the form $u = t'^{\alpha} \psi[\xi, \tau]$ with ψ a periodic function of period T in τ. This is known as *discrete self-similarity* [49, 151], since at times $\tau_n = \tau_0 + nT$, n integer, the solution looks self-similar. A closely related phenomenon is the sequence of eddies described by (5.61), whose periodicity occurs in space. Examples of discrete self-similarity will be described below.

(V) *Strange attractors*   The dynamics on the scale τ are described by a nonlinear (low-dimensional) dynamical system such as the Lorenz equation.

12.2 Periodic orbits: a toy model

We now give a simple example of a dynamical system with periodic orbits. Its singular behavior is once more driven by the hyperbolic growth model (3.1), described by

$$u_t(x,t) = u^2(x,t). \tag{12.1}$$

However, we shall now take u to be coupled to two additional fields v and w, which renders the dynamics more interesting:

$$v_t(x,t) = -u(x,t)w(x,t), \tag{12.2}$$
$$w_t(x,t) = u(x,t)v(x,t). \tag{12.3}$$

The similarity solution to (12.1) is

$$u(x',t') = (t')^{-1}U(\xi,\tau), \quad \xi = \frac{x'}{t'^{1/2}}, \tag{12.4}$$

and the convergence onto it is exponential, as we have seen. Thus we can assume that U has reached its fixed point, and we will neglect the dependence on τ, thus

$$U = \frac{1}{1+c\xi^2}.$$

To determine the dynamics of v and w, we seek similarity solutions of the form

$$v = V(\xi,\tau), \quad w = W(\xi,\tau), \tag{12.5}$$

which leads to the equations

$$V_\tau = -\frac{\xi V_\xi}{2} - UW, \quad W_\tau = -\frac{\xi W_\xi}{2} + UV, \tag{12.6}$$

which are solved by the ansatz

$$V = V_0 \cos[C(\xi)+\tau], \quad W = V_0 \sin[C(\xi)+\tau]. \tag{12.7}$$

Indeed, if (12.7) is inserted into (12.6) then one finds that $C(\xi)$ is determined by

$$\frac{\xi C'(\xi)}{2} = U(\xi) - 1. \tag{12.8}$$

Equation (12.8) has the solution $C(\xi) = -\ln(1+c\xi^2)$, and so we find that

$$V = V_0 \cos\left[-\ln(1+c\xi^2)+\tau\right], \tag{12.9}$$
$$W = V_0 \sin\left[-\ln(1+c\xi^2)+\tau\right]. \tag{12.10}$$

In other words, in the space of self-similar variables the solution traces out a *closed orbit*, which in this simple case is a circle. As the singularity is approached, the orbit lives on smaller and smaller scales in real space. For $t' \to 0$ and x' fixed, the argument in (12.9), (12.10) behaves as follows:

$$- \ln \left[1 + c \left(\frac{x'}{t'^{1/2}} \right)^2 \right] + \tau \approx - \ln cx'^2 + \ln t' + \tau = - \ln cx'^2,$$

which is a constant. This means that the solution (12.9), (12.10) satisfies the matching condition (3.15).

It is worth noting that the similarity equations (12.6) can be written in complex form by introducing $Z \equiv V + iW$, resulting in

$$Z_\tau + \frac{\xi Z_\xi}{2} = iUZ. \tag{12.11}$$

The solution is

$$Z = Z_0 e^{\lambda \tau} e^{iC(\xi)}, \tag{12.12}$$

with an imaginary scaling exponent $\lambda = i$. Note the analogy with the corner eddies (5.58), whose periodic structure is the result of a scaling exponent becoming complex.

However, in general the orbit will be more complicated and is not necessarily described by a complex exponent. A well-known example of periodic behavior occurs in cosmology [49], in the dynamics of the gravitational field at the threshold to black hole formation. In general, the structure of (a single component of) the solution is

$$V = \psi(\phi(\xi) + \tau), \tag{12.13}$$

where ψ is *periodic* in τ with (say) period T.

If one observes V at times $\tau_n = \tau_s + nT$, where $\tau_n = - \ln t'_n$, the resulting sequence of images will appear self-similar, in the same sense as (12.4). For this reason, if the dynamical system lies on a periodic orbit then one speaks of *discrete self-similarity* [80], as mentioned in Section 12.1. However, for each time shift τ_s the spatial picture of this self-similar evolution will be different. In practice, discrete self-similarity is very difficult to detect from numerical or experimental data. A detailed numerical analysis of the periodic orbits in the above cosmological problem was reported in [150]. This involved a search for periodic solutions in function space, with the period T an unknown.

Finally, we can also investigate the *stability* of the periodic solution of the toy model studied above. Applying the complex formulation (12.11), we use the ansatz

$$Z = e^{i\tau} e^{\nu \tau} P(\xi). \tag{12.14}$$

Here we have separated the factor $e^{i\tau}$ describing the periodic time dependence; ν is the eigenvalue describing the stability. Inserting (12.14) into (12.11), we find the eigenvalue equation

$$(\nu + i)P + \frac{\xi}{2}P' = \frac{iP}{1 + c\xi^2}. \tag{12.15}$$

The solution of this equation is

$$P = P_0 \frac{e^{iC(\xi)}}{\xi^{2\nu}} \tag{12.16}$$

and, imposing regularity at the origin, we find the eigenvalue sequence

$$\nu = 0, -1/2, -1, \dots \tag{12.17}$$

The vanishing eigenvalue corresponds to a perturbation in the direction of the orbit, but all the other eigenvalues are negative. This demonstrates the stability of the periodic solutions found for our toy model: they are limit cycles [65].

12.3 Discrete self-similarity in the implosion of polygonal shocks

A more physical example of discrete self-similarity appears in the dynamics of imploding shock waves. We found that the case of a collapsing cylindrical wave is described accurately by (11.114), (11.115) within geometrical shock dynamics. In reality the initial shape of the shock wave will never be perfectly cylindrical but will be at least somewhat deformed, as seen in the experimental sequence in Fig. 12.1. In the course of time the shock front appears to develop sharp, corner-like, features. Indeed, simulations of geometrical shock dynamics, (11.108), (11.109), show that an initially smooth shock front tends to develop a corner in finite time [193]. To understand the origin of this behavior, note that (11.108), (11.109) can be written in the characteristic form

$$\left(\frac{\partial}{\partial\lambda} \pm c\frac{\partial}{\partial\sigma}\right)\left(\theta \pm \int \frac{dM}{Ac}\right) = 0, \tag{12.18}$$

where the nonlinear wave speed is

$$c = \sqrt{-\frac{M}{AA'}} \propto M^{\nu+1}. \tag{12.19}$$

The wave propagation is thus similar to (11.34), with a wave speed that increases monotonically with the speed of the front. As a result, shocks will form in the direction of σ, along the shock front, producing a jump of θ. But

Figure 12.1 Converging cylindrical shock waves with Mach number 1.5 in air. From [216] with kind permission from Springer Science and Business Media.

this means that the shock front has formed a corner, called a *shock-shock* to emphasize the underlying mechanism for corner formation. As a result of the tendency of a shock front to form corners, an initially smooth initial condition with N-fold symmetry tends to approach a polygonal shape [193].

The motion resulting from a polygonal initial condition is shown in Fig 12.2, using the example of a hexagon. It is evident that the shape of the front undergoes a periodic evolution, but on a progressively smaller scale, as the shock front focuses. First, a hexagon rotated by $\pi/6$ is created; in a second step a smaller hexagon with the original orientation is reproduced. It is apparent that corners are created by the *collision* of two vertices of a previous incarnation of the polygon. For example, the corner at C comes from the collision of A and A'.

The process of such a collision of vertices at A and A', occurring at C, is illustrated in Fig 12.3. We can work out the geometry of the reflected shock using the rules of geometric shock dynamics. Let the Mach number of the original shock along CB be M_1 and that of the newly created shock along QP be M_2. Owing to the symmetry along the broken line CQ this line can be viewed as a solid wall, from which the incoming shock wave is reflected. The creation of a new piece of shock front QP, which connects the shock to the wall, is known as a *Mach stem*. Some small disagreements exist between shock reflection in the geometric theory and in the full shock theory based on the two-dimensional Euler equation. For example, in the full theory the Mach

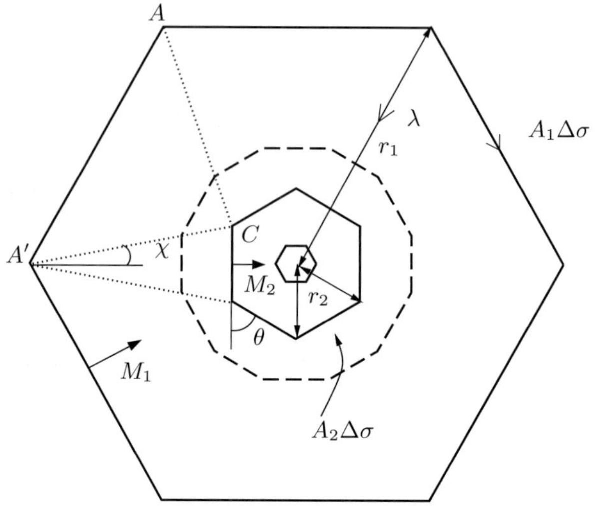

Figure 12.2 An imploding hexagonal shock; the angle between two sides is $\theta = \pi/N$, $N = 6$. After some time Δt_1 the hexagon transforms into another hexagon, which is rotated from the first by an angle of π/N radians. After another time interval Δt_2 the shock front is transformed back into the original hexagonal shape, but on a smaller scale. The broken curve shows the shock front in between two hexagonal shapes. The open arrowheads give the orientations of the space and time axes.

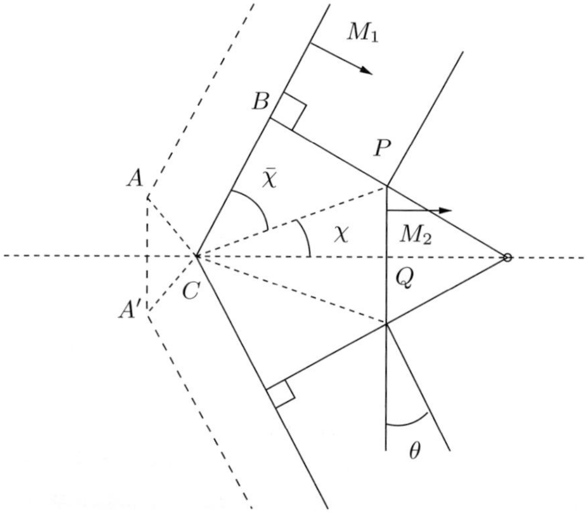

Figure 12.3 The collision of two vertices A and A' at C. In the process a new shock front, with Mach number M_2, is created.

stem is slightly curved and disappears altogether above a critical angle, while it is always there in the geometric theory [222].

Now, the distance along the ray BP is $\overline{BP} = M_1 \Delta \lambda$, where $\Delta \lambda$ is the "time" elapsed in (λ, σ) coordinates during the motion of the vertex from C to P; similarly, $\overline{CQ} = M_2 \Delta \lambda$. The distance CB along the shock front is $\overline{CB} = A_1 \Delta \sigma$, where $\Delta \sigma$ is the interval traveled by the vertex in (λ, σ) coordinates; analogously, $\overline{QP} = A_2 \Delta \sigma$. Expressing $\overline{CP}$ in two different ways using the Pythagorean theorem, we have

$$(M_1 \Delta \lambda)^2 + (A_1 \Delta \sigma)^2 = (M_2 \Delta \lambda)^2 + (A_2 \Delta \sigma)^2$$

and thus

$$c \equiv \frac{\Delta \sigma}{\Delta \lambda} = \sqrt{\frac{M_2^2 - M_1^2}{A_1^2 - A_2^2}}. \tag{12.20}$$

The angle χ can be computed from

$$\tan \chi = \overline{PQ}/\overline{QC} = \frac{A_2 \Delta \sigma}{M_2 \Delta \lambda} = \frac{A_2 c}{M_2},$$

and $A_i \propto M_i^{-\nu}$ according to (11.112). Thus if we set $\mu = M_2/M_1$ and use (12.20), we find that

$$\tan \chi = \sqrt{\frac{1 - \mu^{-2}}{\mu^{2\nu} - 1}}. \tag{12.21}$$

Using virtually identical arguments, we also find that

$$\tan \bar{\chi} = \sqrt{\frac{1 - \mu^{-2\nu}}{\mu^2 - 1}}. \tag{12.22}$$

Since the exterior angle θ between two sides of the polygon is given by

$$\theta = \frac{\pi}{2} - \chi - \bar{\chi},$$

with some elementary manipulations we find from (12.21) and (12.22) that

$$\cos \theta = \frac{\mu^\nu + \mu}{\mu^{\nu+1} + 1}. \tag{12.23}$$

Equation (12.23) allows to compute μ, the ratio of successive Mach numbers, given the angle θ formed by the normals to the two shock fronts at a corner. If we start with a regular polygon with N sides ($N = 6$ in Fig. 12.2) and a constant Mach number M_1 everywhere then instantaneously new shocks will appear at the corners, propagating at Mach number M_2. For a polygon with N sides,

$$\cos \frac{\pi}{N} = \frac{\mu^\nu + \mu}{\mu^{\nu+1} + 1}, \tag{12.24}$$

according to (12.23). If we take, for instance, air, with $\gamma = 7/5$, then $\nu = 5.0743$ from (11.113). For a square, $N = 4$ and $\mu = 1.545$ and for a hexagon $N = 6$ and $\mu = 1.289$; these values are obtained from (12.24). With the values of μ and M_1 we can compute the Mach number $M_2 = \mu M_1$ of the new shock. These new shocks grow in length while the old ones, of Mach number M_1, disappear. Eventually, we come back to the same polygon, but rotated by an angle π/N and smaller.

The distances r_1, r_2 of the corners of the old and new polygons respectively to their common center (see Fig. 12.2) are related by (cf. (11.112))

$$\frac{r_2}{r_1} = \frac{A_2}{A_1} = \left(\frac{M_1}{M_2}\right)^\nu = \frac{1}{\mu^\nu}, \tag{12.25}$$

since the length of the shock front between two vertices is $A_i \Delta \sigma$. The times $\lambda_1' = \lambda_0 - \lambda_1$, $\lambda_2' = \lambda_0 - \lambda_2$ at which the old and new polygons are formed, relative to the time of collapse, λ_0, satisfy

$$\frac{\lambda_2'}{\lambda_1'} = \frac{M_1}{M_2} \frac{r_2}{r_1} = \frac{1}{\mu^{\nu+1}}, \tag{12.26}$$

since M_1 and M_2 are constant between collisions.

The same process repeats itself with the new polygon, to form an even smaller polygon with the same orientation as the first. Since the two processes are self-similar, a sequence of nested polygons of sizes r_i is created such that

$$\frac{r_{i+2}}{r_i} = \frac{1}{\mu^{2\nu}}, \tag{12.27}$$

using (12.25). As for the corresponding times, (12.26) leads to, setting $\tau_i = -\ln \lambda_i'$,

$$\frac{\lambda_{i+2}'}{\lambda_i'} = e^{\tau_i - \tau_{i+2}} = \frac{1}{\mu^{2\nu+2}},$$

and therefore the process is periodic in $\tau = -\ln \lambda'$, with period

$$T = \tau_{i+2} - \tau_i = \ln \mu^{2\nu+2}. \tag{12.28}$$

This characterizes the evolution as discretely self-similar, and we can recast the full geometric shock model (11.108), (11.109) with (11.112) in self-similar form; $\lambda' = \lambda_0 - \lambda$ is the "time" to breakup. Inserting the similarity description

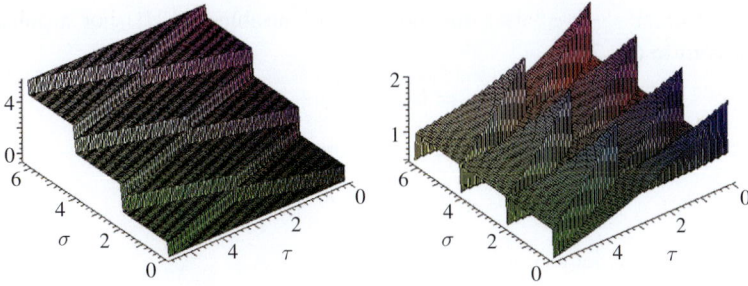

Figure 12.4 The profiles $|\Theta|(\tau, \sigma)$ (left) and $\Psi(\tau, \sigma)$ (right) for $N = 4$.

$$\theta(\lambda, \sigma) = \Theta(\tau, \sigma), \tag{12.29}$$

$$M(\lambda, \sigma) = (\lambda')^{-1/\nu+1} \Psi(\tau, \sigma), \tag{12.30}$$

into (11.108), (11.109), one obtains

$$\frac{\partial \ln \Psi}{\partial \tau} + \frac{\Psi^{\nu+1}}{\nu} \frac{\partial \Theta}{\partial \sigma} = \frac{1}{\nu + 1}, \tag{12.31}$$

$$\frac{\partial \Theta}{\partial \tau} - \Psi^{\nu+1} \frac{\partial \ln \Psi}{\partial \sigma} = 0. \tag{12.32}$$

If there is a corner in the shape of the front, Θ and Ψ have to satisfy the jump conditions (12.21) and (12.23).

The above analysis shows that for each N-polygon there exists a τ-periodic solution of (12.31), (12.32) with period $T = 2(\nu + 1) \ln \mu$, where μ is determined by (12.24). In Fig 12.4 we represent $|\Theta|(\tau, \sigma)$ and $\Psi(\tau, \sigma)$ for the case $N = 4$ (a square). The variable σ runs from 0 to 2π, τ runs over one period T, where $T = 5.28$. Full numerical simulations of the two-dimensional compressible Euler equations show that solutions are attracted to a polygonal shape [24] and that the geometrical theory describes the solution well. However, whether the full Euler equations have solutions which display discrete self-similarity exactly remains an open problem.

12.4 Chaos

To give an explicit example of a system of PDEs exhibiting *chaotic* dynamics, consider the structure of the system (12.1)–(12.3). It can be generalized to produce *any* low-dimensional dynamics near the singularity, as follows:

$$u_t(x, t) = u^2(x, t), \tag{12.33}$$

$$u_t^{(i)}(x, t) = 2u F_i(\{u^{(i)}\}), \quad i = 1, \ldots, n, \tag{12.34}$$

where $F : \mathbb{R}^n \rightarrow \mathbb{R}^n$ is a function of the variables $u^{(i)}$. Using an ansatz analogous to (12.7),

$$u^{(i)} = U^{(i)}\,[C(\xi) + \tau],\qquad(12.35)$$

and choosing $C(\xi) = -\ln(1 + c\xi^2)$, one obtains the system

$$U_\tau^{(i)} = F_i\left\{U^{(i)}\right\}.\qquad(12.36)$$

To be specific, let us consider $n = 3$ and

$$F_1 = \sigma(u^{(2)} - u^{(1)}),\quad F_2 = \rho u^{(1)} - u^{(2)} - u^{(1)}u^{(3)},\quad F_3 = u^{(1)}u^{(2)} - \beta u^{(3)},$$
$$(12.37)$$

so that (12.36) becomes the Lorenz system [202]. Here σ, ρ, and β are constants. As before, for $t' \rightarrow 0$ the variable τ goes to infinity and near the singularity one is exploring the long-time behavior of the dynamical system (12.36). In the case of (12.37), and for sufficiently large ρ, the resulting dynamics will be chaotic.

In particular, taking $\sigma = 10$, $\rho = 28$, and $\beta = 8/3$, as was done by Lorenz [147], the maximal Lyapunov exponent is 0.906. The initial conditions with which (12.35) is to be solved depend on ξ. Thus the chaotic dynamics will follow a completely different trajectory for each initial space point. As a result it is very difficult to detect self-similar behavior of this type as such, even if data arbitrarily close to the singularity time is taken. This is illustrated in Fig 12.5, where the rescaled profile $U^{(1)}$ is shown at constant intervals of the logarithmic time τ. Clearly, the spatial structure of the singularity appears to be very different in the three panels of Fig. 12.5, and the underlying dynamics is not easy to decipher. A possible approach to this challenge was proposed in [180]: that chaotic motion is characterized by unstable periodic orbits, for which one could search numerically.

Finally, we mention an interesting example of chaotic self-similarity, which occurs for the so-called Gauss sums [23]

$$S_L(\tau) = \sum_{n=1}^{L} \exp\left(i\pi\tau n^2\right).\qquad(12.38)$$

An example is plotted in Fig. 12.6 for the case $\tau = 1/e$; one observes an intricate superposition of structures on many scales. A way to understand the self-similarity of the structure is to derive the approximate renormalisation transformation [23]

$$S_L(\tau_0) \approx \frac{\exp(i\pi/4)}{\tau_0^{1/2}} K^{[1/\tau_0]+1} S_{[L\tau_0]}(\tau_1(\tau)),\qquad(12.39)$$

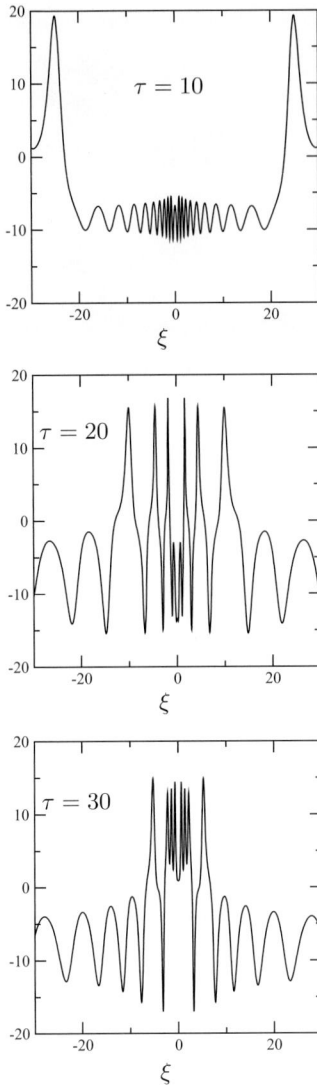

Figure 12.5 Three "snapshots" of the self-similar profile $U^{(1)}(C(\xi)+\tau)$, (12.35), for the Lorenz model.

which describes how the sum behaves under a *change of scale*. Here K is the complex conjugation operator and $\tau_1(\tau_0)$ is the map

$$\tau_1(\tau_0) = \begin{cases} (1/\tau_0) \bmod 1, & [1/\tau_0] \text{ even,} \\ 1 - (1/\tau_0) \bmod 1, & [1/\tau_0] \text{ odd.} \end{cases} \quad (12.40)$$

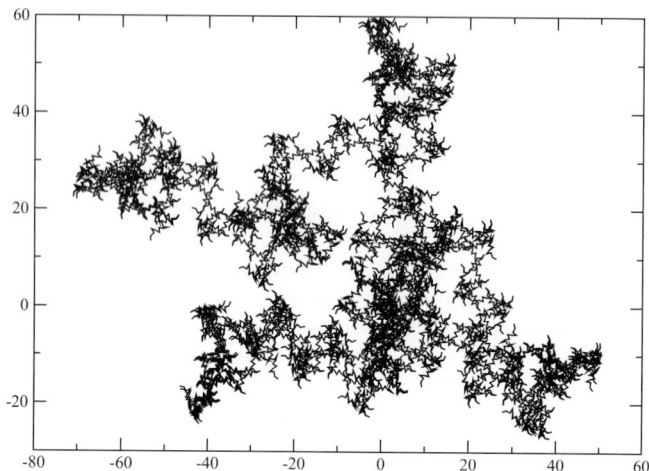

Figure 12.6 The sum $S_L(\tau)$ for $\tau = 1/e$, plotted for $L = 1$ to $L = 10\,000$.

Here $[x]$ denotes the largest integer not greater than x. If τ is a fixed point of (12.40), such as is the case for the quadratic irrational $\tau = \sqrt{101} - 10$ (see Exercise 12.7), the pattern renormalizes itself up to a change in scale $\tau^{-1/2}$ at each iteration. This corresponds to the generation of a fractal, which for $L \to \infty$ is exactly self-similar under a magnification of $\tau^{-1/2}$. If, however, τ_0 has some generic transcendental value, (12.40) is a chaotic map and the rescalings are accordingly random, as illustrated in Figure 12.6.

Exercises

12.1 Instead of (12.1), consider

$$u_t(x, t) = u^p(x, t),$$

with $p > 1$. Derive the similarity equations and solve them.

12.2 Consider the PDE

$$t^2 \frac{\partial^2 u}{\partial t^2} + \frac{\partial^2 u}{\partial x^2} + \gamma u = 0,$$

which has solutions of the form

$$u = t^\alpha e^{ikx}, \tag{12.41}$$

with complex $\alpha = \alpha_1 + i\alpha_2$. Find the real and imaginary parts α_1 and α_2. For which values of γ can (12.41) be represented as

$$u = t^{\alpha_1} F(x),$$

and for which values is the solution discretely self-similar, so that

$$u = t^{\alpha_1} F(x + \alpha_2 \ln t)?$$

12.3 In the limit $N \rightarrow \infty$, a polygon approaches a circle. Confirm that the discretely self-similar dynamics approaches the self-similar collapse $R \propto t'^{\beta}$, with β given by (11.117), for the case $D = 2$. In doing so, show that the difference $\lambda_2 - \lambda_1$ between the times at which polygons are formed goes to zero for $N \rightarrow \infty$.

12.4 Study the stability of a collapsing cylindrical shock.
   (i) Show that (11.110) leads to the nonlinear elliptic equation

$$\frac{\partial^2 Y}{\partial \lambda^2} - \frac{\partial^2 Y^{-1}}{\partial \sigma^2} = 0, \tag{12.42}$$

   where $Y = \sqrt{v} \chi M^{-v-1}$.
   (ii) Find the function $Y(\lambda, \sigma)$ corresponding to (11.115), and verify that it satisfies (12.42).
   (iii) Introduce perturbations of the form

$$Y = a\lambda' + \epsilon f(\tau, \sigma),$$

   where $\tau = -\ln \lambda'$, and show that, for small ϵ,

$$\left(\frac{\partial^2}{\partial \tau^2} + \frac{\partial}{\partial \tau} - \frac{1}{a^2} \frac{\partial^2}{\partial \sigma^2} \right) f = 0. \tag{12.43}$$

   Derive the corresponding dispersion relation and discuss the stability of the shock.

12.5 Instead of (11.108), consider a regularized equation of the form

$$\frac{\partial \theta}{\partial \sigma} = \frac{1}{M} \frac{\partial A}{\partial \lambda} + \eta \frac{\partial^2 M^{-1-v}}{\partial \sigma^2}, \tag{12.44}$$

where η is a regularization parameter. Alternatively, consider

$$\frac{\partial \theta}{\partial \lambda} = -\frac{1}{A} \frac{\partial M}{\partial \sigma} + \eta \frac{\partial^3 M^{-1-v}}{\partial \sigma^3}. \tag{12.45}$$

Discuss the evolution of a circular shock that is slightly perturbed and show that, for sufficiently small perturbations and a suitable sign of η, the shock front returns to a circular shape. Can one expect the appearance of polygonal shocks?

12.6 Convince yourself that, far from the singular point, the chaotic self-similar solution (12.35), (12.36) obeys the proper matching condition (3.15).

12.7 Show that the values

$$\tau_k^e = \sqrt{k^2 + 1} - k \quad \text{(even)},$$

$$\tau_k^o = k + 1 - \sqrt{k^2 + 2k} \quad \text{(odd)},$$

where $k = 1, 2, \ldots$ are fixed points of the map (12.40). Hint: show that $k < \sqrt{k^2 + 1} < k + 1$ and $k < \sqrt{k^2 + 2k} < k + 1$.

13

Vortices

In previous chapters we described how singularities develop in various systems of partial differential equations. In many cases these singularities persist and become a permanent feature of the underlying field, as is the case for shock waves. Shock waves represent a planar discontinuity of a hydrodynamic field such as the density. Vortices form another important type of persistent singularities [169]; they correspond to singularities of the phase. A vortex is shown schematically in Fig. 13.1. For example, in a bathtub vortex or the tornado shown in Fig. 13.2 (left), the arrows would represent the direction of flow. At the center of a vortex both the direction of motion and the phase become indeterminate. Two other physical examples are the vortices in type-II superconductors in a magnetic field (see Fig. 13.2 (right)) and dislocations in crystal lattices.

As suggested by the images in Fig. 13.2, a vortex is associated with a (in some limit) vanishingly small length scale compared with the system size. As a result, we expect the vortex to exhibit a universal scaling structure. For example, according to (4.47) a line vortex in fluid mechanics has the power law form

$$\mathbf{v} = \frac{\Gamma}{2\pi} \frac{\mathbf{e}_\theta}{r}, \tag{13.1}$$

where r is the distance from the core and Γ is the circulation. It is clear from (13.1) that the velocity goes to infinity at the core, but also that its direction is indefinite. When a vortex structure dominates a flow, the description of the field simplifies drastically and reduces to the analysis of vortices and their mutual interaction. The nature of these interactions depends on the (nonlinear) dynamics of the system under consideration, and has to be determined on a case-by-case basis, as we will see below.

However, some important features of vortices can be understood by considering the kinematics of a complex wave field alone, disregarding the detailed dynamics underlying it. Consider the complex wave field

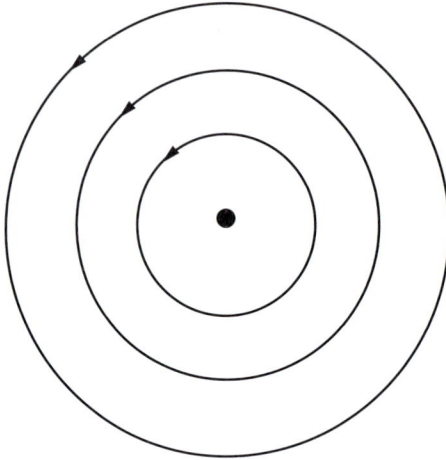

Figure 13.1 The flow lines in a vortex.

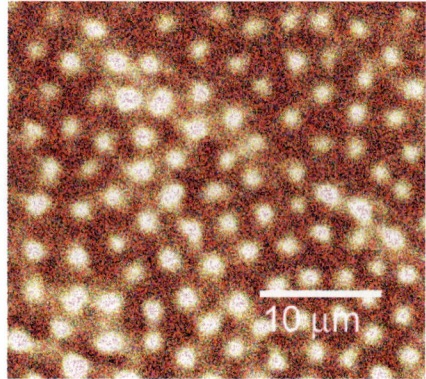

Figure 13.2 Vortices formed in two very different contexts and with very different length scales: on the left, a tornado in the atmosphere, with a typical length scale of 1 km and, on the right, vortices forming an Abrikosov lattice in a type-II superconductor, with a typical length scale of 1 μm. Left-hand image, courtesy of TV Tropes Foundation; right-hand image, the Nobel Prize in Physics – Popular information: Nobelprize.org. Nobel Media AB 2014. Web. 1 Dec 2014. http://www.nobelprize.org/nobel_prizes/physics/laureates/2003/popular.html.

$$\psi(\mathbf{r}, t) = \rho(\mathbf{r}, t)e^{i\chi(\mathbf{r},t)}, \qquad (13.2)$$

determined by the amplitude $\rho(\mathbf{r}, t)$ and phase $\chi(\mathbf{r}, t)$ [20]. The physically observable wave field corresponds to the real part of ψ. A vortex or wave dislocation would then correspond to the places where $\rho(\mathbf{r}, t)$ is zero and, as

Figure 13.3 Cotidal lines of spring tides, labeled by the hour of occurrence around the world, after [167]. The lines are organized around a small number of vortices, which determine the overall structure and where the amplitude goes to zero.

a result, where the phase is indeterminate: thus a vortex is a singularity of the phase.

As an example, Fig. 13.3 shows a map of the lines of equal tide around the world. It illustrates nicely how the existence of only a few vortices determines the structure of the whole wave field and the sequence of arrival of the tides. Namely, along a closed loop around a vortex, the arrival time varies by a full 12 hours. Moreover, one can deduce immediately that there is no tidal motion at the position of the vortices. The topological nature of the problem implies that the assembly of vortices is structurally stable: a small rearrangement of the land mass will only result in a small shift in the position of the vortices, but will leave the phase structure intact. Therefore, phase singularities are often referred to as "topological singularities" or "topological defects."

An even better way to characterize the topological nature of such a singularity is to consider the integral of the phase. If one integrates the phase χ in (13.2) around a closed loop containing the vortex core, one obtains

$$\oint \chi = \pm 2\pi n, \tag{13.3}$$

where the integer n is called the charge. The requirement (13.3) ensures that ψ is continuous as one goes around the loop. The crucial observation is that the value of the charge is independent of the form of the loop as long as only one vortex is enclosed. If there are several then the integral will give the sum of all the charges enclosed. This also shows generically that there will always be vortices: the line integral of χ over the boundary of the flow will be generically nonzero, which implies that there must be vortices in the interior.

13.1 Point vortices in inviscid fluid flow

Let us begin with the simplest case, that of a straight-line vortex in fluid
mechanics, which can be represented by a point singularity in the plane; see
Fig. 13.1. We assume that the fluid motion is potential everywhere except at
these isolated singularities. Since we are looking for a swirling motion, the
velocity field is of the form $\mathbf{v} = A\mathbf{e}_\theta$. If one integrates along a closed curve
C which contains the vortex, one thus picks a finite circulation Γ, defined by
(4.34); otherwise the circulation is zero. In view of Stokes' theorem (4.35) this
illustrates nicely that all the vorticity is concentrated at the point $r = 0$. Choosing C to be a circle of radius r and moving in the anticlockwise direction, the
circulation is $\Gamma = 2\pi Ar$ and we thus find (13.1) for the velocity field. From
Kelvin's theorem we know that the circulation around a loop convected by the
flow cannot change. Thus the circulation of a point vortex must be a conserved
quantity.

The velocity is related to the potential ϕ by

$$v_\theta = \frac{\Gamma}{2\pi r} = \frac{1}{r}\frac{\partial \phi}{\partial \theta},$$

and thus

$$\phi = \frac{\Gamma}{2\pi}\theta + \text{const.}$$

As described by (4.44), ϕ is the real part of a complex potential w so we must
have

$$w = \frac{\Gamma}{2\pi i}\ln z, \tag{13.4}$$

as given earlier in Example 4.9. The imaginary part of (13.4) is the stream
function

$$\psi = -\frac{\Gamma}{2\pi}\ln r, \tag{13.5}$$

which is constant around circles centered at the position of the vortex, as
expected.

13.1.1 Vortex motion

Using the velocity field for a single point vortex, it is easy to determine the
motion of N vortices in the plane. If the position of the ith vortex is (x_i, y_i)
and its strength is Γ_i then, according to (13.1), the velocity at (x, y) produced
by the ith vortex is

$$\mathbf{v}_i = \frac{\Gamma_i}{2\pi}\frac{(-(y - y_i), x - x_i)}{(x - x_i)^2 + (y - y_i)^2}.$$

Since each vortex is convected by the velocity field produced by all the other vortices, we arrive at the dynamical system

$$\frac{dx_j}{dt} = \sum_{i \neq j} \frac{\Gamma_i}{2\pi} \frac{-(y_j - y_i)}{(x_j - x_i)^2 + (y_j - y_i)^2}, \tag{13.6}$$

$$\frac{dy_j}{dt} = \sum_{i \neq j} \frac{\Gamma_i}{2\pi} \frac{x_j - x_i}{(x_j - x_i)^2 + (y_j - y_i)^2}. \tag{13.7}$$

The system of equations (13.6), (13.7) possesses a Hamiltonian structure, with Hamiltonian

$$H = \frac{-1}{4\pi} \sum_{i \neq j} \Gamma_i \Gamma_j \ln |\mathbf{x}_i - \mathbf{x}_j|. \tag{13.8}$$

It is easy to confirm that the Hamilton equations

$$\frac{dx'_j}{dt} = \frac{\partial H}{\partial y'_j}, \quad \frac{dy'_j}{dt} = -\frac{\partial H}{\partial x'_j}$$

are equivalent to (13.6), (13.7), with

$$(x'_j, y'_j) = \sqrt{|\Gamma_j|}(x_j, \operatorname{sign}(\Gamma_j) y_j).$$

The problem of solving the ideal fluid equations in the plane has therefore been reduced to solving a low-dimensional Hamiltonian dynamical system, which entails a much lower level of difficulty than solving the full PDE (4.31).

Example 13.1 (Counterrotating vortices) As a first illustration, consider two counterrotating vortices of equal strength Γ; see Fig. 13.4. According to (13.1) one vortex convects the other, at a speed

$$U_{\text{conv}} = \frac{\Gamma}{2\pi h} \tag{13.9}$$

and in a direction perpendicular to the line connecting the two vortices. As illustrated in Fig. 13.5 (left), both vortices thus move at the same speed in the same direction, so their distance h remains constant. This is of course consistent with the fact that the Hamiltonian (13.8) is conserved under the motion, so $\mathbf{r} = \mathbf{x}_1 - \mathbf{x}_2$ must be of constant length. $\qquad\square$

Example 13.2 (Corotating vortices) By contrast, if the two vortices have the *same* sense of circulation, then they move in opposite directions but at the same speed. Again, their instantaneous direction of motion is perpendicular to the line connecting them, and hence their distance cannot change in time. This

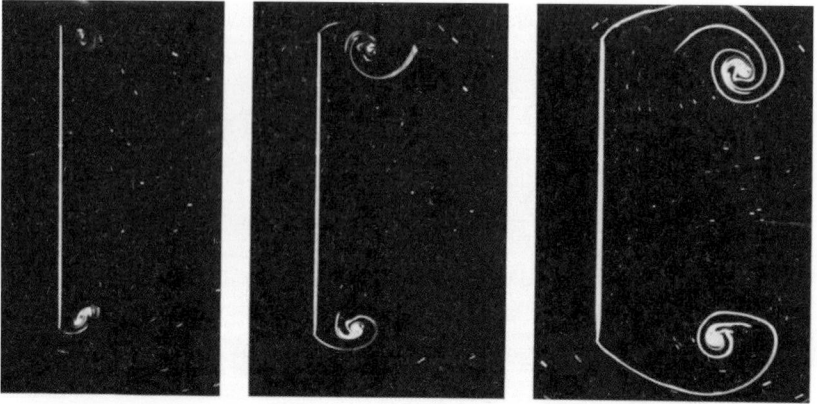

Figure 13.4 Two counterrotating vortices of equal strength created by separation from the edges of a paddle. The vortices move in a straight line, keeping their distance constant [205, 212].

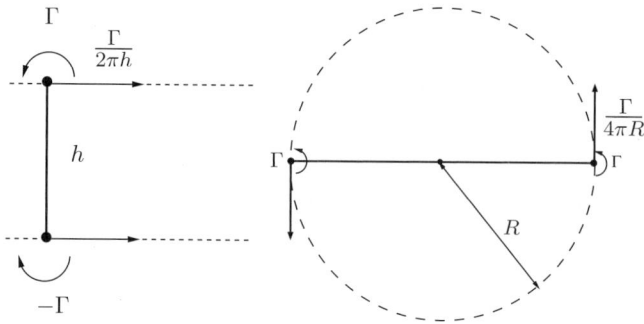

Figure 13.5 The motion of two vortices of equal strength. If the vortices are counterrotating (left), they move in a straight line parallel to each other; if they are corotating, they circle each other (right).

implies that the two vortices rotate around a common center, being located at opposite sides of a circle of radius R. Their speed is again given by (13.9), with $h = 2R$, and so their *angular* frequency is

$$\Omega = \frac{\Gamma}{4\pi R^2}.$$

$$(13.10)$$

Incidentally, (13.10) is generalized easily to the case of N vortices of equal strength (see Exercise 13.1), spaced equally around a circle. In that case the above formula is simply to be multiplied by $N - 1$.   □

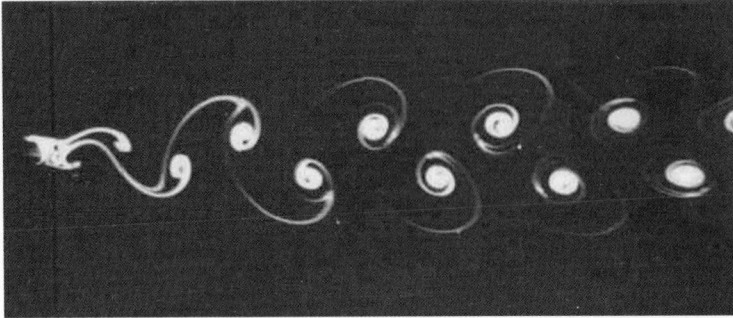

Figure 13.6 A von Kármán vortex street formed by the flow behind a circular cylinder at Reynolds number Re $=$ 105 [212].

However, perhaps the most common vortex configuration is the von Kármán vortex street, which forms behind obstacles in a uniform stream; see Fig. 13.6. Vortices are shed alternately from either side of the obstacle. This leads to two *staggered* arrays of vortices, in which each vortex sits in the middle between two vortices of the other row. Because of the overall conservation of vorticity, the vortices must be counterrotating to keep the total circulation zero.

These simple ideas are incorporated into the following model: vortices are located at points

$$(x_m, y_m) = (ma, 0), \quad (x'_m, y'_m) = ((m + 1/2)a, -b),$$

the primed and unprimed coordinates denoting the upper and lower rows, respectively, in Fig. 13.7; $m = 0, \pm1, \pm2, \ldots$ The distance a between two vortices in the same row and the vertical distance b between the two rows are in principle arbitrary. The point vortices at $y = 0$ are rotating clockwise with strength Γ while those located at $y = -b$ rotate anticlockwise with the same strength. Our aim is to show that this pattern is stationary in time up to a constant speed of convection U. For this to be the case, the velocity field created at *each* vortex position must be $-U\mathbf{e}_x$. If the flow speed incident on the body is U_{in}, the pattern will be convected at speed $U_{\text{in}} - U$ without changing its shape.

Since the street extends infinitely in both directions, we can choose a representative vortex arbitrarily, for example that at $(a/2, -b)$ in the lower row, and compute the velocity field generated by all the other vortices at that point. It is also useful to notice that the contributions of all the other vortices *in the same row* cancel out each other exactly. This is seen by considering opposite pairs of vortices located at equal distances from the target vortex, which generate equal and opposite velocities. It is thus sufficient to sum over the contributions from the top row.

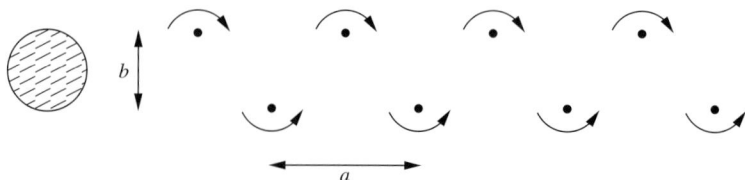

Figure 13.7 A simple model of a von Kármán vortex street.

The complex potential w of each vortex is given by (13.4), so the sum over the top row is

$$w(z) = \frac{i\Gamma}{2\pi} \sum_{m=-\infty}^{\infty} \ln(z - ma).$$

For $m \neq 0$, each logarithmic contribution can be rewritten as follows:

$$\ln(z - ma) = \ln\left(1 - \frac{z}{ma}\right) + \ln(-ma),$$

and since the potential is defined only up to a constant, we can disregard the constant $\ln(-ma)$. Thus the potential created by the top row of vortices can be written

$$
\begin{aligned}
w_{\text{top}}(z) &= \frac{i\Gamma}{2\pi}\left\{\ln z + \sum_{m=1}^{\infty} \ln\left(1 - \frac{z}{ma}\right) + \ln\left(1 + \frac{z}{ma}\right)\right\} \\
&= \frac{i\Gamma}{2\pi}\left\{\ln z + \sum_{m=1}^{\infty} \ln\left[1 - \left(\frac{z}{ma}\right)^2\right]\right\} \\
&= \frac{i\Gamma}{2\pi} \ln\left\{z \prod_{m=1}^{\infty}\left[1 - \left(\frac{z}{ma}\right)^2\right]\right\}.
\end{aligned}
$$

But a well-known result of complex function theory states that (see [1], (4.3.89))

$$z \prod_{m=1}^{\infty}\left[1 - \left(\frac{z}{ma}\right)^2\right] = \sin\frac{\pi}{a}z,$$

and so we obtain

$$w_{\text{top}}(z) = \frac{i\Gamma}{2\pi} \ln \sin\frac{\pi z}{a}. \tag{13.11}$$

Differentiating, we obtain the complex velocity $u - iv$ at the location of the vortex at $z = a/2 - ib$:

$$\frac{dw}{dz} = u - iv = \frac{i\Gamma}{2a}\cot\left(\frac{\pi}{2} - i\frac{\pi b}{a}\right) = \frac{i\Gamma}{2a}\tan i\frac{\pi b}{a} = -\frac{\Gamma}{2a}\tanh\frac{\pi b}{a},$$
(13.12)

which indeed has a u-component only, and the relationship between the convection speed U and the parameters of the vortex street is

$$\frac{1}{2a}\tanh\frac{\pi b}{a} = \frac{U}{\Gamma}.$$
(13.13)

The key feature of von Kármán's staggered configuration of vortices is its stability. For example, if the rows were not staggered (Exercise 13.2), the configuration could not be stable: a vortex in the upper row that becomes shifted to the right is then convected less by the vortex immediately below and thus falls behind even more. In the staggered configuration, by contrast, there is a trade-off between the two vortices just below, whose velocity fields compensate each other. More formally, one has to consider perturbations to the vortex positions in the form of a Fourier series [131]. Thus if z_m is the perturbation to the mth vortex in the upper row and z_n' the perturbation to the nth vortex in the lower row, we write for a single Fourier mode

$$z_m(t) = \gamma(t)\cos m\theta, \quad z_n'(t) = \gamma'(t)\cos\left(n + \frac{1}{2}\right)\theta,$$
(13.14)

where γ, γ' are complex numbers. The positions of the vortices in the upper and lower rows are then given by

$$ma - Ut + z_m(t), \quad \left(n + \frac{1}{2}\right)a - Ut - ib + z_n'(t),$$
(13.15)

respectively. In Exercise 13.3 the lowest Fourier mode, $\theta = \pi$, is considered and it can be shown that the von Kármán vortex street is only stable if $r = b/a$ assumes a critical value given by

$$\frac{b}{a} = \frac{1}{\pi}\text{arccosh}\sqrt{2} = 0.280\,55\ldots$$
(13.16)

A more detailed calculation [131] shows that under the condition (13.16) the street is indeed stable against *all* perturbations. Measuring the ratio for the experiment shown in Fig. 13.6, we find $b/a \simeq 0.36$–0.42, in fair agreement with (13.16). Inserting (13.16) into (13.13), it follows that

$$\frac{U}{\Gamma} = \frac{1}{a\sqrt{8}}.$$
(13.17)

Unfortunately, our analytical approach does not provide any hint as to the values of a and b relative to the dimension, or shape, of the body. This issue needs to be studied separately [103]. Once a is determined, formula (13.17) provides the velocity U of the vortices as a function of the circulation Γ. Since the circulation can be estimated roughly as $2U_{in}a$ (by integrating the velocity U_{in} of the flow incident on the body over the boundary of an a-periodic cell around one vortex), we conclude from (13.17) that the velocity of the vortices U is smaller than the velocity of the incident flow U_{in}, so that the vortices will trail behind the body.

13.2 Vortex filaments

In the previous section we looked at straight vortex filaments oriented in parallel, producing a two-dimensional flow. Now we generalize to the case where the vorticity is concentrated along a line of arbitrary shape, called a vortex filament. Let us parameterize such a curve with the arc length s; the geometry of the filament is shown in Fig. 13.8. The tangent vector is found by differentiating $\mathbf{r}$ along the curve:

$$\mathbf{t} = \frac{\partial \mathbf{r}}{\partial s} \equiv \mathbf{r}_s. \tag{13.18}$$

The rate of change of $\mathbf{t}$ is a measure of the curvature $\kappa(s)$, and the resulting vector is normal to the curve:

$$\mathbf{t}_s = \kappa \mathbf{n}. \tag{13.19}$$

This means that $\mathbf{n}$ and the binormal

$$\mathbf{b} = \mathbf{t} \times \mathbf{n} \tag{13.20}$$

span the plane perpendicular to the local direction of the filament (see Fig. 13.8).

To complete the geometrical description of a curve we need to specify the derivatives of the basis vectors $\mathbf{t}$, $\mathbf{n}$, and $\mathbf{b}$. One can show that in addition to (13.19) (see for example [62]) one has

$$\mathbf{n}_s = -\kappa \mathbf{t} + \tau \mathbf{b}, \quad \text{and} \quad \mathbf{b}_s = -\tau \mathbf{n}. \tag{13.21}$$

The second scalar function, $\tau(s)$, is called the torsion of the curve. It describes how $\mathbf{b}$ winds itself around the centerline. The three equations in (13.19) and (13.21) are known as the *Frenet–Serret formulas*.

The velocity generated by a vortex can be calculated from a line integral along the filament, in exactly the same way as the magnetic field of a

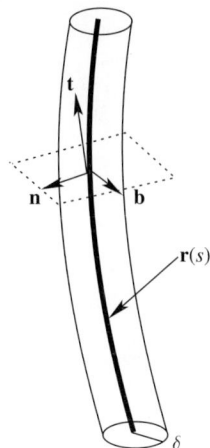

Figure 13.8 The geometry of a vortex filament. The position along the line is described by the position vector $\mathbf{r}(s)$, where s is the arc length, and the strength of the vortex is Γ. The tangent vector $\mathbf{t}$ points in the direction of the filament. The normal and binormal vectors $\mathbf{n}$ and $\mathbf{b}$ lie in a plane perpendicular to the filament.

current-carrying wire is calculated using the Biot–Savart law. In the electrical analogy (4.35) is Ampère's law; the strength Γ of the vortex filament plays the role of the electrical current I through the wire, and must be constant along the filament for consistency. In view of Kelvin's theorem, Γ must also be constant in time. Thus in the present context the Biot–Savart law has the form

$$\mathbf{v}(\mathbf{r}) = \frac{\Gamma}{4\pi} \int \frac{\mathbf{t}(s') \times [\mathbf{r} - \mathbf{r}(s')]}{|\mathbf{r} - \mathbf{r}(s')|^3} ds'. \tag{13.22}$$

Other vortex filaments will be convected by the velocity field (13.22). In addition, a filament is convected by its *own* velocity field, for which we need $\mathbf{v}(\mathbf{r})$ for $\mathbf{r} = \mathbf{r}(s)$ on the filament. However, the integral (13.22) becomes divergent for such a choice owing to the singular nature of the vorticity distribution. As $\mathbf{r}(s')$ approaches $\mathbf{r}$, $\mathbf{r} - \mathbf{r}(s')$ becomes parallel to $\mathbf{t}$ and the numerator in (13.22) goes to zero, so that a non-integrable singularity of size $O(|\mathbf{r} - \mathbf{r}(s')|^{-2})$ remains in the integrand. Thus one has to account for the fact, that, in reality, the vorticity is smeared out over an area of size $\delta S = \pi \delta^2$; see Fig. 13.8. We model this fact by evaluating the velocity at a distance δ from the centerline, rendering the result finite.

We choose a fixed $\mathbf{r}$ at a distance δ from the centerline of the vortex filament and calculate $\mathbf{v}(\mathbf{r})$ to leading order as $\delta \to 0$. To that end we split up the integral into two contributions I_1 and I_2 (see Fig. 13.9),

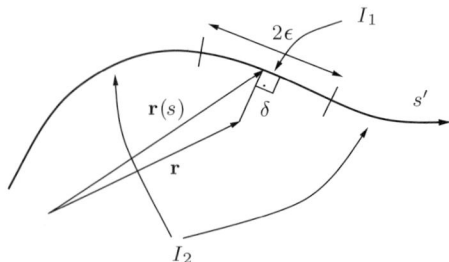

Figure 13.9 The local geometry of a vortex line $\mathbf{r}(s')$ used to derive the self-induction approximation (13.27). Here s marks the position closest to the fixed vector $\mathbf{r}$, and δ is the distance between $\mathbf{r}$ and $\mathbf{r}(s)$. The region of integration is divided into I_1, comprising the points s' closer to s than ϵ, and its complement I_2.

$$I_{1/2} = \frac{\Gamma}{4\pi} \int_{|s-s'|\lessgtr\epsilon} \frac{\mathbf{t}(s') \times [\mathbf{r} - \mathbf{r}(s')]}{|\mathbf{r} - \mathbf{r}(s')|^3} ds', \qquad (13.23)$$

where $\mathbf{r}(s)$ is the point on the filament closest to $\mathbf{r}$. The parameter ϵ is assumed to be small compared with the radius of curvature R of the filament yet large compared with δ: $\delta \ll \epsilon \ll R$.

Thus, in computing the local contribution I_1 we can approximate the shape of the filament by a Taylor expansion:

$$\mathbf{r}(s) = \mathbf{r}(s') + \frac{d\mathbf{r}}{ds}(s')(s - s') + \frac{1}{2}\frac{d^2\mathbf{r}}{ds^2}(s')(s - s')^2 + O((s - s')^3)$$

$$= \mathbf{r}(s') + \mathbf{t}(s')(s - s') + \frac{1}{2}\kappa\mathbf{n}(s')(s - s')^2 + O((s - s')^3).$$

Since $\mathbf{r}$ is evaluated at a distance δ from the filament (in a direction perpendicular to $\mathbf{t}$; see Fig. 13.9), we can use the estimate

$$|\mathbf{r} - \mathbf{r}(s')|^3 = \left(\delta^2 + |s - s'|^2\right)^{3/2} + O(|s - s'|^4).$$

Then, since we are assuming that $|s - s'| < \epsilon$ and $\epsilon \gg \kappa^{-1} = R$, the integrand in (13.22) can be approximated as

$$-\frac{1}{2}\frac{\kappa |s - s'|^2 \mathbf{b}}{\left(\delta^2 + |s - s'|^2\right)^{3/2}} + \text{lower-order terms.} \qquad (13.24)$$

If one neglects the lower-order terms in (13.24) and considers the limit $\epsilon/\delta \gg 1$, then explicitly

$$I_1 \simeq -\frac{\Gamma\kappa}{8\pi} \int_{s-\epsilon}^{s+\epsilon} \frac{(s - s')^2 ds'}{[\delta^2 + (s - s')^2]^{3/2}} \mathbf{b} \simeq \frac{\Gamma\kappa \ln(\epsilon/\delta)}{4\pi}\mathbf{b}. \qquad (13.25)$$

The integral I_2, however, is dominated by contributions from regions close to $s' = s \pm \epsilon$ (say $(s - s_0, s - \epsilon)$ and $(s + \epsilon, s + s_0)$), but where (13.24) is still valid because $\epsilon \ll R$. However, we will now neglect δ^2 relative to the second term in the denominator of (13.24), which is of order ϵ^2. The contributions from points on the filament far from $\mathbf{r}$ are bounded, given the fast decay of the Biot–Savart kernel ($\sim |\mathbf{r}|^{-2}$) at infinity. Hence we obtain

$$I_2 \simeq -\frac{\Gamma \kappa}{8\pi} \left(\int_{s+\epsilon}^{s+s_0} \frac{ds}{|s - s'|} + \int_{s-s_0}^{s-\epsilon} \frac{ds}{|s - s'|} \right) \mathbf{b} \simeq -\frac{\Gamma \kappa \ln \epsilon}{4\pi} \mathbf{b}. \qquad (13.26)$$

Therefore, on adding (13.25) and (13.26) the dependence on ϵ cancels out and one obtains the following leading-order contribution for the velocity:

$$\mathbf{v} = -\frac{\Gamma \kappa \ln \delta}{4\pi} \mathbf{b}, \qquad (13.27)$$

which becomes singular as $\delta \to 0$. Equation (13.27) is known as the *self-induction approximation* for isolated vortex filaments. We will take a constant finite value of δ as representing the "radius" of the vortex filament.

Computing the cross product of (13.19) with $\mathbf{t} = \mathbf{r}_s$ and using (13.20), one finds $\kappa \mathbf{b} = \mathbf{r}_s \times \mathbf{r}_{ss}$. In particular, $\mathbf{v}$ is orthogonal to the filament and using (13.27) we arrive at the evolution equation

$$\frac{\partial \mathbf{r}}{\partial t} = \frac{\Gamma \ln \delta^{-1}}{4\pi} \mathbf{r}_s \times \mathbf{r}_{ss} \qquad (13.28)$$

for $\mathbf{r}(s, t)$. Equation (13.28) describes the dynamics of a singular distribution of vorticity lying on a smooth curve $\mathbf{r}(s)$.

Example 13.3 (Stability of a filament) The binormal flow equation (13.28) for a vortex filament has a stationary solution corresponding to a straight filament $\mathbf{r}_s = s\mathbf{e}_3$. Now consider small perturbations of the form

$$\mathbf{r}(s, t) = \mathbf{r}_0 + \epsilon \mathbf{g}(s, t), \qquad (13.29)$$

with $\mathbf{g}$ perpendicular to $\mathbf{e}_3$: $\mathbf{g}(s, t) = g_1 \mathbf{e}_1 + g_2 \mathbf{e}_2$. It follows that

$$\mathbf{r}_s \times \mathbf{r}_{ss} = \epsilon \left(\mathbf{e}_3 + \epsilon \mathbf{g}_s \right) \times \mathbf{g}_{ss} = \epsilon \mathbf{e}_3 \times \mathbf{g}_{ss} + O(\epsilon^2),$$

so that, at linear order,

$$\mathbf{g}_t = a \left(\mathbf{e}_3 \times \mathbf{g}_{ss} \right), \qquad (13.30)$$

where $a = \Gamma \ln \delta^{-1} / (4\pi)$.

The cross products of (13.30) with $\mathbf{e}_2$ and $\mathbf{e}_1$ yield the following equations for the components of $\mathbf{g}$:

$$g_{1,t} = -ag_{2,ss}, \quad g_{2,t} = ag_{1,ss}. \qquad (13.31)$$

The solution to (13.31) is of the form

$$g_1 = Ae^{iks+\omega(k)t}, \quad g_2 = Be^{iks+\omega(k)t}, \tag{13.32}$$

which results in the equations

$$A\omega(k) = ak^2 B, \quad B\omega(k) = -ak^2 A.$$

There exists a nontrivial solution for A and B provided that the determinant vanishes, giving the dispersion relation

$$\omega^2(k) + a^2 k^4 = 0. \tag{13.33}$$

Hence $\omega(k) = \pm ak^2 i$, so perturbations will neither grow nor decay in time. This is characteristic of systems with a Hamiltonian structure, where energy is conserved. The nonlinearity of (13.33) points to the presence of dispersive phenomena, where the waves forming an initial wave packet travel with different speeds. The real solutions are of the form

$$g_1(s,t) = \cos(ks \pm ak^2 t), \quad g_2(s,t) = \sin(ks \pm ak^2 t), \tag{13.34}$$

representing helical propagation of the perturbation $\mathbf{g}$. $\qquad\square$

13.2.1 Corner singularity of a vortex filament

We now ask whether there are solutions of (13.28) for which the curve itself may develop singularities in finite time; for simplicity, we absorb the prefactor into a change in time scale. A natural candidate for such a solution, given the scalings of (13.28) and the constraint that $|\mathbf{t}| = |\mathbf{r}_s| = 1$, is (see [102])

$$\mathbf{r}(s,t) = t'^{1/2}\mathbf{G}(\xi), \quad \xi = \frac{s}{t'^{1/2}}, \tag{13.35}$$

where we have put $t' = t_0 - t$. Inserting the similarity form (13.35) for the similarity profile $\mathbf{G}(\xi)$ into (13.28) leads to

$$\frac{1}{2}\mathbf{G}(\xi) - \frac{\xi}{2}\mathbf{G}_\xi(\xi) = \mathbf{G}_\xi(\xi) \times \mathbf{G}_{\xi\xi}(\xi). \tag{13.36}$$

From the matching condition (3.16) we conclude that the behavior of the filament must be linear at infinity, and so

$$\mathbf{r}(s,t) = \mathbf{G}_0^{\pm} s, \quad s \to \pm\infty. \tag{13.37}$$

Given that the problem is rotationally invariant, the only parameter that matters is the angle θ between the directions $\mathbf{G}_0^+$ and $\mathbf{G}_0^-$.

Taking derivatives of (13.36) with respect to ξ and setting $\mathbf{T} = \mathbf{G}_\xi$ we get

$$-\frac{\xi}{2}\mathbf{T}_\xi = \mathbf{T} \times \mathbf{T}_{\xi\xi}. \tag{13.38}$$

We now rewrite (13.38) using the self-similar version $\mathbf{T}_\xi = \bar{\kappa}\mathbf{N}$ and $\mathbf{N}_\xi = -\bar{\kappa}\mathbf{T} + \bar{\tau}\mathbf{B}$ of the Frenet–Serret formulas (13.19), (13.21); $\bar{\kappa}$ and $\bar{\tau}$ are rescaled versions of the curvature and torsion, respectively:

$$\kappa(s,t) = t'^{-1/2}\bar{\kappa}(\xi), \quad \tau(s,t) = t'^{-1/2}\bar{\tau}(\xi). \tag{13.39}$$

Thus we obtain

$$-\frac{\xi}{2}\bar{\kappa}\mathbf{N} = \mathbf{T} \times (\bar{\kappa}_\xi\mathbf{N} - \bar{\kappa}^2\mathbf{T} + \bar{\kappa}\bar{\tau}\mathbf{B}) = \bar{\kappa}_\xi\mathbf{B} + \bar{\kappa}\bar{\tau}\mathbf{T} \times (\mathbf{T} \times \mathbf{N})$$

$$= \bar{\kappa}_\xi\mathbf{B} - \bar{\kappa}\bar{\tau}\mathbf{N}, \tag{13.40}$$

where we have used the orthonormality conditions in (13.44) below. Scalar-multiplying (13.40) by $\mathbf{B}$ and $\mathbf{N}$, respectively, one finds that

$$\bar{\kappa}(\xi) = a, \quad \bar{\tau}(\xi) = \frac{\xi}{2}, \tag{13.41}$$

which is the solution to the similarity problem in terms of the curvature and torsion.

The solution (13.41) is particularly simple, but finding the curve $\mathbf{G}(\xi)$ is not. To this end one has to solve the equation

$$\mathbf{G}_\xi = \mathbf{T}(\xi), \tag{13.42}$$

where the tangent vector $\mathbf{T}(\xi)$ obeys the Frenet–Serret system (13.13), (13.21). Thus we have

$$\begin{pmatrix} \mathbf{T} \\ \mathbf{N} \\ \mathbf{B} \end{pmatrix}_\xi = \begin{pmatrix} 0 & a & 0 \\ -a & 0 & \xi/2 \\ 0 & -\xi/2 & 0 \end{pmatrix} \begin{pmatrix} \mathbf{T} \\ \mathbf{N} \\ \mathbf{B} \end{pmatrix}, \tag{13.43}$$

together with the conditions

$$|\mathbf{T}| = |\mathbf{N}| = |\mathbf{B}| = 1, \quad \mathbf{T}\cdot\mathbf{N} = \mathbf{T}\cdot\mathbf{B} = \mathbf{B}\cdot\mathbf{N} = 0. \tag{13.44}$$

The system (13.42)–(13.44) was solved numerically in [102], where some analytical results are also provided. A unique solution is found for any angle θ between the two asymptotic directions of the filament. For one particular angle, a sequence of self-similar solutions is shown in Fig. 13.10. The vortex filament remains smooth until it develops into a corner at $t = t_0$.

It is important to remark that, although the binormal flow equation (13.28) produces vortex filaments that look very similar to those observed in real situations, one must be aware of the limitations of the model: (i) it ignores the

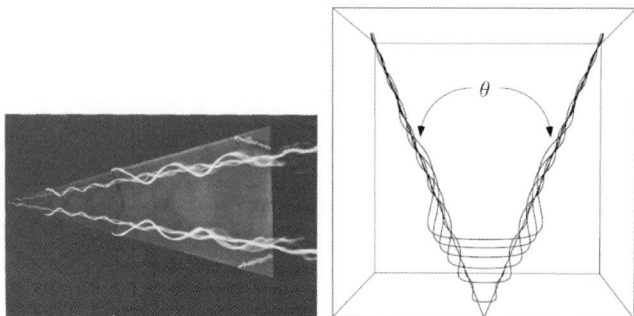

Figure 13.10 Production of vortex filaments by a delta wing (left) and the development of singularities in finite time in the solution to the binormal flow equation (13.28) (right).

stretching and twisting of vortex filaments; (ii) the self-induction approximation (13.27) introduces a cutoff parameter δ, whose meaning is unclear. A more complete description should match to the inner core structure of the filament and provide a value of the parameter δ as a result.

13.3 Vortex sheets

In the preceding section we studied singular distributions of vorticity along a line, while the rest of the flow was assumed to be potential. A natural extension of this concept is the vortex sheet, where it is assumed that the vorticity is concentrated on sheets. Indeed, vortex sheets can be regarded as the limiting case of a row of many vortex filaments. Once more, the hope is to reduce the dynamics to an equation for the motion of the sheet instead of having to compute the velocity field everywhere. To construct a vortex sheet we begin with the simplest case, that of straight vortex filaments (which can be viewed as point vortices in two dimensions), arranged along a straight line; see Fig. 13.11.

If the distance between the vortices is $\delta\xi$, the complex potential of the arrangement is

$$w(z) = \frac{\Gamma}{2\pi i} \sum_{-\infty}^{\infty} \ln(z - n\delta\xi). \tag{13.45}$$

If we take the circulation to scale as $\Gamma = \sigma\delta\xi$ and if the distance between the vortices goes to zero, (13.45) turns into the integral

$$w(z) = \frac{\sigma}{2\pi i} \int_{-\infty}^{\infty} \ln(z - \xi)d\xi. \tag{13.46}$$

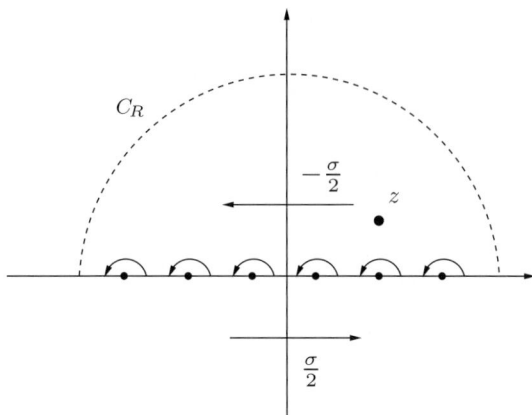

Figure 13.11 A vortex sheet can be described as a row of vortices. It corresponds to a shear flow with all the shear concentrated along the sheet.

The complex velocity becomes

$$u - iv = \frac{dw}{dz} = \frac{\sigma}{2\pi i} \int_{-\infty}^{\infty} \frac{d\xi}{z - \xi}, \qquad (13.47)$$

and we need to evaluate this. If we construct a closed loop by integrating along the real axis and then along a far circle C_R (see Fig. 13.11), we obtain

$$\int_{-\infty}^{\infty} \frac{d\xi}{\xi - z} + \lim_{R \to \infty} \int_{C_R} \frac{d\xi}{\xi - z} = \begin{cases} 2\pi i, & \Im\{z\} > 0, \\ 0, & \Im\{z\} < 0. \end{cases}$$

Here we have used the fact that, according to Cauchy's theorem, the integral is $2\pi i$ if z is inside the loop and vanishes if z is outside, i.e. if $\Im\{z\} < 0$. Now

$$\lim_{R \to \infty} \int_{C_R} \frac{d\xi}{\xi - z} = \lim_{R \to \infty} \int_0^{\pi} \frac{i R e^{i\theta} d\theta}{R e^{i\theta} - z} = \int_0^{\pi} i \, d\theta = i\pi,$$

and so we finally have

$$u - iv = \begin{cases} -\sigma/2, & y > 0, \\ \sigma/2, & y < 0. \end{cases} \qquad (13.48)$$

In other words, at the vortex sheet there is a sudden jump σ in the tangential velocity. It can be regarded as an extreme case of shear flow, with all the shear concentrated in the sheet.

Now we will generalize to a sheet that is curved, but we will remain in the framework of two-dimensional flow. In that case the vortex sheet of Fig. 13.11, which lies on the x-axis, turns into a curve C in the plane. In complex notation this curve can be written as $z = z(\alpha, t)$, where α is a parameter. The local

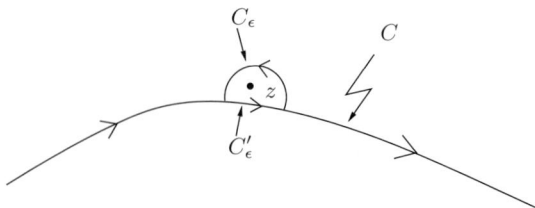

Figure 13.12 The path of integration used to extract the local contribution to the velocity field generated by a vortex sheet

strength $\sigma(\alpha)$ can of course vary, corresponding to a variation in the strengths of the vortices which make up the sheet. Thus (13.47) becomes

$$u - iv = \frac{dw(z,t)}{dz} = \frac{1}{2\pi i} \int_C \frac{\sigma(\alpha',t)d\alpha'}{z - z(\alpha',t)}, \tag{13.49}$$

where we have included a possible time dependence for both z and σ. Notice that $d\alpha' = dz(\alpha',t)/z_{\alpha'}(\alpha',t)$ so that (13.49) is in fact the Cauchy integral

$$u - iv = \frac{1}{2\pi i} \int_C \frac{\sigma}{z_\alpha} \frac{dz'}{z - z'}. \tag{13.50}$$

If the vortex sheet moves, following the velocity field that it creates, then the vortices making up the sheet are convected by the flow, which implies that σ has to be constant along particle paths.

To calculate the motion of the vortex sheet, we analyze (13.50) as z approaches the sheet. Once more there are local contributions which are singular, but they are not divergent. To compute them, we use the procedure indicated in Fig. 13.12. First, the integral along C is split into a local contribution from C_ϵ' and a contribution from the part of C exterior to it:

$$\int_C \frac{\sigma}{z_\alpha} \frac{dz'}{z - z'} = \int_{C_\epsilon'} \frac{\sigma}{z_\alpha} \frac{dz'}{z - z'} + \mathrm{PV} \int_C \frac{\sigma}{z_\alpha} \frac{dz'}{z - z'}.$$

The letters PV in front of the integral denote the principal value, which means that a region of size ϵ on either side of the singular point is excluded, and ϵ is subsequently sent to zero.

To compute the integral along C_ϵ' we close the loop over a semicircle C_ϵ, of radius ϵ, parameterized as $z' = z + \epsilon e^{i\theta}$. Since σ and z_α can be expanded about the center of the circle, for small ϵ we have

$$\int_{C_\epsilon'} \frac{\sigma}{z_\alpha} \frac{dz}{z - z'} = \frac{\sigma}{z_\alpha} \int_{C_\epsilon'} \frac{dz}{z - z'}$$

and, using Cauchy's theorem,

$$\int_{C'_\epsilon} \frac{dz'}{z'-z} = 2\pi i - \int_{C_\epsilon} \frac{dz'}{z'-z} = 2\pi i - i \int_0^\pi d\theta = \pi i. \qquad (13.51)$$

Had z been approaching the vortex sheet from below, C_ϵ would close the loop from below and the result of (13.51) would have been $-i\pi$, since the path would turn in the opposite direction. Therefore from (13.49) we find

$$\lim_{z \to C^\pm} \frac{dw(z,t)}{dz} = \mp \frac{\sigma(\alpha,t)}{2z_\alpha(\alpha,t)} + (\tilde{u} - i\tilde{v})(z(\alpha,t),t), \qquad (13.52)$$

where

$$(\tilde{u} - i\tilde{v})(z(\alpha,t),t) = \frac{1}{2\pi i} \mathrm{PV} \int_C \frac{\sigma(\alpha',t)d\alpha'}{z(\alpha,t) - z(\alpha',t)}. \qquad (13.53)$$

Since $z_\alpha(\alpha,t)$ is the complex representation of a vector that is tangent to C, so is $1/z_\alpha^*(\alpha,t) = z_\alpha(\alpha,t)/|z_\alpha(\alpha,t)|^2$. Therefore the complex conjugate of the first term on the right of (13.52) can be viewed as a velocity which is tangential but discontinuous across the sheet. It corresponds to (13.48) in the case of a straight sheet.

The average of the discontinuous tangential velocity is zero but, even if one were to choose a different interpretation, a tangential velocity does not affect the motion of the sheet so we need to consider the second contribution to (13.52) only. If α is the label of a vortex advected by the velocity field $\tilde{u} - i\tilde{v}$, the equation of motion becomes

$$\frac{\partial z^*(\alpha,t)}{\partial t} = \frac{1}{2\pi i} \mathrm{PV} \int_C \frac{\sigma(\alpha',t)d\alpha'}{z(\alpha,t) - z(\alpha',t)}. \qquad (13.54)$$

In addition $\sigma(\alpha,t)$ is independent of t, owing to the fact that an individual vortex moves with the fluid and carries its own vorticity. If we define $\mu(\alpha,t)$ such that $\mu_\alpha = \sigma$ then μ will also be constant along Lagrangian paths. We can thus parameterize the curve with μ and write

$$\frac{\partial z^*(\mu,t)}{\partial t} = \frac{1}{2\pi i} \mathrm{PV} \int_C \frac{d\mu'}{z(\mu,t) - z(\mu',t)}. \qquad (13.55)$$

Equation (13.55) is known as the *Birkhoff–Rott equation*; it governs the evolution of the vortex sheet.

Example 13.4 (Prandtl spiral) The Prandtl spiral is a similarity solution of the Birkhoff–Rott equation (13.55), which can be parameterized in terms of the polar angle θ in the form

$$z = t^\alpha f(\theta)e^{i\theta}, \quad \mu = t^\beta g(\theta), \qquad (13.56)$$

where f, g are real. Inserting (13.56) into (13.55), the relation between the similarity exponents is $\beta = 2\alpha - 1$ and the profiles $f(\theta)$, $g(\theta)$ satisfy the following complex integro-differential equation:

$$e^{-i\theta}\left(\alpha f + (1 - 2\alpha)\frac{g}{g'}(f' - if)\right) = \frac{1}{2\pi i}\mathrm{PV}\int_{-\infty}^{\infty}\frac{g'(\theta')d\theta'}{f(\theta)e^{i\theta} - f(\theta')e^{i\theta'}}.$$

(13.57)

We seek solutions to (13.57) of the form

$$f = Ae^{a\theta}, \quad g = Be^{b\theta},$$

which, after the substitution $\theta' - \theta = \varphi$, leads to

$$A^2 e^{(2a-b)\theta}\left(\alpha + \frac{1 - 2\alpha}{b}(a - i)\right) = \frac{Bb}{2\pi i}\mathrm{PV}\int_{-\infty}^{\infty}\frac{e^{b\varphi}}{1 - e^{(i+a)\varphi}}d\varphi;$$

for this to be a solution we must have $b = 2a$. However, in that case (and for $a > 0$, say) the integral diverges, as the vorticity is unbounded at infinity [189]. We avoid this problem by formally assuming a priori that $0 < b < a$, in which case the integral is convergent. Using contour deformation and the residue theorem, the integral becomes

$$\mathrm{PV}\int_{-\infty}^{\infty}\frac{e^{b\varphi}}{1 - e^{(i+a)\varphi}}d\varphi = -\frac{i\pi}{i + a}\frac{1 + e^{2\pi ib/(i+a)}}{1 - e^{2\pi ib/(i+a)}}.$$

(13.58)

Now the result (13.58) can be analytically continued to the relevant case $b = 2a$, to yield the complex equation

$$\frac{A^2}{Ba}\left(\alpha(a + i) + \frac{a^2 + 1}{2a}(1 - 2\alpha)\right) = -\frac{1 + e^{4\pi ia/(i+a)}}{1 - e^{4\pi ia/(i+a)}}.$$

(13.59)

The right-hand side of (13.59) is

$$-\frac{1 + e^{4\pi ia/(i+a)}}{1 - e^{4\pi ia/(i+a)}} = \frac{e^{2\mu} - 1 - 2ie^{\mu}\sin\mu}{1 - 2e^{\mu}\cos\mu + e^{2\mu}},$$

where $\mu = 4\pi a^2/(a^2 + 1)$. Eliminating A^2/B between the real and imaginary parts of (13.59), one finally obtains

$$\alpha = \left(\frac{2}{1 + a^2} - \frac{2a}{1 + a^2}\frac{\sinh[4\pi a/(1 + a^2)]}{\sin[4\pi a/(1 + a^2)]}\right)^{-1},$$

(13.60)

and the vortex sheet has the shape

$$x = At^{\alpha}e^{a\theta}\cos\theta, \quad y = At^{\alpha}e^{a\theta}\sin\theta.$$

(13.61)

The curve (13.61) is an infinite logarithmic spiral; see Fig. 13.13. As a function of time, the size of the spiral increases as t^{α}.   □

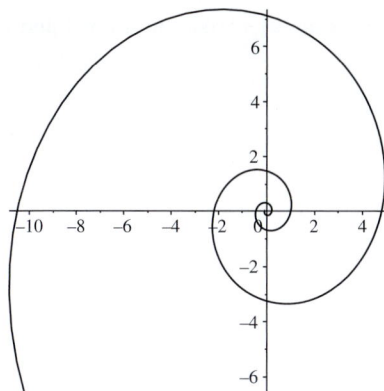

Figure 13.13 The Prandtl spiral (13.61) for $a = 1/4$ and $A = t = 1$. From (13.60) it follows that $\alpha \approx 0.759$.

Figure 13.14 Two layers of air move relative to one another to produce a Kelvin–Helmholtz instability [66]. The flow is visualized by clouds in the transition region, which produce a spiral pattern known as billow clouds. (Image by Kristin Bartee.)

13.3.1 Linear instability of vortex sheets

If two streams of fluid are flowing against each other in a low-viscosity environment an instability occurs, as illustrated in Fig. 13.14. In the idealized case where the flow changes discontinuously, as in a vortex sheet, this is known as the Kelvin–Helmholtz instability [66]. We will begin by performing this linear stability analysis of a vortex sheet using the formulation (13.55).

The base state solution is $z(\mu, t) = \mu$, corresponding to a planar surface; see Fig. 13.11. This solution is time independent since

$$
\begin{aligned}
\mathrm{PV} \int_{-\infty}^{\infty} \frac{d\mu'}{\mu - \mu'} &= \lim_{R \to \infty} \lim_{\epsilon \to 0} \left(\int_{-R}^{\mu - \epsilon} \frac{d\mu'}{\mu - \mu'} + \int_{\mu + \epsilon}^{R} \frac{d\mu'}{\mu - \mu'} \right) \\
&= \lim_{R \to \infty} \lim_{\epsilon \to 0} \left(\ln \frac{\mu + R}{\epsilon} - \ln \frac{R - \mu}{\epsilon} \right) \\
&= \lim_{R \to \infty} \ln \frac{\mu + R}{R - \mu} = 0.
\end{aligned} \tag{13.62}
$$

Now we probe the linear stability by adding a small perturbation ϵ ($\epsilon \ll 1$):

$$z(\mu, t) = \mu + \epsilon \zeta(\mu, t).$$

At leading order in ϵ the linearized form of (13.55) becomes

$$\frac{\partial \zeta^*(\mu, t)}{\partial t} = -\frac{1}{2\pi i} \text{PV} \int_{-\infty}^{\infty} \frac{\zeta(\mu, t) - \zeta(\mu', t)}{(\mu - \mu')^2} d\mu'. \tag{13.63}$$

Integrating by parts, one finds that

$$\begin{aligned} 0 &= \lim_{\epsilon \to 0} \frac{\zeta(\mu) - \zeta(\mu')}{\mu - \mu'} \Big|_{\mu'=\mu-\epsilon}^{\mu'=\mu+\epsilon} \\ &= \text{PV} \int_{-\infty}^{\infty} \frac{\zeta(\mu, t) - \zeta(\mu', t)}{(\mu - \mu')^2} d\mu' - \text{PV} \int_{-\infty}^{\infty} \frac{\zeta_\mu(\mu')d\mu'}{\mu - \mu'}, \end{aligned}$$

and so

$$\text{PV} \int_{-\infty}^{\infty} \frac{\zeta(\mu, t) - \zeta(\mu', t)}{(\mu - \mu')^2} d\mu' = \text{PV} \int_{-\infty}^{\infty} \frac{\zeta_\mu(\mu')d\mu'}{\mu - \mu'}. \tag{13.64}$$

Thus, using (13.64), equation (13.63) turns into

$$\frac{\partial \zeta^*(\mu, t)}{dt} = -\frac{1}{2\pi i} \text{PV} \int_{-\infty}^{\infty} \frac{\zeta_\mu(\mu', t)d\mu'}{\mu - \mu'}. \tag{13.65}$$

Noting that the *Hilbert transform* [148] is defined as

$$H(f)(x) = \frac{1}{\pi} \text{PV} \int_{-\infty}^{\infty} \frac{f(x')dx'}{x - x'}, \tag{13.66}$$

equation (13.65) can be written compactly as

$$\zeta_t^* = -\frac{1}{2i} H(\zeta_\mu). \tag{13.67}$$

Now we assume sinusoidal perturbations of the form

$$\zeta(\mu, t) = e^{\omega(k)t} \left(a_k e^{ik\mu} + a_{-k} e^{-ik\mu} \right), \quad k > 0, \tag{13.68}$$

and make use of the identity

$$H\left(e^{ik\mu}\right) = \frac{1}{\pi} \text{PV} \int_{-\infty}^{\infty} \frac{e^{ik\mu'} d\mu'}{\mu - \mu'} = -i\frac{k}{|k|} e^{ik\mu}, \tag{13.69}$$

which follows on using the residue theorem. Inserting (13.68) into (13.65), one obtains

$$\omega a_k^* e^{-ik\mu} + \omega a_{-k}^* e^{ik\mu} = -\frac{|k|}{2i} \left(a_k e^{ik\mu} + a_{-k} e^{-ik\mu} \right),$$

and thus

$$\omega a_k^* = -\frac{|k|}{2i} a_{-k}, \quad \omega a_{-k}^* = -\frac{|k|}{2i} a_k.$$

Taking the complex conjugate of the first equation and inserting this into the second, we obtain the dispersion relation

$$\omega(k) = \pm \frac{|k|}{2}. \tag{13.70}$$

This relation implies that perturbations grow exponentially and render the vortex sheet unstable. Even worse, the shorter the wavelength (the larger is k), the faster the growth, which is the hallmark of an *ill-posed* differential problem (in the sense of Hadamard).

In reality there must be a mechanism that limits the growth at short wavelengths. For example, the shear layer will have a finite thickness δ. If this is taken into account then the growth rates remain finite and the maximum growth rate occurs at a wavenumber $k \approx \delta^{-1}$ [189], although we will not elaborate on this calculation here. In Section 13.3.3 we describe how the ill-posedness can be removed by introducing surface tension.

13.3.2 Moore's singularity of vortex sheets

Now we consider the evolution of a curved vortex sheet, illustrated in Fig. 13.15. Moore [157] showed numerically that, regardless of the initial conditions, a singularity develops in finite time. The singularity is very weak and only appears in the *curvature* of the vortex sheet, so that the tangent remains continuous. As a result we can use the linearized evolution (13.67) and apply it to the tangent vector to describe the singularity. By taking the μ derivative in (13.67) and writing $\eta = \xi_\mu$, we arrive at

$$\eta_t^* = \frac{1}{2i} H(\eta_\mu). \tag{13.71}$$

We now show that (13.71) leads to the blowup of η_μ and η_t, while all the nonlinear corrections to (13.71) remain bounded [59]. By setting

$$\eta(\mu, t) = \varphi(\mu, t) + i\psi(\mu, t)$$

we can decompose equation (13.71) into the two equations

$$\varphi_t = \frac{1}{2} H(\psi_\mu), \quad \psi_t = \frac{1}{2} H(\varphi_\mu). \tag{13.72}$$

Since $H(H(f)) = -f$ [148], the relations (13.72) are precisely the Cauchy–Riemann equations for $\Phi = \varphi - iH\psi$ in the complex variable $\sigma = \mu + it/2$.

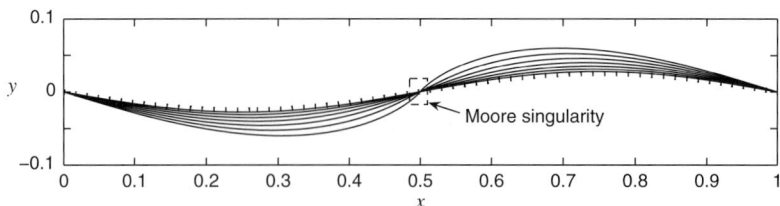

Figure 13.15 Evolution of a vortex sheet, starting from the initial profile $y = 0.1 \sin 2\pi x$ (dotted line), with constant vortex sheet strength σ. The interface is shown at constant time intervals.

Specifying the initial configuration for the vortex sheet thus corresponds to prescribing Φ on the real axis, which determines it completely. By analytic continuation, one can determine the time t_0 at which a singularity first appears. The smoothness of η in a time interval $[0, t_0)$ is associated with the smoothness of the Riemann sheet $\Re\{\Phi(\sigma)\}$ in the strip $0 \leq \Im\{\sigma\} < t_0/2$. However, one should keep in mind that, strictly speaking, the linearized equation (13.71) is a true description of the full nonlinear problem only in the neighborhood of t_0.

A generic way for a singularity to occur is for the Riemann sheet to fold at some $\sigma_0 = \mu_0 + i t_0/2$ (we can take $\mu_0 = 0$ without loss of generality). In other words, at $t = t_0$ the mapping $\sigma(\Phi)$ becomes non-invertible:

$$\frac{d\sigma(\Phi)}{d\Phi}\bigg|_{\Phi_0 = \Phi(\sigma_0)} = 0$$

and hence

$$\sigma = \sigma_0 + C(\Phi - \Phi_0)^n + \cdots \tag{13.73}$$

if $\sigma'(\Phi_0) = \cdots = \sigma^{(n-1)}(\Phi_0) = 0$, $\sigma^{(n)}(\Phi_0) \neq 0$. Note that (13.73) is a complex version of (7.26), describing the singularities of a viscous pinching thread or, even more simply, of (3.19).

As in those problems one can expect that generically $n = 2$, which means that

$$\Phi(\sigma) \sim \Phi_0 + A\left[(t_0 - t) + 2i\mu\right]^{1/2} = \Phi_0 + A t'^{1/2}(1 + is)^{1/2} \tag{13.74}$$

as $t \to t_0$, with A a complex number. As usual we have put $t' = t_0 - t$ and introduced a similarity variable,

$$s = \frac{2\mu}{t'}.$$

Writing (13.74) in terms of real and imaginary parts, using

$$(1 + is)^{1/2} = (1 + s^2)^{1/4}\left(\cos\frac{\arctan s}{2} + i \sin\frac{\arctan s}{2}\right), \tag{13.75}$$

we obtain

$$\Phi = \varphi - iH\psi$$
$$\sim \Phi_0 + At'^{1/2}(1+s^2)^{1/4}\left(\cos\frac{\arctan s}{2} + i\sin\frac{\arctan s}{2}\right).$$

Since the imaginary part of a function which is analytic in the upper half-plane is the Hilbert transform of its real part [110], we conclude that

$$(1+s^2)^{1/4}\sin\frac{\arctan s}{2} = -H\left((1+s^2)^{1/4}\cos\frac{\arctan s}{2}\right),$$

so that

$$\varphi + i\psi \sim \Phi_0 + At'^{1/2}(1+s^2)^{1/4}\cos\frac{\arctan s}{2}. \qquad (13.76)$$

Thus the Cartesian coordinates of the tangent along the sheet are

$$x_\mu = \Re\{\eta\} = \Re\{\Phi_0\},$$
$$y_\mu = \Im\{\eta\} = \Im\{\Phi_0\} + At'^{1/2}(1+s^2)^{1/4}\cos\frac{\arctan s}{2}.$$

Since $y_\mu = 0$ at the singularity, the curvature is $\kappa = -y_{\mu\mu}/x_\mu^2$ for $t' \to 0$, and

$$\kappa = -\frac{A}{\Re\{\Phi_0\}^2}\frac{(1+s^2)^{-1/4}}{t'^{1/2}}\sin\frac{\arctan s}{2}. \qquad (13.77)$$

Locally, μ is proportional to the spatial distance x from the singularity, so it follows from (13.77) that the curvature has the similarity description

$$\kappa(x,t) = (t')^{-1/2}K(\xi), \quad \xi = \frac{x}{t'}. \qquad (13.78)$$

The curvature profile has the universal form

$$K(\xi) = C\frac{\sin\left[(\arctan a\xi)/2\right]}{[(a\xi)^2+1]^{1/4}}, \qquad (13.79)$$

where C and a are constants. Therefore, by rescaling the curvature profile in the x-direction with κ_m^2 and dividing the result by κ_m, we should see convergence to the self-similar solution. In Fig. 13.16 we have rescaled the curvature profiles of Fig. 13.15 in this fashion and very good agreement with the theoretical prediction (13.79) can be seen.

The construction of (13.78), (13.79) relied on the assumption that we could approximate the nonlinear Birkhoff–Rott equation (13.55) by its linearized version (13.67). Now we can verify this assumption by integrating (13.78) over the singularity. From the scaling it is clear that in the limit $t' \to 0$ the slope will experience a jump proportional to $t'^{1/2}$, which goes to zero. This justifies

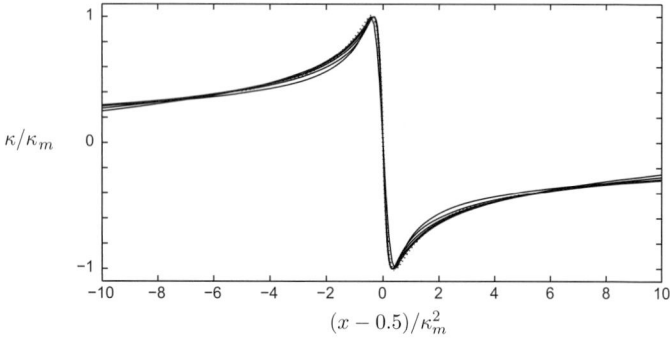

Figure 13.16 Rescaled curvature profiles (see Fig. 13.15) near $x = 0.5$, where Moore's singularity takes place. The theoretical prediction (13.79) is shown as the dotted line.

our use of the linear approximation. More detailed estimates of the difference between linear and nonlinear evolution are given in [59].

Since the evolution problem for vortex sheets is such that at the linear level all modes grow exponentially, one needs to concentrate on a class of initial data such that the Fourier coefficients a_k decay exponentially fast as $|k| \to \infty$, so that one can find a solution for at least a finite period of time. This seems to be a very stringent condition, since it implies that the initial sheet has infinitely many continuous derivatives. Nevertheless, physical effects such as surface tension or viscosity may introduce a cutoff in the spectrum that damps out high frequencies, so that assuming a fast decay of the coefficients a_k is not unreasonable.

We have argued that (13.79) is the generic form of Moore's singularity, since it is not to be expected that several coefficients of the expansion (13.73) vanish at the same time. In principle, however, other singularities are possible. For example, if $n = 3, 4, \ldots$ in (13.73), leaving the details of the calculation as Exercise 13.6 one finds that

$$\kappa(x, t) = (t')^{1/n-1} K(\xi), \quad \xi = \frac{x}{t'}, \tag{13.80}$$

with

$$K(\xi) = C \frac{\sin\left[(\arctan a\xi)/n\right]}{\left((a\xi)^2 + 1\right)^{1/2-1/(2n)}}. \tag{13.81}$$

Of course, singularities associated with holomorphic functions can be even more general than (13.74), and include poles, branch points of any order, and even fractal boundaries [41, 203]. However, the initial data need to have a very specific form to yield these behaviors. This means that it is possible to introduce small perturbations into the initial data to generate the generic form (13.74) of folding of the Riemann sheet, and hence Moore's singularity, in agreement with numerical findings. A similar situation arises for the complex Burgers' equation and more general first-order systems. For the complex Burgers' equation it can be shown [40] that square root singularities similar to those in (13.74) are indeed generic among all possible complex singularities. A very similar example of such a complex map will be discussed in the next chapter, when we study singularities in Hele-Shaw flow.

13.3.3 Continuation of Moore's singularity

To continue Moore's singularity beyond t_0, one has to introduce a regularization. A physical way to do this is to include a small amount of surface tension, whose effect will become significant when the curvature becomes large. We consider *two* immiscible fluids of equal density, with surface tension between them, moving relative to each other. Thus, according to (4.40),

$$\frac{\partial}{\partial t}(\phi_1 - \phi_2) + \left(\frac{\nabla\phi_1 + \nabla\phi_2}{2}\right)\cdot(\nabla\phi_1 - \nabla\phi_2) + \frac{\gamma}{\rho}\kappa = 0, \qquad (13.82)$$

where ϕ_1, ϕ_2 are the velocity potentials in either phase and κ is the interface curvature.

Now, as before let us take α as a Lagrangian (particle) label, so that

$$\dot{\mathbf{x}}(\alpha, t) = \mathbf{v},$$

where $\mathbf{v}$ is the fluid velocity. If we *define* $\mathbf{v}$ as the average velocity $(\nabla\phi_1 + \nabla\phi_2)/2$ across the interface, we obtain from (13.82) that

$$\frac{\partial}{\partial t}[\phi_1(\alpha, t) - \phi_2(\alpha, t)] = \left.\frac{\partial}{\partial t}\right|_{\mathbf{x}}(\phi_1 - \phi_2) + \mathbf{v}\cdot(\nabla\phi_1 - \nabla\phi_2) = -\frac{\gamma}{\rho}\kappa. \quad (13.83)$$

As before, the ambiguity that exists in the definition of the tangential velocity does not matter, as it does not affect the interface motion but only leads to a reparameterization within the interface.

Taking the α-derivative of (13.83), we find on the one hand

$$\frac{\partial^2}{\partial t\partial\alpha}(\phi_1 - \phi_2) + \frac{\gamma}{\rho}\kappa_\alpha = 0.$$

On the other hand, multiplying (13.52) by z_α we obtain

$$\frac{\partial w}{\partial\alpha} = z_\alpha\frac{\partial w}{\partial z} = \mp\frac{\sigma(\alpha, t)}{2} + \frac{\partial\tilde{w}}{\partial\alpha},$$

where the second term on the right is continuous across the interface. Taking the difference across the interface we have

$$\frac{\partial}{\partial \alpha}(w_1 - w_2) = \frac{\partial}{\partial \alpha}(\phi_1 - \phi_2) = \sigma(\alpha, t), \qquad (13.84)$$

since the expression on the right is purely real. Thus, inserting (13.84) into (13.83), we arrive at the following evolution equation for the vortex sheet strength:

$$\frac{\partial \sigma}{\partial t}(\alpha, t) = -\frac{\gamma \kappa_\alpha}{\rho}. \qquad (13.85)$$

Together with (13.54), the relation (13.85) is the equation of motion for the vortex sheet with surface tension.

As before, we consider a linearized version of the vortex sheet equation, putting

$$\sigma(\alpha, t) = \sigma_0 + \epsilon \tilde{\sigma}(\alpha, t), \quad z(\alpha, t) = \alpha + \epsilon \zeta(\alpha, t),$$

and inserting these expressions into (13.54):

$$\frac{\partial \zeta^*(\alpha, t)}{\partial t} = \frac{1}{2\pi i} \mathrm{PV} \int_{-\infty}^{\infty} \frac{\tilde{\sigma}(\alpha', t)d\alpha'}{\alpha - \alpha'} - \frac{\sigma_0}{2\pi i} \mathrm{PV} \int_{-\infty}^{\infty} \frac{\zeta(\alpha, t) - \zeta(\alpha', t)}{(\alpha - \alpha')^2} d\alpha'.$$

Using (13.64), this turns into

$$\frac{\partial \zeta^*}{\partial t} = \frac{1}{2i} H(\tilde{\sigma}) - \frac{\sigma_0}{2i} H(\zeta_\alpha), \qquad (13.86)$$

where H is the Hilbert transform (13.66). If we write

$$\zeta = \varphi + i\psi$$

and linearize around a flat interface, the curvature becomes

$$\kappa \simeq -\psi_{\alpha\alpha}.$$

Then, from (13.85), (13.86) we obtain the following system for φ, ψ, and $\tilde{\sigma}$:

$$\varphi_t = -\frac{\sigma_0}{2} H(\psi_\alpha), \qquad (13.87)$$

$$\psi_t = \frac{1}{2} H(\tilde{\sigma}) - \frac{\sigma_0}{2} H(\varphi_\alpha), \qquad (13.88)$$

$$\tilde{\sigma}_t = \frac{\gamma}{\rho} \psi_{\alpha\alpha\alpha}. \qquad (13.89)$$

Taking the time derivative of (13.88) and inserting into it φ_t and $\tilde{\sigma}_t$, we find

$$\psi_{tt} = \frac{\gamma}{2\rho} H(\psi_{\alpha\alpha\alpha}) - \frac{\sigma_0^2}{4} \psi_{\alpha\alpha}.$$

Decomposing into Fourier modes by writing

$$\psi = e^{\omega t} \left(a_k e^{ik\alpha} + a_k^* e^{-ik\alpha} \right),$$

and using (13.69), one finds the dispersion relation

$$\omega^2(k) = -\frac{\gamma}{2\rho} |k|^3 + \frac{\sigma_0^2}{4} k^2. \tag{13.90}$$

First, note that the term proportional to surface tension resolves the ill-posedness of the original dispersion relation (13.70). In particular, there is no longer an instability for $|k| \to 0$. Surface tension introduces an new length scale, $l = \gamma/(\sigma_0^2 \rho)$, so if one rescales (13.90) according to

$$k' = kl, \quad \omega' = \omega l / \sigma_0, \tag{13.91}$$

the dispersion relation becomes parameter-free. Thus, in a boundary layer of size l around the singularity and a period of time of size $\tau = l/\sigma_0$ around t_0, the effects of surface tension modify the singular solution (13.77) and possibly regularize it. Whatever occurs beyond t_0 will be determined by modifying the universal self-similar solution at the scale l. Hence, one can expect that the solution beyond t_0 will be a universal quantity, when nondimensionalized using l. Specifically, we expect the maximum curvature (i.e. at the tip of the spiral) to scale as

$$\kappa_m = \frac{1}{l} f \left(\frac{|t'| \sigma_0}{l} \right). \tag{13.92}$$

A careful analysis [59] shows that f saturates at some maximum value, which sets the size of the spiral arm (cf. Fig. 13.14). Its size will be of order κ_m^{-1}, i.e. proportional to l. In Fig. 13.17 we represent numerical solutions with two different values of the Weber number $We = \epsilon/l$, where ϵ is a typical length scale of the problem, for instance the initial amplitude of the wave. As predicted by the above scaling argument, the size of the spiral arm that develops during the post-singularity evolution is inversely proportional to We.

13.4 Vortices in the Ginzburg–Landau equation

The fluid vortex (13.1) has a pure power law form, but we now turn to a system whose description is complicated by the presence of an intrinsic length scale ϵ. The Ginzburg–Landau equation for a complex wave field Ψ in D dimensions reads

$$\frac{\partial \Psi}{\partial t} = \epsilon^2 \Delta \Psi + \Psi - |\Psi|^2 \Psi; \tag{13.93}$$

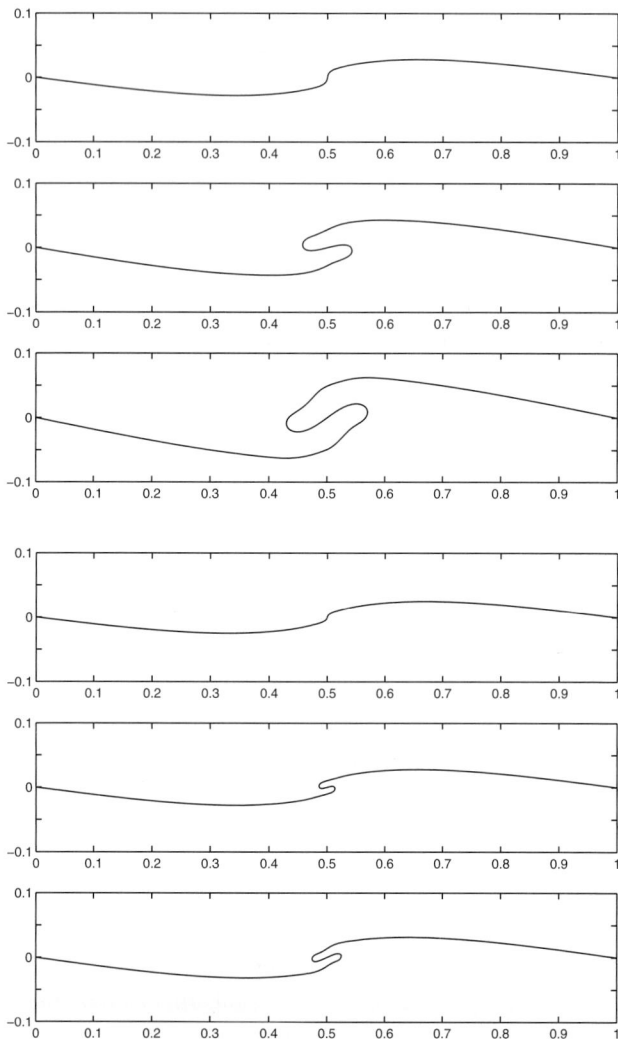

Figure 13.17 Upper sequence: vortex sheet with We = 200 at three different increasing times. In the first frame high values of the curvature develop at the center and an arm develops (second frame) and revolves over itself (third frame). Lower sequence: the same process with We = 400.

the small parameter ϵ is known as the coherence length. The field Ψ represents an order parameter, i.e. a complex number whose modulus describes the transition between two pure phases, one phase corresponding to $|\Psi| = 0$ and the other to $|\Psi| = 1$. In the Ginzburg–Landau theory of superconductors, electron pairs (so-called *Cooper pairs*) form a superfluid and $|\Psi|^2$ represents the fraction of electrons that have condensed into the superfluid phase. The coherence length corresponds to the approximate spatial dimension of the Cooper pairs, which is of order 100 nm. Equivalently, ϵ is the typical length over which the order parameter changes.

For the following, it is instructive to note that (13.93) is the gradient flow of the free energy, which in D dimensions is given by

$$F_\epsilon\left[\Psi\right] = \int \left(\frac{1}{2}|\nabla\Psi|^2 + \frac{1}{4\epsilon^2}(1 - |\Psi|^2)^2\right) d^D x. \tag{13.94}$$

To demonstrate this, we compute the variation of (13.94) under a small perturbation $\delta\Psi$:

$$\begin{aligned} &F_\epsilon\left[\Psi + \delta\Psi\right] \\ &= \int \left(\frac{1}{2}\nabla(\Psi + \delta\Psi) \cdot \nabla(\Psi^* + \delta\Psi^*)\right. \\ &\quad \left. + \frac{1}{4\epsilon^2}(1 - (\Psi + \delta\Psi)(\Psi^* + \delta\Psi^*))^2\right) d^D x \\ &= F_\epsilon\left[\Psi\right] + \int \left(\nabla\Psi \cdot \nabla(\delta\Psi^*) - \frac{1}{\epsilon^2}\delta\Psi^*\Psi(1 - |\Psi|^2)\right) d^D x + O\left(|\delta\Psi|^2\right) \\ &= F_\epsilon\left[\Psi\right] + \int \left(-\Delta\Psi - \frac{1}{\epsilon^2}\Psi(1 - |\Psi|^2)\right)\delta\Psi^* d^D x + O\left(|\delta\Psi|^2\right). \end{aligned}$$

In other words, (13.93) can be written as:

$$-\epsilon^2 \frac{\delta F_\epsilon\left[\Psi\right]}{\delta\Psi^*} = \epsilon^2\Delta\Psi + \Psi(1 - |\Psi|^2) = \frac{\partial\Psi}{\partial t}, \tag{13.95}$$

so the motion is in the direction of the gradient of the free energy. Multiplying (13.95) by the conjugate of $\delta F_\epsilon\left[\Psi\right]/\delta\Psi^*$ and integrating over space, we obtain

$$\frac{dF_\epsilon\left[\Psi\right]}{dt} = -\epsilon^2 \int_{\mathbb{R}^D} \left|\frac{\delta F_\epsilon\left[\Psi\right]}{\delta\Psi^*}\right|^2 d^D x.$$

This means that the Ginzburg–Landau equation is purely dissipative: the free energy can only decrease. Stationary solutions of the equation correspond to minimizers of the free energy.

13.4.1 Structure of stationary vortices

We now specialize to solutions in the plane ($D = 2$), where points with $|\Psi| = 0$ correspond to vortices and thus to phase singularities; see Fig. 13.2 (right). During the fast evolution of a system from a completely disordered state to an ordered state (with $|\Psi| = 1$), the local incompatibilities of different orderings

Table 13.1 *The first five constants a_n which determine the asymptotic behavior (13.99) of $\rho(\xi)$, as computed in [3].*

n	a_n
1	0.583 189 495 860 3
2	0.153 099 102 859 5
3	0.026 183 420 716 8
4	0.003 327 173 400 7
5	0.000 336 593 940 9

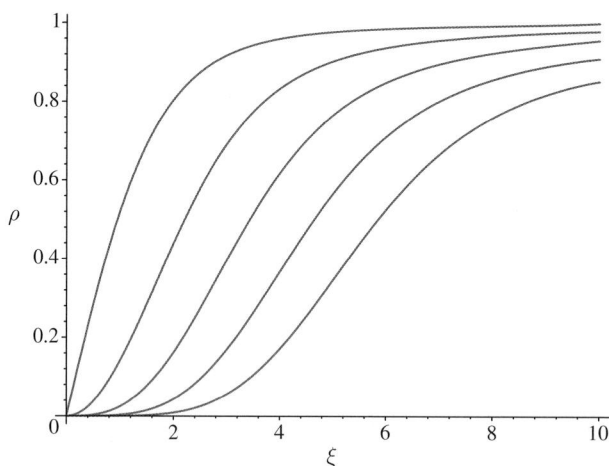

Figure 13.18 The shape of the vortex core for $n = 1, \ldots, 5$ (from left to right), as determined numerically from a solution of (13.97) using the constants a_n quoted in Table 13.1. It can be shown [3] that solutions are ordered in such a way that $\rho_n(\xi) > \rho_{n+1}(\xi)$ for all ξ.

create vortices. We will see that the vortices contain most of the free energy of the system.

A static vortex has a local structure that can be described, using polar coordinates in two dimensions, in the similarity form

$$\Psi(r, \theta) = \rho(\xi)e^{in\theta}, \quad \xi = \frac{r}{\epsilon}, \tag{13.96}$$

which has the same structure as (13.2). Therefore, by integrating the phase along a closed loop around the vortex core one obtains the value $2n\pi$, so that a quantized *topological charge* (also called the *degree* or *winding number*) can be assigned to each of the solutions (13.96).

The substitution of (13.96) into (13.93) leads to the following equation for $\rho(\xi)$:

$$\frac{1}{\xi}\left(\xi\rho'\right)' - \frac{n^2}{\xi^2}\rho + \rho - \rho^3 = 0, \tag{13.97}$$

to be satisfied along with the condition that the vortex connects a disordered state ($|\Psi| = 0$) at $\xi = 0$ with an ordered state ($|\Psi| = 1$) as $\xi \to \infty$:

$$\rho(0) = 0, \quad \lim_{\xi\to\infty} \rho(\xi) = 1. \tag{13.98}$$

Equation (13.97) with the conditions (13.98) is known to have a unique solution for each n [46, 109]. Each solution is monotonic and has asymptotics

$$\rho(\xi) \underset{\xi\to 0}{\sim} a_n \xi^n + O(\xi^{n+2}), \quad \rho(\xi) \underset{\xi\to\infty}{\sim} 1 - \frac{n^2}{2\xi^2} + O(\xi^{-4}), \tag{13.99}$$

which depend on the winding number n. The constant a_n has to be determined numerically [3].

Inserting (13.96) into (13.94), the free energy of a vortex contained in a ball of radius R is

$$F_\epsilon = 2\pi \int_0^{R/\epsilon} \left[\frac{1}{2}\left(\rho'^2 + \frac{n^2}{\xi^2}\rho^2\right) + \frac{1}{4}(1-\rho^2)^2 \right] \xi\, d\xi, \tag{13.100}$$

where we have used

$$|\nabla\Psi|^2 = \frac{1}{\epsilon^2}\left(\rho'^2 + \frac{n^2}{\xi^2}\right).$$

Examining the asymptotics (13.99), one observes that the second term in the integrand of (13.100) produces a logarithmically diverging contribution for $\epsilon \to 0$. Hence we can use the asymptotics (13.99) for $\xi \to \infty$ and neglect contributions from the origin:

$$F_\epsilon \simeq \pi \int_1^{R/\epsilon} \frac{n^2}{\xi}\rho^2 d\xi \simeq \pi n^2 \ln\frac{R}{\epsilon}. \tag{13.101}$$

The logarithmically diverging contribution (13.101) coming from the neighborhood of each vortex is called the *self-energy*.

Naively, one would think that a minimizer would simply be $\Psi = 1$. But, as we mentioned above, there are constraints of a topological nature that may force the presence of vortices: if the problem is posed in a bounded domain Ω then one needs to impose the boundary condition

$$\Psi = g \quad \text{on} \quad \partial\Omega.$$

Thus, the degree or winding number of g (the integral over the phase along the boundary) must equal the sum of the degrees of the vortices in Ω.

13.4.2 The renormalized energy

A difficulty of dealing with the free energy arises from the fact that the self-energy of a single vortex diverges logarithmically. Thus, in order to capture the interaction energy W between vortices, one has to separate the self-energy from the renormalized energy W [25], so that

$$F_\epsilon [\Psi] = \pi \left(\sum_{i=1}^{N} n_i^2 \right) \ln \frac{1}{\epsilon} + W(a_1, a_2, \ldots, a_N) + O(\epsilon), \qquad (13.102)$$

where N is the number of vortices in Ω, n_i is the degree of the ith vortex, and a_i is a complex number representing its location. Notice that minimizing the logarithmic contribution to the energy requires that either all vortices have degree $+1$ or all have degree -1. Hence if deg $g > 0$ then there will be $k = \deg g$ vortices of degree $+1$ and then there will be if deg $g < 0$ $|k|$ vortices of degree -1. We will now show that the renormalized energy has the general form

$$W(a_1, a_2, \ldots, a_k) = -\pi \sum_{i \neq j} n_i n_j \ln |a_i - a_j| + B(g, \partial\Omega), \qquad (13.103)$$

where B contains terms depending on the particular form of the boundary. For solutions extending to the whole plane, $\Omega = \mathbb{R}^2$, only the first term on the right-hand side plays a role in the selection of vortex configurations with a minimal free energy.

To proceed, we use a generalization of the similarity solution (13.96) which allows for an arbitrary phase φ as a function of the angle:

$$\Psi(r, \theta) = \rho \left(\frac{r}{\epsilon} \right) e^{i\varphi(r,\theta)}. \qquad (13.104)$$

We will consider superpositions of such vortices, each centered at a point in the plane (r being in each case the distance to the center of the vortex), so that the distance d between any pair of these points is $d \gg \epsilon$. Our goal is to determine the locations of the centers of each vortex and the phase φ such that the free energy (13.94) is minimized.

Inserting (13.104) into (13.94), one obtains the free energy:

$$F_\epsilon [\Psi] = \int_\Omega \left(\frac{1}{2} |\nabla\rho|^2 + \frac{1}{2} \rho^2 |\nabla\varphi|^2 + \frac{1}{4\epsilon^2} (1 - \rho^2)^2 \right) d^2x. \qquad (13.105)$$

Performing variations with respect to ρ and ϕ in exact analogy to the calculation leading to (13.95), one arrives at the following Euler–Lagrange system:

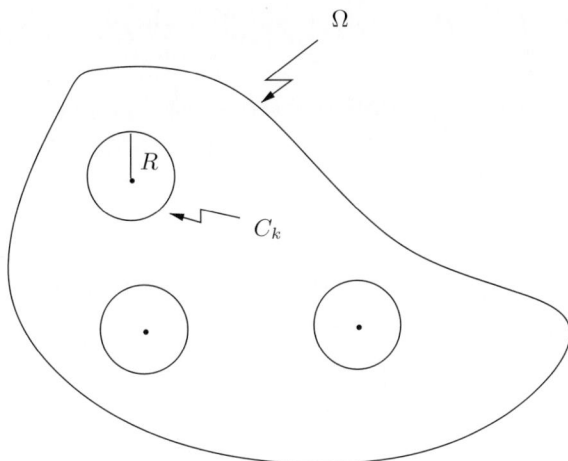

Figure 13.19 Vortices in a domain Ω. The free energy is separated into the self-energy contribution (13.101) coming from a disk of radius R around each vortex and the contribution coming from the exterior of the disks.

$$\Delta\rho + (1 - |\nabla\varphi|^2 - \rho^2)\rho = 0, \qquad (13.106)$$

$$\Delta\varphi + \frac{2}{\rho}\nabla\rho \cdot \nabla\varphi = 0. \qquad (13.107)$$

We now verify (13.103) by decomposing the energy (13.105) into an integral over the union of disks C_k of radius R around each vortex k, and the integral over their exterior (see Fig. 13.19). The first contribution F_{int} is the self-energy and the second contribution F_{ext} will produce the renormalized energy W. The interior contribution is

$$
\begin{aligned}
F_{\text{int}} &= \int_{\cup C_k} \left(\frac{1}{2}|\nabla\rho|^2 + \frac{1}{2}\rho^2 |\nabla\varphi|^2 + \frac{1}{4\epsilon^2}(1 - \rho^2)^2 \right) d^2x \\
&= \sum_k \int_{C_k} \left(\frac{1}{2}|\nabla\rho|^2 + \frac{1}{2}\rho^2 |\nabla\varphi|^2 + \frac{1}{4\epsilon^2}(1 - \rho^2)^2 \right) d^2x \\
&\simeq \sum_k \pi n_k^2 \ln \frac{R}{\epsilon}.
\end{aligned}
$$

The last line follows since the vortices are well separated and R is chosen to be small compared with the distance between vortices. This is the sum of the self-energies (13.101).

The domain outside each C_k is $\Omega \setminus \cup C_k$; see Fig. 13.19. According to (13.99), in this domain we have $\rho \sim 1 + O(R^{-2})$ and thus

$$F_{\text{ext}} = \int_{\Omega \setminus \cup C_k} \left(\frac{1}{2} |\nabla \rho|^2 + \frac{1}{2} \rho^2 |\nabla \varphi|^2 + \frac{1}{4\epsilon^2} (1 - \rho^2)^2 \right) d^2 x$$

$$\simeq \int_{\Omega \setminus \cup C_k} \frac{1}{2} \rho^2 |\nabla \varphi|^2 \, d^2 x \simeq \int_{\Omega \setminus \cup C_k} \frac{1}{2} |\nabla \varphi|^2 \, d^2 x. \qquad (13.108)$$

Moreover, (13.107) reduces to

$$\Delta \varphi = 0$$

in the exterior of $\cup C_k$, so that, by linearity, the phase φ is the sum of the phases of each individual vortex. This means that, at a point (x, y),

$$\varphi = \sum_k \theta_k = \sum_k n_k \Re\{-i \ln(z - z_k)\},$$

where $z = x + iy$. Notice that φ is the real part of a complex potential

$$\Phi = \varphi + i\psi = -i \sum_j n_j \ln(z - a_j). \qquad (13.109)$$

Since φ is harmonic, we deduce the identity

$$\nabla \cdot [(\nabla \varphi)\varphi] = |\nabla \varphi|^2,$$

and so from Gauss' theorem, (13.108) can be rewritten as

$$F_{\text{ext}} \simeq \frac{1}{2} \int_{\cup_k \partial C_k \cup \partial \Omega} \varphi \nabla \varphi \cdot \mathbf{n} ds, \qquad (13.110)$$

where the domain is the boundary of all the C_k as well as of Ω. The Cauchy–Riemann equations for the complex potential Φ are $\varphi_x = -\psi_y$ and $\psi_x = \varphi_y$, while the components of the tangent vector are $\mathbf{t} = (t_1, t_2) = (-n_2, n_1)$. Thus

$$\nabla \varphi \cdot \mathbf{n} = \varphi_x n_1 + \varphi_y n_2 = -\psi_x t_1 + \psi_y t_2 = -\nabla \psi \cdot \mathbf{t},$$

which we can use in (13.110). The integral over a closed curve vanishes,

$$\oint \nabla \cdot (\psi \varphi) \mathbf{t} ds = 0,$$

and so we can integrate by parts to find

$$F_{\text{ext}} \simeq -\frac{1}{2} \int_{\cup_k \partial C_k \cup \partial \Omega} \varphi \nabla \psi \cdot \mathbf{t} ds = \frac{1}{2} \int_{\cup_k \partial C_k \cup \partial \Omega} \psi \nabla \varphi \cdot d\mathbf{l}. \qquad (13.111)$$

The complex potential on the boundary of C_k, given by $a_k + Re^{i\theta}$, is

$$\Phi = -i \sum_j n_j \ln \left(a_k - a_j + Re^{i\theta} \right)$$

$$\approx -i \sum_{j \neq k} n_j \ln \left(a_k - a_j \right) - i n_k \ln R + n_k \theta, \qquad (13.112)$$

assuming that R is chosen small compared with the distance between vortices. But this means that the imaginary part ψ of (13.112) on ∂C_k assumes the constant value

$$\psi_k = -\sum_{j \neq k} n_j \ln|a_k - a_j|,$$

disregarding $\ln R$, which does not depend on vortex position. From the real part of (13.112) we find $\varphi_k = n_k \theta$, and so

$$\nabla \varphi_k = \frac{n_k}{R} \mathbf{e}_\theta.$$

We can split (13.111) into contributions from the different boundaries:

$$F_{\text{ext}} \simeq \frac{1}{2} \sum_k \psi_k \int_{\partial C_k} \nabla \varphi_k \cdot d\mathbf{l} + \frac{1}{2} \int_{\partial \Omega} \psi \nabla \varphi \cdot d\mathbf{l}, \qquad (13.113)$$

where the second contribution comes from the interaction of the vortices with the boundary of Ω. If this boundary is sufficiently far from the vortices, say at a distance L, then the collection of vortices is viewed by the points on the boundary as a single vortex of degree $N = \sum_k n_k$, so that

$$\frac{1}{2} \int_{\partial \Omega} \psi \nabla \varphi \cdot d\mathbf{l} \simeq -\pi N^2 \ln L$$

provides a constant contribution to the free energy. The first integral on the right-hand side of (13.113) is then the interaction energy W as given by (13.103), using

$$\int_{\partial C_k} \nabla \varphi_k \cdot d\mathbf{l} = 2\pi n_k.$$

For a disk of unit radius, vortices of degree $+1$, and $g = e^{ik\theta}$, Ignat *et al.* [117] deduced the exact formula

$$W(a_1, a_2, \ldots, a_k) = -\pi \sum_{i < j} \ln|a_i - a_j| - \pi \sum_{i,j} \ln\left|1 - a_i a_j^*\right|$$

and found that, for $k = 2$, the minimizers consist of two vortices located symmetrically with respect to the center and for $k = 3$ they consist of equilateral triangles of vortices.

13.4.3 Dynamics of Ginzburg–Landau vortices

In large domains it may take a long time for a set of vortices to settle in an equilibrium configuration. If the domain is infinite, it can be the case that such an equilibrium configuration does not exist and the vortices keep moving until they collide and annihilate (provided that they have opposite degrees or winding numbers) or dissipate. Under these circumstances, it is important to describe their dynamics by establishing their mutual-interaction rules. An essential assumption is that vortices are well separated entities, i.e. that two vortices are separated by a distance at least $O(1)$ compared with their size ϵ. Then the vortices will keep their shape and the evolution merely consists of the movement of their centers with velocities

$$v_i = \frac{da_i}{dt}.$$

In order to find a dynamical law relating the velocity of each vortex to the locations of all the vortices, we use the gradient flow (13.95), with F_ϵ approximated by (13.102). An infinitesimal displacement of the vortex locations from $(a_1, a_2, \ldots, a_N)$ to the new locations $(a_1, a_2, \ldots, a_N) + (\delta a_1, \delta a_2, \ldots, \delta a_N)$ produces a change in energy

$$\delta F_\epsilon [\Psi] = \sum_i \nabla_{a_i} W(a_1, a_2, \ldots, a_N) \cdot \delta a_i + O(\epsilon). \qquad (13.114)$$

For ease of notation, in this subsection inner products (indicated by a dot) are written as products of *real* vectors. Thus the complex number δa_i in (13.114) is to be interpreted as a two-dimensional real vector. Furthermore, the gradient $\nabla_{a_i} W$ could be interpreted as the complex derivative of W with respect to a_i.

Since the vortices are well separated, the variation of the wave function under a shift δa_i of the vortex positions is

$$\delta \Psi = \sum_i \nabla \Psi_i \cdot \delta a_i, \qquad (13.115)$$

where Ψ_i denotes the local solution (13.96) centered at a_i.

Multiplying the equation of motion (13.95) by $\delta \Psi^*$, and integrating over space, we obtain

$$\left\langle \frac{\partial \Psi}{\partial t}, \delta \Psi \right\rangle = \epsilon^2 \delta F_\epsilon [\Psi], \qquad (13.116)$$

where we define the angle brackets as follows:

$$\langle A, B \rangle = \int A B^* d^2 x. \qquad (13.117)$$

Equation (13.116) is thus

$$\left\langle \frac{\partial \Psi}{\partial t}, \delta \Psi \right\rangle = \int \left(\sum_i v_i \cdot \nabla \Psi_i \right) \left(\sum_j \nabla \Psi_j^* \cdot \delta a_j \right) d^2 x,$$

using (13.115). The same calculation as in (13.101) (integrating around a vortex core from ϵ to R) gives

$$\int \nabla \Psi_i \cdot \nabla \Psi_j^* d^2 x = 2\pi \delta_{ij} n_i^2 \ln(1/\epsilon) + O(1).$$

The result is zero if $i \neq j$ since the vortices do not overlap. Hence we conclude that

$$\left\langle \frac{\partial \Psi}{\partial t}, \delta \Psi \right\rangle = \sum_i 2\pi n_i^2 \ln(1/\epsilon) v_i \cdot \delta a_i + O(1). \tag{13.118}$$

Inserting (13.118) and (13.114) into (13.116), we obtain

$$\frac{|\ln \epsilon|}{\epsilon^2} v_i = -\frac{1}{2\pi n_i^2} \nabla_{a_i} W(a_1, a_2, \dots, a_N), \tag{13.119}$$

since the position of each vortex can be varied independently.

For the whole plane we have

$$W(a_1, a_2, \dots, a_N) = -\pi \sum_{j \neq k} n_j n_k \ln |a_j - a_k|$$

and thus

$$\nabla_{a_i} W = -\pi \sum_{j \neq k} n_j n_k \left(\frac{\delta_{ji}(a_j - a_k)}{|a_j - a_k|^2} + \frac{\delta_{ki}(a_k - a_j)}{|a_j - a_k|^2} \right)$$

$$= -2\pi \sum_j n_i n_j \frac{a_i - a_j}{|a_i - a_j|^2}. \tag{13.120}$$

It follows that the motion of each vortex is governed by

$$\frac{da_i}{dT} = \frac{1}{n_i} \sum_{j \neq i} \frac{n_j(a_i - a_j)}{|a_i - a_j|^2}, \tag{13.121}$$

with a new time scale

$$T = \frac{\epsilon^2}{|\ln \epsilon|} t.$$

Example 13.5 (Motion of Ginzburg–Landau vortices) If we consider only two vortices, located at a_1 and a_2, we obtain from (13.121)

$$\frac{d(a_1 - a_2)}{dT} = \left(\frac{n_2}{n_1} + \frac{n_1}{n_2} \right) \frac{a_1 - a_2}{|a_1 - a_2|^2} = \left(\frac{n_1^2 + n_2^2}{n_1 n_2} \right) \frac{a_1 - a_2}{|a_1 - a_2|^2}. \tag{13.122}$$

Multiplying (13.122) by $(a_1 - a_2)^*$ and its complex conjugate by $a_1 - a_2$, one finds

$$\frac{d\,|a_1 - a_2|^2}{dT} = 2\left(\frac{n_1^2 + n_2^2}{n_1 n_2}\right).$$

Integrating, we obtain

$$|a_1 - a_2|^2(T) = |a_1 - a_2|^2(0) + 2\left(\frac{n_1^2 + n_2^2}{n_1 n_2}\right)T. \tag{13.123}$$

Therefore two vortices whose degree is of opposite sign attract, while those having the same sign repel. The mathematical proof of this result has been the subject of several studies (see [142], [190]). Matched asymptotic expansion techniques lead to an identical result ([160], [171]), while corrections to the velocity that are higher order in $\mu = 1/|\ln \epsilon|$ can be obtained as well [4]. Note that the behavior of Ginzburg–Landau vortices is markedly different from the dynamics of vortices in an inviscid fluid. Namely, (13.6), (13.7) form a Hamiltonian system while the Ginzburg–Landau equation is dissipative. However, a seemingly small modification turns (13.93) into a Hamiltonian problem, as we will see now. $\qquad\square$

13.5 Nonlinear Schrödinger equation

The nonlinear Schrödinger equation

$$i\frac{\partial \Psi}{\partial t} = \epsilon^2 \Delta \Psi + \Psi - |\Psi|^2\, \Psi \tag{13.124}$$

differs from the Ginzburg–Landau equation in that the left-hand side has as a factor the imaginary unit. It also admits vortices as stationary solutions, but their interaction and motion is radically different. We saw that the Ginzburg–Landau equation can be written as a gradient flow and that free energy is dissipated. Equation (13.124), however, is a Hamiltonian PDE, with Hamiltonian

$$H(\Psi) = \epsilon^2 \int \left(\frac{1}{2}|\nabla \Psi|^2 + \frac{1}{4\epsilon^2}(1 - |\Psi|^2)^2\right)d^2x; \tag{13.125}$$

we note that the equation of motion (13.124) can be written as

$$i\frac{\partial \Psi}{\partial t} = -\frac{\delta H(\Psi)}{\delta \Psi^*}; \tag{13.126}$$

the evolution is such that the energy $H(\Psi)$ is conserved. The energy (13.125) is the same as (13.105) for the Ginzburg–Landau problem, and so the vortex solutions are the same. However, while vortices in the Ginzburg–Landau equation tend to an equilibrium, vortices in the nonlinear Schrödinger equation keep moving and thus do not settle into any equilibrium configuration.

To derive the equation of motion we can proceed along the same lines as for the Ginzburg–Landau problem. Considering variations $(\delta a_1, \delta a_2, \ldots, \delta a_N)$ in the vortex positions, from (13.126) we obtain

$$\left\langle i\frac{\partial \Psi}{\partial t}, \delta\Psi \right\rangle = \sum_i \nabla_{a_i} W(a_1, a_2, \ldots, a_N) \cdot \delta a_i. \tag{13.127}$$

Assuming as before that the vortices are well separated, the left-hand side of (13.127) becomes

$$\int i\left(\sum_i v_i \cdot \nabla\Psi_i\right)\left(\sum_j \nabla\Psi_j^* \cdot \delta a_j\right) d^2x$$

$$= \sum_i \left(\int i\nabla^\perp\Psi_i \cdot \nabla\Psi_i^* d^2x\right) v_i^\perp \delta a_i,$$

where we have used $v_i \cdot \nabla = v_i^\perp \cdot \nabla^\perp$, with $(a,b)^\perp = (-b, a)$; in complex notation, $(a,b)^\perp$ is the same as multiplication with the imaginary unit or a counterclockwise rotation by $90°$. Inserting the vortex solution (13.96) and transforming to polar coordinates, we obtain

$$i\nabla^\perp\Psi_i \cdot \nabla\Psi_i^* = 2\Im\left\{\frac{\partial\Psi_i^*}{\partial x}\frac{\partial\Psi_i}{\partial y}\right\} = 2n_i\frac{\rho\rho'}{\xi\epsilon^2},$$

and thus finally

$$\int i\nabla^\perp\Psi_i \cdot \nabla\Psi_i^* d^2x = 2\pi n_i \int_0^\infty \left(\rho^2\right)' d\xi = 2\pi n_i,$$

in view of the boundary conditions (13.98).

Thus, using (13.120) for the right-hand side, (13.127) becomes the following evolution equation for the vortex centers:

$$\frac{da_i^\perp}{dT} = -\sum_{j\neq i} \frac{n_j(a_i - a_j)}{|a_i - a_j|^2}, \tag{13.128}$$

with $T = \epsilon^2 t$.

Example 13.6 (Motion of vortices in a Bose–Einstein condensate) Note that the evolution given by (13.128) is identical to the evolution (13.6), (13.7) of vortices in Euler flow, but with *quantized* vortex line strengths $\Gamma_i = 2\pi n_i$. If we consider two vortices located at a_1 and a_2, respectively, then

$$\frac{da_1^\perp}{dT} = -\frac{n_2(a_1 - a_2)}{|a_1 - a_2|^2}, \quad \frac{da_2^\perp}{dT} = -\frac{n_1(a_2 - a_1)}{|a_1 - a_2|^2}.$$

By subtracting these equations we get

$$\frac{d(a_1 - a_2)}{dT} = i(n_1 - n_2)\frac{a_1 - a_2}{|a_1 - a_2|^2}, \qquad (13.129)$$

switching to complex notation. The solution of (13.129) is

$$a_1 - a_2 = 2Re^{i\Omega T},$$

which describes the counterclockwise rotation of the two vortices around each other at angular frequency Ω; see Fig. 13.5. The angular frequency is

$$\Omega = \frac{n_1 - n_2}{4R^2},$$

in agreement with (13.10) for two vortices of equal and opposite circulation. □

Exercises

13.1 Consider a system of N vortices at the vertices of a regular polygon. Show that the polygon rotates, without change of shape, with angular velocity

$$\Omega = \Gamma\frac{(N-1)}{4\pi R^2},$$

where Γ is the strength of each vortex and R is the radius of the circle on which the vortices lie. It was shown in [104] that for $N \leq 6$ such an arrangement of vortices is stable.
   (i) Write down the positions of the vortices in complex notation.
   (ii) Show that the total complex potential of the vortices is

$$w(z) = \frac{i\Gamma}{2\pi}\ln\left(z^N - R^N\right).$$

   (iii) Evaluate the velocity at the position of the vortex at $z = R$, subtracting the singular contribution from the vortex itself.
13.2 Calculate the convection speed U for a vortex street which is not staggered, i.e. whose vortices are located at

$$(x_m, y_m) = (ma, 0), \quad (x'_m, y'_m) = (ma, -b)$$

and where the vortices of the upper row are rotating clockwise, those of the lower row anticlockwise.

13.3 (*) Consider the stability of the von Kármán vortex street, using perturbations of the form (13.15), (13.14) [153]. It is sufficient to consider the motion of the vortex $m = 0$ near the origin, whose complex velocity is

$$-U + \frac{\bar{z}_0}{dt};$$

here U is given by (13.13).

(i) Show that the complex velocity at the position z_0 of the vortex $m = 0$ is

$$\frac{i\Gamma}{2\pi} \sum_{m=1}^{\infty} \left(\frac{1}{z_0 - z_m - ma} + \frac{1}{z_0 - z_{-m} + ma} \right)$$

$$- \frac{i\Gamma}{2\pi} \sum_{n=0}^{\infty} \left(\frac{1}{z_0 - z'_n - \left(m + \frac{1}{2}\right)a + ib} \right.$$

$$\left. + \frac{1}{z_0 - z'_{-n-1} + \left(m + \frac{1}{2}\right)a + ib} \right).$$

(ii) Using the decomposition (13.14), expand for small values of γ and γ' to show that

$$\frac{d\bar{\gamma}}{dt} = -\frac{i\Gamma}{\pi a^2} \left(A\gamma + C\gamma' \right), \qquad (13.130)$$

and, by symmetry,

$$\frac{d\bar{\gamma}'}{dt} = \frac{i\Gamma}{\pi a^2} \left(A\gamma' + C\gamma \right), \qquad (13.131)$$

where

$$A = \sum_{m=1}^{\infty} \frac{1 - \cos m\theta}{m^2} - \frac{\left(n + \frac{1}{2}\right)^2 - r^2}{\left[\left(n + \frac{1}{2}\right)^2 + r^2\right]^2}$$

and

$$C = \sum_{n=0}^{\infty} \frac{\left[\left(n + \frac{1}{2}\right)^2 - r^2\right] \cos\left(n + \frac{1}{2}\right)\theta}{\left[\left(n + \frac{1}{2}\right)^2 + r^2\right]^2}.$$

Argue that, for the mean velocity to cancel, the identity

$$\sum_{n=0}^{\infty} \frac{1}{(n + \frac{1}{2})^2 + r^2} = \frac{\pi}{2r} \tanh r\pi \qquad (13.132)$$

must hold.

(iii) Using (13.130) and (13.131) show that the vortex configuration is unstable if $A^2 > C^2$; otherwise it is stable.

(iv) Use the particular mode $\theta = \pi$ to demonstrate that $A = 0$ is a necessary condition for stability. Take the derivative of (13.132) with respect to r, and use the Fourier expansion of $\theta(2\pi - \theta)$ to do the necessary sums. Thus show that (13.16) must be satisfied.

13.4 A vortex patch consists of a bounded region $\Omega(t)$ in the plane that has constant vorticity ω_0 at any time. By introducing the stream function $\psi(x, y, t)$ such that $\mathbf{v} = (-\psi_y, \psi_x)$, it follows from the definition of vorticity that ψ satisfies

$$\Delta\psi = \omega_0, \quad x \in \Omega(t),$$
$$\Delta\psi = 0, \quad x \notin \Omega(t).$$

(i) Using the Biot–Savart law show that the contour $z(\alpha, t)$ of $\Omega(t)$ satisfies the following integro-differential equation:

$$\frac{dz(\alpha, t)}{dt} = -\frac{\omega_0}{2\pi} \int_0^{2\pi} \ln|z(\alpha, t) - z(\alpha', t)| \, z_\alpha(\alpha', t) d\alpha',$$

where α is an arbitrary parameterization of the contour.

(ii) Study the stability of a circular vortex patch.

13.5 (*) A Kirchhoff vortex is an elliptical vortex patch with vorticity ω_0 and semiaxes a and b [131]:

$$\frac{x^2}{a^2} + \frac{y^2}{b^2} = 1.$$

Show that the elliptical shape is maintained during the evolution but that the ellipse rotates with angular velocity

$$\Omega = \frac{ab}{(a + b)^2}\omega_0.$$

To demonstrate this result, consider the velocity field $\mathbf{v} = (u, v)$ given by

$$(u, v) = \frac{\omega_0}{(a + b)} (-ay, bx)$$

inside the ellipse, and

$$(u, v) = \frac{\omega_0\, ab}{\sqrt{a^2 + \lambda} + \sqrt{b^2 + \lambda}} \left(-\sqrt{b^2 + \lambda},\, \sqrt{a^2 + \lambda}\right)$$

outside the ellipse, where λ is such that

$$\frac{x^2}{a^2 + \lambda} + \frac{y^2}{b^2 + \lambda} = 1.$$

Show that $\mathbf{v}$ is incompressible, is continuous across the boundary of the ellipse, and has a constant vorticity ω_0 inside the ellipse. From the velocity field at the boundary of the ellipse deduce that it rotates with the angular velocity Ω.

13.6 Consider a singularity of higher order in (13.73), with $n > 2$ an integer. Show that this corresponds to curvature singularities of the form (13.80), (13.81).

13.7 The Kaden spiral originates from a semi-infinite vortex sheet with initial condition

$$z(\mu, 0) = \left(\frac{\mu}{2}\right)^2, \quad 0 \le \mu < \infty.$$

(i) Show that the self-similar solution

$$z(\mu, t) = t^{2/3} \varphi\left(\frac{\mu}{t^{1/3}}\right)$$

for a vortex sheet represents a spiral developing at the edge of an initially flat vortex sheet whose center moves following the trajectory $z_c = \varphi_0 t^{2/3}$. The asymptotic conditions on φ are $\varphi(\xi) \sim \xi^2/4$ as $\xi \to \infty$, and $\varphi(\xi) \to \varphi_0$ as $\xi \to 0$.

(ii) Derive the equation to be satisfied by $\varphi(\xi)$.

13.8 Consider the generalized Ginzburg–Landau equation

$$\frac{\partial \Psi}{\partial t} = \epsilon^2 \Delta \Psi + \Psi - |\Psi|^p\, \Psi \qquad (13.133)$$

with $p > 1$.

(i) Derive the corresponding free energy.

(ii) Derive the similarity equation for the vortex core and its asymptotics.

(iii) Show that the leading-order contribution to the free energy of a vortex with degree (winding number) n is

$$F_\epsilon \simeq -\frac{2\pi}{p} n^2 \ln \epsilon.$$

14

Cusps and caustics

In Chapter 1 (see Fig. 1.5), we discussed the remarkable similarity that exists between the cusp singularities that form on the surface of a viscous liquid and the cusps formed by caustic lines in optics. We will now examine the common geometrical structure that lies behind this universality.

14.1 Viscous free surface cusps

Let us analyze the viscous flow equations near a cusp, in the spirit of the local analysis of Chapter 5 [124]. We will treat the stationary viscous flow equation (4.48) and solve for possible shapes of the free surface, which we parameterize as $h(y)$ near the cusp; see Fig. 14.1. A crucial simplification comes from the insight that from a distance the two surfaces of the cusp become almost parallel. This means that on a large scale the cusp may be modeled as a *cut* in an infinite two-dimensional domain, as shown on the right of Fig. 14.1. As a result, the viscous flow equations can now be solved in a known domain. The boundary condition is one of vanishing stress on the free surface, i.e. along the positive y-axis.

The geometry of the cusp is that of a *crack* that has opened up in the fluid. Not only that, however; the viscous flow equation (4.49) governing elastic deformations in a two-dimensional solid is the same as (4.65). As to the boundary conditions, they are also those of vanishing stress on the crack surface. We will return to the crack problem in more detail in the next chapter. The geometry of Fig. 14.1 (right) is that of a corner flow with opening angle $\alpha = 2\pi$. This means that we are seeking solutions of the form (5.47), and in polar coordinates the velocity field is given by

$$v_r = \frac{1}{r}\frac{\partial \psi}{\partial \theta}, \quad v_\theta = -\frac{\partial \psi}{\partial r}, \tag{14.1}$$

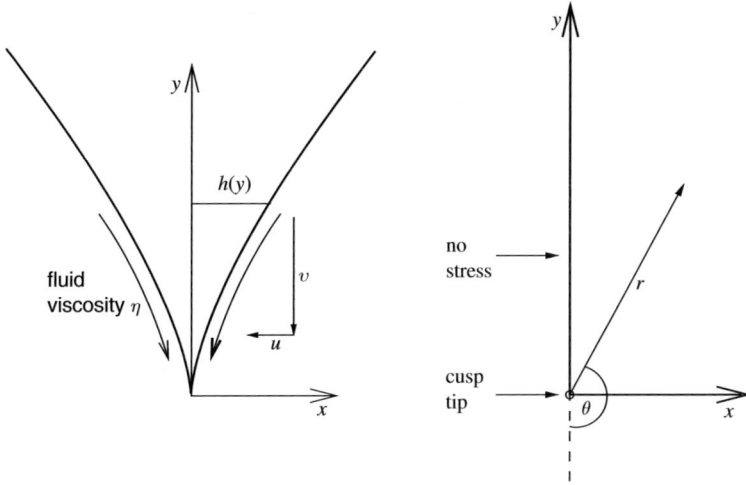

Figure 14.1 Left: a cusp singularity in a viscous fluid and the flow around the free surface. Right: a coarse-grained version of the same problem, that of flow in a slit plane. The boundary condition on the upper half of the y-axis is one of no stress.

where ψ is the stream function. We will restrict ourselves to solutions symmetric with respect to the plane of the cusp, so that (5.54) simplifies to

$$\psi = r^\lambda \left(A \sin \lambda\theta + C \sin (\lambda - 2)\theta \right). \tag{14.2}$$

The exponent λ is determined from a solvability condition, so this is an example of self-similarity of the second kind. In Fig. 14.1 we have chosen the cusp to lie along $\theta = \pi$, so that (14.2) obeys automatically the boundary condition $v_\theta = 0$ on the line of symmetry $\theta = 0$. By contrast, the velocity perpendicular to the face of the crack is nonzero. Indeed, in view of the near field picture Fig. 14.1 (left), our goal is to calculate the velocity field v_θ pointing inwards toward the cusp, from which we can compute the shape of the free surface.

Along the cusp, i.e. for $\theta = \pi$, we impose boundary conditions of zero shear and zero normal stress, which translate into (see Appendix A)

$$\sigma_{r\theta} = \eta \left(\frac{1}{r} \frac{\partial v_r}{\partial \theta} + \frac{\partial v_\theta}{\partial r} - \frac{v_\theta}{r} \right) = 0, \quad \sigma_{\theta\theta} = 2\eta \left(\frac{1}{r} \frac{\partial v_\theta}{\partial \theta} + \frac{v_r}{r} \right) - p = 0,$$
$$\tag{14.3}$$

respectively. The pressure can be eliminated from (14.3) by taking the r-derivative of the second equation and using the r-component of the Stokes equation,

$$\frac{\partial p}{\partial r} = \eta \left(\frac{\partial^2 v_r}{\partial r^2} + \frac{1}{r^2} \frac{\partial^2 v_r}{\partial \theta^2} + \frac{1}{r} \frac{\partial v_r}{\partial r} - \frac{2}{r^2} \frac{\partial v_\theta}{\partial \theta} - \frac{v_r}{r^2} \right), \tag{14.4}$$

so that, in terms of the stream function, equations (14.3) become

$$\frac{1}{r^2} \frac{\partial^2 \psi}{\partial \theta^2} + \frac{1}{r} \frac{\partial \psi}{\partial r} - \frac{\partial^2 \psi}{\partial r^2} = 0, \quad \frac{3}{r} \frac{\partial^3 \psi}{\partial r^2 \partial \theta} + \frac{1}{r^3} \frac{\partial^3 \psi}{\partial \theta^3} - \frac{3}{r^2} \frac{\partial^2 \psi}{\partial r \partial \theta} + \frac{4}{r^3} \frac{\partial \psi}{\partial \theta} = 0. \tag{14.5}$$

Evaluating equations (14.5) at π, one finds the two solvability conditions

$$(\lambda - 1)[\lambda A + (\lambda - 2)C]\sin \lambda \pi = 0, \quad \lambda(\lambda - 1)(\lambda - 2)(A + C)\cos \lambda \pi = 0. \tag{14.6}$$

Solutions to (14.6) have to obey either $\sin \lambda \pi = 0$ or $\cos \lambda \pi = 0$, while the other equation yields a relation between the two amplitudes A and C. The smallest value of λ which does not lead to a diverging velocity at the tip is $\lambda = 1$; this solves the first equation in (14.6). In that case the solution (14.2) is simply a constant velocity field $\mathbf{V} = (u, v) = (0, V)$, streaming along the axis of the cusp, which does not "feel" the presence of the crack. To this we now add a singular part, which is a genuine contribution from the cusp. The dominant contribution comes from the next largest value of λ, $3/2$, which solves the second equation in (14.6) and yields $C = 3A$ from the first equation. Thus we arrive at the singular solution

$$\psi = -4Ar^{3/2} \sin^3 \frac{\theta}{2}. \tag{14.7}$$

If one also includes downward streaming, the leading-order velocity field along the crack (π) is given by

$$u = -v_\theta = \frac{\partial \psi}{\partial r} = -6Ay^{1/2}, \quad v = -V. \tag{14.8}$$

Taking the derivative of (14.8), one finds that the stresses *diverge* as $r^{-1/2}$, as a function of the distance r from the crack tip. In elasticity it is the presence of these large stresses near the crack tip which are responsible for crack propagation, as we will discuss in detail in Section 15.3.

Now we turn to the shape of the free surface, as illustrated in Fig. 14.1 (left). We use the flow field (14.8), observing that a stationary free surface will be a streamline of the flow. Therefore we have

$$\frac{\partial h}{\partial y} = \frac{u}{v} \approx \frac{6A}{V} y^{1/2}. \tag{14.9}$$

Integrating this equation, we finally arrive at

$$h \approx \frac{4A}{V} y^{3/2}. \tag{14.10}$$

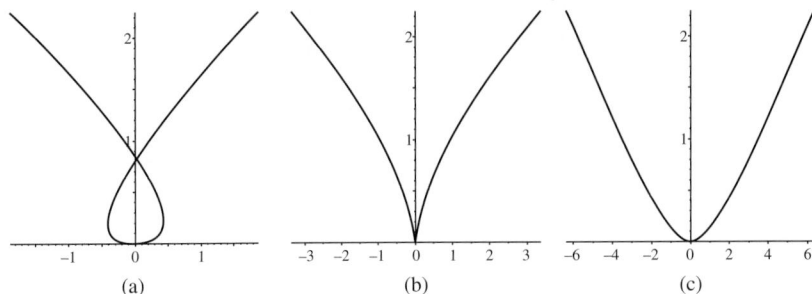

Figure 14.2 A cusp can be understood as arising when a loop in a curve is opened. We plot (14.12) for $a = 1$ and for $\epsilon = -1, 0$, and 1, from left to right.

This is precisely the power law (1.3); to obtain the prefactor A, matching to an outer solution would be required.

Could this simple result have been arrived at with less calculation? For example, according to Exercise 1.10 the tip of a cycloid has exactly the same scaling as the viscous cusp, suggesting a purely geometrical origin of the observed power law. To investigate this possibility, consider a generic curve in the plane that self-intersects, forming a loop; see Fig. 14.2(a). Now, as this loop is pulled open there is a transitional stage where a cusp is formed; see Fig. 14.2(b). Assuming the deformation is performed in a smooth fashion, what is the form of the cusp?

Let us consider the local behavior of a smooth curve near (x_0, y_0) by expanding in φ, which parameterizes the curve:

$$x = x_0 + a_1\varphi, \quad y = y_0 + a_2\varphi. \tag{14.11}$$

The parameters x_0 and y_0 can be eliminated by a translation, and $a_2 = 0$ after a suitable rotation. For a critical point to occur at $\varphi = 0$, a_1 must also vanish. Otherwise the curve would be a straight line locally. Thus we put $a_1 = \epsilon$, where $\epsilon = 0$ is the critical point. For the curve not to become degenerate, we must expand to a higher order. Thus, to leading order we have $y = \varphi^2/2$, normalizing the prefactor to 1/2 by rescaling φ. In the other coordinate we must go to a yet higher order, since the choice $x = \epsilon\varphi + b\varphi^2$ would once more result in a degenerate curve at the critical point $\epsilon = 0$. Thus we obtain a third-order polynomial $x = \epsilon\varphi + b\varphi^2/2 + a\varphi^3/3$. Using a shift in φ, and another rotation, the coefficient b can be eliminated, so we finally obtain the universal form

$$x = \epsilon\varphi + a\varphi^3/3, \quad y = \varphi^2/2. \tag{14.12}$$

This is the generic form of a curve in the neighborhood of a critical point that is given by

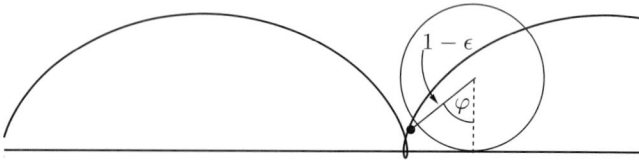

Figure 14.3 A trochoid is the trajectory of a point fixed on (or external to) a rolling disk, shown here for $\epsilon < 0$.

$$\frac{\partial \mathbf{x}}{\partial \varphi} = 0. \tag{14.13}$$

The curve (14.12) has the similarity form

$$x = \epsilon^{3/2} X_c(\sigma), \qquad y = \epsilon Y_c(\sigma), \tag{14.14}$$
$$X_c = \pm\sigma + a\sigma^3/3, \qquad Y_c = \sigma^2/2, \tag{14.15}$$

with similarity variable $\sigma = \varphi/\epsilon^{1/2}$. The positive sign corresponds to an open curve, see Fig. 14.2(c), and the negative sign to a self-intersecting curve, Fig. 14.2(a).

It follows from the above reasoning that given a family of smooth mappings $\mathbf{x}(\epsilon, \varphi)$, singularities are generically of the cusp form (14.12).

Example 14.1 (Trochoid) The trajectory shown in Fig. 14.3 comes from the superposition of the translation and the rotation of a rolling disk:

$$x = \varphi - (1 - \epsilon)\sin\varphi, \quad y = 1 - (1 - \epsilon)\cos\varphi. \tag{14.16}$$

In Exercise 1.10 we saw that for $\epsilon = 0$ a cusp is formed at $\varphi = 0$, where $x_\varphi = y_\varphi = 0$. Expanding about $\varphi = 0$ for finite ϵ we obtain

$$x = \epsilon\varphi - (1 - \epsilon)\varphi^3/6 + O(\varphi^5), \quad y = \epsilon - (1 - \epsilon)\varphi^2/2 + O(\varphi^4),$$

the first equation of which suggests that $\varphi^2 \propto \epsilon$, in order to balance $\epsilon\varphi$ with φ^3. Expanding consistently, we obtain

$$x = \epsilon\varphi - \varphi^3/6, \quad y = \epsilon - \varphi^2/2, \tag{14.17}$$

which is the same curve as (14.12), up to a translation and rescaling of the axes. $\qquad\square$

Remarkably, the free surface singularity (14.10), obtained by solving Stokes' equation, also exhibits the generic two-thirds power law found from (14.12) if we put $\epsilon = 0$. By imposing the stress-free conditions (14.3) we have neglected surface tension, producing an infinite curvature at the tip of the cusp.

If, however, the surface tension is included, then the cusp tip will be rounded, corresponding to $\epsilon > 0$ in (14.12), with a positive curvature $\kappa = \epsilon^{-1}$ at the apex. An exact solution of the Stokes equation with surface tension [122], based on a conformal mapping between the fluid domain and the unit disk, indeed yields precisely the similarity form (14.14) of the tip, as the following example shows:

Example 14.2 (Jeong–Moffatt solution) If a double vortex (oriented vertically downwards) is placed underneath a free surface inside a viscous fluid in two dimensions, the free surface can be parameterized by the smooth mapping

$$x = a\cos\theta + \frac{(a+1)\cos\theta}{1+\sin\theta}, \quad y = a(1+\sin\theta). \tag{14.18}$$

For the curve to be singular we must have $x_\theta = y_\theta = 0$, so we compute

$$x_\theta = -a\sin\theta - \frac{a+1}{1+\sin\theta}, \quad y_\theta = a\cos\theta. \tag{14.19}$$

From the second equation we get $\theta = \pi/2$, so a singularity occurs for $0 = x_\theta(\pi/2) = -3a/2-1/2$, or $a = -1/3$. We put $a = -1/3+\epsilon$ and $\theta = \pi/2+\varphi$ and expand about this point in φ, to obtain

$$x = -\frac{3\epsilon}{2}\varphi - \frac{\varphi^3}{12}, \quad y = -\frac{2}{3} + 2\epsilon + \frac{\varphi^2}{6}, \tag{14.20}$$

which is equivalent to (14.12) up to a translation. $\square$

The conformal mapping found by Jeong and Moffatt automatically leads to the smooth mapping (14.18) when restricted to the unit circle. A singularity arises because the map becomes non-invertible on the boundary, which is condition (14.13). Thus our reasoning shows that a free surface singularity must be of the generic form (14.12) *unless* higher derivatives of the curve vanish as well. We will present an explicit example of singularity formation by complex mapping for a simpler case in Section 14.3 below.

The only features that the geometrical argument cannot predict are the radius of curvature ϵ and its dependence on surface tension. Instead of reproducing the full calculation [122], which is quite intricate, we merely give a scaling argument, due to Hinch, which explains the *exponential* dependence on surface tension. Namely, the effect of surface tension is to generate a point force of strength 2γ (the two sides of the cusp pull with force γ per unit length in the z-direction), at a distance ϵ inside the cusp. Thus the force is smeared out over the size ϵ of the cusp. This generates an upward velocity $-2\gamma\ln\epsilon/(4\pi\eta)$

(see Exercise 7.10). For the tip to be stationary the upward velocity has to be compensated by the streaming velocity V, leading to

$$\epsilon \approx e^{-2\pi\,\mathrm{Ca}} \tag{14.21}$$

where $\mathrm{Ca} = \eta V / \gamma$ is the *capillary number*.

14.2 Singularity theory

The unifying mathematical theory which aims to classify all singularities of maps $\mathbf{f} : \mathbb{R}^n \to \mathbb{R}^m$ whose individual components are smooth is known as *singularity theory* [9, 10, 90, 97]. It includes catastrophe theory [176], which focuses on the critical points of scalar-valued functions. Here, by a singularity we mean a point in $\mathbb{R}^n$ (which we can take as the origin) at which the Jacobian of $\mathbf{f}$ vanishes; in the case of a curve $\mathbf{x}(\varphi)$ this amounts to the condition (14.13). If $\mathbf{f}$ were *not* singular then, by a change of coordinates, we could straighten the map to turn it into an identity (when restricted appropriately).

The crucial idea is to classify maps up to smooth, invertible deformations. This means that two maps $\mathbf{f}$, $\mathbf{g}$ are called *equivalent* (the so-called left–right or $\mathcal{A}$-equivalence) if there are two smooth invertible maps (or diffeomorphisms) ϕ, ψ such that, locally,

$$\psi \circ \mathbf{f} = \mathbf{g} \circ \phi. \tag{14.22}$$

Thus, we need only consider maps up to arbitrary smooth deformations of both their target spaces and their domains.

One important consequence is that each equivalence class defined by (14.22) is essentially represented by a polynomial. Thus the singular cusp map

$$\mathbf{x}(\varphi) = (\varphi^3, \varphi^2), \tag{14.23}$$

called a *germ*, is $\mathcal{A}$-equivalent to all smooth maps $\mathbb{R} \to \mathbb{R}^2$ for which the first two derivatives of one component vanish and for which the first derivative of the other component also vanishes. Classification proceeds by the order of the polynomials. Clearly, (14.23) is equivalent to e.g. $\mathbf{x}(\varphi) = (3\varphi^2, 2\varphi^3)$, which differs only by a rescaling and interchange of the axes.

If we add a small perturbation to the germ (14.23), it will in general no longer be singular. As suggested by our previous calculation, this *unfolding* (14.12) of the cusp has a universal structure and is characterized (up to $\mathcal{A}$-equivalence) by a small number of *unfolding parameters*. The minimum number of parameters needed is called the *codimension* of the singularity, which is one in the case of the cusp. To be specific, consider perturbing (14.23) by powers of any order. First, powers of higher order than those appearing in (14.23) can be eliminated up to $\mathcal{A}$-equivalence. Second, consider perturbations of lower order:

$$\mathbf{x}(\varphi) = (\epsilon_1\varphi + \epsilon_2\varphi^2 + \varphi^3, \eta_1\varphi + \varphi^2). \tag{14.24}$$

By choosing in (14.22) the right transformation $\phi(\varphi) = \varphi + \eta_1/2 \equiv \tilde{\varphi}$, η_1 is eliminated:

$$\mathbf{x}(\tilde{\varphi}) = (\tilde{\epsilon}_1\tilde{\varphi} + \tilde{\epsilon}_2\tilde{\varphi}^2 + \tilde{\varphi}^3, \tilde{\varphi}^2),$$

with $\tilde{\epsilon}_2 = \epsilon_2 - 3\eta_1/2$ and $\tilde{\epsilon}_1 = \epsilon_1 - \epsilon_2\eta_1 + 3\eta_1^2/4$, ignoring shifts in the origin.

Then in a second step we choose in (14.22) the left transformation $\boldsymbol{\psi}(\mathbf{x}) = (x - \tilde{\epsilon}_2 y, y) \equiv \tilde{\mathbf{x}}$, which eliminates all but one parameter:

$$\tilde{\mathbf{x}}(\tilde{\varphi}) = (\epsilon\tilde{\varphi} + \tilde{\varphi}^3, \tilde{\varphi}^2). \tag{14.25}$$

The final parameter ϵ cannot be eliminated, however, since this would require ψ to contain a term $\sqrt{x}$, which would be singular at the origin. Thus using $\mathcal{A}$-equivalence we arrive at the same result as (14.12), except that the amplitude a has been scaled away. The unfolding (14.25) is known as the *miniversal unfolding* of the cusp, to indicate that (i) any unfolding can be represented in this way and (ii) it contains the smallest possible number of parameters.

A classification of the singular germs of plane curves up to fourth order of the lowest power is given in [36]; (14.23) is a representative of the family A_{2k} of singularities given by $\varphi \mapsto (\varphi^{2k+1}, \varphi^2)$. The next most complicated germ is the *swallowtail* singularity E_6 (of which we will see examples below), which is the first member of the family E_{6k}:

$$\mathbf{x}(\varphi) = (\varphi^3, \varphi^4). \tag{14.26}$$

Proceeding as before, we consider the general unfolding

$$\mathbf{x}(\varphi) = (\eta_1\varphi + \eta_2\varphi^2 + \varphi^3, \epsilon_1\varphi + \epsilon_2\varphi^2 + \epsilon_3\varphi^3 + \varphi^4). \tag{14.27}$$

Using the transformation $\phi(\varphi) = \varphi + \eta_2/3 \equiv \tilde{\varphi}$ we can eliminate η_2 and then set $\tilde{\mathbf{x}} \equiv \boldsymbol{\psi}(\mathbf{x}) = (x, y - \tilde{\epsilon}_3 x)$. This leads to the universal unfolding (dropping the tildes, and using a different normalization for convenience)

$$\mathbf{x}(\varphi) = (\epsilon_s\varphi + \varphi^3/3, \eta_1\varphi + \eta_2\varphi^2/2 + \varphi^4/4), \tag{14.28}$$

which shows that the swallowtail germ (14.26) has codimension 3.

An example of the unfolding (14.28) is shown in Fig. 14.4, which illustrates that the germ (14.26) bifurcates into two different cusp singularities, one of which is open while the other has an intersection. There is a special case for $\epsilon_s < 0$, namely when $\eta_2 = \epsilon_s$ and $\eta_1 = 0$, for which both cusps are tuned

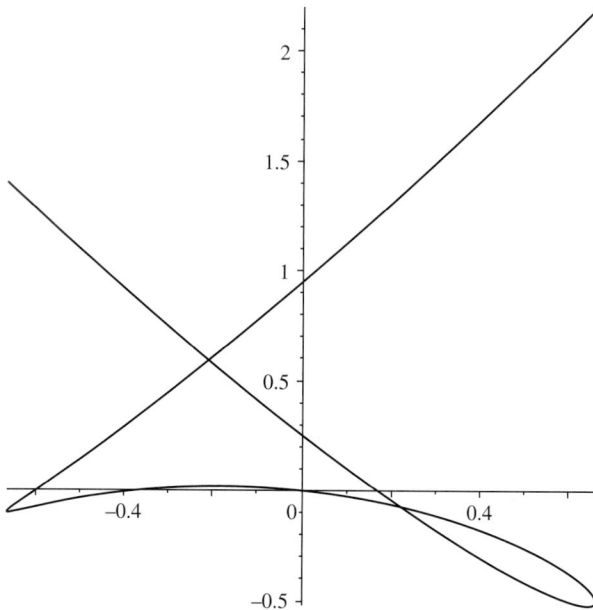

Figure 14.4 The unfolding of the swallowtail (14.28) for parameters $\epsilon = -1$, $\eta_2 = -1.1$, and $\eta_1 = 0.2$.

exactly to the singularity. We will encounter this case below (see (14.73)) as the singularities of a wave front. The resulting figure, shown in Fig. 14.7, is known as the swallowtail. Classifications of the singularities of space curves and surfaces are found in [91] and [156], respectively.

14.3 Hele-Shaw flow

As a time-dependent version of cusp formation, let us consider the singularities in Hele-Shaw flow, as described by (6.33), (6.34). This problem was first considered in the 1940s by Polubarinova-Kochina and Galin [174, 173, 89, 101]. The geometry is illustrated in Fig. 14.5: an initially smooth patch of viscous fluid is surrounded by air, occupying a region Ω, with boundary $\partial\Omega$. The flow is driven by a sink of strength $m < 0$ placed inside the fluid, with the radial velocity field (6.35) and pressure distribution (6.36). As the fluid is sucked out and the fluid patch shrinks, its boundary forms in finite time a cusp singularity, whose structure we aim to investigate.

Apart from the singularity (6.36), the pressure is a harmonic function $\Delta p = 0$ within Ω. The boundary condition on the free surface $\partial\Omega$ is

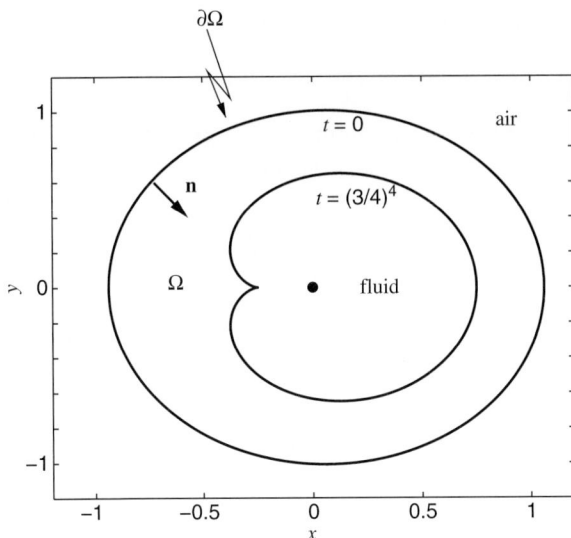

Figure 14.5 The formation of a cusp in the Hele-Shaw problem with suction at the origin, using the exact solution (14.41) with coefficients (14.42), (14.43). The initial conditions are $a_1(0) = 1$, $a_2(0) = 1/16$.

$$p = 0, \tag{14.29}$$

corresponding to a constant pressure in the air surrounding the fluid. Scalar-multiplying (6.33) by the inward normal $\mathbf{n}$, we obtain the normal velocity at the boundary,

$$u_n = -\nabla p \cdot \mathbf{n}, \tag{14.30}$$

where we have chosen units of speed so as to scale the prefactor to unity. Comparing (6.35) and (6.36) this implies that $h^2 = 12\eta$.

We solve the problem by looking for a complex function f that is conformal within Ω and maps the unit circle $|\xi| = 1$ onto $\partial\Omega$:

$$z = f(\xi, t), \quad \xi = re^{i\theta}, \tag{14.31}$$

where points in the plane are written as $z = x + iy$. The pressure is represented as

$$p = \Re\{\Phi(z)\},$$

where Φ is a function that is holomorphic except for the point where the source is located, where it has a logarithmic singularity. Since f is holomorphic inside the fluid, Φ must have a logarithmic singularity as a function of ξ as well and the solution satisfying the boundary condition (14.29) is

$$\Phi(f(\xi,t)) = \ln \xi \quad \text{inside} \quad \Omega, \tag{14.32}$$

choosing units of pressure to normalize the prefactor in (6.36) to unity, i.e. setting $6\eta m/(\pi h^2) = 1$; this means that $m = -2\pi$.

Now let us represent the boundary as $f(s,t)$, where s is the arc length along $\partial\Omega$. Then f_s is the tangent vector and if_s is the normal vector in complex notation; the inner product can be written as $\mathbf{a} \cdot \mathbf{b} = \Re\{\bar{a}b\}$. Then we find that

$$\nabla p \cdot \mathbf{n} = \Re\left\{\frac{\partial\Phi}{\partial z}if_s\right\} = \Re\left\{\frac{\partial\Phi}{\partial\xi}\frac{1}{f_\xi}if_s\right\} = \Re\left\{\frac{1}{\xi f_\xi}if_s\right\}, \tag{14.33}$$

using (14.32) in the last step. The representation $z(s,t)$ of the boundary corresponds to $\xi = e^{i\theta(s,t)}$ on the circle. Thus $f_s = f_\xi \xi_s = f_\xi \xi i\theta_s$, which, when inserted into (14.33), yields

$$-\nabla p \cdot \mathbf{n} = \theta_s. \tag{14.34}$$

Furthermore,

$$u_n = \mathbf{u} \cdot \mathbf{n} = \Re\left\{\bar{f}_t if_s\right\} = -\theta_s \Re\left\{\bar{f}_t \xi f_\xi\right\}, \quad |\xi| = 1. \tag{14.35}$$

Inserting (14.34) and (14.35) into the equation for interface motion (14.30), we arrive at

$$\Re\left\{\bar{f}_t \xi f_\xi\right\} = -1 \quad \text{for} \quad |\xi| = 1, \tag{14.36}$$

which is the Polubarinova–Galin equation in dimensionless form. From (14.31) it follows that $f_\theta = f_\xi \xi_\theta = i\xi f_\xi$, so we finally obtain

$$\Im\left\{f_\theta \bar{f}_t\right\} = -1, \quad |\xi| = 1. \tag{14.37}$$

Example 14.3 (Shrinking circular bubble) According to (14.31) the ansatz $f = a(t)\xi$ with $a \in \mathbb{R}$ describes a circular bubble of radius a. Inserting it into (14.36) we obtain $\Re\{\dot{a}\bar{\xi}\xi a\} = \dot{a}a|\xi|^2 = \dot{a}a = -1$, and so $a^2(t) = a_0^2 - 2t$. This corresponds to a circular bubble of initial radius a_0 whose area decreases uniformly at a rate $m = -2\pi$, as required by volume conservation. $\square$

If the initial bubble is not perfectly circular then the boundary may form a cusp, as seen in Fig. 14.5. From our considerations on Fig 14.2(a) we know that this is associated with the boundary curve becoming non-invertible. Thus we are interested in when the conformal map $f(\xi,t)$ becomes non-invertible on the boundary of the circle; suppose that this happens at some critical time t_0. As discussed in Section 13.3.2, the simplest kind of non-invertibility corresponds to $f'(\xi, t)$ becoming zero at some boundary point ξ_0 at the time t_0 of formation of the singularity. This leads to the generic quadratic behavior

$$f(\xi, t_0) = b_0 + b_2(\xi - \xi_0)^2 \tag{14.38}$$

near ξ_0. Without loss of generality we can assume that the singularity occurs at $\xi_0 = 1$, by making a complex rotation. In addition we may take $b_0 = 0$ and $b_2 = 1$, corresponding to a shift in cusp position, a rotation, and a change of scale. Now, for $\xi = e^{i\theta}$ on the circle we have

$$f = \left(e^{i\theta} - 1\right)^2 = -\theta^2 - i\theta^3 + O(\theta^4),$$

which is the cusp $\mathbf{x} = (-\theta^2, -\theta^3)$ to leading order.

Higher-order zeros of the first derivative,

$$f(\xi, t_0) = b_0 + b_n(\xi - \xi_0)^n, \tag{14.39}$$

with $n > 2$ are also possible, see (13.73) (but they are not the only types of non-invertibility [56]); each leads to a different type of singularity. However, as we discussed in the context of Moore's singularity, any small perturbation to the initial data would produce quadratic terms, making (14.38) the generic behavior.

Let us study the approach to the cusp singularity using a particular solution to (14.36) of the form [174, 173, 89, 114]

$$z = f(\xi, t) = a_1(t)\xi + a_2(t)\xi^2, \tag{14.40}$$

where a_1 and a_2 are real coefficients. In terms of the angle θ, (14.40) is written as

$$f = a_1(t)e^{i\theta} + a_2(t)e^{2i\theta}. \tag{14.41}$$

Inserting (14.41) into (14.37), one finds that

$$\Im\left\{f_\theta \bar{f}_t\right\} = a_1\dot{a}_1 + 2a_2\dot{a}_2 + (2\dot{a}_1a_2 + a_1\dot{a}_2)\cos\theta = -1.$$

Thus, for (14.40) to be a solution of (14.37) the coefficients a_1 and a_2 have to satisfy the following pair of ODEs:

$$a_1\dot{a}_1 + 2a_2\dot{a}_2 = -1, \tag{14.42}$$
$$a_1\dot{a}_2 + 2a_2\dot{a}_1 = 0. \tag{14.43}$$

In Exercise 14.2 can be derived the corresponding ODEs for a polynomial of arbitrary order. Integrating (14.43) leads to

$$a_1^2(t)a_2(t) = B, \tag{14.44}$$

while from (14.42) we obtain

$$\frac{a_1^2(t)}{2} + a_2^2(t) = C - t, \tag{14.45}$$

where B, C are constants of integration.

Example 14.4 (Area conservation) In fact the relation (14.45) follows from volume conservation or from the conservation of area in a two-dimensional setting. For any $f(\xi)$ that maps the unit disk conformally, the area A of the image Ω of the unit disk is given by [42]

$$A = \pi \sum_{n=1}^{\infty} \frac{|f^{(n)}(0)|^2}{(n-1)!n!}. \tag{14.46}$$

In the case of (14.40), $f^{(1)}(0) = a_1$ and $f^{(2)}(0) = 2a_2$ and so

$$A = \pi \left(a_1^2 + 2a_2^2 \right) = A_0 + mt.$$

Remembering that $m = -2\pi$, this is equivalent to (14.45).   □

Equations (14.44), (14.45) represent a particular exact solution to the Hele-Shaw problem. The constants of integration are determined by the initial conditions, through

$$C = \frac{a_1^2(0)}{2} + a_2^2(0), \quad B = a_2(0)a_1^2(0), \tag{14.47}$$

and $a_1(t)$ is determined by solving

$$\frac{a_1^2(t)}{2} + \frac{B^2}{a_1^4(t)} = C - t. \tag{14.48}$$

The resulting evolution is shown in Fig. 14.5 for the initial conditions $a_1(0) = 1$ and $a_2(0) = 1/16$, for which a singularity develops.

We have seen that this happens when $f'(\xi_0) = 0$ at some point ξ_0 along the circle; we can always choose $\xi_0 = -1$ by a complex rotation. The condition $f'(\xi_0) = 0$ implies $a_1 + 2a_2\xi_0 = 0$, so we have to check whether during the evolution the condition

$$\xi_0 = -\frac{a_1(t_0)}{2a_2(t_0)} = -1 \tag{14.49}$$

is satisfied at a singularity time t_0. Using (14.44) with the condition (14.49), we find

$$a_1(t_0) = (2B)^{1/3}, \quad a_2(t_0) = \frac{1}{2}(2B)^{1/3}, \tag{14.50}$$

using B as a convenient parameter. Then from (14.48) one can compute the singularity time as

$$t_0 = C - \frac{3}{4}(2B)^{2/3} = \frac{1}{2}a_1^2(0) + a_2^2(0) - \frac{3}{4}[2a_2(0)a_1^2(0)]^{2/3},$$

in terms of the initial conditions.

Now we expand in the small parameter $t' = t_0 - t$ to find the critical behavior around the singularity. Inserting the ansatz $a_1 = (2B)^{1/3} + \delta$ into (14.48), we find $3\delta^2 \simeq t'$ and thus

$$a_1(t) \simeq (2B)^{1/3} + \frac{1}{\sqrt{3}}t'^{1/2}, \quad a_2(t) \simeq \frac{B^{2/3}}{2^{1/3}} - \frac{B^{1/3}}{2^{2/3}\sqrt{3}}t'^{1/2} \qquad (14.51)$$

close to the singularity. In terms of the parameter B, the singularity occurs at

$$x_0 \equiv x(\pi, t_0) = -\left(\frac{B}{4}\right)^{1/3}, \quad y_0 \equiv y(\pi, t_0) = 0$$

in space.

To find the spatial structure we expand $f(\xi)$ in $\varphi = \pi - \theta$. Using

$$\xi = e^{i\theta} = -1 + i\varphi + \frac{\varphi^2}{2} - \frac{i\varphi^3}{6} + O(\varphi^4),$$

and recognizing that φ is of order $t'^{1/4}$, we arrive at

$$x(\theta, t) - x_0 \simeq -\frac{2}{\sqrt{3}}t'^{1/2} - \frac{B^{1/3}}{2^{2/3}}\varphi^2, \qquad (14.52)$$

$$y(\theta, t) - y_0 \simeq -\sqrt{3}t'^{1/2}\varphi - \frac{B^{1/3}}{2^{2/3}}\varphi^3, \qquad (14.53)$$

which is the desired local solution for which we have been looking.

As expected, (14.52), (14.53) have the same self-similar structure as (14.12), with $\epsilon \propto t'^{1/2}$. To make this explicit, we rotate by 90° and rescale one axis:

$$\tilde{x} = -\frac{(2B)^{1/6}}{\sqrt{3}}y, \quad \tilde{y} = -x. \qquad (14.54)$$

Then, in similarity variables, the cusp can be written in the universal form

$$\tilde{x} = |t'|^{3\gamma/2}X_c, \quad \tilde{y} = |t'|^{\gamma}Y_c, \qquad (14.55)$$

with X_c and Y_c defined by (14.14). Putting

$$X_c = \frac{\tilde{y}}{t'^{3/4}}, \quad Y_c = \frac{\tilde{y} - (B/4)^{1/3}}{t'^{1/2}} - \frac{2}{\sqrt{3}}, \quad \sigma = \frac{(2B)^{1/6}\varphi}{2\sqrt{3}t'^{1/4}}, \qquad (14.56)$$

one obtains (14.55), with $a = \sqrt{3}/2$ in (14.15) and the scaling exponent $\gamma = 1/2$.

14.4 Optical caustics

We begin with the underlying wave physics, described by the wave equation

$$\Delta\phi - c^{-2}\phi_{tt} = 0, \tag{14.57}$$

where c is the wave speed. For simplicity we confine ourselves to a scalar field ϕ. A monochromatic wave of angular frequency ω is described by

$$\phi = \Phi e^{-ikct}, \tag{14.58}$$

where $k = \omega/c$ is the wave number. Inserting (14.58) into (14.57) we obtain the Helmholtz equation,

$$\Delta\Phi + k^2\Phi = 0, \tag{14.59}$$

which we will use later to study diffraction patterns on the scale of the wavelength.

To begin with, however, we are interested in geometrical features much larger than the wavelength, which corresponds to studying the limit of large k. To that end we try the ansatz

$$\Phi = \Phi_0(\mathbf{x}) \exp\left[ik\psi(\mathbf{x})\right], \tag{14.60}$$

where $\psi(\mathbf{x})$ is the phase of the wave and $\Phi_0(\mathbf{x})$ is its amplitude. Inserting (14.60) into (14.59) yields

$$\left\{k^2\Phi_0\left[1 - (\nabla\psi)^2\right] + ik\left(2\nabla\Phi_0 \cdot \nabla\psi + \Phi_0\Delta\psi\right) + \Delta\Phi_0\right\} e^{ik\psi(\mathbf{x})} = 0. \tag{14.61}$$

At leading order, for $k \to \infty$ the quadratic term dominates and we obtain a *nonlinear* equation for the phase:

$$|\nabla\psi|^2 = 1, \tag{14.62}$$

which is known as the *eikonal equation*. At next-to-leading order we obtain a linear equation for the wave amplitude Φ_0:

$$2\nabla\Phi_0 \cdot \nabla\psi + \Phi_0\Delta\psi = 0. \tag{14.63}$$

A wave front is by definition an isosurface (in the two-dimensional case an isoline) of the phase ψ, defined by

$$\psi(\mathbf{x}) - ct = 0. \tag{14.64}$$

Thus, from the eikonal equation (14.62) it follows that the normal to the front is

$$\mathbf{n} = \frac{\nabla\psi}{|\nabla\psi|} = \nabla\psi.$$

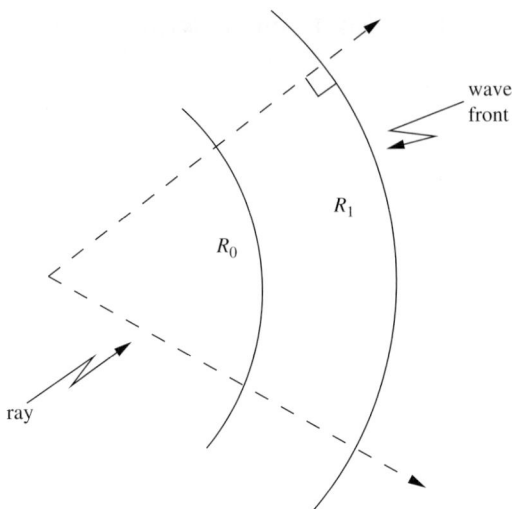

Figure 14.6 A wave front propagates locally like a sequence of concentric circles, rays emanate from the center as described by (14.66).

For simplicity, let us confine ourselves from now on to the two-dimensional case of line-like wavefronts, so the front $\mathbf{x}(\varphi, t)$ is parameterized by a single scalar parameter φ. Differentiating (14.64) with respect to time, one finds

$$c = \frac{\partial \mathbf{x}}{\partial t} \nabla \psi = \dot{\mathbf{x}} \cdot \mathbf{n}.$$

Any component of $\dot{\mathbf{x}}$ tangential to the wave front will not affect its motion. Thus one can always find a parameterization such that

$$\frac{\partial \mathbf{x}}{\partial t} = c\mathbf{n}, \tag{14.65}$$

i.e. the tangential component of $\dot{\mathbf{x}}$ vanishes. The trajectory $\mathbf{x}(\varphi, t)$ at constant φ, which is always normal to the wave front, is called a *ray*. Equation (14.65) is a nonlinear evolution equation for the wave front, whose singularities we will study in order to describe the caustic curve seen in Fig. 1.5. This approach is different from, but of course closely related to, the usual method of studying caustics using catastrophe theory [162]; see Exercise 14.8.

In principle the normal $\mathbf{n}$ in (14.60) is allowed to depend on time for each ray φ (and indeed, this will be the case where c is not constant but is allowed to depend on position). However, for constant c the normal $\mathbf{n}$ is in fact constant for each ray, as illustrated in Fig. 14.6. Namely, the wave front curve can be approximated to second order by a circle; thus the ray can be viewed as originating from the center of the circle or being focused into it. In either case the ray will continue to go straight, without changing direction. But this means that (14.65) can be integrated to give

$$\mathbf{x}(\varphi, t) = \mathbf{x}(\varphi, 0) + c\mathbf{n}(\varphi, 0)t. \qquad (14.66)$$

It is also clear from Fig. 14.6 that the distance between two rays increases as the radius of curvature of the wave front. Since the wave energy, which is proportional to Φ_0^2, is conserved we must have

$$\frac{\Phi_0(\mathbf{x}_1)}{\Phi_0(\mathbf{x}_0)} = \left(\frac{R_0}{R_1}\right)^{1/2}. \qquad (14.67)$$

The same result can also be derived from a more detailed analysis of (14.63) [222]. We mention that (14.67) corresponds to a particular type of relationship (11.111) in the geometrical wave theory developed earlier:

$$\frac{A}{A_0} = \left(\frac{M_0 - 1}{M - 1}\right)^2. \qquad (14.68)$$

Namely, A/A_0 is the fractional change in the length of the wave front between two rays and $M - 1$ (the departure of the Mach number from the linear wave speed) is a measure of the wave's intensity.

Now we analyze the evolution of a wave front as a function of time, which is a nontrivial exercise in spite of the apparent simplicity of (14.66). Let us assume for ease of notation that the initial condition for $t = 0$ is a graph for which $\mathbf{x} = (\varphi, h(\varphi))$. Then (14.66) can be written as

$$x = \varphi - \frac{h'}{\sqrt{1 + h'^2}}ct, \quad y = h(\varphi) + \frac{ct}{\sqrt{1 + h'^2}}, \qquad (14.69)$$

which is an exact solution of the differential equation (14.65), as can be checked by direct computation.

The wave front described by (14.69) is not to be confused with the caustic shown in Fig. 1.5. Rather, a caustic is a place where rays meet, thus increasing the intensity. This means that the conditions for a caustic are $dx/d\varphi = 0$, $dy/d\varphi = 0$: at a caustic the shape of the wave front $\mathbf{x}(\varphi, t)$ must also become singular. Using (14.69) it can be checked that the two conditions are equivalent. Solving for the time at which singularity occurs one finds that

$$ct_c = \frac{(1 + h'^2)^{3/2}}{h''} \equiv \frac{1}{\kappa}, \qquad (14.70)$$

where κ is the local curvature of the initial wave front. This result is very intuitive: the singularity occurs exactly at the focus of a local circle inscribed into the wave front. The first singularity occurs at a time t_0 corresponding to the maximum of the curvature, for which there is optimal focusing:

$$ct_0 = \frac{1}{\max\{\kappa\}} \equiv r_0. \qquad (14.71)$$

The universal structure of this singularity is obtained by expanding the initial wave front $h(\varphi)$ about the point of maximum curvature, taken to be at $\varphi = 0$. By a rotation we can ensure that the focus is along the y-axis, so we have $h(\varphi) = 0$. For $h''(0)$ to be a maximum at $\varphi = 0$, the third derivative must also vanish, resulting in

$$h = a_1\varphi^2 + a_2\varphi^4 + \dots \qquad (14.72)$$

For the maximum curvature to occur at $\varphi = 0$ the coefficients must obey $a_1^3 > a_2$, and we conclude from (14.70) that $r_0 = 1/(2a_1)$. For the dimensionless time distance to the singularity we take $t' = c(t_0 - t)/r_0$. Expanding the solution (14.69) in t', we obtain

$$\begin{aligned} x/r_0 &= t'\theta + a\theta^3/3, \\ y/r_0 &= -t' + t'\theta^2/2 + a\theta^4/4, \end{aligned} \qquad (14.73)$$

where $a = 3(a_1^3 - a_2)/(2a_1^3) > 0$ and $\theta = \varphi/r_0$. The y-axis has been shifted down by r_0, so that the singularity occurs at the origin. The corresponding shape of the wave front is shown in Fig. 14.7 for $t' > 0$ (before the singularity), $t' = 0$ (at the singularity), and $t' < 0$ (after the singularity). After the singularity the curve self-intersects and forms a characteristic shape known as the *swallowtail* in catastrophe theory [176]. The solution (14.73) can be written in similarity form as

$$x = |t'|^{3/2}X, \quad y = |t'|^2 Y, \qquad (14.74)$$

$$X = \pm\Sigma + a\Sigma^3/3, \quad Y = \pm\Sigma^2/2 + a\Sigma^4/4, \qquad (14.75)$$

where $\Sigma = \theta/|t'|^{1/2}$. The positive sign applies to the time $t' > 0$ before the singularity and the negative sign to $t' < 0$.

The swallowtail singularity (14.73) is a special case of the more general unfolding (14.28) of the higher-order singularity germ (14.26). If one puts $\eta_1 = 0$ then the figure becomes symmetric, as in Fig. 14.7. If in addition one puts $\eta_2 = \epsilon_s = t'$, (14.28) becomes (14.73) exactly, up to a shift and a rescaling of the axes. One can show (see Exercise 14.9) that for $\eta_2 = \epsilon_s + \epsilon$ the local solution near the tip has the standard cusp form (14.12). Clearly, the eikonal equation has an underlying symmetry which ensures that $\epsilon = 0$.

To calculate the shape of the caustic from the wave front (14.73), observe that the caustic is the line swept out by the singularities of the wave front; see

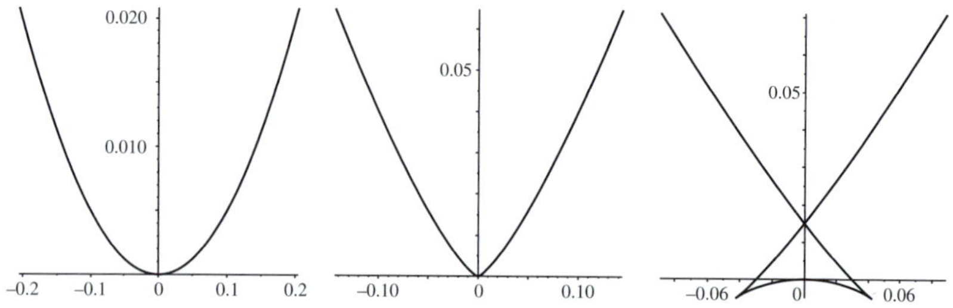

Figure 14.7 Two cusp singularities being born out of a swallowtail singularity. From left to right, (14.73) with $a = 1$ and $t' = 1, 0$, and -1. The curves have been shifted so that the point $\theta = 0$ is at the origin

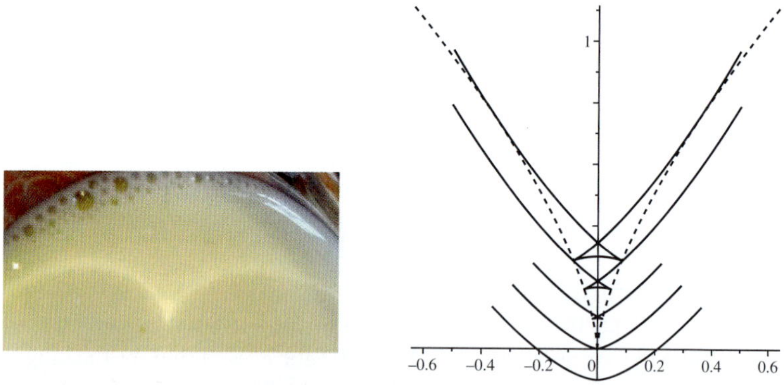

Figure 14.8 A caustic in a glass of milk: on the left, the physical reality; on the right, the evolution of the wave front as it passes through the swallowtail singularity. The caustic forms a two-thirds cusp (broken line), which is the curve traced out by the tips of the swallowtail curve (14.73) (image by Ann Eggers).

Fig. 14.8 (right). From the first equation in (14.73) the tip condition $\partial x / \partial \theta = 0$ corresponds to $\theta_c = \pm\sqrt{-t'_c/a}$. Inserting this value into (14.73), one finds the position of the caustic (x_c, y_c) as function of θ_c. This yields, to leading order in θ_c,

$$\frac{x_c}{r_0} = -\frac{2a}{3}\theta_c^3, \quad \frac{y_c}{r_0} = a\theta_c^2 + O(\theta_c^4), \tag{14.76}$$

which is a generic two-thirds cusp. As shown in Fig. 14.8 (right), the tip of the cusp lies at the position of the swallowtail singularity, $t' = 0$. As t' becomes negative (after the swallowtail singularity), the caustic is traced out. The universal form of the wave front singularity (14.73) yields the leading-order equation for the caustic (14.76).

The fact that the caustic is another generic singular curve is no accident. In most treatments of optical singularities one derives the caustic curves directly,

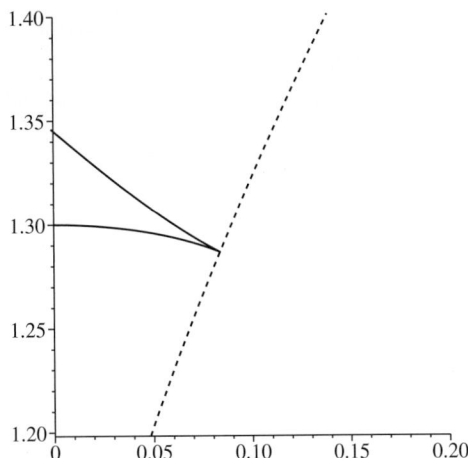

Figure 14.9 The local geometry near a caustic line; shown is a detail of Fig. 14.8 (right). The caustic is the broken line. The wave front has a cusp singularity at the position of the caustic and is oriented perpendicularly to the caustic. The curvature of the wave front diverges as the caustic is approached, leading to a diverging field strength.

using Fermat's principle: the path taken by a ray is that of least time (or least distance, if the index of refraction is constant) [162]. Thus, by studying the critical points of a distance function using catastrophe theory, one can classify the singularities of caustics. The singularities of caustics are called *Lagrange singularities*, while the singularities of wave fronts, in which our interest lies, are called *Legendre singularities* [8]. General relationships exist between the two types of singularity [8].

Returning to the shape of the wave front as described by (14.73) in more detail, at $t' = 0$ the wave front is given by $y/r_0 = (3x/r_0)^{4/3}/(4a^{1/3})$, corresponding to a rather mild singularity. For $t' < 0$ the wave front develops two tips where it meets the caustic (see Fig. 14.9). To examine its shape we use the similarity description (14.74), (14.75) (applying the negative sign). There are two singular points with $X_\Sigma = 0 = -1 + a\Sigma^2$, and so $\Sigma_c = \pm a^{-1/2}$. Inserting this into (14.75), the cusp singularities lie at $X_c = \mp 2/(3\sqrt{a})$, $Y_c = -1/(4a)$. To find out what happens at these points, we shift their position to the origin and introduce the rotation

$$\begin{pmatrix} \bar{Y} \\ \bar{X} \end{pmatrix} = \begin{pmatrix} 1 & \mp\sqrt{a} \\ \pm\sqrt{a} & 1 \end{pmatrix} \begin{pmatrix} X - X_c \\ Y - Y_c \end{pmatrix}; \tag{14.77}$$

we then find that

$$\begin{pmatrix} \bar{Y} \\ \bar{X} \end{pmatrix} = \begin{pmatrix} -2as^3/3 + O(s^4) \\ (a+1)s^2 + O(s^3) \end{pmatrix}, \tag{14.78}$$

where $s = \mp a^{-1/2} + \Sigma$. Thus locally the two tips have the form of cusps with the usual two-thirds exponent.

To find the orientation of the cusp relative to the caustic, note that the caustic can be parameterized by θ_c in the form $\mathbf{x}(\theta_c, t_c(\theta_c))$. Thus

$$\frac{d\mathbf{x}}{d\theta_c} = \frac{\partial \mathbf{x}}{\partial \theta_c} + \frac{\partial \mathbf{x}}{\partial t} \frac{\partial t_c}{\partial \theta_c} \tag{14.79}$$

is a vector tangential to the caustic. Now, the critical condition (14.13) for the singularity of the wave front stipulates that $\partial \mathbf{x}/\partial \theta_c = 0$, so (14.79) implies that $\mathbf{x}_t \equiv \partial \mathbf{x}/\partial t$ is also tangential to the caustic. But $\mathbf{x}_t$ is the direction of a ray and is thus perpendicular to the cusp direction. This means that the orientation of the cusp is locally *perpendicular* to the caustic, as illustrated in Fig. 14.9. In fact wave fronts can be constructed as *evolvents* of the caustic curves [8]. Evolvents or involutes are the curves that are produced as a taut string is rolled off another curve (the evolute, which is the caustic in our case). In particular, the evolvent wave front always has a cusp when it meets the evolute (caustic), and is perpendicular to it.

It follows from these observations that the field intensity diverges as the caustic is approached. We have shown that if ρ is a coordinate along the caustic, and σ is the coordinate perpendicular to it, then the wave front is described locally by the cusp equation $\rho^2 = A\sigma^3$. This means that the curvature of the wave front diverges as $\partial^2\rho/\partial\sigma^2 \propto 1/\sqrt{\sigma}$ as a function of the distance σ from the caustic. According to (14.67) the light intensity $|\Phi_0|^2$ is proportional to the curvature, which thus diverges as $\sigma^{-1/2}$, where σ is the distance from the caustic. This explains the brightness of the caustic, which would go to infinity in the limit $\lambda \to 0$. As we will see now, this divergence dissolves into a diffraction pattern if the wave nature of light is taken into account.

14.5 The wavelength scale

To compute the diffraction pattern mentioned at the end of the previous section, we will use the geometry of the wave front as established above. We observe that the caustic comes from a focusing of rays originating from a smooth wave front described by (14.73) for any $t' \to 0$. According to Fresnel's principle, the diffraction pattern is produced by a superposition of point sources distributed over the wave front [94]. More formally, this description can be derived using Kirchhoff's formula [94, 199] for the wave field $\Phi(\mathbf{x}_0)$, which follows from Helmholtz's equation (14.59) in two dimensions.

Namely, let Γ be a curve enclosing $\mathbf{x}_0$ which is the incoming wave front producing the field at $\mathbf{x}_0$. The Green function G for Helmholtz's equation is the wave field coming from a point source described by a δ-function $\delta(\mathbf{r})$ at the origin:

$$\left(\triangle + k^2\right)G(\mathbf{r}) = \delta(\mathbf{r}). \tag{14.80}$$

Now we use Green's second identity,

$$\int (G\triangle\Phi - \Phi\triangle G)d^2x_1 = \int_\Gamma \left(G\frac{\partial\Phi}{\partial n} - \Phi\frac{\partial G}{\partial n}\right)d\sigma_1,$$

where $\partial/\partial n$ denotes the derivative in the direction of the outward normal $\mathbf{n}$. Inserting (14.59) and (14.80) on the left-hand side one obtains Kirchhoff's representation

$$\Phi(\mathbf{x}_0) = \int_\Gamma \left(\Phi(\mathbf{x}_1)\frac{\partial G(\mathbf{r})}{\partial n} - G(\mathbf{r})\frac{\partial\Phi(\mathbf{x}_1)}{\partial n}\right)d\sigma_1, \tag{14.81}$$

where $\mathbf{r} = \mathbf{x}_0 - \mathbf{x}_1$.

The second term in (14.81) drops out if, instead of the free space Green function G, we use the Green function G_1 which obeys the boundary condition $G_1(\mathbf{r}) = 0$ for $\mathbf{x}_1$ on Γ. Then we obtain

$$\Phi(\mathbf{x}_0) = \int_\Gamma \Phi(\mathbf{x}_1)\frac{\partial G_1(\mathbf{r})}{\partial n}d\sigma_1. \tag{14.82}$$

To calculate G_1 we use two approximations. First, we assume that we are in the short-wave limit $kr \gg 1$, in which case G has the asymptotic form [69]:

$$G(\mathbf{r}) \approx \frac{1}{4i}\sqrt{\frac{2}{i\pi}}\frac{e^{ikr}}{\sqrt{kr}}, \tag{14.83}$$

where $r = |\mathbf{r}|$. Second, we assume that the wave front Γ is relatively flat (i.e. its radius of curvature is large compared with r), so that we can use the method of images to satisfy the boundary condition on Γ; see Fig. 14.10. If $\mathbf{x}_0'$ is the position of the mirror image of $\mathbf{x}_0$, reflected in the wave front, then

$$G_1(\mathbf{r}) \approx \frac{1}{4i}\sqrt{\frac{2}{i\pi}}\left(\frac{e^{ikr}}{\sqrt{kr}} - \frac{e^{ikr'}}{\sqrt{kr'}}\right), \tag{14.84}$$

where $r' = |\mathbf{x}_0' - \mathbf{x}_1|$.

Now using the notation of Fig. 14.10, we calculate the normal derivative on Γ as

$$\frac{\partial}{\partial n}\frac{e^{ikr}}{\sqrt{r}}\bigg|_{z_1=0} = -\frac{\partial}{\partial z_1}\frac{e^{ikr}}{\sqrt{r}}\bigg|_{z_1=0}$$

$$= -\frac{ikz_0}{r}\frac{e^{ikr}}{\sqrt{r}} + O(r^{-3/2}) \approx -\frac{ik\mathbf{n}\cdot\mathbf{r}}{r}\frac{e^{ikr}}{\sqrt{r}},$$

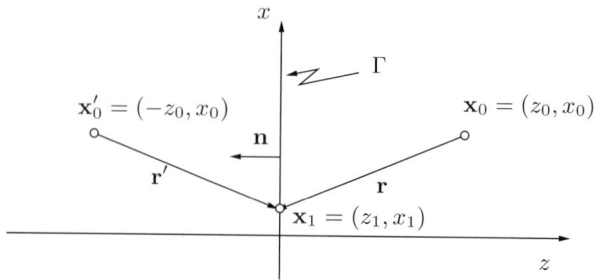

Figure 14.10 Calculation of G_1 using the method of images

since $r = \sqrt{(x_0 - x_1)^2 + (z_0 - z_1)^2}$. The normal derivative of the image gives the same, up to a minus sign, producing a factor 2 overall. Thus, finally the expression (14.82) for the wave field becomes

$$\Phi(\mathbf{x}_0) = \sqrt{\frac{k}{2\pi i}} \int_\Gamma \Phi(\mathbf{x}_1) \frac{\mathbf{n} \cdot \mathbf{r}}{r} \frac{e^{ikr}}{\sqrt{r}} d\sigma_1. \tag{14.85}$$

This representation corresponds to a distribution of wave sources of strength $\Phi(\mathbf{x}_1)$ along the wave front Γ; see Fig. 14.11. The "obliquity factor" $(\mathbf{n} \cdot \mathbf{r})/r$ expresses the fact that the direction of emission is not the same as the direction of observation $\mathbf{r}$. For the present case of a local diffraction pattern coming from a distant wave front the direction of observation can be taken as normal to the wave front and the obliquity factor can be neglected. In addition, instead of the *closed* curve Γ, the main contribution to the wave pattern near the caustic comes from the central part of the incoming wave front, described by (14.73). The field strength $\Phi(\mathbf{x}_1) \equiv \Phi_1$ is taken as constant at the incoming wave front Γ, so we obtain the simplified expression

$$\Phi(\mathbf{x}_0) = \Phi_1 \sqrt{\frac{k}{2\pi i}} \int_\Gamma \frac{e^{ikr}}{\sqrt{r}} d\sigma_1. \tag{14.86}$$

Now the only quantity determining the diffraction pattern near the cusp is the distance r between a point $\mathbf{x}_1 = (x, y)$ on the wave front and a point of observation $\mathbf{x}_0 = (\bar{x}, \bar{y})$. We would like to compute

$$r = \sqrt{(x - \bar{x})^2 + (y - \bar{y})^2}, \tag{14.87}$$

where x, y is given by (14.73), as shown in Fig. 14.11. All lengths are computed in units of r_0, and the distance between the incoming wave front and the caustic is taken to be unity, i.e. we put $t' = 1$ in (14.73). To order the size of all the terms, we use the parameter θ in (14.73). Since $(\bar{x}, \bar{y})$ lies near the cusp

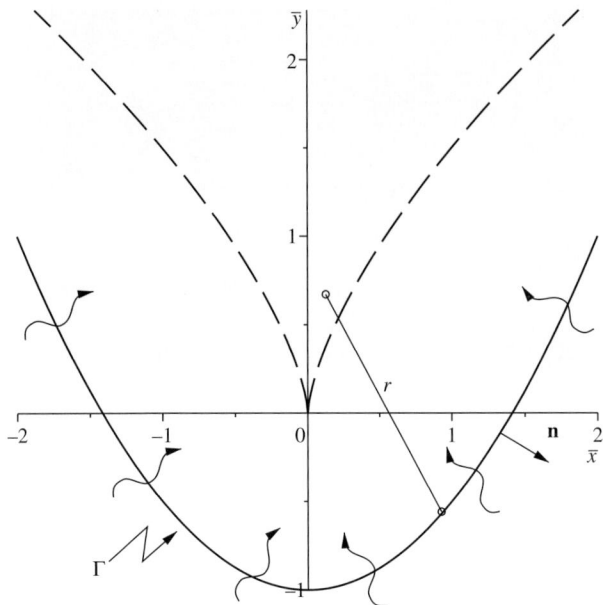

Figure 14.11 To calculate the field strength at a point $(\bar{x}, \bar{y})$ in the neighborhood of the caustic (broken line), one has to integrate over the incoming wave front Γ. The field is indicated by the wavy arrows. The phase shift is determined by the distance r between $(\bar{x}, \bar{y})$ and a point (x, y) on the wave front.

caustic (dashed line), it follows from (14.76) that $\bar{x} = O(\theta^3)$ and $\bar{y} = O(\theta^2)$. With this in mind we put

$$\bar{x} = \xi\theta^3, \quad \bar{y} = \zeta\theta^2, \tag{14.88}$$

and expand r in a power series in θ:

$$r = 1 + \zeta\theta^2 + \left(\frac{a}{12} + \frac{1}{8} - \xi - \frac{\zeta}{2}\right)\theta^4 + O(\theta^6). \tag{14.89}$$

Rewriting (14.89) in terms of the fixed position $(\bar{x}, \bar{y})$, and putting $\bar{a} = a/12 + 1/8$, we obtain

$$r = 1 + \bar{y} + \frac{\bar{a}\theta^4}{12} - \bar{x}\sigma - \frac{\bar{y}}{2}\theta^2. \tag{14.90}$$

The integral in (14.86) is to be performed over $\theta \approx \sigma_1$. Thus, with the substitution $s = \theta(\bar{a}k)^{1/4}$, we finally obtain the similarity solution

$$\Phi \approx e^{ik(1+\bar{y})}(\bar{a}k)^\alpha \, \Xi\left(\frac{\bar{x}}{(\bar{a}k)^{\beta_x}}, \frac{\bar{y}}{(\bar{a}k)^{\beta_y}}\right), \tag{14.91}$$

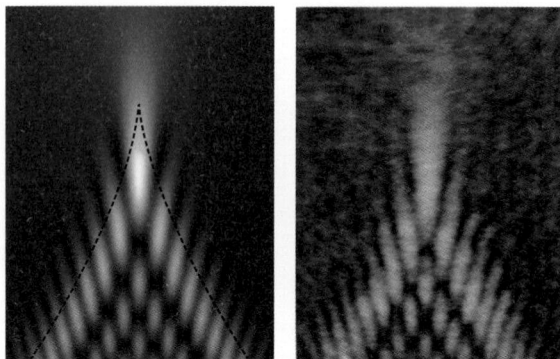

Figure 14.12 Gray-scale image of the wave amplitude near the cusp caustic (left), as calculated from (14.91)–(14.93); the broken line shows the caustic. For comparison, an experimental image is shown (right). Image courtesy of M. Berry.

with similarity exponents $\alpha = 1/4$, $\beta_x = 1/2$, and $\beta_y = 1/4$; the similarity profile Ξ has two arguments and is written concisely as

$$\Xi(\xi, \eta) = \frac{\Phi_1}{\sqrt{2\pi \bar{a}i}} P\left(-\frac{\eta}{2}, -\xi\right),\qquad (14.92)$$

where

$$P(u, v) = \int_{-\infty}^{\infty} e^{i(s^4 + us^2 + vs)} ds \qquad (14.93)$$

is known as *Pearcey's function* [166]. Up to a rapidly oscillating phase factor, in the small-wavelength limit $k \to \infty$ the diffraction pattern is determined by the similarity function (14.92). The spatial similarity exponent differs depending on whether one is moving in the direction of the cusp ($\bar{y}$), or perpendicular to it ($\bar{x}$). Figure 14.12 shows $|\Phi|$ near the cusp [21], as well as diffraction fringes measured experimentally. It can be seen that the singularity which, according to ray theory, would occur along the cusp dissolves into an intricate diffraction pattern, in particular within the cusp. The pattern is caused by interference between the rays, of which there are three inside the cusp [21]. The diffraction pattern near caustic singularities of higher order can be analyzed as well, most elegantly using catastrophe theory [20]. Each singularity is associated with a unique set of scaling exponents characterizing the scaling of the diffraction pattern in space.

Exercises

14.1 Consider the eikonal equation (14.65), and let the wave front be represented in complex notation as

$$z(\varphi, t) = x(\varphi, t) + iy(\varphi, t).$$

If $t = t_1 + it_2$ and $n = n_1 + in_2$ are the tangent and normal vectors, show that $t = z_s$ and $n = iz_s$, where s is the arc length. Thus show that the eikonal equation can be written as

$$\Im\{z_\sigma \bar{z}_t\} = -c\frac{ds}{d\varphi}. \qquad (14.94)$$

14.2 Show that the ansatz (14.40) can be generalized to an arbitrary polynomial

$$f(\xi, t) = \sum_{n=1}^{N} a_n \xi^n,$$

and derive an equation of motion for $\{a_n(t)\}$. To this end note that, for $\xi = e^{i\theta}$, the left-hand side of (14.37) results in a Fourier series.

14.3 Generalize Example 14.3 to an arbitrary mapping $f(\xi, t)$ which is analytic on the unit disk, by showing that the Hele-Shaw dynamics (14.36) satisfies the volume conservation $\dot{A} = m$, where A is the area of the fluid domain and $m = -2\pi$ is the flux. Hint: use the result of the previous exercise to show that, for $N \to \infty$,

$$\Re\left\{\sum_{n=1}^{\infty} na_n\bar{\dot{a}}_n\right\} = -1,$$

and use the area formula (14.46).

14.4 What type of cusp singularity is found for (14.39), with integer $n > 2$?

14.5 Look for solutions to the Hele-Shaw problem such that the conformal mapping of the disk onto the fluid domain is given by the rational function

$$f(\xi, t) = a(t)\frac{\xi[1 - b(t)\xi]}{1 - c(t)\xi},$$

with

$$f(\xi, 0) = \frac{\xi(4 - \sqrt{5/2}\xi)}{2\xi - \sqrt{10}},$$

where $a(t), b(t), c(t)$ are to be found. Show that the resulting solution is such that the boundary of the domain reaches the sink (located at the origin) in finite time and at that time becomes the region $|z + 1| < 1$.

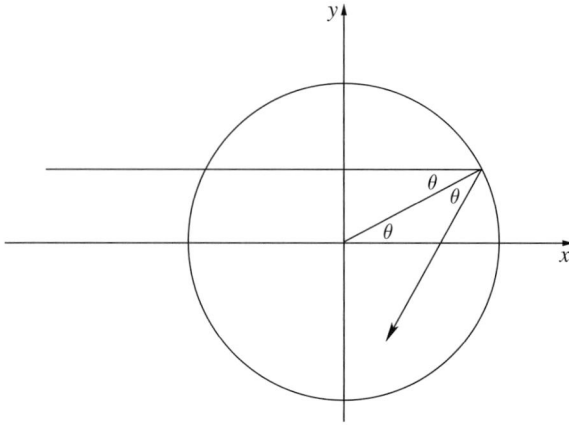

Figure 14.13 A ray reflected by a circle.

14.6   (*) By using the formulae for the limit of a Cauchy integral when the
integration curve C is approached (see Section 14.3), show that the
Polubarinova–Galin equation

$$\Re\{\bar{f}_t \xi f_\xi\} = -\frac{Q}{2\pi},$$

for the evolution of the conformal mapping $z = f(\xi, t)$ in the Hele-
Shaw problem, is equivalent to the following Löwner–Kufarev type
evolution equation:

$$f_t(\xi, t) = -\xi f_\xi(\xi, t) \frac{Q}{4\pi^2} \int_0^{2\pi} \frac{1}{|f_\xi(\xi, t)|^2} \frac{e^{i\theta} + \xi}{e^{i\theta} - \xi} d\theta.$$

14.7   Show that the conformal mapping

$$f(\xi, t) = \frac{Q}{2\pi\lambda} t - \ln\xi + 2(1 - \lambda)\ln(1 + \xi),$$

with $\xi = e^{i\theta}$, $\theta \in [0, 2\pi)$, and $\lambda \in (0, 1)$, is a solution for the
Polubarinova–Galin equation. Interpret the solution as a finger (the
so-called Saffman–Taylor finger) which moves with speed $Q/(2\pi\lambda)$.

14.8   A simple way to find a caustic line is to characterize it as the *enve-
lope* of rays. This means that two neighboring rays almost coincide and
produce a brighter line.
   (i) Consider a beam of parallel rays incident on a unit circle parame-
       terized by the angle θ; see Fig. 14.13. Show that if the law of equal
       angles is obeyed, the equation for the reflected ray is

       $$(y - \sin\theta)\cos 2\theta = (x - \cos\theta)\sin 2\theta.$$

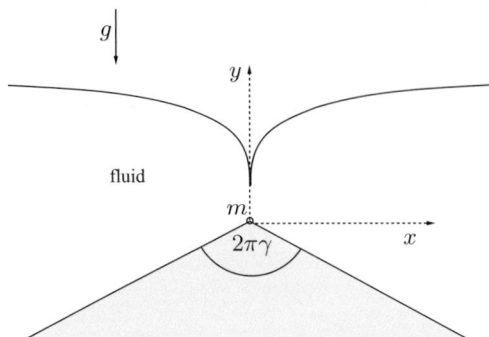

Figure 14.14 A source with strength m is placed at the top of a ridge with opening angle $2\pi\gamma$.

(ii) From the condition that the ray is stationary with respect to θ, derive the parametric equations for the caustic,

$$x = \cos\theta - \frac{1}{2}\cos\theta\cos 2\theta, \quad y = \sin\theta - \frac{1}{2}\cos\theta\sin 2\theta. \quad (14.95)$$

(iii) Show that the tip of the caustic is the usual cusp singularity. In fact, the curve (14.95) is a close relative of the cycloid of Exercise 1.10, in that it is generated by a circle of radius $1/2$ rolling off the interior of the unit circle, a so-called nephroid [176].

14.9 Analyze the generalized swallowtail (14.28) for $\epsilon_s < 0$, assuming symmetry, i.e. $\eta_1 = 0$.
   (i) Show that for $\eta_2 = \epsilon_s$ the curve becomes singular at $\varphi = \pm\sqrt{-\epsilon_s}$.
   (ii) Putting $\varphi = \pm\sqrt{-\epsilon_s} + \psi$ and expanding, show that the singularities are cusps oriented in the direction $(\pm\sqrt{-\epsilon_s}, -\epsilon_s)$.
   (iii) Now taking $\eta_2 = \epsilon_s + \epsilon$, show that the cusp unfolds locally into the form (14.12).

14.10 An exact solution for the special case of free surface potential flow *with gravity* can be found for the geometry shown in Fig. 14.14 [191]; however, surface tension has been neglected. A detailed calculation shows that the free surface is described by the differential equation [70]

$$\frac{dz}{dl} = -i\left(\frac{2}{3}\right)^{1/3}\frac{1-l}{l^{1/3}(1+l)^{4/3}}, \quad (14.96)$$

where $l = \exp(-i\theta)$, $-\pi \le \theta \le \pi$. The units of length are chosen as $(m^2/18\pi^2 g)^{1/3}$. Show that at the line of symmetry the free surface makes a generic cusp, and compute the coefficients.

14.11 (*) Consider a fluid drop with the free surface shown in Fig. 14.15. The potential flow inside the drop is driven by a double vortex; the surface of the drop is a streamline. We neglect surface tension, so the flow speed q

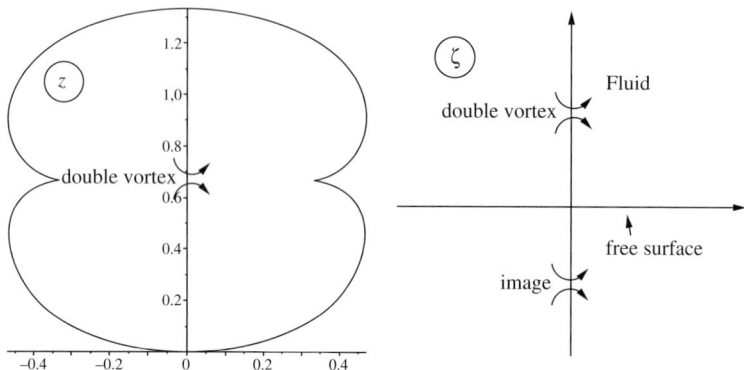

Figure 14.15 The mapping $z(\zeta)$ maps the right-hand part of the interior of the drop to the upper half of the complex ζ-plane. The real ζ-axis corresponds to the free surface of the drop, the double vortex in the interior of the drop is mapped to $\zeta = i$. An image singularity is located at $-i$.

on the surface is constant; and choose units such that $q = 1$. This means that we have to find a complex potential $w(z)$ which is holomorphic inside the drop save for a singularity $w \sim 1/(z - z_0)$ at the position z_0 of the double vortex. On the surface of the drop, $\Im\{w\} = \text{const}$ and $|dw/dz| = 1$.

(i) Following Hopkinson [112], we solve the problem by constructing a conformal map $z(\zeta)$ which maps the inside of the drop (the z-plane) to the upper half-plane (the ζ-plane); see Fig. 14.15. We choose $z(\zeta)$ such that $z_0 = z(i)$. Now, the complex velocity in the ζ-plane reads

$$\frac{dw}{d\zeta} = \frac{dw}{dz}\frac{dz}{d\zeta}. \tag{14.97}$$

Show that by adding an image vortex at $\zeta = -i$ we obtain

$$\frac{dw}{d\zeta} = \frac{1}{(\zeta - i)^2} + \frac{1}{(\zeta + i)^2} = 2\frac{\zeta^2 - 1}{(\zeta^2 + 1)^2}, \tag{14.98}$$

where $w(\zeta)$ satisfies the boundary conditions.

(ii) Noticing that

$$\Omega \equiv \ln\frac{dz}{dw} \tag{14.99}$$

is purely imaginary on the real ζ-axis, argue that

$$\Omega = 2 \ln \frac{\zeta - i}{\zeta + i}.$$ (14.100)

Demonstrate that this implies

$$\frac{dz}{d\zeta} = 2 \frac{\zeta^2 - 1}{(\zeta + i)^4}.$$ (14.101)

(iii) Show that the shape of the drop reads, in parametric form,

$$x = -\frac{2\zeta}{3} \frac{3\zeta^4 - 2\zeta^2 + 3}{(\zeta^2 + 1)^3}, \qquad y = \frac{4}{3} \frac{3\zeta^4 + 1}{(\zeta^2 + 1)^3},$$ (14.102)

where ζ is real. Analyze (14.102) to show that the drop surface has two cusps, of the generic form (14.12) with $\epsilon = 0$.

14.12 (*) Consider a drop like that of the preceding exercise but which in addition to the double vortex is driven by a vortex of strength m relative to the double vortex. Thus the complex potential has the local form $w \sim 1/(z - z_0) - im \ln(z - z_0)$ near the singularity.

(i) Show that

$$\frac{dw}{d\zeta} = \frac{2(1 + m)(\zeta^2 - \gamma^2)}{(\zeta^2 + 1)^2},$$

where $\gamma = \sqrt{(1 - m)/(1 + m)}$. In the following, we restrict ourselves to the case $m < 1$, so γ is real.

(ii) Using (14.100) show that

$$\frac{dw}{d\zeta} = \frac{2(1 + m)(\zeta^2 - \gamma^2)}{(\zeta + i)^4}$$

and thus that the drop shape is given by

$$x = -2(1 + m)\zeta \frac{3\zeta^4 - (1 + \gamma^2)\zeta^2 + 3\gamma^2}{3(\zeta^2 + 1)^3},$$ (14.103)

$$y = 2(1 + m)\frac{6\zeta^4 + 3(1 - \gamma^2)\zeta^2 + 1 + \gamma^2}{3(\zeta^2 + 1)^3},$$ (14.104)

where ζ is real.

(iii) Demonstrate that, setting $m = 1 + \epsilon_1$, the local behavior of the boundary near $(x, y) = (0, 4/3)$ is described by (14.28) with $\eta_2 = \epsilon_s$.

14.13 A space–time curve (in arbitrary space dimensions) is described by a vector $\mathbf{x}(\varphi, t)$, where φ is a real parameter that measures the position along the curve. A cosmic string in flat space–time is described by the equations [213]:

$$\dot{\mathbf{x}} \cdot \mathbf{x}' = 0, \tag{14.105}$$

$$\dot{\mathbf{x}}^2 + \mathbf{x}'^2 = 1, \tag{14.106}$$

$$\ddot{\mathbf{x}} = \mathbf{x}''. \tag{14.107}$$

Here the dot refers to differentiation with respect to time and the prime to differentiation with respect to φ.

(i) Confirm that the wave equation (14.107) is solved by

$$\mathbf{x}(\varphi, t) = \frac{1}{2} [\mathbf{a}(\varphi - t) + \mathbf{b}(\varphi + t)], \tag{14.108}$$

where

$$\mathbf{a}' = \mathbf{b}' = 1,$$

but otherwise $\mathbf{a}$ and $\mathbf{b}$ are arbitrary smooth functions.

(ii) Show that the curvature of the string becomes infinite if $\mathbf{a}' = -\mathbf{b}'$.

(iii) Assume that a singularity occurs at $\varphi = t = 0$, so $\mathbf{a}, \mathbf{b}$ have the local expansions

$$\mathbf{a} = \mathbf{a}_0' \varphi + \frac{\mathbf{a}_0''}{2} \varphi^2 + \frac{\mathbf{a}_0'''}{6} \varphi^3 + \dots,$$

$$\mathbf{b} = \mathbf{b}_0' \varphi + \frac{\mathbf{b}_0''}{2} \varphi^2 + \frac{\mathbf{b}_0'''}{6} \varphi^3 + \dots.$$

Use these expansions to show that

$$\mathbf{x}(\varphi, 0) = \frac{\mathbf{a}_0'' + \mathbf{b}_0''}{4} \varphi^2 + \frac{\mathbf{a}_0''' + \mathbf{b}_0'''}{12} \varphi^3 + \dots,$$

i.e. the curve momentarily develops a cusp, if $\mathbf{a}_0'' + \mathbf{b}_0'' \neq 0$.

14.14 In two dimensions a general solution to (14.105)–(14.107) can be found in the form [113]

$$x'(\varphi, t) = \cos(f - g) \cos(f + g), \tag{14.109}$$

$$z'(\varphi, t) = \cos(f - g) \sin(f + g), \tag{14.110}$$

$$\dot{x}(\varphi, t) = -\sin(f - g) \sin(f + g), \tag{14.111}$$

$$\dot{z}(\varphi, t) = \sin(f - g) \cos(f + g), \tag{14.112}$$

where $f = f(\varphi + t)$ and $g = g(\varphi - t)$ are any two (smooth) functions and φ is a parameter; the primes and the overdots refer to differentiation with respect to φ and t, respectively.

(i) Show that the curvature of the curve $\mathbf{X}(\varphi, t)$ can be written as

$$\kappa(\varphi, t) = \frac{f' + g'}{\cos(f - g)}. \tag{14.113}$$

(ii) (*) Argue that, if the curve described by (14.109)–(14.112) is closed, it must have a singularity around its perimeter. Hint: for the curve to be closed, the range of $f + g$ must be a full period $[0, 2\pi]$; thus show that there exists a time t_0 for which $|f - g| = \pi$.

14.15 To investigate the formation of cusp singularities described by (14.109), (14.110), we expand locally around the point $f - g = \pi/2$, i.e. we set [81]

$$f(\zeta) = \frac{\pi}{4} + f_1\zeta + \frac{f_2}{2}\zeta^2 + O(\zeta^3),$$
$$g(\zeta) = \frac{-\pi}{4} + g_1\zeta + \frac{f_2}{2}\zeta^2 + O(\zeta^3). \tag{14.114}$$

(i) Demonstrate that we must have $f_1 = g_1$ for the singularity to occur first at $t = 0$.

(ii) Show that (14.114) results in a swallowtail singularity of the type (14.28) with $\eta_2 = \epsilon_s$ and $\eta_1 = 0$.

15

Contact lines and cracks

15.1 Driven singularities

In this chapter we discuss two examples of singularities whose motion is controlled by a balance of external driving on a large scale and energy dissipation on a microscopic scale: the first is contact line motion and the second is crack propagation; see Fig. 15.1. As these problems are very different in their physical manifestations they are rarely discussed together, yet their mathematical structures are very similar. In the contact line problem the singularity occurs at the edge of a drop, the so-called contact line. In the crack problem the singularity is at the tip of the crack.

The core of the problem is an intermediate region, which represents the universal singularity and which connects the small and large scales. Toward large scales the intermediate solution matches to an outer solution, which represents the specific geometry of the problem, for example the shape of the elastic body and its loading (in the crack problem) or the shape of the drop (in the contact line problem). Toward small scales the singularity is cut off by phenomena that take place on a microscopic scale and are therefore dependent on the specific system under study.

In both cases it is instructive to consider the problem from the point of view of the energy flux through the system. The energy input is supplied from the large scale and is consumed on the microscopic scale. The balance between the two determines the propagation of the singularity.

15.2 A spreading drop

Three-phase contact lines occur very commonly, for example when a drop of water is attached to a windowpane. In this case the drop is bounded by a line where water, glass, and air meet. Very often these contact lines are observed to

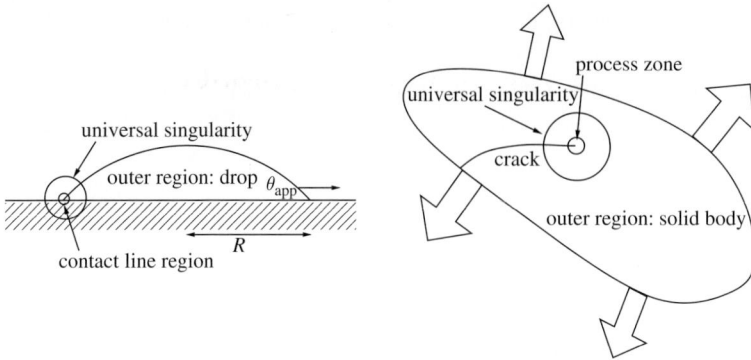

Figure 15.1 Two moving singularities. On the left, a moving contact line at the edge of a spreading drop; on the right, a moving crack. For each problem, a very similar hierarchy of three asymptotic regions can be identified.

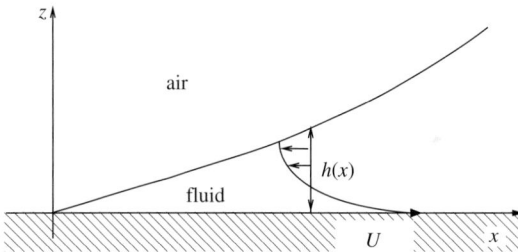

Figure 15.2 Parabolic flow in a wedge of fluid near an advancing contact angle. The substrate moves with speed U to the right.

move; the drop runs down the windowpane or a drop placed on a flat surface spreads. It would seem a straightforward exercise to describe this motion using the equations of fluid motion: inside the fluid drop we solve the Navier–Stokes or Stokes equations, subject to a boundary condition of vanishing shear stress on the free surface and a no-slip condition on the solid. However, solutions to the continuum equations in which the contact line is moving do not exist.

This difficulty arises because at the contact line the velocity field becomes multivalued. This is best appreciated by considering a frame of reference in which the contact line is stationary; see Fig. 15.2. The fluid surface is approximated as a wedge with profile $h(x) \approx \theta x$, θ being the (small) contact angle. Here, and throughout the remainder of this chapter, we will always consider the *lubrication limit*, where $\tan \theta \approx \theta$, and thus will speak of angles and the interface slope interchangeably. If the contact line were to move with speed U to the left, in our frame of reference this is equivalent to the solid moving with speed U to the right. Owing to the no-slip condition, the fluid at the bottom has to move with it. However, the total flux through the wedge has to be zero, so near the free surface the fluid must be moving in the opposite direction (see Fig 15.2). This means that the velocity gradient over the wedge can be estimated as

$$\frac{\partial u}{\partial z} \approx \frac{U}{h(x)} \approx \frac{U}{\theta x}.$$

Bearing in mind that the largest gradients occur in the z-direction normal to the solid, the rate of viscous dissipation per unit volume (4.16) in a fluid of viscosity η is approximately

$$\epsilon \approx \eta \left(\frac{\partial u}{\partial z}\right)^2 \approx \eta \left(\frac{U}{\theta x}\right)^2.$$

Integrating this over the wedge of fluid, one arrives at an estimate of the dissipation per unit time and per unit length of contact line:

$$D_{\text{visc}} \simeq \eta \int_L^R \left(\frac{U}{h}\right)^2 h \, dx = \eta \frac{U^2}{\theta} \ln \frac{R}{L}, \qquad (15.1)$$

where R is an appropriate outer length scale, such as the radius of the spreading drop. In order to get a finite answer we are forced to introduce a very small cutoff length L, associated with the breakdown of the continuum description. We anticipate that L will be approximately the size of an atom, or about a nanometer.

An experiment typical of those that we are aiming to analyze is shown in Fig. 15.3. A drop of silicone oil is placed on a flat and chemically inert surface. The drop spreads since the total surface energy of the system is lower if the solid is covered by silicone oil. We would like to discover the time dependence of the drop radius $R(t)$ by solving the thin film equation (6.29). In view of (15.1) it is clear that this can be done only if the singularity at the contact line can be regularized by introducing a microscopic (molecular) parameter into the description; otherwise, the power needed to move the drop would be infinite.

We can use energy conservation to obtain a first estimate for the rate of spreading, balancing the available surface energy and the energy dissipation (15.1). If the spreading is slow, viscous forces are not important near the center of the drop and the drop shape is determined by capillary equilibrium alone. We will restrict ourselves to the case of small drops, for which gravity is not important and so the drop assumes the shape of a spherical cap, as can be confirmed by an inspection of Fig. 15.3(a). Anticipating the more detailed discussion below, at constant volume V a drop of radius R has the profile

(a)

(b)

Figure 15.3 Two stages of the spreading of a silicone drop on a glass substrate [45]: (a) at $t = 3.25$ s after deposition, (b) at $t = 18.25$ s. The images include the drop's reflection by the solid substrate; the solid surface is thus located at the line of symmetry of the lens-shaped region. The drop volume is $V = 1.7 \times 10^{-4} \, \text{cm}^3$ and the capillary speed is $\gamma/\eta = 10.6$ cm/s. Although the equilibrium contact angle is zero the "apparent" angle is finite and the drop is well fitted by a spherical cap, represented by the curve in (a).

$$h(r) = \frac{2V}{\pi R^2} \left[1 - \left(\frac{r}{R} \right)^2 \right],$$

using the small-angle approximation. This profile makes an angle

$$\theta = -h'(R) = \frac{4V}{\pi R^3}$$

with the substrate, later to be called the *apparent contact angle* since it is determined experimentally by fitting a spherical cap to the macroscopic part of the drop.

The distinction between the apparent and actual contact angles is crucial, since the drop does not in fact end at the apparent contact line determined by such a fit but merges smoothly into a so-called precursor film, which spreads ahead of the drop. Accordingly, the profile angle eventually goes to zero as one moves away from the drop and into the precursor film region. An important consequence is that only the liquid–vapor surface energy matters for this problem, as the relevant part of the solid is covered by liquid already. The available surface free energy is determined by the area of the drop relative to the area of the flat surface covered by the drop:

$$F_s = \gamma \left(2\pi \int_0^R r\sqrt{1 + h'^2(r)}\,dr - \pi R^2 \right). \tag{15.2}$$

In the limit of small slopes $h' \ll 1$ we obtain

$$F_s = \gamma \pi \int_0^R r h'^2(r) dr = \gamma \pi \int_0^R \frac{16 V^2 r^3}{\pi^2 R^8} dr = \frac{4\gamma V^2}{\pi R^4}.$$

To obtain an equation of motion for R, we balance the rate of energy release,

$$-\dot{F}_s = \frac{16\gamma V^2}{\pi R^5} \dot{R},$$

with the total energy dissipation (see (15.1))

$$2\pi R D_{\text{visc}} = \frac{\eta \pi^2 R^4}{2V} \dot{R}^2 \ln \frac{R}{L},$$

where we have inserted the expression for the angle θ and $U = \dot{R}$. The result is the equation of motion

$$\dot{R} = \frac{32\gamma V^3}{\pi^3 \eta R^9} \left(\ln \frac{R}{L} \right)^{-1}. \tag{15.3}$$

Neglecting the slow logarithmic variation and integrating, this leads to the scaling

$$R \propto t^{1/10},$$

known as *Tanner's law* [29].

To perform a more systematic treatment we need to account more carefully for the contact line region, where the macroscopic drop transitions to the precursor film. Thus, as discussed above, the spreading drop problem is split into three different regions. The key is the intermediate region, which "negotiates" between the drop and the contact line and represents the universal singularity. The way in which the singularity is cut off at the contact line depends on the nature of the problem. In the case of complete wetting, to be discussed here, the precursor film provides the cutoff. This film represents the "inner" part of the solution. Finally, the "outer" solution corresponds to the drop itself, which is close to hemispherical in the case of very slow spreading. These regions need to be *matched*; this ensures that information is passed from the outer part of the solution down to the microscopic scales.

15.2.1 Voinov solution: the universal singularity

We address the problem of contact line motion within the framework of the thin film equation (6.29). We are interested in the dynamics close to the contact line, which is dominated, as we have seen, by a very small length scale L. Relative to this scale the radius of curvature of the contact line can be considered as large, and we can use the one-dimensional thin film equation (6.28). Closer to the center of the drop the axial curvature has to be taken into account, as we will do below.

Introducing the distance $x = R - r$ from the contact line, we are in the frame of reference of the contact line; see Fig. 15.2. We are looking for traveling wave solutions $h(x) = h(r, t)$ of the profile, assuming that the dominant time dependence comes from the motion of the contact line while the shape of the profile remains almost constant. With a contact line speed $U = \dot{R}$, the time derivative becomes

$$\frac{\partial h}{\partial t}(r, t) = \frac{\partial}{\partial t} h(R - r) = \dot{R} h'(x) = U h'(x), \tag{15.4}$$

and the one-dimensional thin film equation (6.28) yields

$$3 \operatorname{Ca} h' + \left(h''' h^3 \right)' = 0; \tag{15.5}$$

the capillary number $\operatorname{Ca} = U/v_\eta$ was introduced in (6.12); it measures the relative strength of the viscous and surface tension forces.

The traveling wave equation (15.5) can of course be integrated to give $3 \operatorname{Ca} h + h''' h^3 = \text{const}$. Near the contact line, h becomes microscopic, so on intermediate scales, which we are studying here, we can put $\text{const} = 0$ and obtain

$$\frac{3 \operatorname{Ca}}{h^2} = -h'''. \tag{15.6}$$

This equation represents a balance between the viscous forces, on the left, and the surface tension forces, on the right. The divergence of the viscous forces as h goes to zero is a reflection of the diverging shear rates near the contact line, discussed in the previous section. As a result h''' also has to diverge, indicating that the curvature becomes large and the interface is strongly deformed.

A first observation is that (15.6) is scale invariant, so we can construct a *similarity solution*

$$h = \operatorname{Ca}^{1/3} L H \left(\frac{x}{L} \right), \tag{15.7}$$

where the length L is still to be determined. Inserting (15.7) into (15.6), we find the following equation for the similarity function $H(\xi)$:

$$H''' = -\frac{3}{H^2}. \tag{15.8}$$

We remark that (15.8) has in fact an exact solution expressible in terms of Airy functions; this is described in detail in [68]. Although very useful for a variety of problems, we will not employ it here but rather rely on asymptotic methods to extract the necessary information. The length L in (15.7) will turn out to be the microscopic length at which the contact line singularity is regularized.

To be able to match (15.7) to the drop, which has a macroscopic scale R, we need to find asymptotic solutions to (15.8) in the limit $\xi = R/L \to \infty$. In view of the scale invariance of (15.8), a natural guess is the power law $H = A\xi^{\alpha}$, which leads to

$$A^3\alpha(\alpha - 1)(\alpha - 2)\xi^{\alpha-3} = -3\xi^{-2\alpha}. \tag{15.9}$$

The powers balance for $\alpha = 1$, but this does not give a solution of (15.9) since the prefactor on the left is then zero. We will see below that logarithmic corrections arise from this cancellation. The next possible solution which leads to an asymptotic balance is $\alpha = 2$, since the right-hand side of (15.9) vanishes for large ξ. It is therefore natural to try the ansatz

$$H = a\xi^2 + b\xi + c + \epsilon(\xi), \quad \xi \to \infty, \tag{15.10}$$

in (15.8); this leads to

$$\epsilon''' \approx -\frac{3}{a^2\xi^4}, \quad \xi \to \infty. \tag{15.11}$$

This equation has a decaying solution,

$$\epsilon \approx \frac{1}{2a^2\xi}, \quad \xi \to \infty, \tag{15.12}$$

so (15.10) indeed describes the generic asymptotic behavior for large ξ.

However, the solution (15.10) does not have the right properties to match to the drop. Note that the similarity solution (15.7) has curvature

$$h'' = \frac{Ca^{1/3}}{L}H''\left(\frac{x}{L}\right). \tag{15.13}$$

However, $H''(\infty) = 2a$ according to (15.10) so on the one hand the physical profile $h(x)$ has curvature $h'' \to 2aCa^{1/3}/L$ as $x/L \to \infty$, which is of order $1/L$. On the other hand, the outer solution will have a typical curvature $1/R \ll 1/L$, so the two solutions do not match. The only way (15.13) can be consistent in the limit $L \to 0$ is if H satisfies the *matching condition*

$$H'' \to 0, \quad \xi \to \infty. \tag{15.14}$$

This implies that $a = 0$ in (15.10), but it is clear from the remainder (15.12) that a power law ansatz is not consistent in that case; logarithmic corrections are called for. We will try

$$H = c\xi(\ln\xi)^\beta, \tag{15.15}$$

and look for a leading-order balance. One finds

$$H' = c\left[(\ln\xi)^{\beta} + \beta(\ln\xi)^{\beta-1}\right] \approx c(\ln\xi)^{\beta}$$

and similarly, to leading order,

$$H''' \approx -c\beta\frac{(\ln\xi)^{\beta-1}}{\xi^2}.$$

Thus from (15.6) we find that

$$\frac{3}{c^2(\ln\xi)^{2\beta}} \approx c\beta(\ln\xi)^{\beta-1}, \tag{15.16}$$

from which we deduce that $\beta = 1/3$ and $c = 3^{2/3}$.

This means that, to leading order, the solution of (15.6) which is consistent with the matching condition (15.14) is

$$H(\xi) = \xi(9\ln\xi)^{1/3}. \tag{15.17}$$

To check that this form remains consistent at higher orders, we look for an asymptotic expansion of the form

$$H(\xi) = \xi\,(9\ln\xi)^{1/3}\left(1 + \frac{A}{\ln\xi} + \frac{B}{(\ln\xi)^2} + \cdots\right), \tag{15.18}$$

which satisfies (15.14). Inserting (15.18) into (15.8) and comparing powers of $\ln\xi$, one finds for example $B = -(A^2 + 10/27)$; all other coefficients can also be found in terms of the single parameter A. The origin of this free constant is the *scale invariance* of the similarity equation (15.8): if $H(\xi)$ is a solution, so is $\xi_0 H(\xi/\xi_0)$. In other words, H' is determined only up to an arbitrary rescaling of the independent variable.

This means that the slope can be written in the form

$$H'(\xi) = \left[9\ln(\xi/\xi_0)\right]^{1/3}\left(1 + \frac{a_2}{[\ln(\xi/\xi_0)]^2} + \frac{a_3}{[\ln(\xi/\xi_0)]^3} + \cdots\right), \tag{15.19}$$

with a single adjustable parameter $\xi_0 = \exp(-3A)$. This choice makes the coefficient a_1 disappear; since there is only one free parameter in the expansion (15.18), all the other coefficients are fixed, for example $a_2 = -7/27$, $a_3 = 10/81$, $a_4 = -10/9$. Note that H' goes to infinity as $\xi \to 0$, i.e. as the contact line is approached. This once more reflects the fact that there is no consistent moving-contact-line solution unless a microscopic length scale is introduced.

In terms of the original profile (15.7), the leading-order behavior is now given by

$$h'^3 = \mathrm{Ca}H'^3\left(\frac{x}{L}\right) \approx 9\mathrm{Ca}\ln\frac{x}{\xi_0 L}, \tag{15.20}$$

which is known as the *Voinov solution* [214]. The solution is valid in the range $L \ll x \ll R$, i.e. in an *intermediate* range between a microscopic neighborhood of size L near the contact line, and a macroscopic scale given by the drop radius R. This permits us to connect the inner region (the contact line) to the outer region (the drop) using (15.20). We reiterate that (15.20) obeys the matching condition (15.14), which ensures that it can be connected to an outer solution of negligible curvature.

15.2.2 Regularization: inner region

To be able to identify the microscopic length scale $\xi_0 L$ that appears in (15.20), we have to be more specific about the interaction of the liquid with the solid substrate surrounding the drop. We consider the case where it is energetically favorable for the liquid to cover the substrate, so that the drop is preceded by a thin film of about 10 nm thickness, as we discussed earlier.

The precursor film is produced since it is favored by long-range or van der Waals interactions between molecules. These interactions decay slowly (according to a power law) and thus extend beyond the range of a single interatomic distance (about a nanometer); hence the name. This means there is a thickness-dependent contribution to the free energy of a liquid film of thickness l [29]:

$$V(l) = \frac{A}{12\pi l^2}.$$

(15.21)

If the *Hamaker constant* A is positive then the energy is lowered for thick films, resulting in a repulsive effective interaction between the liquid–vapor and the solid–liquid interfaces. In other words, a repulsive interaction will tend to thicken the liquid film. The energy (15.21) results in an additional contribution to the pressure inside the film, which now reads (see (6.24))

$$p = -\gamma h'' + \frac{dV(h)}{dh} = -\gamma h'' - \frac{A}{6\pi h^3}.$$

(15.22)

With this extra term, the thin film equation (15.6) becomes

$$\frac{3\text{Ca}}{h^2} = -h''' + 3a^2 \frac{h'}{h^4},$$

(15.23)

with microscopic length scale

$$a = \sqrt{\frac{A}{6\pi\gamma}},$$

(15.24)

which is typically below a nanometer. Thus, unless h is very small (i.e. inside the precursor film), the last term on the right of (15.23) is negligible and one returns to (15.6).

Example 15.1 (Slip length) In the case of a perfectly wetting fluid (for which the equilibrium contact angle θ_{eq} vanishes), a emerges as the length scale which cuts off the logarithmic contact line singularity. In the case of partial wetting (i.e. $\theta_{eq} = -h'(R, t) > 0$), there is no mesoscopic precursor film and we can assume that $h(R, t) = 0$ at the edge of the drop. One can regularize the logarithmic singularity by introducing a slip length $\lambda > 0$; see Exercise 6.2. This implies that the fluid is allowed to slip over the substrate when sheared, and the thin film equation is then given by (6.70).

Once more looking for traveling wave solutions $h(x) = h(r, t)$ as function of the distance $x = R - r$ from the contact line, and inserting the time derivative (15.4) into (6.70), we have

$$Uh' + \frac{v_\eta}{3}\left[h'''(h^3 + 3\lambda h^2)\right]' = 0.$$

Integrating, and using the fact that $h(0) = 0$ to eliminate the constant of integration, we obtain

$$\frac{3Ca}{h^2 + 3\lambda h} = -h'''. \tag{15.25}$$

The boundary conditions at the contact line are

$$h(0) = 0, \quad h'(0) = \theta_{eq}. \tag{15.26}$$

□

Exact solutions to (15.23) are not known to exist, but it is easy to confirm that (15.23) has the similarity solution

$$h(x) = \frac{a}{Ca^{1/3}}\phi(Ca^{2/3}x/a), \tag{15.27}$$

which is the same as (15.7), with $L = a/Ca^{2/3}$. The similarity function ϕ depends on the variable $\xi = a^{-1}Ca^{2/3}x$ and satisfies the similarity equation

$$\frac{3}{\phi^2} = -\phi''' + \frac{3\phi'}{\phi^4}. \tag{15.28}$$

To find the inner solution we have to solve (15.28) subject to the matching condition (15.14), which implies the boundary condition $\phi''(\infty) = 0$. The opposite limit, $\xi \to -\infty$, is the one in which the precursor film is reached. If one tries a power law, one finds

$$\phi = -\xi^{-1}, \quad \xi \to -\infty \tag{15.29}$$

from a balance between viscosity and van der Waals forces. Surface tension (the first term on the right of (15.28)) clearly gives a subdominant contribution. Ultimately, we are only interested in the limit $\xi \to \infty$ of the part of the inner solution toward the intermediate region, expected to be of the form (15.19), with $\phi \equiv H$. Before we construct this solution we will return to the slip length model.

Example 15.2 (Inner solution for slip model) To solve the slip similarity equation (15.25), we use the similarity solution

$$h = 3\lambda H\,(\xi)\,,\ \xi = \frac{x\theta_{eq}}{3\lambda}, \tag{15.30}$$

based on the slip length. Then the similarity equation (15.25) becomes

$$H''' = -\frac{\delta}{H^2 + H}, \tag{15.31}$$

with $\delta = 3\mathrm{Ca}/\theta_{eq}^3$. On account of the conditions (15.26) at the contact line, as well as the matching condition (15.14), the boundary conditions to be satisfied by H are

$$H(0) = 0, \quad H'(0) = 1, \quad H''(\infty) = 0. \tag{15.32}$$

Below we will solve (15.31), (15.32) using an expansion in the small parameter δ, i.e. for small capillary numbers. $\qquad\square$

We want to solve the precursor-film similarity solution (15.28) subject to the condition (15.29). This suggests that we should look for an asymptotic solution in terms of a slowly varying base solution $\phi_0 \approx -1/\xi$ with a more rapidly varying perturbation $\delta(\xi)$ superimposed on it:

$$\phi(\xi) = \phi_0(\xi) + \delta(\xi); \tag{15.33}$$

the perturbation satisfies the equation

$$\delta(6\phi_0 + 4\phi_0^3\phi_0''') - 3\delta' + \delta'''\phi_0^4 = 0. \tag{15.34}$$

Since the base solution is not constant, a WKB ansatz, similar to (7.52), is called for. To leading order, we begin with the ansatz

$$\delta(\xi) = \epsilon e^{b\xi^\alpha}. \tag{15.35}$$

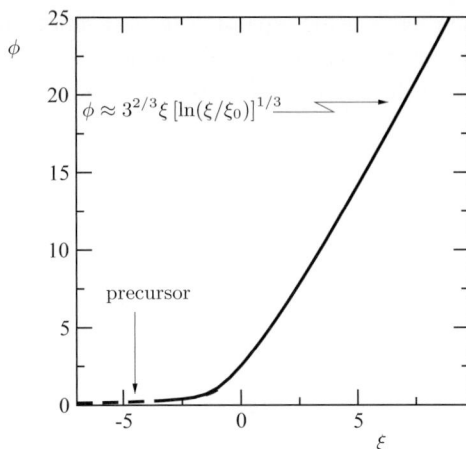

Figure 15.4 The solution of (15.28) corresponding to the inner region; the broken line is the precursor-film solution $\phi \approx -1/\xi$. To the right, ϕ approaches the Voinov solution (intermediate region) with $\xi_0 \approx 0.63$.

The two derivative terms in (15.34) provide the leading-order balance,

$$(\alpha b)^3 \xi^{3\alpha-7} \approx 3\alpha b \xi^{\alpha-1},$$

from which we find $\alpha = 3$ and $b = 1/\sqrt{3}$. Thus for $\xi \to -\infty$ we can use the description

$$\phi = -\xi^{-1} + \epsilon \exp(\xi^3/\sqrt{3}), \qquad (15.36)$$

with ϵ the only free parameter. For a more detailed expansion to higher orders, see [82].

The structure of the solution of (15.28) that we are seeking is illustrated in Fig. 15.4. For $\xi \to \infty$ it must be of the asymptotic form (15.19); our only task is to compute the value of the unknown parameter ξ_0. We can find the solution numerically by the shooting method. An initial condition $\phi(\xi_i), \phi'(\xi_i), \phi''(\xi_i)$ is computed from (15.36) at e.g. $\xi_i = -3$, which contains a single parameter ϵ. One then integrates (15.28) to a very large value (e.g. $\xi_a = 10^{10}$), adjusting ϵ so that the curvature $\phi''(\xi_a)$ vanishes. The free parameter ξ_0 is determined from solving the following nonlinear equation derived from (15.19):

$$\phi'(\xi_f) = [9\ln(\xi_f/\xi_0)]^{1/3} \left(1 + \sum_{i=2}^{\infty} \frac{a_i}{[\ln(\xi_f/\xi_0)]^i}\right);$$

the left-hand side is determined from the numerical solution at some intermediate fitting point ξ_f, with $1 \ll \xi_f \ll \xi_a$. In practice it is enough to take into account the first few terms in the expansion and to confirm that ξ_0 is independent of ξ_f. The result is $\xi_0 \approx 0.63$, so that to leading order the asymptotic solution in the intermediate limit $\xi \to \infty$ is

$$\phi'(\xi) = \left[9\ln(\xi/0.63)\right]^{1/3}. \tag{15.37}$$

Using the similarity solution (15.27) it follows that

$$h'^3(x) = \mathrm{Ca}\left[\phi'\left(\frac{\mathrm{Ca}^{2/3}x}{a}\right)\right]^3, \tag{15.38}$$

so that

$$h'^3(x) \approx 9\,\mathrm{Ca}\ln\frac{\mathrm{Ca}^{2/3}x}{0.63a}, \tag{15.39}$$

which is the intermediate limit of the inner solution. This is indeed the anticipated structure, if in (15.20) one identifies the length scale

$$L\xi_0 = 0.63a\,\mathrm{Ca}^{-2/3}. \tag{15.40}$$

Note that, somewhat surprisingly, this length scale depends on the capillary number and thus on the contact-line speed [82]. Our next task will be to find a corresponding expression based on an analysis of the outer (drop) region, which we will then match to (15.39).

Example 15.3 (Outer asymptotics of the slip model) To derive an expression analogous to (15.39), we solve the slip model (15.31) using the perturbation expansion

$$H(\xi) = H_0(\xi) + \delta H_1(\xi) + O(\delta^2).$$

At order $O(\delta^0)$ (15.31) becomes $H_0''' = 0$. Integrating three times, we have

$$H_0 = a + b\xi + c\xi^2.$$

The boundary conditions (15.32) have to be satisfied for any δ, so they also have to be satisfied by H_0. But this means that $a = 0$, $b = 1$, and $c = 0$, so that $H_0 = \xi$.

At order $O(\delta)$ the equation for H_1 is now

$$H_1''' = -\frac{1}{\xi^2 + \xi}, \tag{15.41}$$

having inserted H_0 on the right-hand side of (15.31). A first integral of (15.41) is

$$H_1'' = \ln(1 + \xi^{-1}) + \text{const},$$

and using $H_1(\infty) = 0$ it follows that const $= 0$. Integrating a second time, we have

$$H_1' = (1 + \xi) \ln(1 + \xi) - \xi \ln \xi + \text{const}.$$

Now, the boundary condition at the contact line is $H_1'(0) = 0$ since $H'(0) = H_0(0) = 1$. It follows that again the constant is 0, and there is no need to find H_1 itself since we are calculating the solution to first order in δ only.

In the limit $\xi \to \infty$ the result is

$$H' = 1 + \delta \left[\ln \xi + 1 + O(\xi^{-1}) \right] + O(\delta^2). \tag{15.42}$$

Using the similarity form (15.30), and raising the result to the third power, we obtain

$$h'^3(x) \approx \theta_{eq}^3 + 9 \, \mathrm{Ca} \ln \frac{x e \theta_{eq}}{3\lambda}, \tag{15.43}$$

which corresponds to (15.39) in the case of the slip model. Note that we have separated the contribution from the equilibrium contact angle θ_{eq}, which vanishes in the case of complete wetting. $\qquad \square$

15.2.3 The drop: outer region

Now we consider the scale of the drop. Away from the contact line, viscous forces are not important; thus the shape of the drop must be an equilibrium surface. Since we are considering small drops below the scale of the capillary length ($R < \ell_c$), gravity is not important either, and the drop is a surface of constant curvature,

$$\triangle h_0 = \frac{1}{r} \frac{\partial}{\partial r} \left(r \frac{\partial h_0}{\partial r} \right) = \text{const} \tag{15.44}$$

(by contrast large drops, for which gravity is important, are "puddles" with a flat top and curved edge). For a given radius R, the volume V of the drop can be calculated by integrating over the footprint A of the drop:

$$V = \int_A h_0 d^2 x = 2\pi \int_0^R r h_0 dr. \tag{15.45}$$

The solution of (15.44) with $h_0(R) = 0$ for a given volume V is

$$h_0(r) = \frac{2V}{\pi R^2} \left[1 - \left(\frac{r}{R} \right)^2 \right]. \tag{15.46}$$

In this approximation, where h_0 denotes the static solution for which all viscous effects are neglected, the drop meets the solid at an *apparent* contact angle

$$\theta_{\text{app}} = \frac{4V}{\pi R^3}.$$ (15.47)

The solution (15.46), shown as the solid curve in Fig. 15.3, is indeed an almost perfect fit on the scale of the drop. Thus, on the basis of such a macro-scopic measurement, one would predict θ_{app} to be the contact angle at which the fluid meets the solid. However, in the equilibrium state the solid is covered by a film, so in reality the contact angle is zero: $\theta_{\text{eq}} = 0$. Since $\theta_{\text{app}} > \theta_{\text{eq}}$, the drop must be in a non-equilibrium state in which it attempts to reach the vanishing equilibrium angle and hence it spreads. The gradual increase in the slope of the profile is described by the intermediate solution (15.39). The mathematical problem that presents itself is that (15.39) describes a slope that is varying continuously rather than approaching a well-defined limit. Thus we must find a correction to (15.46) which matches to (15.39) in the contact line region.

To this end we perform an expansion in the capillary number of which h_0 is the zeroth order:

$$h_{\text{outer}}(r) = h_0(r) + \text{Ca}\, h_1(r) + O(\text{Ca}^2).$$ (15.48)

According to our analysis of the intermediate region, we know that the structure of the solution is determined by (15.20) to all orders in the capillary number. Thus we anticipate that the outer solution can be written in the form

$$h_{\text{outer}}'^3(r) \approx -\theta_{\text{app}}^3 - 9\,\text{Ca} \ln\left(\frac{R-r}{bR}\right),$$ (15.49)

since in the limit of vanishing speed $h_{\text{outer}}' = h_0' = -\theta_{\text{app}}$; $R - r \equiv x$ is the distance from the contact line. The variable x is measured from the contact line into the drop, while r is measured in the opposite direction, which accounts for the minus signs in (15.49). The asymptotic form (15.49) is the inner limit, toward the intermediate region, of the outer solution.

Our goal is to calculate the constant b, for which we use the perturbation expansion (15.48). From (15.48) we obtain

$$h_{\text{outer}}'^3(r) = -\theta_{\text{app}}^3 + 3\,\text{Ca}\, \theta_{\text{app}}^2 h_1'(r),$$ (15.50)

which must match the structure given by (15.49); comparing, we find

$$h_1'(r) = -\frac{3}{\theta_{\text{app}}^2} \ln\left(\frac{R-r}{bR}\right).$$ (15.51)

This determines the structure of the first-order correction in (15.48), which follows by taking into account only the outer, hydrodynamic, features of the problem. As a result, (15.51) encounters a logarithmic singularity for $r = R$.

However, comparing (15.48) and (15.51) for $r < R$, we can determine the constant b.

To describe the drop dynamics on the scale R, it is necessary to use the full two-dimensional description (6.29). The time derivative of h_0 is contained in R alone, and therefore

$$v_\eta^{-1} \frac{\partial h_0}{\partial t} = \mathrm{Ca} \frac{\partial h_0}{\partial R}, \tag{15.52}$$

since $U = \dot{R}$ is the speed of the contact line. Thus to leading (i.e. linear) order in the capillary number, (6.29) becomes

$$-3 \frac{\partial h_0}{\partial R} = \nabla \left[h_0^3 \nabla \Delta h_1 \right] \equiv \mathcal{L} h_1. \tag{15.53}$$

Here we have introduced the linear operator

$$\mathcal{L} = \nabla \cdot \left(q \, \nabla \nabla^2 \right). \tag{15.54}$$

For brevity we have put $h_0^3 \equiv q$.

Now we will show that it is possible to derive an equation for b without solving (15.53) explicitly [170]. To this end we multiply (15.53) by a suitable test function φ, to be found below, and integrate over the footprint A of the drop:

$$\int_A \psi \varphi d^2 x = \int_A \varphi \mathcal{L} h_1 d^2 x, \tag{15.55}$$

where we have defined

$$\psi(r) \equiv -3 \frac{\partial h_0}{\partial R} = 3\theta_{\mathrm{app}} \left[1 - 2 \left(\frac{r}{R} \right)^2 \right]. \tag{15.56}$$

Integrating by parts four times, (15.55) can be rewritten as

$$\int_A \psi \varphi d^2 x = \int_A (\mathcal{L}^\dagger \varphi) h_1 d^2 x + \text{boundary terms}, \tag{15.57}$$

where

$$\mathcal{L}^\dagger = \Delta \left[\nabla \cdot (q \nabla) \right]$$

is the operator adjoint to $\mathcal{L}$. The idea is to choose φ such that the integral over the drop on the right of (15.57) vanishes. Once φ is known we can compute the integral on the left of (15.57), yielding an equation in terms of the boundary alone. Near the boundary, however, we can use the asymptotic description (15.51), in which b is the only unknown parameter.

To find φ, we note that the drop volume is

$$V = \int_A h d^2 x = \int_A h_0 d^2 x + \mathrm{Ca} \int_A h_1 d^2 x, \tag{15.58}$$

and so according to (15.45), h_1 must obey the boundary condition

$$\int_A h_1 d^2x = 0. \tag{15.59}$$

Thus the integral on the right of (15.57) will vanish if φ is chosen such that $\mathcal{L}^\dagger \varphi = \text{const.}$ According to (15.44) this is achieved if

$$\nabla \cdot (q \, \nabla \varphi) = h_0 + c. \tag{15.60}$$

The constant c is necessary since, from (15.60),

$$\int_A (h_0 + c) d^2x = \int_A \nabla \cdot (q \nabla \varphi) d^2x = \int_{\partial A} q(\mathbf{n} \cdot \nabla)\varphi d\sigma = 0. \tag{15.61}$$

This follows because $q = h_0^3 \to 0$ everywhere on the boundary ∂A and $\mathbf{n}$ is the normal to ∂A. The left-hand side of (15.61) is $V + Ac$, and thus $c = -V/A = -\theta_{\text{app}} R/4$.

To solve (15.60) and thus find φ, it is useful to choose cylindrical coordinates. Then (15.60) becomes

$$\frac{(rq\varphi')'}{r} = h_0 - \frac{\theta_{\text{app}} R}{4} \tag{15.62}$$

and integrating (15.62) twice, we find

$$\varphi(r) = \int_0^r \frac{1}{r' h_0^3(r')} \int_0^{r'} r'' \left(h_0(r'') - \frac{\theta_{\text{app}} R}{4} \right) dr'' dr' = \frac{r^2}{2\theta_{\text{app}}^2 (R^2 - r^2)}. \tag{15.63}$$

With (15.63) in hand we have achieved our goal: the left-hand side of (15.57) is expressed in terms of boundary values alone. Since $\varphi(r)$ has a singularity at the edge of the drop, the integral

$$\int \psi \varphi r \, dr$$

on the left-hand side of (15.57) diverges logarithmically as $r \to R$. Thus we need to compute the integrals in (15.57) up to $r = R - \epsilon$ and then investigate the limit $\epsilon \to 0$. To facilitate the calculation we put $\theta_{\text{app}} = 1$ and $R = 1$ in the next few steps, which is possible since the two are independent parameters from the point of view of the outer problem. It then follows from (15.47) that $V = \pi/4$ and, from (15.51), we also know that

$$h_1' \approx -3 \ln(1 - r) + 3 \ln b$$

near the edge of the drop. Beginning with the left-hand side of (15.57) (divided by 2π), we find

$$\int_0^{1-\epsilon} r\psi\varphi \, dr = \frac{3}{2}\int_0^{1-\epsilon} \frac{r^3(1-2r^2)}{1-r^2}\,dr$$

$$= \frac{3}{4}\left[\left(-2+r^2+r^4+\ln(1-r^2)\right)\right]_0^{1-\epsilon}$$

$$= \frac{3}{4}\left(\ln\epsilon + \ln 2 + 2\right) + O(\epsilon). \qquad (15.64)$$

To compute the boundary terms on the right-hand side of (15.57), we start with the right-hand side of (15.55) and perform a sequence of partial integrations. Note that

$$\mathcal{L}h_1 = \nabla \cdot (q\,\nabla\Delta h_1) = \frac{1}{r}\left(rq\Delta h_1'\right)',$$

and so the right-hand side of (15.55), divided by 2π, is

$$\int_0^{1-\epsilon} \left(rq\Delta h_1'\right)' \varphi\, dr = -\int_0^{1-\epsilon} rq\Delta h_1'\varphi'\, dr + \left[rq\Delta h_1'\varphi\right]_0^{1-\epsilon}$$

$$= \int_0^{1-\epsilon} \Delta h_1\left(rq\varphi'\right)'\, dr + \left[rq\Delta h_1'\varphi - \Delta h_1 rq\varphi'\right]_0^{1-\epsilon}. \qquad (15.65)$$

Using $\Delta h_1 = (rh_1')'/r$, the integral on the right is transformed by two further partial integrations:

$$\int_0^{1-\epsilon} \frac{(rh_1')'}{r}\left(rq\varphi'\right)'\, dr = -\int_0^{1-\epsilon} rh_1'\left(\frac{(rq\varphi')'}{r}\right)'\, dr + \left[h_1'\left(rq\varphi'\right)'\right]_0^{1-\epsilon}$$

$$= \int_0^{1-\epsilon} h_1\left[r\left(\frac{(rq\varphi')'}{r}\right)'\right]'\, dr + \left[h_1'\left(rq\varphi'\right)' - rh_1\left(\frac{(rq\varphi')'}{r}\right)'\right]_0^{1-\epsilon} + O(\epsilon). \qquad (15.66)$$

By the way that we have constructed φ, we know that

$$0 = \mathcal{L}^\dagger\varphi = \Delta\left[\nabla \cdot (q\nabla\varphi)\right] = \frac{1}{r}\left[r\left(\frac{(rq\varphi')'}{r}\right)'\right]'$$

and so the integral on the right of (15.66) vanishes, as it must. Thus in summary we find, collecting the boundary terms from (15.65) and (15.66),

$$\int_0^{1-\epsilon} \left(rq\Delta h_1'\right)' \varphi\, dr$$

$$\approx \left[rq\Delta h_1'\varphi - \Delta h_1 rq\varphi' + h_1'h_1'\left(rq\varphi'\right)' - rh_1\left(\left(rq\varphi'\right)'/r\right)'\right]_0^{1-\epsilon}$$

$$= rq\Delta h_1'\varphi - \Delta h_1 rq\varphi' + h_1'r\left(h_0 - \frac{1}{4}\right) - rh_1 h_0'\bigg|_{r=1-\epsilon}, \qquad (15.67)$$

using the fact that all terms are proportional to r, and so the contribution from the lower limit vanishes. We have also used the facts that, according to (15.62), $(rq\varphi')' = r(h_0 - 1/4)$ and $h_1(1) = 0$. To analyze (15.67) near the boundary, we use that, for $r = 1 - \epsilon$,

$$\varphi \approx \frac{1}{4\epsilon}, \quad q \approx \epsilon^3, \quad h_1' \approx -3\ln\epsilon + 3\ln b,$$

$$\Delta h_1 \approx \frac{3}{\epsilon} \quad \text{and} \quad \Delta h_1' \approx \frac{3}{\epsilon^2}, \tag{15.68}$$

and so

$$\int_0^{1-\epsilon} \left(rq\Delta h_1'\right)' \varphi \, dr \approx \epsilon^3 \frac{3}{\epsilon^2}\frac{1}{4\epsilon} - \epsilon^3\frac{3}{\epsilon}\frac{1}{4\epsilon^2} + 3(\ln\epsilon - \ln b)\frac{1}{4} = \frac{3}{4}(\ln\epsilon - \ln b). \tag{15.69}$$

Comparing (15.64) and (15.69) we find, of course, the same logarithmic divergence in ϵ and the constant terms give $2 + \ln 2 = -\ln b$, so finally the constant b in (15.49) is found to be given by

$$b = \frac{1}{2e^2}. \tag{15.70}$$

As a result, (15.49) becomes

$$h'^3(x) \approx \theta_{\text{app}}^3 + 9\,\text{Ca}\ln\frac{2e^2 x}{R}, \tag{15.71}$$

with $x = R - r$ and $h_{\text{outer}}'(r) = -h'(x)$.

15.2.4 The global problem: matching

To obtain the global solution to the spreading problem, we compare the outer limit (15.39) of the inner solution to the inner limit (15.71) of the outer solution. The key feature of the third power of the slope is that it exhibits the universal x-dependence $9\text{Ca}\ln x$ (cf. (15.20)) in the intermediate region. We obtain

$$9\,\text{Ca}\ln\frac{\text{Ca}^{2/3}x}{0.63a} = \theta_{\text{app}}^3 + 9\,\text{Ca}\ln\frac{2e^2 x}{R},$$

which indeed matches, with apparent contact angle

$$\theta_{\text{app}}^3 = 9\,\text{Ca}\ln\frac{R\,\text{Ca}^{2/3}}{0.63 \times 2e^2 a}. \tag{15.72}$$

Now (15.72) can be used together with (15.47) to obtain

$$\left(\frac{4V}{\pi R^3}\right)^3 = 9\,\text{Ca}\ln\frac{R\,\text{Ca}^{2/3}}{0.63 \times 2e^2 a} \equiv 9B\text{Ca}. \tag{15.73}$$

The scaling of this equation agrees with our earlier result (15.3) based on an energy balance. However, the prefactors are different since our estimate (15.1) is insufficiently accurate. Since $\text{Ca} = \dot{R}/v_\eta$, (15.73) is an equation of motion for the drop radius R. As long as $a \ll R\text{Ca}^{2/3}$ the logarithm can be taken to be a constant; for example, in the drop-spreading experiment of Fig. 15.3, $R \approx 1$ mm and $\text{Ca} \approx 4 \times 10^{-5}$, giving $B \approx 5.7$ if one uses $a \approx 2\text{Å}$ [211]. In this approximation,

$$R(t) \approx \left[\frac{10\gamma}{9B\eta} \left(\frac{4V}{\pi} \right)^3 t \right]^{1/10}, \tag{15.74}$$

which is Tanner's law.

Example 15.4 (Drop with slip) In the case of partial wetting with slip, we compare (15.71) with (15.43) instead, obtaining

$$\theta_{\text{app}}^3 + 9\,\text{Ca}\ln\frac{2e^2 x}{R} = \theta_{\text{eq}}^3 + 9\,\text{Ca}\ln\frac{x e \theta_{\text{eq}}}{3\lambda}.$$

Thus the apparent contact angle can be identified as

$$\theta_{\text{app}}^3 = \theta_{\text{eq}}^3 + 9\,\text{Ca}\ln\frac{R\theta_{\text{eq}}}{6e\lambda}. \tag{15.75}$$

In the same way as before, this can be combined with (15.47) to obtain an equation of motion for $R(t)$:

$$\dot{R} = \frac{v_\eta}{9}\left(\ln\frac{R\theta_{\text{eq}}}{6e\lambda} \right)^{-1} \left[\left(\frac{4V}{\pi R^3} \right)^3 - \theta_{\text{eq}}^3 \right],$$

which describes the relaxation of the radius R toward its equilibrium value,

$$R_{\text{eq}} = \left(\frac{4V}{\pi\theta_{\text{eq}}} \right)^{1/3}. \qquad\qquad \square$$

15.3 A moving crack

In Fig. 15.5 we show the optical birefringence pattern in the neighborhood of a straight crack loaded in the direction perpendicular to the crack. In the vicinity of the crack tip the stress and strain fields become very large, as can be seen from the experimental birefringence lines. It is the concentration of stress near the tip which drives the propagation of the crack, once a threshold is passed. In the limit where the size of the crack tip vanishes the stresses become infinite, but in reality they are cut off on a small scale, for example the length of the

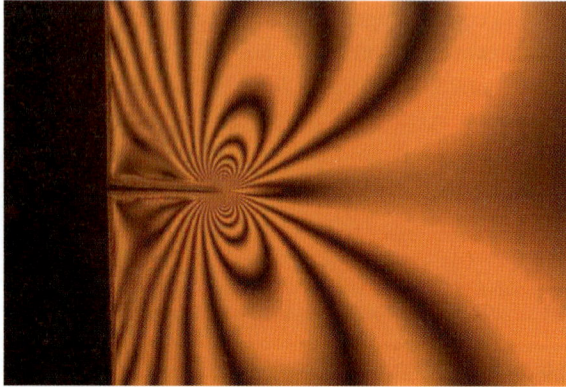

Figure 15.5 A 7.25-mm-long loaded crack in polycarbonate, as seen under monochromatic light. (Image courtesy of DoITPoMS, University of Cambridge: http://www.doitpoms.ac.uk/.) The lines of constant stress are concentrated near the crack tip, where a singularity in the equations of linear elasticity appears.

atomic bonds of the solid. This is the idealized case of brittle materials such as Plexiglas or silicon, on which we will focus here. In the case of softer materials, the cutoff is provided by plastic deformations of the material near the tip; for more details, see [88].

As indicated in Fig. 15.1, this implies a structure of the crack problem that is identical to the moving-contact-line problem of the previous section. Owing to the scale separation between atomic scales and the size of the body, there exists a universal, scale-invariant, solution close to the crack tip which is the analogue of the Voinov solution. After studying this solution, we will show how the flux of energy to the microscopic scale can be described by a formalism known as the J-integral. Finally, we explain the matching to the large-scale behavior of the body. We will follow closely the treatment presented in [136].

15.3.1 Universal tip singularity

We begin by analyzing the local behavior of stresses near the crack tip, on scales much larger than the atomic scale but much smaller than the size of the body. Even if the crack is curved, sufficiently close to the tip it can be considered straight, as shown in Fig. 15.6. We confine ourselves to planar deformations, so in the bulk of the material we have to solve the biharmonic equation (4.65):

$$\triangle^2 \phi = 0.$$

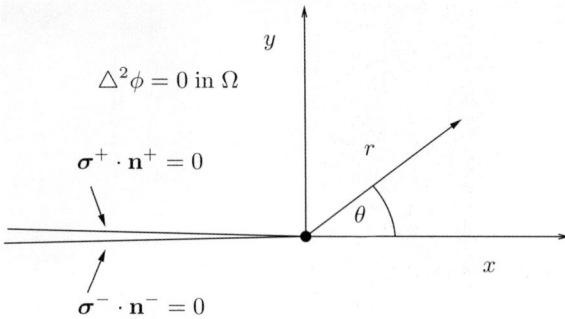

Figure 15.6 Geometry of the crack tip. The faces of the crack are stress free (cf. Fig. 14.1).

The defining characteristic of the crack faces (denoted by $\pm$ superscripts) is that they are stress-free (cf. (4.62)):

$$\sigma_{ij}^+ n_j^+ = \sigma_{ij}^- n_j^- = 0.$$

We will solve this problem using the local expansion methods of Chapter 5; in fact the problem is almost identical to the fluid cusp problem of Section 14.1, except that here we will not assume the fluid to be incompressible and will not impose any symmetry constraints on the crack.

The calculation is done most easily in polar coordinates, using the representation (4.66) for the stress tensor. The boundary conditions on the crack surfaces are

$$\sigma_{\theta\theta}^{\pm} = \frac{\partial^2 \phi}{\partial r^2}(r, \pm\pi) = 0, \quad \sigma_{r\theta} = -\frac{\partial}{\partial r}\left(\frac{1}{r}\frac{\partial\phi}{\partial\theta}\right)(r, \pm\pi) = 0. \quad (15.76)$$

To find a local solution to the biharmonic equation we use the ansatz

$$\phi = r^\lambda f(\theta)$$

as we did before for viscous flow, with solution (5.54).

The first condition in (15.76) gives $(\lambda - 1)\lambda r^{\lambda-2} f(\pm\pi) = 0$, which implies that $f(\pm\pi) = 0$, as the cases $\lambda = 1$ or 0 are to be excluded as we shall see. The second condition leads to $(\lambda - 1)r^{\lambda-2} f'(\pm\pi) = 0$, and so $f'(\pm\pi) = 0$. Inserting this into (5.54) it follows directly that

$$(A + C)\cos\lambda\pi \pm (B + D)\sin\lambda\pi = 0,$$
$$\pm[A\lambda + C(\lambda - 2)]\sin\lambda\pi + [B\lambda + D(\lambda - 2)]\cos\lambda\pi = 0,$$

or

$$
\left.
\begin{aligned}
(A + C)\cos\lambda\pi &= 0,\\
(B + D)\sin\lambda\pi &= 0,\\
[B\lambda + D(\lambda - 2)]\cos\lambda\pi &= 0,\\
[A\lambda + C(\lambda - 2)]\sin\lambda\pi &= 0.
\end{aligned}
\right\}
\qquad (15.77)
$$

It is clear that unless $\lambda = n/2$, where n is an integer, the four equations are linearly independent and the only solution is $A = B = C = D = 0$. An additional requirement the solution has to fulfill is that the elastic energy (4.55) is finite. The strain ϵ scales as $r^{\lambda-2}$, and so F behaves as $r^{2(\lambda-2)}$ near the crack tip. The volume element is $2\pi r$ in two dimensions; thus for (4.55) to be finite the exponent must satisfy $2(\lambda - 2) + 1 > -1$, or $n > 2$. Since the local behavior is dominated by the *lowest* possible exponent, it follows that $n = 3$. This means that near the crack tip the stresses diverge as $1/\sqrt{r}$ and the deformations behave as $\sqrt{r}$. This result is the exact analogue of the $\sqrt{r}$ behavior of typical *flow velocities* near a fluid cusp; see (14.8).

For $n = 3$ the coefficients in (15.77) are given by $\sin(3\pi/2) = 1$ and $\cos(3\pi/2) = 0$, and so $D = -B$ and $C = 3A$. Thus, introducing constants $K_I = A\sqrt{2\pi}$ and $K_{II} = B\sqrt{2\pi}$, one finds from (4.66) that

$$
\sigma = \frac{K_I}{4\sqrt{2\pi r}}\Sigma^I + \frac{K_{II}}{4\sqrt{2\pi r}}\Sigma^{II},
\qquad (15.78)
$$

the tensors Σ^I, Σ^{II}, being defined by the angular dependences

$$
\Sigma_{rr}^I = 5\cos\frac{\theta}{2} - \cos\frac{3\theta}{2}, \quad \Sigma_{rr}^{II} = -5\sin\frac{\theta}{2} + 3\sin\frac{3\theta}{2}, \quad (15.79)
$$

$$
\Sigma_{\theta\theta}^I = 3\cos\frac{\theta}{2} + \cos\frac{3\theta}{2}, \quad \Sigma_{\theta\theta}^{II} = -3\sin\frac{\theta}{2} - 3\sin\frac{3\theta}{2}, \quad (15.80)
$$

$$
\Sigma_{r\theta}^I = \sin\frac{\theta}{2} + \sin\frac{3\theta}{2}, \quad \Sigma_{r\theta}^{II} = \cos\frac{\theta}{2} + 3\sin\frac{3\theta}{2}. \quad (15.81)
$$

The problem being linear, the solution is a superposition of two independent modes, called Mode I and Mode II; their amplitudes K_I and K_{II} are called *stress intensity factors*. Their values depend on the body geometry, the crack configuration, and, linearly, on the external traction. However, the propagation of the crack is determined by microscopic features of the material, as we will see in the next section. Thus K_I and K_{II} contain all the necessary local information as far as the continuum description of the problem is concerned. Apart from that, the spatial structure of the singular solution is universal. This means that (15.78) is the exact analogue of the Voinov solution (15.20).

The singular deformation fields corresponding to (15.78) are of the form

$$\mathbf{u} = \frac{K_I}{4\mu}\sqrt{\frac{r}{2\pi}}\mathbf{f}^I + \frac{K_{II}}{4\mu}\sqrt{\frac{r}{2\pi}}\mathbf{f}^{II}. \tag{15.82}$$

Inserting this into (4.52) and (4.53), we have for both modes

$$\Sigma_{rr} = f_r + \frac{2v}{1-2v}\left(\frac{3f_r}{2} + f_\theta'\right),$$

$$\Sigma_{\theta\theta} = 2\left(f_\theta' + f_r\right) + \frac{2v}{1-2v}\left(\frac{3f_r}{2} + f_\theta'\right),$$

$$\Sigma_{r\theta} = f_r' - \frac{f_\theta}{2}.$$

Integrating these equations and using (15.79)–(15.81), we obtain

$$\begin{aligned} f_r^I &= (5-8v)\cos\frac{\theta}{2} - \cos\frac{3\theta}{2}, \\ f_r^{II} &= (-5+8v)\sin\frac{\theta}{2} + 3\sin\frac{3\theta}{2}, \end{aligned} \tag{15.83}$$

$$\begin{aligned} f_\theta^I &= (-7+8v)\sin\frac{\theta}{2} + \sin\frac{3\theta}{2}, \\ f_\theta^{II} &= (-7+8v)\cos\frac{\theta}{2} + 3\cos\frac{3\theta}{2}. \end{aligned} \tag{15.84}$$

Of particular interest is the deformation field for $\theta = \pm\pi$, shown schematically in Fig. 15.7, since it determines the shape of the crack opening. From (15.82) and (15.83), (15.84) one finds for Mode I

$$u_x(r, \pm\pi) = -u_r(r, \pm\pi) = 0,$$

$$u_y(r, \pm\pi) = -u_\theta(r, \pm\pi) = \pm\frac{2(1-v)K_I}{\mu}\sqrt{\frac{r}{2\pi}},$$

so that the jumps across the crack become

$$[u_x] = 0, \quad [u_y] = \frac{4(1-v)K_I}{\mu}\sqrt{\frac{r}{2\pi}}.$$

For Mode II we have

$$u_x(r, \pm\pi) = -u_r(r, \pm\pi) = \pm\frac{2(1-v)K_{II}}{\mu}\sqrt{\frac{r}{2\pi}},$$

$$u_y(r, \pm\pi) = -u_\theta(r, \pm\pi) = 0,$$

with jumps

$$[u_x] = \frac{4(1-v)K_{II}}{\mu}\sqrt{\frac{r}{2\pi}}, \quad [u_y] = 0.$$

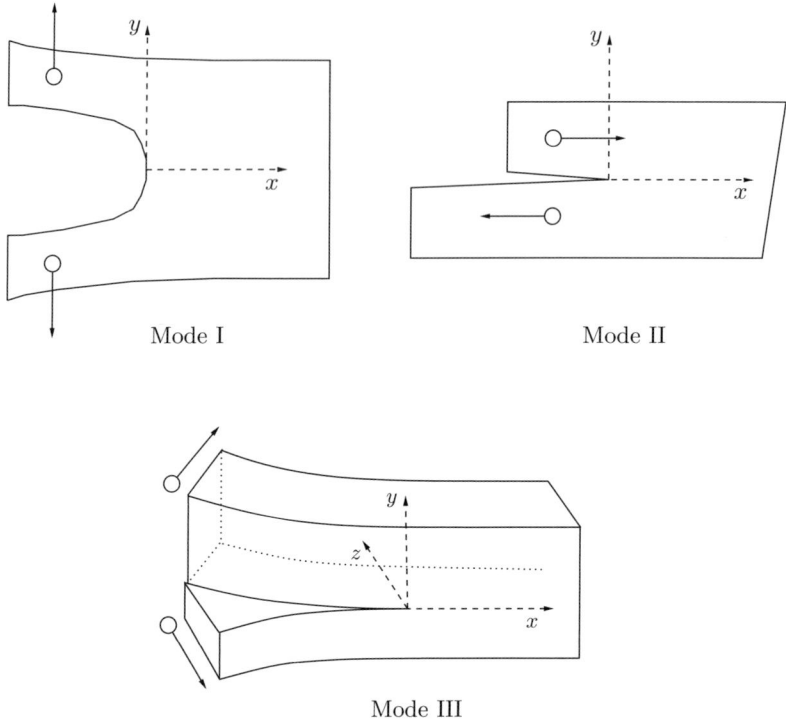

Mode I   Mode II

Mode III

Figure 15.7 Schematic of Mode I, Mode II, and Mode III cracks in plane elasticity.

In Mode I the crack is opening in the direction perpendicular to itself. The deformation in the direction of the crack is continuous along the crack. In Mode II, however, the crack is sheared and the deformation in the normal direction is continuous; see Fig. 15.7. If one allows for out-of-plane deformations, there also exists a third mode (a Mode III crack, see Fig. 15.7), which is treated in Exercises 15.6 and 15.8 (see 15.107). In this mode only u_z has nonzero components, and since it depends on x and y and not on z the mode can still be described within two-dimensional elasticity. Thus, in Mode III the crack opens by a shearing motion in the direction perpendicular to the plane.

Example 15.5 (Slit of Mode I type) Consider a crack of length $2a$ in a thin but otherwise infinite sheet; see Fig. 15.8. Far from the crack the sheet is pulled apart with a constant traction $\sigma_{yy} = \sigma$, producing a pure Mode I deformation. Westergaard [220] gave a formula for the stresses in terms of the complex function

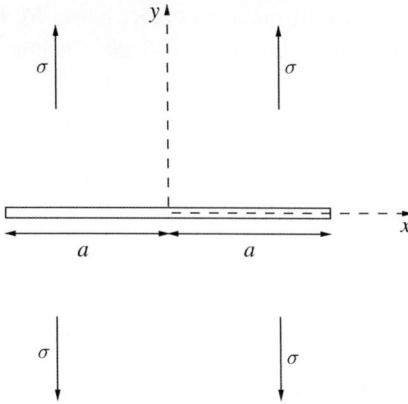

Figure 15.8 A crack of length $2a$ in a two-dimensional elastic medium. A traction σ is applied at infinity, so that $\sigma_{yy} = \sigma$ far from the crack.

$$f(z) = \frac{\sigma}{\sqrt{1 - a^2/z^2}}. \tag{15.85}$$

In particular, σ_{yy} can be obtained as

$$\sigma_{yy} = \Re\{f(z)\} + y\Im\{f'(z)\}.$$

In the plane of the crack, at $y = 0$ and for $|x| \geq a$, we obtain

$$\sigma_{yy} = \Re\left\{\frac{\sigma}{\sqrt{1 - a^2/x^2}}\right\} = \sigma\frac{|x|}{\sqrt{x^2 - a^2}}.$$

Near the crack tip, as $x \to a$,

$$\sigma_{yy} = \frac{\sigma a}{\sqrt{2a}}\frac{1}{\sqrt{r}} + O(1),$$

where $r = x - a$. Along the line $\theta = 0$, corresponding to $y = 0$, we have $\sigma_{yy} = \sigma_{rr}$. Thus from (15.78) and (15.79) we find the Mode I stress intensity factor

$$K_I = \frac{\sigma a}{\sqrt{2a}}\sqrt{2\pi} = \sigma\sqrt{\pi a}. \tag{15.86}$$

This value cannot be computed by local analysis; it is determined by solving the full elastic problem. $\qquad\qquad\qquad\qquad\qquad\qquad\qquad\square$

15.3.2 Inner problem: the fracture energy

Next we describe how to incorporate the microscopic aspect of the theory into the continuum description. This corresponds to the inner solution, for example a precursor film in the case of moving contact lines. We follow the simplest description, due to Griffith, who introduced the fracture energy Γ per unit length of crack length. Often Γ is approximated as $\Gamma = 2\gamma$, where γ is the surface tension of the solid and the factor 2 comes from the two sides of the crack. More properly, the theory should be written in terms of irreversible thermodynamics, taking entropic effects into account. In the interest of simplicity, however, we will present the argument in terms of energies alone. We also restrict ourselves to *quasistatic* situations, which include the conditions under which a crack tip starts to advance as well as the slow motion of crack tips. We will discuss in a little while what the conditions are for the quasistatic approximation to hold true.

Let us consider a solid body Ω to which external forces are applied, such that the applied traction (the force per unit area) $\mathbf{T}_a$ remains constant as the crack tip advances infinitesimally. As the length of the crack changes by $d\ell$, energy conservation requires that the work W_{ext} done by the external forces equals the change in elastic energy W_{el} plus the work required to open the crack:

$$\delta W_{\text{ext}} = dW_{\text{el}} + 2\gamma d\ell. \tag{15.87}$$

Here

$$\delta W_{\text{ext}} = \int_{\partial\Omega} \mathbf{T}_a \cdot \delta\mathbf{u}\, ds$$

and

$$W_{\text{el}} = \int_{\Omega} F d^2x,$$

where F is the energy density given by (4.56).

To compute the work done by the external forces we introduce

$$\phi = \int_{\partial\Omega_T} \mathbf{T}_a \cdot \mathbf{u}\, ds, \tag{15.88}$$

where $\partial\Omega_T$ is the part of the boundary where the external forces are applied. The crack can be seen as an internal boundary $\partial\Omega_c$ and the boundary may be clamped along $\partial\Omega_u$, so that $\partial\Omega = \partial\Omega_T + \partial\Omega_u + \partial\Omega_c$. But $\mathbf{T} = 0$ on $\partial\Omega_c$ and $\mathbf{u} = 0$ on $\partial\Omega_u$, so neither contributes. Since the traction is held constant over $\partial\Omega_T$, we finally have

$$\dot{\phi} = \int_{\partial\Omega_T} \mathbf{T}_a\dot{\mathbf{u}}\, ds. \tag{15.89}$$

Now choosing time as the parameter as a function of which the crack advances, (15.87) can be written as

$$-\dot{W}_{\text{el}} + \dot{\phi} = 2\gamma\dot{\ell}. \tag{15.90}$$

This suggests that we introduce the total potential energy $\mathcal{P}$, where

$$\mathcal{P} = W_{\text{el}} - \phi, \tag{15.91}$$

and the energy release rate

$$G \equiv -\frac{d}{d\ell}(W_{\text{el}} - \phi) = -\frac{d\mathcal{P}}{d\ell}. \tag{15.92}$$

In terms of these quantities, (15.90) can be written in the form

$$G\dot{\ell} = 2\gamma\dot{\ell}. \tag{15.93}$$

If $\dot{\ell} > 0$ (the crack length is increasing), one must have

$$G = 2\gamma. \tag{15.94}$$

This equality says that the available elastic energy goes into the energy necessary to open the crack. However, if $G < 2\gamma$, then $\dot{\ell} = 0$ and the crack does not move.

We will see next that G can in fact be understood in terms of an *energy flux* into the crack tip. This observation implies that G can be calculated in an arbitrarily small region around the tip, so that it is determined by the stress intensity factors alone.

15.3.3 The *J*-integral

Consider the body Ω in Fig. 15.9. As we will show explicitly below, the energy release rate G is indeed determined by the stress intensity factors in the neighborhood of the tip alone. This means we can replace the problem of calculating G for the body Ω by that for a hypothetical body Ω' localized around the crack tip. The boundary of Ω' *excluding* the crack will be denoted by Γ; here we will set boundary conditions that keep the deformation constant (clamped) at a value which corresponds to the original problem. Using (15.92) it follows that

$$-G\dot{\ell} = \frac{d\mathcal{P}}{dt} = \dot{W}_{\text{el}} \tag{15.95}$$

since $\dot{\mathbf{u}} = 0$ on Γ and so according to (15.89) there is no contribution from $\dot{\phi}$.

We take the crack, assumed straight, to advance in the x_1-direction. To keep the part of the boundary represented by the crack constant, we use a frame of reference in which the crack tip is stationary. Thus if the actual velocity of the crack tip is $\mathbf{U} = \dot{\ell}\mathbf{e}_1$, in such a frame of reference the boundary of Ω' will move with velocity $-\mathbf{U}$ to the left. Thus we obtain from (15.95)

$$-G\dot{\ell} = \frac{d}{dt}\int_{\Omega'} F d^2x = \int_{\Omega'} \frac{\partial F}{\partial t} d^2x - \int_{\Gamma} F(\mathbf{U} \cdot \mathbf{n}) ds, \tag{15.96}$$

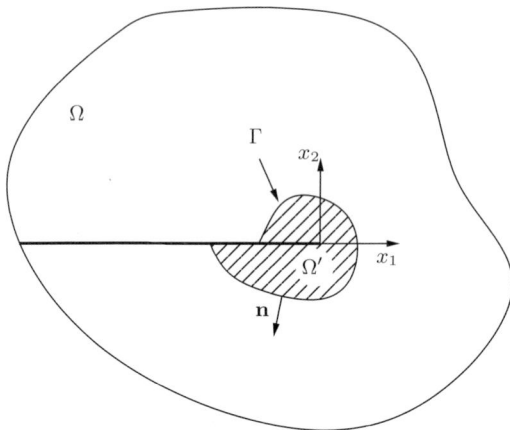

Figure 15.9 The J-integral involves a local region Ω' (hatched) around the crack tip. The boundary Γ of Ω' excludes the crack surface.

where $\mathbf{n}$ is the normal to $\partial\Omega'$. There is no contribution from the crack surface in the second integral, since $\mathbf{U}\cdot\mathbf{n}=0$, crucially using the fact that the crack is straight. Now, the change in the elastic energy density is

$$\frac{\partial F}{\partial t}=\sigma_{ij}\frac{\partial\epsilon_{ij}}{\partial t}=\sigma_{ij}\frac{\partial}{\partial t}\frac{\partial u_i}{\partial x_j},$$

where we have used $\sigma_{ij}=\sigma_{ji}$ and the definition (4.52) of ϵ_{ij} in the second step. Using Gauss' theorem and $\partial_j\sigma_{ij}=0$, it follows that

$$\int_{\Omega'}\frac{\partial F}{\partial t}d^2x=\int_{\Omega'}\sigma_{ij}\frac{\partial}{\partial t}\frac{\partial u_i}{\partial x_j}d^2x=\int_{\Omega'}\frac{\partial}{\partial x_j}\left(\sigma_{ij}\frac{\partial u_i}{\partial t}\right)d^2x$$
$$=\int_{\Gamma}\sigma_{ij}\frac{\partial u_i}{\partial t}n_j ds,$$

where as before there is no contribution from the crack surface.

Since the boundary Γ is clamped, the only change in u_i comes from the fact that the values of u_i are swept to the right at speed $\dot\ell$ with respect to the boundary, and so

$$\frac{\partial u_i}{\partial t}=\dot\ell\frac{\partial u_i}{\partial x_1}.$$

In addition $\mathbf{U}\cdot\mathbf{n}=\dot\ell n_1$, so that (15.96) becomes

$$G=J\equiv\int_{\Gamma}\left(Fn_1-\sigma_{ij}\frac{\partial u_i}{\partial x_1}n_j\right)ds,\qquad(15.97)$$

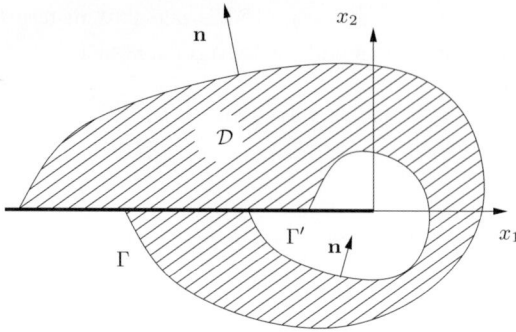

Figure 15.10 The J-integral is independent of Γ. The domain $\mathcal{D}$ is shown as hatched.

known as *Rice's J-integral*. The physical meaning of J is that it is an instantaneous energy flux through the contour Γ toward the crack tip. From that it is clear that J should be independent of the path Γ, as we show now.

Namely, let Γ' be an alternative contour, and so

$$J - J' = \int_{\Gamma \cup \Gamma'} \left(Fn_1 - \sigma_{ij} \frac{\partial u_i}{\partial x} n_j \right) ds; \qquad (15.98)$$

see Fig. 15.10. Here $\mathbf{n}$ is the outward normal of the domain $\mathcal{D}$; note that the normal vector on Γ' has the opposite sign with respect to the convention adopted in Fig. 15.9. There is no contribution from the crack, since the integrand vanishes there as $n_1 = 0$ and $\sigma_{ij} n_j = 0$. Thus (15.98) can be thought of as an integral over the boundary of $\mathcal{D}$. Using first Gauss' theorem and then $\partial_j \sigma_{ij} = 0$ in the second step, we have

$$J - J' = \int_{\mathcal{D}} \left[\frac{\partial F}{\partial x_1} - \frac{\partial}{\partial x_j} \left(\sigma_{ij} \frac{\partial u_i}{\partial x_1} \right) \right] d^2 x$$

$$= \int_{\mathcal{D}} \left(\frac{\partial F}{\partial \epsilon_{ij}} \frac{\partial \epsilon_{ij}}{\partial x_1} - \sigma_{ij} \frac{\partial^2 u_i}{\partial x_j \partial x_1} \right) d^2 x.$$

But $\partial F / \partial \epsilon_{ij} = \sigma_{ij}$, and so the integrand becomes

$$\frac{1}{2} \sigma_{ij} \frac{\partial}{\partial x_1} \left(\frac{\partial u_j}{\partial x_i} + \frac{\partial u_i}{\partial x_j} \right) - \sigma_{ij} \frac{\partial}{\partial x_1} \left(\frac{\partial u_i}{\partial x_j} \right) = 0,$$

on account of $\sigma_{ij} = \sigma_{ji}$. Thus in summary $J - J' = 0$, and we have proven the path independence of the J-integral.

In particular, it follows that a possible path is a circle around the crack tip whose radius R goes to zero. This means that one can calculate J on the basis of the asymptotic behavior (15.78), (15.82) and thus in terms of the stress intensity factors alone. To that end we convert the kernel

$$K \equiv F n_1 - \sigma_{ij} \frac{\partial u_i}{\partial x_1} n_j$$

to polar coordinates, using

$$\frac{\partial}{\partial x_1} = \cos\theta \frac{\partial}{\partial r} - \frac{\sin\theta}{r} \frac{\partial}{\partial \theta}, \quad \mathbf{n} = \mathbf{e}_r,$$

and

$$\frac{\partial \mathbf{u}}{\partial x_1} = \frac{\partial}{\partial x_1}(\mathbf{e}_r u_r + \mathbf{e}_\theta u_\theta) = \left(\frac{\partial u_r}{\partial x_1} + \frac{\sin\theta}{r} u_\theta\right)\mathbf{e}_r + \left(\frac{\partial u_\theta}{\partial x_1} - \frac{\sin\theta}{r} u_r\right)\mathbf{e}_\theta.$$

Furthermore, we have that

$$F = \frac{\lambda}{2}(\epsilon_{rr} + \epsilon_{\theta\theta})^2 + \mu\left(\epsilon_{rr}^2 + \epsilon_{\theta\theta}^2 + 2\epsilon_{r\theta}^2\right),$$

where

$$\epsilon_{rr} = \frac{\partial u_r}{\partial r}, \quad \epsilon_{\theta\theta} = \frac{1}{r}\frac{\partial u_\theta}{\partial \theta} + \frac{u_r}{r}, \quad \text{and} \quad \epsilon_{r\theta} = \frac{\partial u_\theta}{\partial r} - \frac{u_\theta}{r} + \frac{1}{2r}\frac{\partial u_r}{\partial \theta}.$$

A somewhat lengthy but elementary calculation yields

$$K = \frac{1-\nu^2}{\pi RE}\left[(K_I^2 - K_{II}^2)\sin^2\theta + K_{II}^2 + K_I K_{II}\sin 2\theta\right],$$

where we have used (15.82). Thus

$$G = J = R\int_{-\pi}^{\pi} K d\theta = \frac{1-\nu^2}{E}\left(K_I^2 + K_{II}^2\right), \quad (15.99)$$

which is known as the *Irwin relationship;* it shows explicitly that the energy release rate G can be written in terms of local quantities alone.

15.3.4 The global problem

However, G can also be calculated in terms of global properties such as the work of the external forces. This can be made explicit by the following calculation. The elastic energy is determined by the work of the external forces:

$$W_{\text{el}} = \frac{1}{2}\int_{\partial\Omega} \mathbf{T}\cdot\mathbf{u}\,ds = \frac{1}{2}\int_{\partial\Omega_T} \mathbf{T}_a\cdot\mathbf{u}\,ds + \frac{1}{2}\int_{\partial\Omega_u} \mathbf{T}\cdot\mathbf{u}_a\,ds.$$

Using (15.88) we thus obtain

$$\mathcal{P} = W_{\text{el}} - \phi = \frac{1}{2}\int_{\partial\Omega_u} \mathbf{T}\cdot\mathbf{u}_a\,ds - \frac{1}{2}\int_{\partial\Omega_T} \mathbf{T}_a\cdot\mathbf{u}\,ds.$$

To find G, according to (15.92) we have to take the derivative with respect to the crack length ℓ. We know that $\partial \mathbf{u}/\partial \ell = 0$ on $\partial \Omega_u$ and that $\partial \mathbf{T}/\partial \ell = 0$ on $\partial \Omega_T$ and so

$$G = -\frac{\partial \mathcal{P}}{\partial \ell} = \frac{1}{2} \int_{\partial \Omega} \left(\mathbf{T} \cdot \frac{\partial \mathbf{u}}{\partial \ell} - \frac{\partial \mathbf{T}}{\partial \ell} \cdot \mathbf{u} \right) ds. \qquad (15.100)$$

To evaluate the integral on the right of (15.100), one has to solve the global elastic problem with given boundary conditions and in the presence of a crack of length ℓ. However, the crucial point is that the calculation does not involve any microscopic information; the global and the local problem have been separated by introducing the powerful concept of a local singular solution around the crack tip.

There are two situations in which the present quasistatic theory applies:

(i) The initiation of crack motion   If $G < 2\gamma$ initially and the loading changes so that $G = 2\gamma$, the crack will start to move.
(ii) The quasistatic motion of cracks   If $\partial G/\partial \ell < 0$ at constant load then the system is stable: after having started to move at $G = 2\gamma$, the crack will stop moving immediately. Only if the loading increases will the crack resume its motion. The length of the crack will increase so as to keep $G = 2\gamma = $ constant during the motion.

If, however, $\partial G/\partial \ell > 0$, then the system is unstable: as the crack advances, more and more energy becomes available to keep it moving. Our present quasistatic description does not apply to this case. In fact, an unstable crack will advance typically at a speed similar to that of the speed of sound in the material. To describe this, (4.58), which includes inertial terms, needs to be employed. Note that stability depends on the way a sample is tested; see Fig. 15.11. In Exercise 15.9 it can be shown that $\partial G/\partial \ell$ changes sign according to the compliance (the inverse of the spring constant k_s) of the spring.

Example 15.6 (Crack in a strip) We end by giving an example which unites the global and local aspects of the problem. Consider an infinite strip of width $2h$ (see Fig. 15.12), at the centerline of which is a semi-infinite crack. The strip is being pulled apart in the transverse direction. We will calculate the critical deformation δ at which the crack tip will start to advance. To this end we choose a contour Γ (the dotted line) far from the crack tip and use the J-integral. At the boundaries $x_2 = \pm h$ the deformations are $u_1 = 0$ and $u_2 = \pm \delta$, respectively.

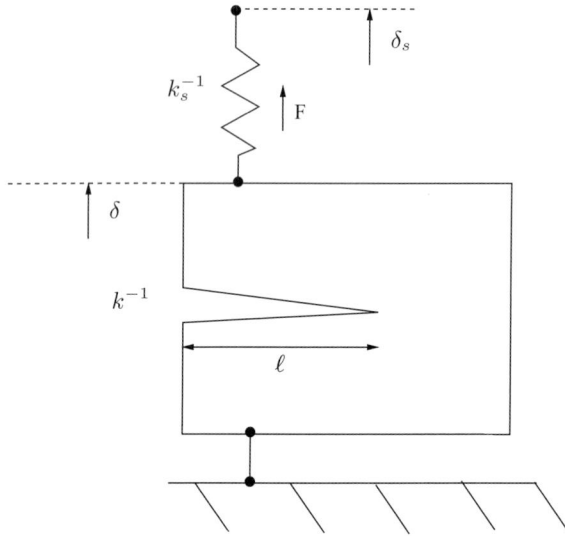

Figure 15.11 A sample with a crack of length ℓ being tested by deforming a spring of compliance k_s^{-1} attached to it. The compliance of the sample is $k^{-1}(\ell)$.

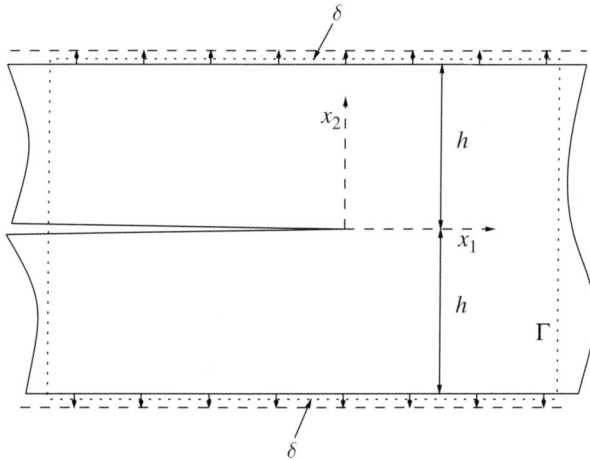

Figure 15.12 A semi-infinite crack advancing in a strip.

At the horizontal boundaries, $n_1 = 0$ and $\partial u_i/\partial x_1 = 0$ and so the integrand of the J-integral vanishes. On the far left, the material is relaxed on account of the crack, and so $\boldsymbol{\sigma} = 0$ and the integrand is zero once again. On the far right $\boldsymbol{\epsilon}$ is uniform, with $\epsilon_{22} = \delta/h$, and all the other components vanish. In addition $\partial u_j/\partial x_1 = 0$, so that the integrand at the right-hand border is

$$Fn_1 = \frac{\lambda}{2}(\epsilon_{11} + \epsilon_{22})^2 + \mu\left(\epsilon_{11}^2 + \epsilon_{22}^2\right) = \left(\frac{\lambda}{2} + \mu\right)\frac{\delta^2}{h^2}$$

$$= E\frac{1-\nu}{2(1+\nu)(1-2\nu)}\frac{\delta^2}{h^2},$$

and the J-integral becomes

$$G = J = 2hFn_1 = E\frac{(1-\nu)\delta^2}{(1+\nu)(1-2\nu)h}. \tag{15.101}$$

Thus, using (15.94), the critical deformation, at which the crack will begin to move, is

$$\delta_{cr} = \sqrt{\frac{2(1+\nu)(1-2\nu)}{(1-\nu)}}\sqrt{\frac{\gamma h}{E}}. \tag{15.102}$$

□

Exercises

15.1 Consider the viscous surface-tension-driven spreading of a drop, as described by the lubrication equation (6.29); we can choose units such that

$$\frac{\partial h}{\partial t} = -\frac{1}{3}\nabla\cdot\left(h^3\nabla\Delta h\right).$$

(i) Use the similarity solution

$$h(r,t) = t^{-\alpha}H(\xi), \quad \xi = \frac{r}{t^\beta}$$

to show that to achieve a (formal) solution, we must have $\alpha = 1/5$ and $\beta = 1/10$, consistent with Tanner's law $R \propto t^{1/10}$. Hint: the solution must be such that the drop volume is constant.

(ii) Show that the similarity profile must satisfy [32]

$$\frac{3\xi}{10} = H^2\left(H'' + \frac{H'}{\xi}\right)'. \tag{15.103}$$

(iii) (*) Analyze (15.103) near a hypothetical touchdown point ξ_0 at which $H(\xi_0) = 0$. Use the asymptotic form (15.17) to show that there is no such solution for which $H > 0$ for $\xi < \xi_0$.

15.2 Using (6.82) and (6.83), respectively, find the equivalent of Tanner's law for the spreading of drops of a power law fluid in two dimensions, i.e.

$$h(x,t) = \frac{1}{t^\alpha}f\left(\frac{x}{t^\beta}\right),$$

and axisymmetric drops in three dimensions, i.e.

$$h(r, t) = \frac{1}{t^\alpha} f\left(\frac{r}{t^\beta}\right).$$

15.3 (*) In Example 15.3 we calculated H_1', solving (15.31) in a perturbative fashion. Using the operator method of Section 15.2.3, this can also be achieved without computing H_1 explicitly. As in (15.53), write the first-order problem as

$$\psi = H_1''' \equiv \mathcal{L}H_1. \tag{15.104}$$

Clearly, $\mathcal{L}^\dagger H_0 = 0$; thus evaluate

$$\int_0^\ell H_0 \mathcal{L} H_1 d\xi$$

for $\ell \to \infty$, to show that $H_1' \approx \ln \xi + 1$ in agreement with (15.42).

15.4 The slip model (15.25) can also be applied to the case of zero contact angle, $\theta_{eq} = 0$. Show that there is a similarity solution of the form

$$h(x) = \lambda\phi\left(\frac{Ca^{1/3}x}{\lambda}\right),$$

and write down the similarity equation for ϕ, including the boundary conditions. Conclude that the solution far from the contact line is of the form (15.20) with

$$L = \lambda Ca^{-1/3},$$

where the constant ξ_0 in (15.20) has to be computed from a numerical solution of the similarity equation [111].

15.5 (*) Repeat the analysis of Section 15.2.3 for the equivalent two-dimensional problem: the spreading of a two-dimensional ridge of width $2R$.

(i) Show that the static ridge shape is

$$h_0(x) = \frac{\theta_{app}R}{2}\left[1 - \left(\frac{x}{R}\right)\right].$$

(ii) Show that, in deriving the analogue of (15.55), the following test function φ must be chosen:

$$\varphi = \frac{2x^2}{3\theta_{app}^2(R^2 - x^2)}.$$

(iii) Finally, show that the constant b in (15.51) is $\sqrt{2}/e^2$, rather than the value given by (15.70) for the axisymmetric case.

15.6 In a Mode III crack, deformations are exclusively in the out-of-plane direction and depend on the plane coordinates x_1, x_2 alone; the equation of elastic equilibrium is (4.69). Show that the local deformation field near the crack tip is

$$u_3 = \frac{2K_{III}}{\mu}\sqrt{\frac{r}{2\pi}}\sin\frac{\theta}{2},\qquad(15.105)$$

where K_{III} is, by definition, the stress intensity factor for a Mode III crack. Also, show that

$$\sigma_{13} = -\frac{K_{III}}{\sqrt{2\pi r}}\sin\frac{\theta}{2},\quad \sigma_{23} = \frac{K_{III}}{\sqrt{2\pi r}}\cos\frac{\theta}{2}.\qquad(15.106)$$

15.7 Consider the out-of-plane deformations of a plate, with a slit of width $2a$ in the x-direction, subjected to a constant shear stress $\sigma_{23} = \tau$. Use the complex potential from Exercise 4.15, together with the Joukowsky transformation

$$z = f(\zeta) = \zeta + \frac{R^2}{\zeta},$$

to show that the complex potential around the slit is

$$w(z) = -i\tau\sqrt{z^2 - a^2}.$$

Thus use $dw/dz = \sigma_{13} - i\sigma_{23}$ to show that the stress intensity factor of a slit under constant shear stress is

$$K_{III} = \tau\sqrt{\pi a}.$$

15.8 Use the J-integral (15.97) and the results of the preceding exercise to show that the Irwin relationship for a Mode III crack is

$$G = J = \frac{K_{III}^2}{2\mu}.\qquad(15.107)$$

15.9 Consider the test rig shown in Fig. 15.11: through a spring with spring constant k_s a force F is applied to the test sample. The total deformation of the rig is δ_s and the deformation of the sample is δ. In the case of a point force F acting on the boundary,

$$\int_{\partial\Omega} \mathbf{T}\cdot\frac{\partial\mathbf{u}}{\partial\ell}\,ds = F\frac{\partial\delta}{\partial\ell},$$

where δ is the displacement. Likewise,

$$\int_{\partial\Omega} \frac{\partial\mathbf{T}}{\partial\ell}\cdot\mathbf{u}\,ds = \delta\frac{\partial F}{\partial\ell}.$$

(i) The behavior of the sample can be summarized by the relation

$$\delta = k^{-1}(\ell)F,$$

where $k^{-1}(\ell)$ is the compliance of the spring. Show that

$$G = \frac{F^2}{2}\frac{dF}{d\ell}.$$

(ii) Analyze the test rig to show that $\delta_s = k^{-1}F + k_s^{-1}F$. Thus demonstrate that, at constant δ,

$$\frac{dG}{d\ell} = F^2 \left[\frac{1}{2}\frac{d^2k^{-1}}{d\ell^2} - \frac{1}{k^{-1} + k_s^{-1}} \left(\frac{dk^{-1}}{d\ell} \right)^2 \right].$$

This relation implies that in general the sign of $dG/d\ell$ will depend on the stiffness k_s of the spring.

Appendix A Vector calculus

Here we recall results from vector calculus used throughout the book. For practical calculations it is useful to represent vector operators in coordinate systems adapted to the symmetry of the problem studied.

Cartesian coordinates

We denote the Cartesian basis vectors as $\mathbf{e}_x = (1, 0, 0)$, $\mathbf{e}_y = (0, 1, 0)$, and $\mathbf{e}_z = (0, 0, 1)$; they are mutually orthogonal (e.g. $\mathbf{e}_x \cdot \mathbf{e}_z = 0$), normalized ($\mathbf{e}_x \cdot \mathbf{e}_x = 1$), and form a right-handed coordinate system (i.e. $\mathbf{e}_x \times \mathbf{e}_y = \mathbf{e}_z$). Then a vector can be represented as

$$\mathbf{v} = (u, v, w) = u\mathbf{e}_x + v\mathbf{e}_y + w\mathbf{e}_z.$$

Let $\phi(\mathbf{r})$ be a scalar field and $\mathbf{v}(\mathbf{r})$ a vector field, both of which depend on position $\mathbf{r} = (x, y, z)$. The standard vector operators can be written in terms of the nabla operator

$$\nabla = \mathbf{e}_x \frac{\partial}{\partial x} + \mathbf{e}_y \frac{\partial}{\partial y} + \mathbf{e}_z \frac{\partial}{\partial z}. \tag{A.1}$$

They are the *gradient*,

$$\nabla \phi = \frac{\partial \phi}{\partial x} \mathbf{e}_x + \frac{\partial \phi}{\partial y} \mathbf{e}_y + \frac{\partial \phi}{\partial z} \mathbf{e}_z, \tag{A.2}$$

the *divergence*,

$$\nabla \cdot \mathbf{v} = \frac{\partial u}{\partial x} + \frac{\partial v}{\partial y} + \frac{\partial w}{\partial z}, \tag{A.3}$$

and the *curl*,

$$\nabla \times \mathbf{v} = \left(\frac{\partial w}{\partial y} - \frac{\partial v}{\partial z} \right) \mathbf{e}_x + \left(\frac{\partial u}{\partial z} - \frac{\partial w}{\partial x} \right) \mathbf{e}_y + \left(\frac{\partial v}{\partial x} - \frac{\partial u}{\partial y} \right) \mathbf{e}_z. \tag{A.4}$$

Important operators derived from this are the *Laplacian*

$$\triangle \phi \equiv \nabla \cdot \nabla \phi = \frac{\partial^2 \phi}{\partial x^2} + \frac{\partial^2 \phi}{\partial y^2} + \frac{\partial^2 \phi}{\partial z^2}, \tag{A.5}$$

and the *convective derivative*

$$(\mathbf{v} \cdot \nabla)\phi = u\frac{\partial \phi}{\partial x} + v\frac{\partial \phi}{\partial y} + w\frac{\partial \phi}{\partial z}. \tag{A.6}$$

The Laplacian and the convective derivative of a vector are defined to act on each component independently:

$$\triangle \mathbf{v} = (\triangle u, \triangle v, \triangle w), \quad (\mathbf{v} \cdot \nabla)\mathbf{v} = ((\mathbf{v} \cdot \nabla)u, (\mathbf{v} \cdot \nabla)v, (\mathbf{v} \cdot \nabla)w).$$

The Cartesian components of the rate-of-deformation tensor $\mathbf{D}$, which can be defined in a coordinate-free manner as

$$\mathbf{D} = \nabla\mathbf{v} + (\nabla\mathbf{v})^T, \tag{A.7}$$

are given by (4.9). Its analogue in linear elasticity is the strain tensor (4.52).

Cylindrical coordinates

Cylindrical coordinates r, θ, z are defined by $\mathbf{r} = (r\cos\theta, r\sin\theta, z)$, and so the unit vectors are

$$\mathbf{e}_r = (\cos\theta, \sin\theta, 0), \quad \mathbf{e}_\theta = (-\sin\theta, \cos\theta, 0), \quad \mathbf{e}_z = (0, 0, 1);$$

they form an orthogonal and right-handed coordinate system. In this basis the nabla operator is represented as

$$\nabla = \mathbf{e}_r\frac{\partial}{\partial r} + \frac{\mathbf{e}_\theta}{r}\frac{\partial}{\partial \theta} + \mathbf{e}_z\frac{\partial}{\partial z}, \tag{A.8}$$

and vectors are written as

$$\mathbf{v} = v_r\mathbf{e}_r + v_\theta\mathbf{e}_\theta + v_z\mathbf{e}_z. \tag{A.9}$$

In cylindrical coordinates the differential operators introduced above are

$$\nabla\phi = \frac{\partial \phi}{\partial r}\mathbf{e}_r + \frac{1}{r}\frac{\partial \phi}{\partial \theta}\mathbf{e}_\theta + \frac{\partial \phi}{\partial z}\mathbf{e}_z, \tag{A.10}$$

$$\nabla \cdot \mathbf{v} = \frac{1}{r}\frac{\partial(rv_r)}{\partial r} + \frac{1}{r}\frac{\partial v_\theta}{\partial \theta} + \frac{\partial v_z}{\partial z}, \tag{A.11}$$

$$\nabla \times \mathbf{v} = \frac{1}{r}\begin{vmatrix} \mathbf{e}_r & r\mathbf{e}_\theta & \mathbf{e}_z \\ \partial/\partial r & \partial/\partial \theta & \partial/\partial z \\ v_r & rv_\theta & v_z \end{vmatrix}, \tag{A.12}$$

$$\Delta\phi = \frac{\partial^2\phi}{\partial r^2} + \frac{1}{r}\frac{\partial\phi}{\partial r} + \frac{1}{r^2}\frac{\partial^2\phi}{\partial\theta^2} + \frac{\partial^2\phi}{\partial z^2}, \tag{A.13}$$

$$(\mathbf{v}\cdot\nabla)\mathbf{v} = \left(v_r\frac{\partial v_r}{\partial r} + \frac{v_\theta}{r}\frac{\partial v_r}{\partial\theta} + v_z\frac{\partial v_r}{\partial z} - \frac{v_\theta^2}{r} \right)\mathbf{e}_r$$

$$+ \left(v_r\frac{\partial v_\theta}{\partial r} + \frac{v_\theta}{r}\frac{\partial v_\theta}{\partial\theta} + v_z\frac{\partial v_\theta}{\partial z} + \frac{v_r v_\theta}{r} \right)\mathbf{e}_\theta \tag{A.14}$$

$$+ \left(v_r\frac{\partial v_z}{\partial r} + \frac{v_\theta}{r}\frac{\partial v_z}{\partial\theta} + v_z\frac{\partial v_z}{\partial z} \right)\mathbf{e}_z,$$

$$\Delta\mathbf{v} = \left(\Delta v_r - \frac{2}{r^2}\frac{\partial v_\theta}{\partial\theta} - \frac{v_r}{r^2} \right)\mathbf{e}_r + \left(\Delta v_\theta + \frac{2}{r^2}\frac{\partial v_r}{\partial\theta} - \frac{v_\theta}{r^2} \right)\mathbf{e}_\theta + \Delta v_z\mathbf{e}_z. \tag{A.15}$$

In cylindrical coordinates the rate of deformation tensor (A.7) is

$$D_{rr} = 2\frac{\partial v_r}{\partial r}, \quad D_{r\theta} = \frac{1}{r}\frac{\partial v_r}{\partial\theta} + \frac{\partial v_\theta}{\partial r} - \frac{v_\theta}{r},$$

$$D_{\theta\theta} = \frac{2}{r}\frac{\partial v_\theta}{\partial\theta} + \frac{2v_r}{r}, \quad D_{\theta z} = \frac{\partial v_\theta}{\partial z} + \frac{1}{r}\frac{\partial v_z}{\partial\theta}, \tag{A.16}$$

$$D_{zz} = 2\frac{\partial v_z}{\partial z}, \quad D_{zr} = \frac{\partial v_z}{\partial r} + \frac{\partial v_r}{\partial z}.$$

Spherical coordinates

Spherical coordinates r, θ, φ are defined by

$$\mathbf{r} = (r\sin\theta\cos\varphi, r\sin\theta\sin\varphi, r\cos\theta),$$

and the unit vectors are

$$\mathbf{e}_r = (\sin\theta\cos\varphi, \sin\theta\sin\varphi, \cos\theta),$$

$$\mathbf{e}_\theta = (\cos\theta\cos\varphi, \cos\theta\sin\varphi, -\sin\theta),$$

$$\mathbf{e}_\varphi = (-\sin\varphi, \cos\varphi, 0);$$

again, they are orthogonal and right-handed. The nabla operator is represented as

$$\nabla = \mathbf{e}_r\frac{\partial}{\partial r} + \frac{\mathbf{e}_\theta}{r}\frac{\partial}{\partial\theta} + \frac{\mathbf{e}_\varphi}{r\sin\theta}\frac{\partial}{\partial\varphi}, \tag{A.17}$$

and vectors are written as

$$\mathbf{v} = v_r\mathbf{e}_r + v_\theta\mathbf{e}_\theta + v_\varphi\mathbf{e}_\varphi. \tag{A.18}$$

The differential operators are

$$\nabla\phi = \frac{\partial\phi}{\partial r}\mathbf{e}_r + \frac{1}{r}\frac{\partial\phi}{\partial\theta}\mathbf{e}_\theta + \frac{1}{r\sin\theta}\frac{\partial\phi}{\partial\varphi}\mathbf{e}_\varphi, \tag{A.19}$$

$$\nabla \cdot \mathbf{v} = \frac{1}{r^2} \frac{\partial (r^2 v_r)}{\partial r} + \frac{1}{r \sin \theta} \frac{\partial (\sin \theta \, v_\theta)}{\partial \theta} + \frac{1}{r \sin \theta} \frac{\partial v_\varphi}{\partial \varphi}, \tag{A.20}$$

$$\nabla \times \mathbf{v} = \frac{1}{r^2 \sin \theta} \begin{vmatrix} \mathbf{e}_r & r \mathbf{e}_\theta & r \sin \theta \, \mathbf{e}_\varphi \\ \partial/\partial r & \partial/\partial \theta & \partial/\partial \varphi \\ v_r & r v_\theta & r \sin \theta \, v_\varphi \end{vmatrix}, \tag{A.21}$$

$$\Delta \phi = \frac{1}{r^2} \frac{\partial}{\partial r} \left(r^2 \frac{\partial \phi}{\partial r} \right) + \frac{1}{r^2 \sin \theta} \frac{\partial}{\partial \theta} \left(\sin \theta \frac{\partial \phi}{\partial \theta} \right) + \frac{1}{r^2 \sin^2 \theta} \frac{\partial^2 \phi}{\partial \varphi^2}, \tag{A.22}$$

$$(\mathbf{v} \cdot \nabla)\mathbf{v} = \left(v_r \frac{\partial v_r}{\partial r} + \frac{v_\theta}{r} \frac{\partial v_r}{\partial \theta} + \frac{v_\varphi}{r \sin \theta} \frac{\partial v_r}{\partial \varphi} - \frac{v_\theta^2 + v_\varphi^2}{r} \right) \mathbf{e}_r$$
$$+ \left(v_r \frac{\partial v_\theta}{\partial r} + \frac{v_\theta}{r} \frac{\partial v_\theta}{\partial \theta} + \frac{v_\varphi}{r \sin \theta} \frac{\partial v_\theta}{\partial \varphi} + \frac{v_r v_\theta}{r} - \frac{v_\varphi^2 \cot \theta}{r} \right) \mathbf{e}_\theta$$
$$+ \left(v_r \frac{\partial v_\varphi}{\partial r} + \frac{v_\theta}{r} \frac{\partial v_\varphi}{\partial \theta} + \frac{v_\varphi}{r \sin \theta} \frac{\partial v_\varphi}{\partial \varphi} + \frac{v_r v_\varphi}{r} + \frac{v_\theta v_\varphi \cot \theta}{r} \right) \mathbf{e}_\varphi, \tag{A.23}$$

$$\Delta \mathbf{v} = \left(\Delta v_r - \frac{2 v_r}{r^2} - \frac{2}{r^2 \sin \theta} \frac{\partial v_\theta \sin \theta}{\partial \theta} - \frac{2}{r^2 \sin \theta} \frac{\partial v_\varphi}{\partial \varphi} \right) \mathbf{e}_r$$
$$+ \left(\Delta v_\theta + \frac{2}{r^2} \frac{\partial v_r}{\partial \theta} - \frac{v_\theta}{r^2 \sin^2 \theta} - \frac{2 \cos \theta}{r^2 \sin^2 \theta} \frac{\partial v_\varphi}{\partial \varphi} \right) \mathbf{e}_\theta$$
$$+ \left(\Delta v_\varphi + \frac{2}{r^2 \sin \theta} \frac{\partial v_r}{\partial \varphi} + \frac{2 \cos \theta}{r^2 \sin^2 \theta} \frac{\partial v_\theta}{\partial \varphi} - \frac{v_\varphi}{r^2 \sin^2 \theta} \right) \mathbf{e}_\varphi. \tag{A.24}$$

The components of the rate of deformation tensor (A.7) are

$$D_{rr} = 2 \frac{\partial v_r}{\partial r}, \quad D_{r\theta} = \frac{1}{r} \frac{\partial v_r}{\partial \theta} + \frac{\partial v_\theta}{\partial r} - \frac{v_\theta}{r},$$

$$D_{\theta\theta} = \frac{2}{r} \frac{\partial v_\theta}{\partial \theta} + \frac{2 v_r}{r}, \quad D_{\theta\varphi} = \frac{1}{r \sin \theta} \frac{\partial v_\theta}{\partial \varphi} + \frac{1}{r} \frac{\partial v_\varphi}{\partial \theta} - \frac{v_\varphi \cot \theta}{r}, \tag{A.25}$$

$$D_{\varphi\varphi} = \frac{2}{r \sin \theta} \frac{\partial v_\varphi}{\partial \varphi} + \frac{2 v_r}{r} + \frac{2 v_\theta \cot \theta}{r}, \quad D_{\varphi r} = \frac{\partial v_\varphi}{\partial r} + \frac{1}{r \sin \theta} \frac{\partial v_r}{\partial \varphi} - \frac{v_\varphi}{r}.$$

Appendix B Index notation and the summation convention

Suppose that we have two vectors $\mathbf{u} = (u_1, u_2, u_3)$ and $\mathbf{v} = (v_1, v_2, v_3)$, so that the dot product is

$$\mathbf{u} \cdot \mathbf{v} = \sum_{i=1}^{3} u_i v_i.$$

However, knowing that the result should be a scalar, we can drop the summation symbol on the understanding that repeated indices imply summation. Thus, using the so-called *summation convention*, we can write the dot product more compactly as

$$\mathbf{u} \cdot \mathbf{v} = u_i v_i. \tag{B.1}$$

The Kronecker delta is defined by

$$\delta_{ij} = \begin{cases} 1, & i = j, \\ 0, & i \neq j, \end{cases} \tag{B.2}$$

so, using the summation convention, $\delta_{ij} a_j = a_i$ since

$$\delta_{ij} a_j \equiv \sum_{j=1}^{3} \delta_{ij} a_j = a_i.$$

Furthermore,

$$\delta_{ii} = \sum_{i=1}^{3} \delta_{ii} = \sum_{i=1}^{3} 1 = 3. \tag{B.3}$$

The summation convention is particularly powerful in dealing with double sums. For example, depending on which index is summed first we find that

$\delta_{ij} u_i v_j = u_i v_i = \mathbf{u} \cdot \mathbf{v}$ or $\delta_{ij} u_i v_j = u_j v_j = \mathbf{u} \cdot \mathbf{v}$. Using conventional notation this would be rendered much more clumsily as

$$\delta_{ij} u_i v_j \equiv \sum_{i,j=1}^{3} \delta_{ij} u_i v_j = \sum_{i=1}^{3} u_i \left(\sum_{j=1}^{3} \delta_{ij} v_j \right) = \sum_{i=1}^{3} u_i v_i = \mathbf{u} \cdot \mathbf{v}.$$

To write cross products using the summation convention, we introduce the Levi–Civita symbol ϵ_{ijk} by the two requirements that $\epsilon_{123} = 1$ and that ϵ_{ijk} changes sign if any two indices are interchanged. From this it follows directly that

$$\epsilon_{ijk} = \begin{cases} 1, & (i, j, k) \in \{(1, 2, 3), (3, 1, 2), (2, 3, 1)\}, \\ -1, & (i, j, k) \in \{(3, 2, 1), (1, 3, 2), (2, 1, 3)\}, \\ 0, & i = j \text{ or } j = k \text{ or } k = i. \end{cases} \tag{B.4}$$

It is easy to check that now the cross product $\mathbf{w} = \mathbf{u} \times \mathbf{v}$ can be written in components as

$$w_i = \epsilon_{ijk} u_j v_k,$$

where again summation over j and k is implied. For example, one can write the triple product efficiently as

$$\mathbf{w} \cdot (\mathbf{u} \times \mathbf{v}) = w_i \epsilon_{ijk} u_j v_k,$$

where the summation is over i, j, and k, so the result is a scalar.

To deal with several cross products, we need to evaluate the product of ϵ_{ijk} with itself:

$$\epsilon_{ijk} \epsilon_{ilm} = \delta_{jl} \delta_{km} - \delta_{jm} \delta_{kl}. \tag{B.5}$$

This expression is easy to check by direct calculation (there is a finite number of possibilities). For example, we can now derive the well-known formula for the vector triple product:

$$\mathbf{w} \times (\mathbf{u} \times \mathbf{v}) = (\mathbf{w} \cdot \mathbf{v})\mathbf{u} - (\mathbf{w} \cdot \mathbf{u})\mathbf{v}. \tag{B.6}$$

Since we are dealing with a vector, we will calculate the ith component of the left-hand side:

$$\begin{aligned}
[\mathbf{w} \times (\mathbf{u} \times \mathbf{v})]_i &= \epsilon_{ijk} w_j (\mathbf{u} \times \mathbf{v})_k = \epsilon_{ijk} w_j \epsilon_{klm} u_l v_m = \epsilon_{kij} \epsilon_{klm} w_j u_l v_m \\
&= (\delta_{il} \delta_{jm} - \delta_{jl} \delta_{im}) w_j u_l v_m = w_j u_i v_j - w_j u_j v_i \\
&= (\mathbf{w} \cdot \mathbf{v}) u_i - (\mathbf{w} \cdot \mathbf{u}) v_i.
\end{aligned}$$

Thus, since we have derived the vector identity for each of its components $i = 1, 2, 3$, (B.6) is proved.

Differential operations

Now we will apply the above conventions to the differential operators introduced in Appendix A, which can be written in terms of the nabla operator (A.1). To save work, we adopt the sleeker notation

$$\nabla \equiv (\partial_1, \partial_2, \partial_3).$$

Then the ith component of the gradient is

$$[\nabla \phi]_i = \partial_i \phi,$$

the divergence is

$$\nabla \cdot \mathbf{v} = \partial_i v_i,$$

and the ith component of the curl is

$$[\nabla \times \mathbf{v}]_i = \epsilon_{ijk} \partial_j v_k.$$

To perform calculations in Cartesian coordinates it is useful to note that $\partial_j x_i = \delta_{ij}$, so that for example $\nabla \cdot \mathbf{r} = \partial_i x_i = \delta_{ii} = 3$, and $\nabla r = \mathbf{r}/r$ since

$$[\nabla r]_i = \partial_i \sqrt{x_j x_j} = \frac{\partial_i x_j^2}{2\sqrt{x_j^2}} = \frac{x_j \partial_i x_j}{r} = \frac{x_j \delta_{ij}}{r} = \frac{x_i}{r}.$$

Finally, a very useful identity in hydrodynamics is

$$\mathbf{v} \times (\nabla \times \mathbf{v}) = \frac{1}{2}\nabla v^2 - (\mathbf{v} \cdot \nabla)\mathbf{v}, \tag{B.7}$$

which is proved by the following calculation:

$$
\begin{aligned}
[\mathbf{v} \times (\nabla \times \mathbf{v})]_i &= \epsilon_{ijk} v_j (\nabla \times \mathbf{v})_k \\
&= \epsilon_{ijk} v_j \epsilon_{klm} \partial_l v_m = \epsilon_{kij} \epsilon_{klm} v_j \partial_l v_m \\
&= (\delta_{il}\delta_{jm} - \delta_{im}\delta_{jl}) v_j \partial_l v_m = v_j \partial_i v_j - v_j \partial_j v_i \\
&= \frac{1}{2}\partial_i v^2 - (\mathbf{v} \cdot \nabla)v_i.
\end{aligned}
$$

Appendix C Dimensional analysis

Dimensional analysis is very intuitive and perhaps does not need much explanation. However, in view of its importance it seems worthwhile to give a brief technical exposition [28]. The idea behind dimensional analysis is that no law of nature can depend on the units we give to the fundamental quantities such as mass, length, time, or charge.

(i) Assume that there are n fundamental units q_i, $i = 1, \ldots, n$, which can be changed by multiplying with a scale factor α_i, giving scaled units $T(q_i) = \alpha_i q_i$. In mechanical problems the fundamental quantities are mass, length, and time, with units kilogram (kg), meter (m), and second (s), respectively. Alternatively, we might use the foot (ft) as our unit of length, which corresponds to a scale factor $\alpha = 0.3048$, so that 1 ft = 0.3048 m.

(ii) A derived quantity Q_j (such as density, velocity, etc.) then transforms under a scale change to

$$T(Q_j) = Q_j \Pi_{i=1}^n \alpha_i^{a_{ji}}.$$

The exponents a_{ji} are called the dimensions of Q_j. Now any physical law

$$f(Q_1, \ldots, Q_k) = 0 \tag{C.1}$$

must be *unit free*, i.e. invariant under all transformations T. Physically, this means that the validity of the law (C.1) does not depend on our choice of units (meters or inches, kilograms or pounds).

The statement that permits us to simplify (C.1), given that $Tf = 0$ for all T, is known as the Buckingham Π-theorem. To formulate it, we have to make one more technical assumption, which accounts for the fact that in a given problem a fundamental unit may not appear at all (in which case the corresponding row in the matrix formed by the exponents a_{ij} is zero) or two units may depend

434

trivially on one another, such as the pressure and its square. In the latter case, the rows of a_{ij} will be linearly dependent. Hence the rank $m \le n$ of a_{ij} is the number of effectively independent units. Now the Buckingham Π-theorem states that (C.1) is *equivalent* to

$$\phi(\Pi_1, \dots, \Pi_{k-m}) = 0, \qquad (C.2)$$

for a suitably defined function ϕ; the new variables Π_i (called dimensionless groups) have no dimensions and are thus invariant under any rescaling of the fundamental units.

In our example of a pinching thread of water, presented in Section 1.1, we have

$$f(h_{\min}, \Delta t, \gamma, \rho) = 0 \qquad (C.3)$$

and thus $k = 4$ unknown quantities and $m = 3$ independent units of length, time, and mass. After applying the Buckingham Π-theorem, there is only a single dimensionless group,

$$\Pi_1 = h_{\min}^{a_1} \rho^{a_2} (\Delta t)^{a_3} \gamma^{a_4}.$$

Since any power of Π_1 will also be dimensionless, we can choose $a_1 = 1$, and the dimensions of Π_1 are then $\text{cm}^{1-3a_2} \text{s}^{a_3-2a_4} \text{g}^{a_2-a_4}$. Requiring Π_1 to be dimensionless leads to $a_2 = 1/3$, $a_4 = a_2 = 1/3$, and $a_3 = 2a_4 = 2/3$, so that

$$\Pi_1 = \frac{h_{\min} \rho^{1/3}}{(\Delta t)^{2/3} \gamma^{1/3}}.$$

The theorem states that (C.3) is equivalent to $\phi(\Pi_1) = 0$, which means that Π_1 must be a constant, in accordance with (1.1).

References

[1] M. Abramowitz and I. A. Stegun. *Handbook of Mathematical Functions*. Dover, 1968.

[2] D. J. Acheson. *Elementary Fluid Dynamics*. Clarendon Press, 1990.

[3] M. Aguareles and I. Baldomà. Structure and Gevrey asymptotics of solutions representing topological defects to some partial differential equations. *Nonlinearity*, 24:2813, 2011.

[4] M. Aguareles, S. J. Chapman, and T. Witelski. Motion of spiral waves in the complex Ginzburg–Landau equation. *Physica D*, 239:348, 2010.

[5] S. Altschuler, S. Angenent, and Y. Giga. Mean curvature flow through singularities for surfaces of rotation. *J. Geom. Anal.*, 5:293, 1995.

[6] S. B. Angenent and J. J. L. Velázquez. Degenerate neckpinches in mean curvature flow. *J. Reine Angew. Math.*, 482:15, 1997.

[7] S. L. Anna and G. H. McKinley. Elasto-capillary thinning and breakup of model elastic liquids. *J. Rheol.*, 45:115, 2001.

[8] V. I. Arnol'd. *Mathematical Methods of Classical Mechanics, Second Edition*. Springer, 1989.

[9] V. I. Arnol'd, V. A. Vasil'ev, V. V. Goryunov, and O. V. Lyashko. Singularity theory I: Local and global theory. In *Dynamical Systems VI*. Springer, 1993.

[10] V. I. Arnol'd, V. A. Vasil'ev, V. V. Goryunov, and O. V. Lyashko. Singularity theory II: Classification and applications. In *Dynamical Syatems VIII*. Springer, 1993.

[11] M. Arrayás, M. A. Fontelos, and J. L. Trueba. Ionization fronts in negative corona discharges. *Phys. Rev. E*, 71:037401, 2005.

[12] H. Ashley and M. Landahl. *Aerodynamics of Wings and Bodies*. Addison-Wesley, 1965.

[13] G. I. Barenblatt. On one class of the one-dimensional problem of non-stationary filtration of a gas in a porous medium. *Prikl. Mat. i Mekh.*, 17:739, 1953.

[14] G. I. Barenblatt. *Similarity, Self-Similarity and Intermediate Asymptotics*. Cambridge University Press, 1996.

[15] G. I. Barenblatt and Y. B. Zel'dovich. Self-similar solutions as intermediate asymptotics. *Ann. Rev. Fluid Mech.*, 4:285, 1972.

[16] O. A. Basaran. Small-scale free surface flows with breakup: drop formation and emerging applications. *AICHE*, 48:1842, 2002.

[17] G. K. Batchelor. *An Introduction to Fluid Dynamics*. Cambridge University Press, 1967.

[18] C. M. Bender and S. A. Orszag. *Advanced Mathematical Methods for Scientists and Engineers*. McGraw-Hill, 1978.

[19] A. J. Bernoff, A. L. Bertozzi, and T. P. Witelski. Axisymmetric surface diffusion: dynamics and stability of self-similar pinch-off. *J. Stat. Phys.*, 93:725, 1998.

[20] M. V. Berry. Singularities in waves and rays. In R. Balian, M. Kleman, and J.-P. Poirier, editors, *Les Houches, Session XXXV*, pp. 453–543. North-Holland, 1981.

[21] M. V. Berry. Rays, wavefronts and phase: a picture book of cusps. In H. Blok, H. A. Ferwerda, and H. K. Kuiken, editors, *Huygens' Principle 1690–1990: Theory and Applications*, pp. 97–111. Elsevier, 1992.

[22] M. V. Berry. Asymptotics, singularities and the reduction of theories. In D. Prawitz, B. Skyrms, and D. Westerståhl, editors, *Proc. 9th Int. Cong. Logic, Method., and Phil. of Sci. IX*, pp. 597–607. Elsevier, 1994.

[23] M. V. Berry and J. Goldberg. Renormalisation of curlicues. *Nonlinearity*, 1:1, 1988.

[24] S. I. Betelu and D. G. Aronson. Focusing of noncircular self-similar shock waves. *Phys. Rev. Lett.*, 87:074501, 2001.

[25] F. Bethuel, H. Brezis, and F. Hélein. *Ginzburg–Landau Vortices*. Birkhäuser, 1994.

[26] S. Bianchini and A. Bressan. Vanishing viscosity solutions of nonlinear hyperbolic systems. *Ann. Math.*, 161:223, 2005.

[27] R. B. Bird, R. C. Armstrong, and O. Hassager. *Dynamics of Polymeric Liquids, Volume I: Fluid Mechanics; Volume II: Kinetic Theory*. Wiley, 1987.

[28] G. Birkhoff. *Hydrodynamics: A Study in Logic, Fact, and Similitude*. Princeton University Press, 1950.

[29] D. Bonn, J. Eggers, J. Indekeu, J. Meunier, and E. Rolley. Wetting and spreading. *Rev. Mod. Phys.*, 81:739, 2009.

[30] A. Boudaoud and S. Chaïeb. Singular thin viscous sheet. *Phys. Rev. E*, 64:050601, 2001.

[31] M. P. Brenner. Droplet breakup and other problems involving surface tension driven flows. Ph.D. thesis, University of Chicago, 1994.

[32] M. P. Brenner and A. L. Bertozzi. Spreading of droplets on a solid surface. *Phys. Rev. Lett.*, 71:593, 1993.

[33] M. P. Brenner, J. Eggers, K. Joseph, S. R. Nagel, and X. D. Shi. Breakdown of scaling in droplet fission at high Reynolds number. *Phys. Fluids*, 9:1573, 1997.

[34] M. P. Brenner, J. R. Lister, and H. A. Stone. Pinching threads, singularities and the number 0.0304 ... *Phys. Fluids*, 8:2827, 1996.

[35] M. P. Brenner, X. D. Shi, and S. R. Nagel. Iterated instabilities during droplet fission. *Phys. Rev. Lett.*, 73:3391, 1994.

[36] J. W. Bruce and T. J. Gaffney. Simple singularities of mappings $\mathbb{C}, 0 \to \mathbb{C}^2, 0$. *J. London Math. Soc.*, 26:465, 1982.

[37] J. M. Burgers. Mathematical examples illustrating relations occurring in the theory of turbulent fluid motion. *Kon. Ned. Akad. Wet., Verh. (Eerste Sectie)*, 17:1, 1939.

[38] J. C. Burton, J. E. Rutledge, and P. Taborek. Fluid pinch-off dynamics at nanometer length scales. *Phys. Rev. Lett.*, 92:244505, 2004.

[39] J. C. Burton and P. Taborek. 2D inviscid pinch-off: an example of self-similarity of the second kind. *Phys. Fluids*, 19:102109, 2007.

[40] R. E. Caflisch, N. Ercolani, T. Y. Hou, and Y. Landis. Multi-valued solutions and branch point singularities for nonlinear hyperbolic and elliptic systems. *Comm. Pure and Appl. Math.*, 46:453, 1993.

[41] R. E. Caflisch and O. F. Orellana. Singular solutions and ill-posedness for the evolution of vortex sheets. *SIAM J. Math. Anal.*, 20:293, 1989.

[42] G. F. Carrier, M. Krook, and C. E. Pearson. *Functions of a Complex Variable.* McGraw-Hill, 1966.

[43] J. R. Castrejón-Pita, A. A. Castrejón-Pita, E. J. Hinch, J. R. Lister, and I. M. Hutchings. Self-similar breakup of near-inviscid liquids. *Phys. Rev. E*, 86:015301(R), 2012.

[44] A. U. Chen, P. K. Notz, and O. A. Basaran. Computational and experimental analysis of pinch-off and scaling. *Phys. Rev. Lett.*, 88:174501, 2002.

[45] J.-D. Chen. Experiments on a spreading drop and its contact angle on a solid. *J. Colloid Interf. Sci.*, 122:60, 1988.

[46] X. Chen, C. M. Elliott, and T. Qi. Shooting method for vortex solutions of a complex-valued Ginzburg–Landau equation. *Proc. Roy. Soc. Edinburgh A*, 124:1075, 1994.

[47] Y.-J. Chen and P. H. Steen. Dynamics of inviscid capillary breakup: collapse and pinchoff of a film bridge. *J. Fluid Mech.*, 341:245, 1997.

[48] R. F. Chisnell. An analytic description of converging shock waves. *J. Fluid Mech.*, 354:357, 1998.

[49] M. W. Choptuik. Universality and scaling in gravitational collapse of a massless scalar field. *Phys. Rev. Lett.*, 70:9, 1993.

[50] A. Chorin and J. E. Marsden. *A Mathematical Introduction to Fluid Mechanics.* Springer, 2000.

[51] C. Clasen, J. Eggers, M. A. Fontelos, J. Li, and G. H. McKinley. The beads-on-string structure of viscoelastic jets. *J. Fluid Mech.*, 556:283, 2006.

[52] I. Cohen, M. P. Brenner, J. Eggers, and S. R. Nagel. Two fluid drop snap-off problem: experiment and theory. *Phys. Rev. Lett.*, 83:1147, 1999.

[53] P. Constantin, T. F. Dupont, R. E. Goldstein, L. P. Kadanoff, M. J. Shelley, and S.-M. Zhou. Droplet breakup in a model of the Hele-Shaw cell. *Phys. Rev. E*, 47:4169, 1993.

[54] S. D. Conte and C. De Boor. *Elementary Numerical Analysis.* McGraw Hill, 1965.

[55] S. Courrech du Pont and J. Eggers. Sink flow deforms the interface between a viscous liquid and air into a tip singularity. *Phys. Rev. Lett.*, 96:034501, 2006.

[56] V. F. Cowling and W. C. Royster. Domains of variability for univalent polynomials. *Proc. Amer. Math. Soc.*, 19:767, 1968.

[57] R. F. Day, E. J. Hinch, and J. R. Lister. Self-similar capillary pinchoff of an inviscid fluid. *Phys. Rev. Lett.*, 80:704, 1998.

[58] P.-G. de Gennes, F. Brochart-Wyart, and D. Quéré. *Capillarity and Wetting Phenomena: Drops, Bubbles, Pearls, Waves.* Springer, 2003.

[59] F. de la Hoz, M. A. Fontelos, and L. Vega. The effect of surface tension on the Moore singularity of vortex sheet dynamics. *J. Nonlinear Sci.*, 18:463, 2008.

[60] W. R. Dean and P. E. Montagnon. On the steady motion of viscous liquid in a corner. *Proc. Camb. Phil. Soc.*, 45:389, 1949.

[61] R. D. Deegan, O. Bakajin, T. F. Dupont, G. Huber, S. R. Nagel, and T. A. Witten. Contact line deposits in an evaporating drop. *Phys. Rev. E*, 62:756, 2000.

[62] M. P. do Carmo. *Differential Geometry of Curves and Surfaces.* Prentice-Hall, 1976.

[63] P. Doshi, I. Cohen, W. W. Zhang, M. Siegel, P. Howell, O. A. Basaran *et al.* Persistence of memory in drop breakup: the breakdown of universality. *Science,* 302:1185, 2003.

[64] J. Douglas Jr. and T. F. Dupont. Alternating-direction Galerkin methods on rectangles. In B. Hubbard, editor, *Numerical Solution of Partial Differential Equations II*, pp. 133–214. Academic Press, 1971.

[65] P. G. Drazin. *Nonlinear Systems.* Cambridge University Press, 1992.

[66] P. G. Drazin and W. H. Reid. *Hydrodynamic Stability.* Cambridge University Press, 1981.

[67] L. Duchemin and J. Eggers. The explicit–implicit–null method: removing the numerical instability of PDEs. *J. Comp. Phys.*, 263:37, 2014.

[68] B. R. Duffy and S. K. Wilson. A third-order differential equation arising in thin-film flows and relevant to Tanner's law. *Appl. Math. Lett.*, 10:63, 1997.

[69] D. G. Duffy. *Green's Functions with Applications.* Chapman & Hall, 2001.

[70] C. R. Dun and G. C. Hocking. Withdrawal of fluid through a line sink beneath a free surface above a sloping boundary. *J. Eng. Math.*, 29:1, 1995.

[71] H. E. Edgerton. *Stopping Time: The Photographs of Harold Edgerton.* Abrams, 1977.

[72] H. Effinger and S. Grossmann. Static structure function of turbulent flow from the Navier–Stokes equation. *Z. Phys. B*, 66:289, 1987.

[73] J. Eggers. Theory of drop formation. *Phys. Fluids*, 7:941, 1995.

[74] J. Eggers. Nonlinear dynamics and breakup of free-surface flows. *Rev. Mod. Phys.*, 69:865, 1997.

[75] J. Eggers. Drop formation – an overview. *ZAMM*, 85:400, 2005.

[76] J. Eggers. Stability of a viscous pinching thread. *Phys. Fluids*, 24:072103, 2012.

[77] J. Eggers. Post-breakup solutions of Navier–Stokes and Stokes threads. *Phys. Fluids*, 26:072104, 2014.

[78] J. Eggers and T. F. Dupont. Drop formation in a one-dimensional approximation of the Navier–Stokes equation. *J. Fluid Mech.*, 262:205, 1994.

[79] J. Eggers and M. A. Fontelos. Isolated inertialess drops cannot break up. *J. Fluid Mech.*, 530:177, 2005.

[80] J. Eggers and M. A. Fontelos. The role of self-similarity in singularities of partial differential equations. *Nonlinearity*, 22:R1, 2009.

[81] J. Eggers and J. Hoppe. Singularity formation for timelike extremal hyper-surfaces. *Phys. Lett. B*, 680:274, 2009.

[82] J. Eggers and H. A. Stone. Characteristic lengths at moving contact lines for a perfectly wetting fluid: the influence of speed on the dynamic contact angle. *J. Fluid Mech.*, 505:309, 2004.

[83] J. Eggers and E. Villermaux. Physics of liquid jets. *Rep. Progr. Phys.*, 71:036601, 2008.

[84] J. Eggers, M. A. Fontelos, D. Leppinen, and J. H. Snoeijer. Theory of the collapsing axisymmetric cavity. *Phys. Rev. Lett.*, 98:094502, 2007.

[85] V. M. Entov and A. L. Yarin. Influence of elastic stresses on the capillary breakup of dilute polymer solutions. *Fluid Dyn.*, 19:21, 1984.

[86] M. A. Fontelos. Break-up and no break-up in a family of models for the evolution of viscoelastic jets. *Z. Angew. Math. Phys.*, 54:84, 2003.

[87] M. A. Fontelos, J. Eggers, and J. H. Snoeijer. The spatial structure of bubble pinch-off. *SIAM J. Appl. Math.*, 71:1696, 2011.

[88] L. B. Freund. *Dynamic Fracture Mechanics*. Cambridge, 1998.

[89] L. A. Galin. Unsteady seepage with a free surface. *Dokl. Akad. Nauk. SSSR*, 47:246, 1945.

[90] C. G. Gibson. *Singular Points of Smooth Mappings*. Pitman, 1979.

[91] C. G. Gibson and C. A. Hobbs. Simple singularities of space curves. *Math. Proc. Camb. Phil. Soc.*, 113:297, 1993.

[92] Y. Giga and R. V. Kohn. Asymptotically self-similar blow-up of semi-linear heat-equations. *Comm. Pure Appl. Math.*, 38:297, 1985.

[93] Y. Giga and R. V. Kohn. Characterizing blowup using similarity variables. *Indiana University Math. J.*, 36:1, 1987.

[94] J. W. Goodman. *Introduction to Fourier Optics*. Roberts & Co., 2004.

[95] J. M. Gordillo and M. A. Fontelos. Satellites in inviscid breakup of bubbles. *Phys. Rev. Lett.*, 98:144503, 2007.

[96] I. S. Gradshteyn and I. M. Ryzhik. *Table of Integrals, Series, and Products*. Academic Press, 1980.

[97] G. M. Greuel, C. Lossen, and E. Shustin. *Introduction to Singularities and Deformations*. Springer, 2007.

[98] W. C. Griffith and W. Bleakney. Shock waves in gases. *Amer. J. Phys.*, 22:597, 1954.

[99] R. E. Grundy. Local similarity solutions for the initial value problem in nonlinear diffusion. *IMA J. Appl. Math.*, 30:209, 1983.

[100] G. Guderley. Starke kugelige und zylindrische Verdichtungsstöße in der Nähe des Kugelmittelpunktes bzw. der Zylinderachse. *Luftfahrtforschung*, 19:302, 1942.

[101] B. Gustafsson and A. Vasil'ev. *Conformal and Potential Analysis in Hele-Shaw Cells*. Birkhäuser, 2006.

[102] S. Gutiérrez, J. Rivas, and L. Vega. Formation of singularities and self-similar vortex motion under the localized induction approximation. *Comm. Partial Diff. Eq.*, 28:927, 2003.

[103] D. A. Hammond and L. G. Redekopp. Global dynamics of symmetric and asymmetric wakes. *J. Fluid Mech.*, 331:231, 1997.

[104] T. H. Havelock. The stability of motion of rectilinear vortices in ring formation. *Phil. Mag.*, 11:617, 1931.

[105] H. S. Hele-Shaw. The flow of water. *Nature*, 58:34, 1898.

[106] D. Henderson, H. Segur, L. B. Smolka, and M. Wadati. The motion of a falling liquid filament. *Phys. Fluids*, 12:550, 2000.

[107] M. A. Herrero and J. J. L. Velázquez. Chemotactic collapse for the Keller–Segel model. *J. Math. Biol.*, 35:177, 1996.

[108] M. A. Herrero and J. J. L. Velázquez. Singularity patterns in a chemotaxis model. *Math. Ann.*, 306:583, 1996.

[109] R.-M. Hervé and M. Hervé. Étude qualitative des solutions réelles d'une équation différentielle liée à l'équation de Ginzburg–Landau. *Ann. Inst. H. Poincaré Anal. Non Linéaire*, 11:427, 1994.

[110] H. Hochstadt. *Integral Equations*. Wiley, 1973.

[111] L. M. Hocking. Rival contact-angle models and the spreading of drops. *J. Fluid Mech.*, 239:671, 1992.

[112] B. Hopkinson. Discontinuous fluid motions involving sources and vortices. *Proc. Lond. Math. Soc.*, 29:142–164, 1898.

[113] J. Hoppe. Conservation laws and formation of singularities in relativistic theories of extended objects. In *Nonlinear Waves, Gakuto International Series: Mathematical Sciences and Applications, Volume 8*. Gakkotosho, 2006.

[114] S. D. Howison. Cusp development in Hele-Shaw flow with a free surface. *SIAM J. Appl. Math*, 46:20, 1986.

[115] C. Huh and L. E. Scriven. Hydrodynamic model of steady movement of a solid/liquid/fluid contact line. *J. Coll. Int. Sci.*, 35:85, 1971.

[116] G. Huisken. Local and global behaviour of hypersurfaces moving by mean curvature. *Proc. Symp. Pure Math.*, 54:175, 1993.

[117] L. Ignat, C. Lefter, and V. D. Radulescu. Minimization of the renormalized energy in the unit ball of $\mathbf{R}^2$. *Nieuw Arch. Wiskd.*, 1:278, 2000.

[118] R. Ishiguro, F. Graner, E. Rolley, and S. Balibar. Coalescence of crystalline drops. *Phys. Rev. Lett.*, 93:235301, 2004.

[119] R. Ishiguro, F. Graner, E. Rolley, S. Balibar, and J. Eggers. Dripping of a crystal. *Phys. Rev. E*, 75:041606, 2007.

[120] J. D. Jackson. *Classical Electrodynamics*. Wiley, 1975.

[121] D. J. Jeffrey and Y. Onishi. The slow motion of a cylinder next to a plane wall. *Quart. J. Mech. Appl. Math.*, 34:129, 1981.

[122] J.-T. Jeong and H. K. Moffatt. Free-surface cusps associated with a flow at low Reynolds numbers. *J. Fluid Mech.*, 241:1, 1992.

[123] F. John. Two-dimensional potential flows with a free boundary. *Comm. Pure Appl. Math.*, 6:497, 1953.

[124] D. D. Joseph, J. Nelson, M. Renardy, and Y. Renardy. Two-dimensional cusped interfaces. *J. Fluid Mech.*, 223:383, 1991.

[125] N. C. Keim, P. Møller, W. W. Zhang, and S. R. Nagel. Breakup of air bubbles in water: memory and breakdown of cylindrical symmetry. *Phys. Rev. Lett.*, 97:144503, 2006.

[126] T. A. Kowalewski. On the separation of droplets. *Fluid Dyn. Res.*, 17:121, 1996.

[127] S. N. Kruzkov. Generalized solutions of the Cauchy problem in the large for first order nonlinear equations. *Dokl. Akad. Nauk. SSSR*, 187:29, 1969.

[128] P. K. Kundu and I. M. Cohen. *Fluid Mechanics, Second Edition*. Academic Press, 2002.

[129] B. Lafaurie, C. Nardone, R. Scardovelli, S. Zaleski, and G. Zanetti. Modelling merging and fragmentation in multiphase flows with SURFER. *J. Comput. Physics*, 113:134, 1994.

[130] G. Lagubeau, M. A. Fontelos, C. Josserand, A. Maurel, V. Pagneux, and P. Petitjeans. Flower patterns in drop impact on thin liquid films. *Phys. Rev. Lett.*, 105:184503, 2010.

[131] H. Lamb. *Hydrodynamics, Sixth Edition*. Cambridge University Press, 1932.

[132] L. D. Landau and E. M. Lifshitz. *Elasticity*. Pergamon, 1970.

[133] L. D. Landau and E. M. Lifshitz. *Fluid Mechanics*. Pergamon, 1987.

[134] L. D. Landau and E. M. Lifshitz. *Electrodynamics of Continuous Media*. Pergamon, 1984.

[135] N. N. Lebedev. *Special Functions and Their Applications*. Prentice-Hall, 1965.

[136] J.-B. Leblond. *Mécanique de la rupture fragile et ductile*. Hermès, 2003.

[137] D. Leppinen and J. Lister. Capillary pinch-off in inviscid fluids. *Phys. Fluids*, 15:568, 2003.

[138] J. Leray. Sur le mouvement d'un liquide visqueux emplissant l'espace. *Acta Math.*, 63:193, 1934.

[139] B. M. Levitan and I. S. Sargsjan. *Sturm–Liouville and Dirac Operators*. Springer, 1990.

[140] H. Li, T. C. Halsey, and A. Lobkovsky. Singular shape of a fluid drop in an electric or magnetic field. *Europhys. Lett.*, 27:575, 1994.

[141] J. Li and M. A. Fontelos. Drop dynamics on the beads-on-string structure of viscoelastic jets: a numerical study. *Phys. Fluids*, 15:922, 2003.

[142] F.-H. Lin and J. X. Xin. On the dynamical law of the Ginzburg–Landau vortices on the plane. *Comm. Pure Appl. Math.*, 52:1189, 1999.

[143] J. R. Lister and H. A. Stone. Capillary breakup of a viscous thread surrounded by another viscous fluid. *Phys. Fluids*, 10:2758, 1998.

[144] E. W. Llewellin, H. M. Mader, and S. D. R. Wilson. The rheology of a bubbly liquid. *Proc. Roy. Soc. London A*, 458:987, 2002.

[145] R. A. London and B. P. Flannery. Hydrodynamics of x-ray induced stellar winds. *Astrophys. J.*, 258:260, 1982.

[146] B. Lopez. Comportement hydrodynamique d'un jet capillaire: influence de la buse d'ejection. Ph.D. thesis, University of Grenoble, 1998.

[147] E. N. Lorenz. Deterministic nonperiodic flow. *J. Atmos. Sci.*, 20:130, 1963.

[148] A. J. Majda and A. L. Bertozzi. *Vorticity and Incompressible Flow*. Cambridge University Press, 2002.

[149] N. Marheineke and R. Wegener. Asymptotic model for the dynamics of curved viscous fibres with surface tension. *J. Fluid Mech.*, 622:345, 2009.

[150] J. M. Martin-Garcia and C. Gundlach. Global structure of Choptuik's critical solution in scalar field collapse. *Phys. Rev. D*, 68:024011, 2003.

[151] J. M. Martin-Garcia and C. Gundlach. Critical phenomena in gravitational collapse. *Living Rev. Rel.*, 10:5, 2007.

[152] J. B. McLeod. The asymptotic behavior near the crest of waves of extreme form. *Trans. Amer. Math. Soc.*, 299:299, 1987.

[153] L. M. Milne-Thompson. *Theoretical Hydrodynamics, Fourth Edition*. Macmillan & Co., 1962.

[154] H. K. Moffatt. Viscous and resistive eddies near a sharp corner. *J. Fluid Mech.*, 18:1, 1963.

[155] H. K. Moffatt and B. R. Duffy. Local similarity solutions and their limitations. *J. Fluid Mech.*, 96:299, 1980.

[156] D. Mond. On the classification of germs of maps from $\mathbb{R}^2 \to \mathbb{R}^3$. *Proc. London Math. Soc.*, 50:333, 1985.

[157] D. W. Moore. Spontaneous appearance of a singularity in the shape of an evolving vortex sheet. *Proc. Roy. Soc. London A*, 365:105, 1979.

[158] M. Moseler and U. Landman. Formation, stability, and breakup of nanojets. *Science*, 289:1165, 2000.

[159] J. D. Murray. *Mathematical Biology*. Springer, New York, 1993.

[160] J. C. Neu. Vortices in complex scalar fields. *Physica D*, 43:385, 1990.

[161] P. Nozières. Shape and growth of crystals. In C. Godrèche, editor, *Solids far from Equilibrium*, pp. 1–154. Cambridge University Press, 1992.

[162] J. Nye. *Natural Focusing and Fine Structure of Light: Caustics and Wave Dislocations*. Institute of Physics Publishing, 1999.

[163] A. Oron, S. H. Davis, and S. G. Bankoff. Long-scale evolution of thin liquid films. *Rev. Mod. Phys.*, 69:931, 1997.

[164] C. Pantano, A. M. Gañán-Calvo, and A. Barrero. Zeroth order, electrohydrostatic solution for electrospraying in cone-jet mode. *J. Aerosol Sci.*, 25:1065, 1994.

[165] D. T. Papageorgiou. On the breakup of viscous liquid threads. *Phys. Fluids*, 7:1529, 1995.

[166] T. Pearcey. The structure of the electromagnetic field in the neighbourhood of a cusp of a caustic. *Phil. Mag.*, 37:311, 1946.

[167] C. L. Pekeris and Y. Accad. Solution of Laplace's equations for the M_2 tide in the world oceans. *Phil. Trans. Roy. Soc. London*, 265:413, 1969.

[168] D. H. Peregrine, G. Shoker, and A. Symon. The bifurcation of liquid bridges. *J. Fluid Mech.*, 212:25, 1990.

[169] L. M. Pismen. *Vortices in Nonlinear Fields: From Liquid Crystals to Superfluids, from Non-equilibrium Patterns to Cosmic Strings*. Oxford, 1999.

[170] L. M. Pismen and J. Eggers. Solvability condition for the moving contact line. *Phys. Rev. E*, 78:056304, 2008.

[171] L. M. Pismen and J. Rubinstein. Motion of vortex lines in the Ginzburg–Landau model. *Physica D*, 47:353, 1991.

[172] P. I. Plotnikov and J. F. Toland. Convexity of Stokes waves of extreme form. *Arch. Rational Mech. Anal.*, 171:349, 2004.

[173] P. Ya. Polubarinova-Kochina. Concerning unsteady motions in the theory of filtration. *Prikl. Matem. Mech.*, 9:79, 1945.

[174] P. Ya. Polubarinova-Kochina. On a problem of the motion of the contour of a petroleum shell. *Dokl. Akad. Nauk USSR*, 47:254, 1945.

[175] S. Popinet. An accurate adaptive solver for surface-tension-driven interfacial flows. *J. Comput. Phys.*, 228:5838, 2009.

[176] T. Poston and I. Stewart. *Catastrophe Theory and Its Applications*. Dover, 1978.

[177] C. Pozrikidis. *Boundary Integral and Singularity Methods for Linearized Flow*. Cambridge University Press, 1992.

[178] W. H. Press, S. A. Teukolsky, W. T. Vetterling, and B. P. Flannery. *Numerical Recipes: The Art of Scientific Computing, Third Edition.* Cambridge University Press, 2007.

[179] W. H. Press, S. A. Teukolski, W. T. Vetterling, and B. P. Flannery. *Numerical Recipes in Fortran; The Art of Scientific Computing Second Edition.* Cambridge University Press, 1992.

[180] A. Pumir and E. D. Siggia. Development of singular solutions to the axisymmtric Euler equations. *Phys. Fluids A,* 4:1472, 1992.

[181] A. Ramos and A. Castellanos. Conical points in liquid–liquid interfaces subjected to electric fields. *Phys. Lett. A,* 184:268, 1994.

[182] Michael Renardy. Self-similar breakup of non-Newtonian fluid jets. In *Rheology Reviews,* pp. 171–196, 2004.

[183] O. Reynolds. On the theory of lubrication and its application to M. Beauchamp Tower's experiments, including an experimental determination of the viscosity of olive oil. *Phil. Trans. Roy. Soc.* 177:157, 1886.

[184] E. Reyssat, E. Lorenceau, F. Restagno, and D. Quéré. Viscous jet drawing air into a bath. *Phys. Fluids,* 20:091107, 2008.

[185] L. F. Richardson. Atmospheric diffusion shown in a distance–neighbor graph. *Proc. Roy. Soc. London A,* 110:709, 1926.

[186] H. Risken. *The Fokker–Planck Equation.* Springer, 1984.

[187] A. Rothert, R. Richter, and I. Rehberg. Transition from symmetric to asymmetric scaling function before drop pinch-off. *Phys. Rev. Lett.,* 87:084501, 2001.

[188] A. I. Ruban and J. S. B. Gajjar. *Fluid Dynamics.* Oxford University Press, 2014.

[189] P. G. Saffman. *Vortex Dynamics.* Cambridge University Press, 1992.

[190] E. Sandier and S. Serfaty. Gamma-convergence of gradient flows with applications to Ginzburg–Landau. *Comm. Pure Appl. Math.,* 57:1627, 2004.

[191] C. Sautreaux. Mouvement d'un liquide parfait soumis à la pesanteur. Détermination des lignes de courant. *J. Math. Pures Appl.,* 7:125, 1901.

[192] K. Schowalter. Quadratic and cubic reaction–diffusion fronts. *Nonlinear Sci. Today,* 4:3, 1995.

[193] D. W. Schwendeman and G. B. Whitham. On converging shock waves. *Proc. Roy. Soc. London A,* 413:297, 1987.

[194] L. I. Sedov. *Similarity and Dimensional Methods in Mechanics.* CRC Press, 1993.

[195] D. Segalman and M. W. Johnson Jr. A model for viscoelastic fluid behavior which allows non-affine deformation. *J. Non-Newtonian Fluid Mech.,* 2:255, 1977.

[196] M. J. Sewell. On Legendre transformations and elementary catastrophes. *Math. Proc. Camb. Soc.,* 82:147, 1977.

[197] X. D. Shi, M. P. Brenner, and S. R. Nagel. A cascade of structure in a drop falling from a faucet. *Science,* 265:157, 1994.

[198] A. Sierou and J. R. Lister. Self-similar solutions for viscous capillary pinch-off. *J. Fluid Mech.,* 497:381, 2003.

[199] I. N. Sneddon. *Elements of Partial Differential Equations.* McGraw-Hill, 1957.

[200] J. H. Snoeijer. Free surface flows with large slopes: beyond lubrication theory. *Phys. Fluids,* 18:021701, 2006.

[201] H. A. Stone, J. R. Lister, and M. P. Brenner. Drops with conical ends in electric and magnetic fields. *Proc. Roy. Soc. Lond. A*, 455:329, 1999.

[202] S. H. Strogatz. *Nonlinear Dynamics and Chaos*. Westview Press, 1994.

[203] C. Sulem, P. L. Sulem, C. Bardos, and U. Frisch. Finite time analyticity for the two and three dimensional Kelvin–Helmholtz instability. *Comm. Math. Phys.*, 80:485, 1981.

[204] S. Taneda. Visualization of separating Stokes flows. *J. Phys. Soc. Japan*, 46:1935, 1979.

[205] S. Taneda and M. Honji. Unsteady flow past a flat plate normal to the direction of motion. *J. Phys. Soc. Japan*, 30:262, 1971.

[206] G. I. Taylor. The formation of a blast wave by a very intense explosion. *Proc. Roy. Soc. London A*, 201:159, 1950.

[207] G. I. Taylor. On scraping viscous fluid from a plane surface. In G. K. Batchelor, editor, *G. I. Taylor: Scientific Papers, Volume IV*, p. 410. Cambridge University Press, 1960.

[208] S. T. Thoroddsen, E. G. Etoh, and K. Takeara. Experiments on bubble pinch-off. *Phys. Fluids*, 19:042101, 2007.

[209] L. Ting and J. B. Keller. Slender jets and thin sheets with surface tension. *SIAM J. Appl. Math.*, 50:1533, 1990.

[210] L. R. G. Treloar. *The Physics of Rubber Elasticity*. Oxford University Press, 1975.

[211] M. P. Valignat, N. Fraysse, A.-M. Cazabat, F. Heslot, and P. Levinson. An ellipsometric study of layered droplets. *Thin Solid Films*, 234:475, 1993.

[212] M. Van Dyke. *An Album of Fluid Motion*. The Parabolic Press, 1982.

[213] A. Vilenkin and E. P. S. Shellard. *Cosmic Strings and Other Topological Defects*. Cambridge University Press, 2000.

[214] O. V. Voinov. Hydrodynamics of wetting [English translation]. *Fluid Dynamics*, 11:714, 1976.

[215] H. von Foerster, P. M. Mora, and L. W. Amiot. Doomsday: Friday, 13 November, AD 2026. *Science*, 132:1291, 1960.

[216] M. Watanabe and K. Takayama. Stability of converging cylindrical shock waves. *Shock Waves*, 1:149, 1991.

[217] R. C. Weast, editor. *Handbook of Chemistry and Physics*. CRC Press, 1978.

[218] F. J. Wegner. Corrections to scaling laws. *Phys. Rev. B*, 5:4529, 1972.

[219] H. Werlé. Hydrodynamic flow visualization. *Ann. Rev. Fluid Mech.*, 5:361, 1973.

[220] H. M. Westergaard. Bearing pressures and cracks. *J. Appl. Mech.*, 6:A49, 1939.

[221] G. B. Whitham. A new approach to problems of shock dynamics. Part I: Two-dimensional problems. *J. Fluid Mech.*, 2:145, 1957.

[222] G. B. Whitham. *Linear and Nonlinear Waves*. Wiley, 1974.

[223] T. P. Witelski and A. J. Bernoff. Self-similar asymptotics for linear and nonlinear diffusion equations. *Stud. Appl. Math.*, 100:153, 1998.

Index